CHIROGENESIS IN CHEMICAL SCIENCE

CHIROGENESIS IN CHEMICAL SCIENCE

Editors

Victor Borovkov
Riina Aav
Tallinn University of Technology, Estonia

World Scientific

NEW JERSEY • LONDON • SINGAPORE • BEIJING • SHANGHAI • HONG KONG • TAIPEI • CHENNAI • TOKYO

Published by

World Scientific Publishing Co. Pte. Ltd.
5 Toh Tuck Link, Singapore 596224
USA office: 27 Warren Street, Suite 401-402, Hackensack, NJ 07601
UK office: 57 Shelton Street, Covent Garden, London WC2H 9HE

British Library Cataloguing-in-Publication Data
A catalogue record for this book is available from the British Library.

CHIROGENESIS IN CHEMICAL SCIENCE

ISBN 978-981-125-921-0 (hardcover)
ISBN 978-981-125-922-7 (ebook for institutions)
ISBN 978-981-125-923-4 (ebook for individuals)

For any available supplementary material, please visit
https://www.worldscientific.com/worldscibooks/10.1142/12915#t=suppl

Typeset by Stallion Press
Email: enquiries@stallionpress.com

Dedication

In memory of Professor Yoshiteru Sakata.

https://doi.org/10.1142/9789811259227_fmatter

Preface

Chirality is a fundamental property of the universe having an enormous impact for different organic/inorganic materials, living organisms, and human beings. The basic principle of chirality is existence of an object in two mirror image forms, which are not superimposable. This phenomenon is widely seen in various fields of knowledge including mathematics, astronomy, physics, chemistry, biology, and ranging in a scale from galaxies to nuclear particles. In chemical science chirality is generally associated with a single molecule or group of molecules belonging to the C_n or D_n point groups with the simplest example of sp^3 carbon atom bonded to four different substituents. The dynamic processes of chirality generation, modulation, transfer, amplification, etc. are termed chirogenesis. This is a fast-growing and interdisciplinary field of research, which is widely seen in many natural (such as DNA double helix, the secondary alpha-helix structure of proteins, lipid membranes, saccharides, heme proteins, and other biological molecular objects) and various artificial systems being of prime importance not only for fundamental science but also for a number of practical applications in such areas as pharmacology and agrochemistry, perfumery and food industry, materials and polymer sciences, enantioselective catalysis and nonlinear optics, nanoscience and nanotechnology, molecular devices and chemical sensors, and others. Therefore, understanding the mechanisms and various influencing factors is of particular significance for smart control and further effective application of chirogenesis in chemistry. This book covers the whole range of chemical fields, where chirogenic phenomena are widely observed including parity violation and self-assembly, supramolecular and host–guest systems, molecular aggregation and polymers, asymmetric synthesis and catalysis, solid-state phase and photochemistry, computer simulation, and materials science.

Chapter 1, by Fujiki, is devoted to one of the most intriguing and fundamental chirogenic issues that is a parity violation. In general, parity is associated with symmetry, while parity violation results in chirality. The latter arises from weak interaction, which is one of the four known universal physical interactions, with the others being electromagnetism, the strong interaction, and gravitation being responsible for parity conservation. The present comprehensive chapter gives an overview of the historical backgrounds and recent advances regarding this phenomenon in cosmology, elemental particle physics, subatomic and atomic physics, various chemistries (inorganic, organic, supramolecular, macromolecular, polymer, and materials), biochemistry, and life science.

In chemistry, there are also different types of interactions including covalent and non-covalent binding, which are responsible for various chirogenic processes as well. In Chapter 2, Ustrnul *et al.* discuss a role of the latter in chirality induction in supramolecular systems. In general, non-covalent interactions have a very dynamic character due to their mostly low energy of binding per individual bond and their utilization is a cornerstone of supramolecular and host–guest chemistry. In connection with this, different types of interactions, their strength, solvent effect, general chirogenic mechanisms, chirality switching phenomena, influence of aggregation, and sensory application, as well as further perspectives of supramolecular chirogenesis, are thoroughly evaluated by using some selected examples.

Aggregation can also be considered as a kind of supramolecular interactions. Therefore, a more detailed analysis of the aggregation impact on the chirogenic processes is carried out in Chapter 3 by Gaeta *et al.* Aggregation and self-assembly processes appear to be a powerful and versatile tool for constructing complex supramolecules by spontaneous organization of appropriate building blocks via non-covalent interactions. For the chirogenesis reason, several strategies have been implemented to investigate how molecular aggregates become chiral and which processes govern the bias for a certain stereochemistry. This chapter first addresses the basic principles regulating the self-assembly phenomena and their chirogenesis aspects from single units to multicomponent systems with particular regard to porphyrin-based assemblies. Moreover, the fundamental strategies for designing and building chirally oriented supramolecular aggregates are taken into account.

In the case of considerably stronger covalent interactions, the chirogenesis is mainly achieved through asymmetric synthesis and catalysis. These chemical aspects are distinctly analyzed in Chapter 4 by Kananovich and Lopp. In particular, the basic principles and major concepts of this area are highlighted by using the most common types and techniques of chirality induction in organic reactions. Two main approaches: (a) diastereoselective synthesis, in which asymmetric induction takes place by utilizing an existing element of chirality in the starting material or with the help of a chiral auxiliary, and (b) enantioselective reactions of prochiral substrates with stoichiometric chiral reagents are clearly highlighted. Furthermore, enantioselective catalysis, in which chiral molecules are produced with the aid of catalytic amount of a chiral inductor, is discussed with particular focus on the most intriguing new directions in the field and chirogenic mechanisms, as well as corresponding green chemistry aspects.

A specific class of chemically linked compounds, being essentially important for chirogenesis, is polymers and macrocycles, the chirogenic properties of which are discussed in Chapter 5 by Puneet *et al*. The synthetic organic polymers and macromolecules possess an added chirogenic advantage in providing a molecular backbone with a desirable type and number of functional groups that may have chiral centers or to be induced. In this chapter, unique and peculiar aspects of chirality transfer and various molecular stages on the homochiral evolution of polymers and macromolecules are explored through the results of selected studies by focusing on providing a fundamental understanding of different branches of chiroptical spectroscopic techniques and their tandem use for a comprehensive investigation of chirogenic phenomena in these chemical structures.

Whilst the majority of chirogenic investigations have been carried out in various solutions, solid-state chirogenesis has scarcely been studied. However, it is of particular value especially for potential practical applications. Therefore, in Chapter 6 Oda *et al*. focused on chirogenesis observed in a solid-state matter, which is based on the solid-state organization of atoms or molecules including crystals and amorphous solids of both organic and inorganic origins, as well as surfaces and nanostructures, by using several representative systems. Importantly, peculiarities of the optical properties of chiral materials in the solid state are comprehensively analyzed by evaluating all polarimetric parameters on the basis of Mueller matrix formalism.

Beside conventional chemical stimuli, physical factors such as light can be a powerful driving force of chirogenesis by applying photoexcitation energy. Hence, Chapter 7 by Singathi *et al.* is devoted to various chirogenic effects in photochemistry. It is well known that light interacts with chromophoric molecules to produce highly energetic short-lived excited-state species with low activation energy barriers and making the process downhill and to produce a little opportunity to differentiate the diastereomeric transition states leading to poor stereo-enhancement in the products. However, a physical or chemical source of chirality can be guided to the reactants in the ground state to drive a desired reaction trajectory so that chirality can be induced or enhanced in the products, and this chapter highlights different strategies to achieve chirogenesis during photochemical reactions.

Another important part of chirogenesis study is rationalization of chirogenic mechanisms, which can be carried out by theoretical modeling. Burk and Burk in Chapter 8 comprehensively review various approaches of theoretical modeling, which is used to gain a versatile information for understanding the chirality induction/transfer mechanisms. Over the years the application of theoretical modeling has steadily grown and been applied to wider aspects of chirality: first to predict the energies of most important conformations, to obtain circular dichroism spectra computationally, to analyze the different interactions present in host–guest and supramolecular complexes, and to rationalize the corresponding electronic transitions. To date, there is a significant growth of the theoretical modeling as computational resources steadily become more available and the methodologies used continue to develop into more robust tools, which can increasingly be used by theoretical researchers and experimentalists.

The last but not the least section of this book is devoted to the utilization of chirogenesis in materials science and other specific applications. Whilst in general chirogenesis is associated with the (supra)molecular level, in Chapter 9 Yi *et al.* explicitly demonstrated that chirogenic phenomena are well translated to the macroscopic world including material systems of various dimensions, such as nanoparticles, nanofibers, nanosurfaces, metal- and covalent organic frameworks, as well as self-assembled soft materials and moiré stacking. This new branch of materials science opens a great promise for developing advanced materials with unique properties and has an impact on the development of a next generation of smart responsive devices and molecular machines on surfaces.

In conclusion, this book comprehensively summarizes all major aspects of the chirogenesis phenomenon in different chemistries and related fields and tailors as for general chemistry readership including graduate and postgraduate students and for researchers specializing in the fields of asymmetry and chirality. Moreover, this book can lay foundation for a successful course of further investigations and applications of chirogenesis and is highly recommended for everyone who is interested in this subject.

Victor Borovkov and Riina Aav

Contents

https://doi.org/10.1142/9789811259227_0001

1 Chirogenesis in Parity Violation and Weak Forces

Michiya Fujiki

Graduate School of Science and Technology,
Nara Institute of Science and Technology,
8916-5 Takayama, Ikoma, Nara 630-0192, Japan.
fujikim@ms.naist.jp

The present comprehensive chapter would stimulate to overview historical backgrounds and recent topics, regarding C (charge conjugation), P (parity), T (time reversal), CP (charge-parity), PT (parity-time), and CT (charge-time) violations in the framework of CPT (charge-parity-time) invariance in cosmology, elementary particle physics, subatomic and atomic physics, chemistry (inorganic and organic small molecules, supramolecules, macromolecules, and polymeric materials), biochemistry, and life science. In my view with a deep understanding, and cumulative knowledge of four fundamental forces in physics, involving P-violating weak force and P-conserving gravitational, electromagnetic, and strong forces, it is possible to state that they are no longer segregated, and are likely to intimately connect to each other.

1.1 Introduction — Seven Symmetries and Four Forces in Nature

In the nineteenth century, a theoretical notion of *unsymmetrical molecules*, proposed in 1827 by Sir John F. W. Herschel, who was an eminent British astronomer, physicist, and chemist [1], stimulated Louis Pasteur, a French giant scientist, to conceive that *Life is a function of the asymmetry of the universe and of the consequence of this fact*, led to the grand conjecture, *L'univers est dissymetrique*, meaning that *The universe is asymmetric*. The grand conjecture became mostly a reality subsequently [2, 3]. True enantiomers of L-amino acids and D-ribose made of the proton (p), neutron (n), and electron (e^-) in a real world are not D-amino acids and L-ribose [4] but hypothetical (D-amino acids)* and (L-ribose)*, built-up of antiproton (anti-p), anti-neutron (anti-n), and positron (e^+), that can exist in the imaginary world as a consequence of parity (P) violation [5]. The only e^+ (positron) on Earth can be artificially generated from radioactive decay of ^{11}C, ^{40}K, ^{13}N, ^{18}F, ^{121}I, etc. However, L-amino acids and (D-amino acids)* as well as D-ribose and (L-ribose)* may no longer be enantiomers in the framework of charge-parity (CP) violation with the arrow of time, meaning a positive entropy [6]. Yet, a definitive answer to the question of biomolecular handedness on Earth remains unresolved and experimental proofs are long-awaited in the realms of chemistry, astrochemistry, astrobiology, and molecular physics [4, 7–10].

On the other hand, several astonishing paradigm shifts occurred in fundamental physics including cosmology, elementary particles, subatoms, and atoms over 100 years [11–17]. The events rely on deep understandings, serendipitous discoveries, and precise experimental setups conducted by thoughtful theorists and experimentalists in the following six major topics: (i) seven fundamental symmetries in physics, that include parity (P, inversion between coordinate space r (x, y, z) and $-r$ (–x, –y, –z)), charge conjugation (C, inversion between matter and antimatter), time reversal (T, inversion between future (t) and past ($-t$)), CP, PT, and CT within the framework of CPT invariance, (ii) electroweak force (EWF) unifying weak force (WF) and electromagnetic force (EMF) among four fundamental physical forces — gravity (GF), EMF, strong force (SF), and WF, formulated by the Standard model [18] or the Glashow–Weinberg–Salam theory, (iii) detecting two massive charged gauge bosons ($W^{\pm}$, 80.4

GeV) and a neutral gauge boson Z^0 (91.2 GeV) associated with massless photon (γ) [19], (iv) detecting massive scalar Higgs boson (H^0, 125.1 GeV), (v) elucidating non-massless characteristics of electron neutrino (ν_e), muon neutrino (ν_μ), and tau neutrino (ν_τ), by observing their neutrino oscillations [20–22], and (vi) in 2020, a possible detection of CP-violation between neutrinos and anti-neutrinos in a lepton family [15]. Several symmetries among the seven did not reveal ideally perfect reflections.

In line with the modern Big Bang scenario [23], the four fundamental forces (GF, EMF, SF, WF) were bifurcated after the Big Bang with the arrow of time [23], as illustrated in Figure 1. Before the Big Bang event, the four forces were united as one force. Firstly, the GF was bifurcated at 10^{-43} s (Planck time) that is equivalent to 10^{32} Kelvin and 10^{19} GeV, generating the SF and electroweak force (EWF). The SF and EWF were bifurcated at 10^{-35} s (10^{27} Kelvin, 10^{14} GeV). Finally, the EWF was bifurcated at 10^{-12} s (10^{15} Kelvin, 10^2 GeV), producing the EMF and WF. The four forces at the present day are coexisting in the whole universe under 10^{-13} GeV (0.1 meV) or 2.73 K or 5×10^{17} s (13.8 billion years).

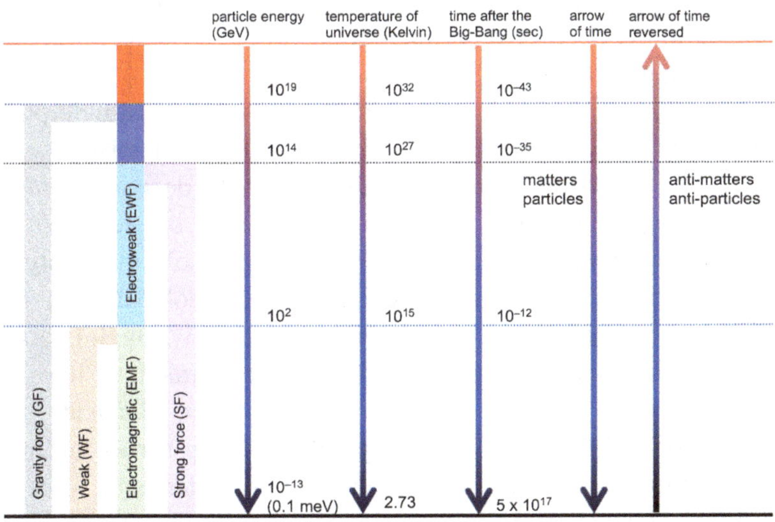

Figure 1. Three bifurcation eras to generate the four fundamental forces (GF, EMF, SF, and WF) after the Big Bang with the arrow of time (T-violation). Modified from Ref. [23].

According to a proposal by Andrey Sakharov in 1967 [6], the origin of asymmetry in baryons, like nucleons and mesons, that are constituted of matters only but not of antimatters, arises from CP-violation that happened once in the past during an extremely rapid expansion of space-time at a nonequilibrium open system in the early universe after the Big Bang. Note that mirror symmetry (energy equality of left and right) in chemistry is equivalent to P-operation in physics, but the chemistry's convention remains accepted as a doubtless principle among the majority of chemists since the historic discovery led by two French scientists, Jean-Baptiste Biot and Louis Pasteur [3, 24]. In the chemists' fraternity, the inherent non-mirror symmetry law, that is, the energy inequality of left and right at the molecular level, is perceived as an extraordinary hypothesis and most of them remain highly skeptical about it, yet, a handful of chemists have dwelled in theoretical and experimental investigations. While they agree with the fact that nature chose L-amino acids and D-ribose on Earth, and not the respective D-amino acids and L-ribose, the conclusive consensus for the origin is still a matter of debate. To prove P-violation in biological substances, an investigation of CP-violation of biological substances is important, while it is extremely difficult to experimentally test CP-violation between left-hand enantiomer and right-hand anti-enantiomer or vice versa owing to the absence of antimatter chemical substances that do not exist on Earth, naturally.

The paradigm shifts with remarkable discoveries in particle physics and cosmology [18, 25] may possibly be connected to chemistry, particularly P- and CP-violations that are monumental to the deep understanding of the nature of atoms, covalently and non-covalently bonded molecules, supramolecules, crystals, colloids, oligomers, macromolecules, polymers, living cells, and the origin of handedness in biomolecules on Earth. Essentially, the Coulombic interactions, van der Waals interactions, hydrogen-bonding interactions, and light-matter interactions have been extensively studied and employed in chemistry, yet, alongside recent development of P-violations in cosmology, elementary particles, and atomic physics, strongly pointing toward the underlying fundamental forces that can be traced down to EMF, and presumably, ν_e as well. Note that, in the mid-nineteenth century, Michael Faraday and James Maxwell unified electronic and magnetic forces as EMF with experiments and theories, respectively [26]. This correlation would be of paramount importance to bridge the gap between fundamental chemistry and particle physics.

All chemical constituents, including the lightest hydrogen, deuteron, and helium, in the universe, are made of p, n, and e^-, that are made up of quarks [12] and leptons [27] as elementary particles. Subsequently, the following questions must attract chemist's attention; (i) how the universe generated these constituents owing to the scenarios of CP- and C-violations with the arrow of time (flowing only from the past to the future, namely, T-violation) and (ii) whether conservation and violation of all the seven symmetries are connected to conservation and/or violation in mirror (P or left-and-right) symmetry of all the chiral chemical constituents?

An open question may be what the antiparticles and antimatters are? Most chemists are not familiar with the antiparticles and antimatters. Noting that the antiparticle of a photon is the photon itself. In the 1940s, Richard Feynman and Ernst Stuckelberg independently proposed the idea of negative energy as particles moving backward in time ($-t$) [28]. They called this state antiparticles and antimatters traveling with the opposite arrows of time, namely, a negative entropy. Heinz-Dieter Zeh, who proposed decoherence — that is now the key in quantum computing — philosophically discussed this topic [29, 30].

Beyond an orthodox stereochemistry in which left-and-right enantiomers are energetically equal, the present chapter highlighted several theories and experimental efforts for a heresy P- and CP-violation stereochemistry, so-called molecular parity violation (MPV) hypothesis, over the past one century. In my view, several experimental results are suggestive proof for the P-violation in several systems, for example, crystals, molecules and oligomers at the ground and photoexcited states in liquids, colloids, and polymers in liquids, molecules under external magnetic field in liquids, though chiral molecules in a gas phase are long-awaiting. To ensure the validity of the experimental results that are testable by anyone, at any places and time, further concrete experimental proof will be needed to be associated with sophisticated MPV theories in realistic condensed matters under collision conditions.

This comprehensive chapter attempts to expedite an intellectual purview on the research gaps between fundamental particle physics to the expanse of cosmology and elegant molecular chemistry to diverse

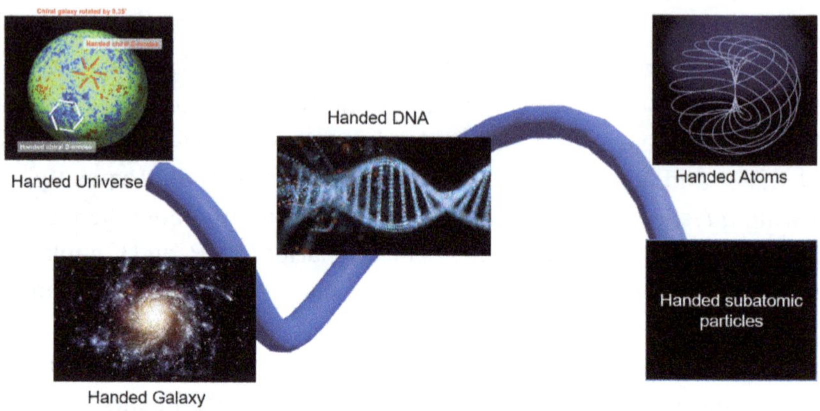

Figure 2. A hierarchical handedness from cosmic microwave radiation, spiral galaxies, weak neural current in atoms and subatomic particles, biomolecules, and biopolymers due to P- and CP-violations after the Big Bang with the arrow of time.

biochemistry. Perhaps these research areas that were once segregated are intimately connected with a common denominator of C-, P-, T-, CP-, PT-, and CT-violations in the CPT theorem illustrated in Figure 2.

1.2 P- and CP-Violations in Cosmology

The standard cosmology teaches that, after the Big Bang occurred 13.7 billion years ago, the universe began expanding [31, 32]. However, in the early twentieth century, most astronomers, even esteemed physicists, such as Albert Einstein, concurred that our universe has a constant cosmological size and is dominated only by the Milky Way. In 1917, Einstein had added a cosmological constant to retain the cosmological size intact while constructing a general theory of relativity based on the static universe hypothesis.

On the other hand, in 1922, Alexander Friedmann considered three scenarios of the universe with time; expansion, constant, and shrinkage, whereas, in 1927, Georges Lemaître was the first to theoretically hypothesize the Big Bang with cold temperature [32]. These two theoretical hypotheses of expanding universe prompted the ground-breaking discovery by Edwin Hubble in 1929, revealing that our universe is actually

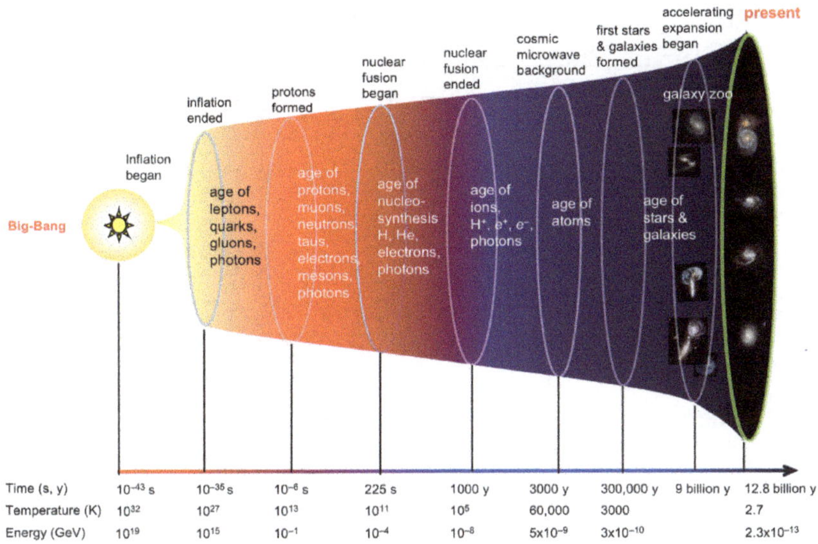

Figure 3. The Big Bang scenario in the universe, followed by a rapid expansion of space with the arrow of time; hypothetic spacetime at each time is represented by the circular sections. (Left) the dramatic expansion occurs in the inflationary epoch; (center) the expansion acceleration occurred at approximately 4 billion years ago. Modified based on Ref. [32].

expanding by observing a cosmological distance from Earth and a recession velocity of several galaxies (called nebulae in those days) [32], as illustrated in Figure 3. His finding was called the Hubble law until recently, but in 2018, it was re-named as Hubble–Lemaître law owing to the great contribution of Lemaître's to modern cosmology.

In 1956, George Gamov predicted cosmic microwave background (CMB) to be around 6 K based on his hot Big Bang hypothesis, inspired by a quantum tunneling in the α-decay process for radioactive elements [32]. The Big Bang hypothesis was confirmed by a serendipitous discovery by Arno Penzias and Robert Wilson, who set up the Dicke phase-sensitive heterodyne radiometer installed to the biggest horn antenna at that time to suppress a background noise [33]. Nowadays, astronomers believe that a remnant light existing in radiofrequencies equals a homogeneous blackbody radiation spectrum of 2.73 K, as shown in Figure 4 [34].

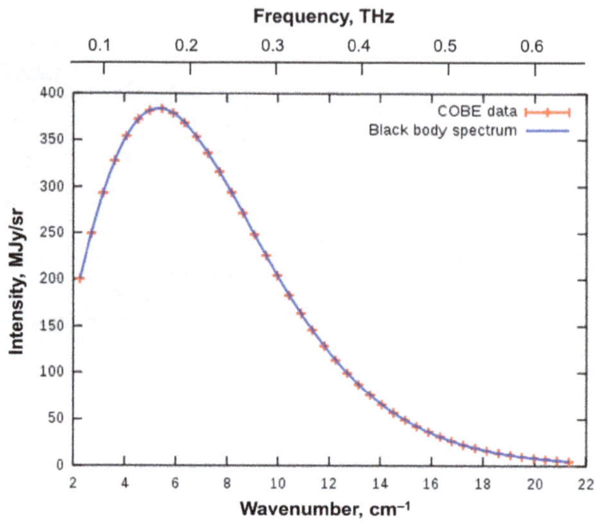

Figure 4. The cosmic microwave background spectrum, measured by the far-infrared absolute spectrophotometer (FIRAS) on the COBE (NASA). The most precisely measured black body spectrum peaking at 2.73 K (160.4 GHz, wavelength 1.87 mm, wavenumber 5.35 cm^{-1}) in nature. Modified from Ref. [34].

Immediately, several researchers postulated theoretically that the early universe originating from 380,000 years after the Big Bang has the primary anisotropy — directional dependency — on the order of 10^{-4}–10^{-5} in Kelvin. To verify this radical idea, the COBE mission endowed with three differential microwave radiometers (DMR, 31.5, 53, and 90 GHz) with a sensitivity of 0.1 mK at the National Aeronautics and Space Administration (NASA) had started [33] (Figure 5). In 1992, COBE confirmed for the first time that the primordial seeds from a growing universe, later called wrinkles, that existed as a result of fluctuation in the early universe using the DMR. Two further missions were started; one is the Wilkinson microwave anisotropy probe (WMAP, 2001–2010, NASA) enabled to precisely detect the anisotropy using five different radio frequencies ranging from 22 GHz to 90 GHz [35] and the other is the Planck Surveyor Satellite (2009–2013, the European Space Agency (ESA)) equipped with nine different radio frequencies ranging from 30 GHz to 857 GHz (Figure 5).

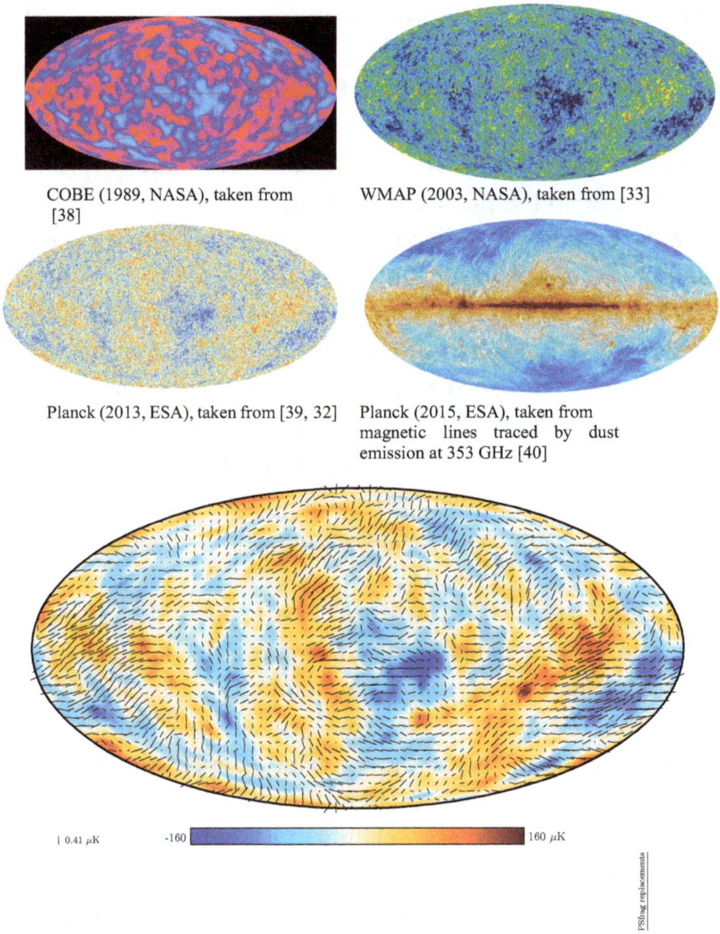

Figure 5. A comparison of five CMB images captured by three space probes followed by analysis [34, 36, 37].

Presumably, chiroptical chemists who are familiar with the high-sensitivity of natural optical rotation dispersion (NORD), natural circular dichroism (NCD), and natural circularly polarized luminescence (NCPL) spectroscopy using AC-modulation techniques with a lock-in amplifier, would immediately recognize that the targeted samples subtracted with the background signals as a function of a wide range of radiofrequency are the key to detect the subtle differences [38–41]. Note that NCD and NORD spectroscopies, which were first observed in 1895 by Aimé Cotton, are nowadays the most powerful tools for characterizing stereochemical structures at the ground state. Contrarily, NCPL spectroscopy is currently a hot topic since it is the only tool available to probe stereochemical structures in the photoexcited state.

In 1998, a serendipitous discovery reported a further acceleration of expansion for the expanding universe that occurred approximately 4 billion years ago [44–46] (Figure 2). Most astronomers hypothesize that our universe will expand eternally and never collapse, eventually approaching an absolute zero Kelvin temperature and an infinite space in the far future. The acceleration of the universe hypothesis arose from invisible *dark energy* (unidentified yet) as a repulsive force over *dark matter* (unknown massive matters) acting as a source of gravitational force that remains otherwise unaccounted for throughout the entire universe.

In the 1930s, Fritz Zwicky was the first to theoretically propose the existence of invisible dark matter, that was recognized as an extraordinary hypothesis based on observations of an anomaly in the luminosities and internal rotations of nebulae in the Coma cluster [47, 48]; the amount of invisible nonluminous matter in the universe is approximately 400 times greater than that of luminous matter. Subsequently, in 1936, Sinclair Smith proposed the great mass existing inter-nebular by analyzing the velocity and numbers of clusters in the Virgo cluster [49].

About 45 years later, in the 1970s and 1980s, Vera Rubin and coworkers showed experimental evidence by carefully analyzing the degree of Doppler red- and blue-shifted lines originating from several atomic lines for a spiral galaxy, Messier 31 (M31), well-known as the Andromeda Galaxy (type Sb, according to Hubble's classification of galaxies [50] and another 20 spiral galaxies (type Sc galaxies) [51–54]; the flat rotation curves in these galaxies indicated the outermost visible stars of the galaxies quickly moving as those close to the galaxy center in support of the radical invisible matter hypothesis. A radical hypothesis was realized by a careful investigation with patience — the questions and answers associated with unexpected possibilities — beyond common sense and textbooks established within the science community and by the authorities.

However, the origins of dark matter and dark energy have endured long-lasting debate. According to a lengthy observation by the Hubble Space Telescope (HST, NASA) and the Planck space observatory (ESA), the ordinary matter we know makes up only 4% of our universe while fractions of dark matter and dark energy constitute 23% and 73%, respectively, as depicted in Figure 6.

This known matter, that accounts for less than 4%, is mostly dominated by hydrogen and helium in abundance; most of all heavier elements, including carbon, oxygen, nitrogen, iron, etc., listed in the periodic table account for only 0.03%, making them rare in our universe. Surprisingly, a total fraction of non-massless, ν_e, ν_μ, and ν_τ reaches 0.3%, which is ten times larger than that of the aforementioned heavier elements. All chemicals and ingredients of life, such as heavy elements, are irreplaceable gifts that were cooked up in an explosion of supernovae and nucleosynthesis, presumably landing on Earth approximately 4 billion years ago. We are now reusing and recycling all elements made of matters, not antimatters, existing crust, and oceans on Earth.

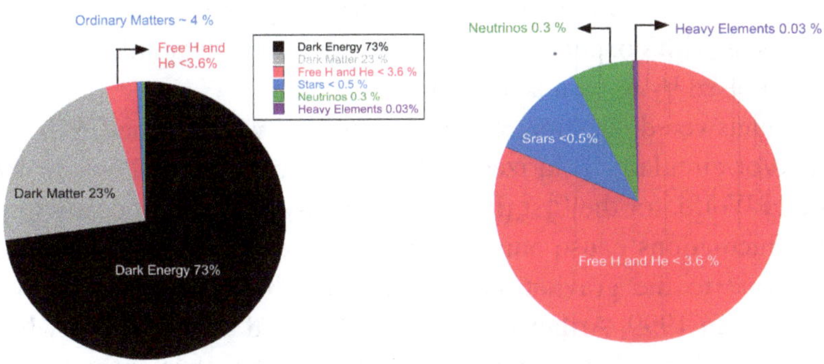

Figure 6. Two charts show the proportions of different components in the universe. Approximately 96% of matters and energy in the universe are governed by dark matter and dark energy. Approximately 4% of detectable matters are hydrogen, helium, carbon, oxygen, silicon, and heavy elements listed in the periodic table. Three neutrinos (ν_e, ν_μ, and ν_τ) contribute to 0.3% mass in the universe. Modified from Ref. [32].

The total mass of all stars, such as the Sun, are undergoing exothermic reactions, including several nuclear fusions, e.g., H + H → D, D + H → T, H + T → He, associated with a 0.7% mass defect. According to the recent cosmology, the Milky Way consists of at least approximately 100 billion stars and 20 billion earth-like planets; Moreover, our universe is composed of at least 100–200 billion galaxies; since the recognition of spiral forms, classified as the Hubble sequence, in a galaxy zoo in the universe, astronomers had a naive question as to whether the spiral direction has no preference or is distributed predominantly one handedness. In 2008, Kate Land and coworkers concluded no clear evidence of spiral handedness among ~37,000 spiral galaxies [55].

However, in 2011, Michael Longo [56] indicated the existence of cosmic P-violation (CPV) by analyzing 15,158 galaxies with redshifts <0.085 that are closer to the Milky Way from a dataset of the Sloan Digital Sky Survey (SDSS) at Apache Point Observatory (New Mexico, USA); a small left-handedness of 7.0% ± 1.3%, expressed as the degree of preference $A = (R - L)/(R + L)$, R and L are clockwise (S-type) and counterclockwise (Z-type) spiral galaxies, respectively, in the universe with a preferred axis. In 2012 and 2016, Lior Shamir reported that the degree of preference A is 1.7%–2.7% with left-handedness by analyzing 126,501 spiral galaxies with redshifts <0.3, far from the Milky Way [57, 58]. However, the discussion of CPV based on analyses of spiral galaxies is controversial with concerns related to the previous data analyses, which might be biased by human selection.

An unanswered question remains open as to whether the CMB is linearly and/or circularly polarized and how one can detect these. In 1967, Sachs and Wolfe, for the first time, theoretically argued that cosmological density fluctuations cause anisotropies in the observed microwave temperature due to the gravitational redshift and other general-relativistic effects [59]. In 1999, Arthur Lue *et al.* [60] posed a question of whether any manifestations of P-violation are possible other than those observed in weak interactions in the $\beta^{\pm}$-decay experiments of the radioactive isotopes. They predicted, theoretically, that a mapping in temperature and polarization of the CMB provided by the Planck Surveyor would provide two classes of P-violating interactions; the first interaction leads to a cosmological birefringence, and the second one leads to an asymmetry in the amplitude between right- and left-handed gravitational waves during the Big Bang.

In an analogy to electrostatics, two polarizations, called E-mode and B-mode, are considered. E-mode, that is, the electric field (E-field), has a vanishing curl field, while B-mode, that is, the magnetic field (B-field), has a vanishing divergence. Note that the B-mode adopts, in principle, left-hand and/or right-hand spiral divergences, that are similar to often employed to characterize enantiopurity of chiral molecules.

Recently, several cosmological research groups reported the degree of hidden CPV by analyzing the big data of temperature anisotropy on the order of 10^{-4}–10^{-6} K in CMB radiation (~2.73 K) as a function of angular scale that is multipole moment [61, 62], as given in Figure 5. The E-modes arise from Thomson scattering in a heterogeneous plasma, whereas the B-modes are generated by two mechanisms; the first one is by gravitational lensing of E-modes and the second one is from gravitational waves arising from cosmic inflation. Nonetheless, detecting the B- and E-modes is an extremely difficult task due to the presence of non-zero curled angles [63–67].

In 2020, using the open datasets of polarized CMB provided by the Planck project (Figure 5), two Japanese astrophysicists, seeking evidence of the CPV, reported a precision data analysis of cross-correlation between the curled B- and curled E-polarizations hidden in the CMB of the Milky Way [68, 69], as illustrated in Figure 7. This was performed

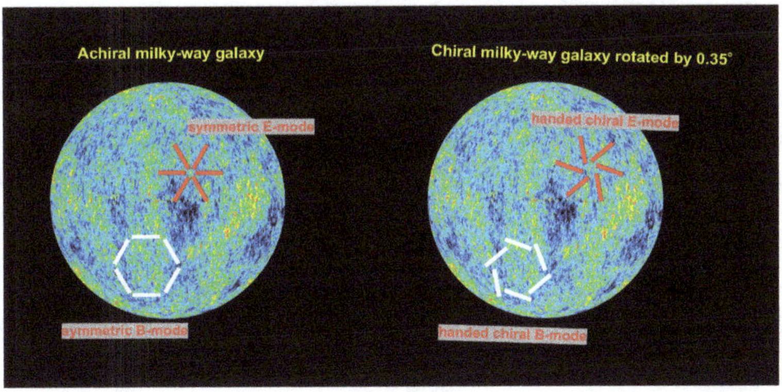

Figure 7. Handedness in the Milky Way by analyzing CMB at ~2.73 K. (Left) Hypothetical symmetric E- and B-modes in the achiral galaxy and (right) asymmetric, curled E- and B-modes in the handed chiral galaxy [68]. Modified from Ref. [69].

by cleverly subtracting a galactic foreground emission induced by cosmic dust that may align to the magnetic field of the galaxy. They showed a handed cosmic birefringence rotation angle of $\beta = 0.35° \pm 0.14°$ at 68% confidence level (CL) by excluding the foreground component of $\beta = 0°$ at 99.2% CL. They conjectured that dark matter and/or dark energy, which are dominant constituents in the universe, are responsible for the CPV. To my understanding, the tiny-handed β-angle, possibly, is equal to $0.39 \pm 0.16\%$ as the degree of circular polarization that corresponds to Kuhn's anisotropy g-value of $8.7 \pm 3.1 \times 10^{-3}$. It is assumed that the CPV unveiled in the Milky Way may commonly exist in other galaxies and in the entire universe. As conjectured by Pasteur [2, 3], one can envision that the CPV could affect the origin of biomolecular handedness, which may correlate to the big question of why nature chose L-amino acids and D-ribose on Earth. The tiny but inherently handed CPV may result in unintended experimental results with a handed chiral bias in chirogenesis and natural chiroptical data [NCD, natural optical rotation (NOR), NORD, NCPL, natural Raman optical activity (NROA), natural vibrational circular dichroism (NVCD)] at any time, location, and laboratory.

The key underlying procedure to detect the hidden CPV is to utilize differences in spontaneous radiation modes from the targeted radiation sources by choosing proper radiation sources as references and background. Similarly, the idea may be applicable to efficiently detect a subtle atomic parity violation (APV) and a subtle MPV as their spontaneous radiation modes of atomic and molecular emitters, as mentioned in the following sections.

Yet, it is obscure whether the physics and chemistry laws are applicable inside of *black holes*. After three years of observation of three orbiting bright stars near Sagittarius A*, among 90 stars of the Milky Way, a supermassive black hole (SBH) was proven to exist at the center of the galaxy, and possibly, might exist at other galaxy centers [70]. In 2019, the Event Horizon Telescope (EHT) collaboration team with a global baseline interferometry array at 1.3 mm wavelength confirmed the SBH at the heart of the Milky Way [71, 72]. The author speculates that the origin of CPV could be connected to SBH with a handed spinning motion.

1.3 P- and CP-Violations in Elementary Particle

Nowadays, cosmological events are visualizable and detectable spectroscopically in a wide range of energy probes, covering γ-ray, X-ray, vacuum-ultraviolet (VUV), UV–Vis, near-infrared (NIR), mid-IR, far-IR, microwaves, and gravitational waves, allowing particle physicists to treat mostly tiny invisible matter that often requires large facilities to enable the generation of extremely high-energy sources.

From the late nineteenth to the twenty-first centuries, physicists identified several particles and epoch-making events, as given chronologically [13, 17, 73–75]; e^- (J.J. Thomson, 1897), p (E. Rutherford, 1909), e^+ (anti-e^-, P. Dirac theoretically postulated, 1928), n (J. Chadwick, 1932), neutrino (ν) (W.E. Pauli, postulated as a massless particle, 1931, E. Amaldi and E. Fermi coined, 1934), ν (C. Cowan and F. Reines detected, 1956), e^+ (D. Skobeltsyn detected, 1923, C. Anderson, 1932, P. Blackett and G. Occhialini, 1932), muon (μ) as electron-like lepton (C. Anderson and S. Neddermeyer, 1936), anti-p (E. Segrè and O. Chamberlain, 1955), anti-n (B. Cork, 1956), anti-ν (M. Goldhaber, L. Grodzins, A. Sunyar, 1958), ν_μ (L. Lederman, M. Schwartz, J. Steinberger, 1962), CP-violation of neutral K-meson (kaon, made of down (d) and anti-strange (anti-s) quarks (J.W. Cronin, V.L. Fitch, 1964), a theory of three quark model (hypothesis, M. Gell-Mann, G. Zweig, 1964), Z-boson (Gargamelle experiment at Conseil Européen pour la Recherche Nucléaire (CERN), 1973), a theory of six quarks, leading to CP-violation (M. Kobayashi, T. Maskawa, 1973), τ as the third electron-like lepton (M. Perl *et al.*, 1975), a theory of great CP-violation upto 20% in B-meson (d and anti-bottom (anti-b) quarks) and anti-B-meson (anti-d and b quarks) (I. Sanda, 1980), oscillation of atmospheric ν_μ-to-ν_τ (T. Kajita *et al.*, at Super-Kamiokande, 1998), ν_τ (Direct Observation of the Nu Tau (DONUT) collaboration at Fermi lab, 2000), CP-violation in B-meson (Belle Collaboration at Kō Enerugī Kasokuki Kenkyū Kikō (KEK), 2001 and BaBar Collaboration at Stanford Linear Accelerator Center (SLAC), 2001), solar neutrino oscillation (A.B. McDonald *et al.* at Sudbury Neutrino Observatory (SNO), 2001 and 2002), anti-neutrino oscillation (Kamioka Liquid Scintillator Anti-Neutrino Detector (KamLAND), 2003), ν_μ-to-ν_τ oscillations (Oscillation Project with Emulsion-tRacking Apparatus (OPERA) at CERN, 2010), detection of Higgs boson (A Toroidal LHC ApparatuS (ATLAS) and Compact Muon Solenoid (CMS) collaborations at CERN, 2012), a precision comparison between *p* (uud) and anti-*p* (anti-u, anti-u, anti-d) (S. Ulmar *et al.*, RIKEN, 2015), oscillation experiment of CP-violation between ν and anti-ν (Tokai-to-Kamioka (T2K) collaboration, 2020).

Since J.J. Thomson discovered the electron in 1897; people long thought that proton (*p*), neutron (*n*), and e^- are elementary particles responsible for constituents of atoms. In 1928, Dirac theoretically predicted anti-e^-, now called the positron, e^+. In 1931, to explain the origin of a broad spectrum in the β-decay reaction of ^{210}Bi ($p = 83$, $n = 127$, Niels Bohr in 1930) [17], Pauli theoretically hypothesized a neutral massless tiny particle with a half-integer spin of 1/2, satisfying the energy conservation law; later, the tiny particle is called neutrino coined by Enrico Fermi [17]. "Anyone cannot detect the neutrino because of massless neutral particle," Pauli thought. The masslessness means that neutrino travels with the same speed as the massless photon. In 1957, Bruno Pontecorvo postulated a possibility of alteration from ν_e to ν_μ after long traveling of ν_e as a solar neutrino [73].

A half-century later, neutrino oscillations between ν_e and ν_μ, and ν_μ and ν_τ were experimentally discovered, leading to the conclusion that ν_e, ν_μ, and ν_τ have tiny masses, <0.22 eV/c^2, <2 MeV/c^2, and <15.5 MeV/c^2, respectively. Note that the rest masses of e^- and *p* are 0.51 MeV/c^2 and 938.3 MeV/c^2, respectively (Figure 8 and Table 1). Thus, the masses of ν_e and anti-ν_e are tiny, but the mass of ν_μ is of the order similar e^- and ν_τ is massive. The three neutrinos are no longer massless beyond the first theoretical hypothesis as the virtual massless particles.

Either *p* or *n* of atomic nucleon is no longer elementary particles while e^- and e^+ are elementary particles. Therefore, *p* (charge +1, spin 1/2) is formed from three quarks; u (charge =2/3, spin 1/2), u (charge =2/3, spin 1/2), d (charge = −1/3, spin −1/2), while *n* (charge 0, spin −1/2) is made of three quarks; u (charge = 2/3, spin 1/2), d (charge = −1/3, spin −1/2), d (charge = −1/3, spin −1/2) (Figure 8 and Table 2). Three π-mesons called pions, involving π^+ (charge = +1, spin 1), π^0 (charge = 0, spin 0), and π^- (charge = −1, spin −1), are made of u/anti-d, u/anti-u (d/anti-d), and d/anti-u, respectively. Noting that π^+ is antiparticle of π, while π^0 is antiparticle of π^0 itself.

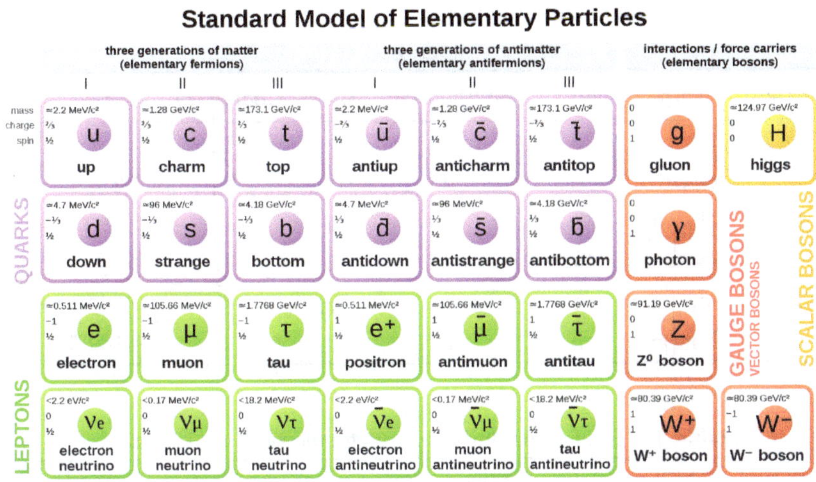

Figure 8. Elementary particles tabulated in the standard model. Six quarks — fermions — belong to hadrons — heavy particles, while six leptons, The quarks, and leptons, half-integer spin 1/2, are classified as fermions. The gluon, photon, and Z^0/$W^{\pm}$ particles, responsible for a carrier of strong nuclear, electromagnetic, and weak nuclear forces, respectively, are integer spin 1, then called vectorial gauge bosons, or simply, gauge bosons. The graviton is speculated to be a massless boson with spin 2 and assumed to be the carrier of gravitational force but is not detected yet. Higgs particles — called a scalar boson with spin 0 — are responsible for providing a mass to all elementary fermions and Z^0/$W^{\pm}$ bosons, excepting massless photons. The chart is taken from Refs. [17, 18, 76].

Table 1. Brief characteristics of leptons and anti-leptons, taken from Refs. [17, 77].

Leptons

Generation	Name	Symbol	Antiparticle	Spin	Charge (e)	Mass(MeV/c^2)
1	Electron	e^-	e^+	½	−1	0.511
	Electron neutrino	ν_e	$\bar{\nu}_e$	½	0	<0.0000022
2	Muon	μ^-	μ^+	½	−1	105.7
	Muon neutrino	ν_μ	$\bar{\nu}_\mu$	½	0	<0.170
3	Tau	τ^-	τ^+	½	−1	1,776.86 ± 0.12
	Tau neutrino	ν_τ	$\bar{\nu}_\tau$	½	0	<15.5

Table 2. Brief characteristics of quarks and anti-quarks [17, 18, 77].

Quarks

Generation	Name	Symbol	Antiparticle	Spin	Charge (e)	Mass(MeV/c^2)
1	up	u	$\bar{u}$	½	$+^2/_3$	$2.2^{+0.6}_{-0.4}$
	down	d	$\bar{d}$	½	$-^1/_3$	$4.6^{+0.5}_{-0.4}$
2	charm	c	$\bar{c}$	½	$+^2/_3$	1,280 ± 30
	strange	s	$\bar{s}$	½	$-^1/_3$	96^{+8}_{-4}
3	top	t	$\bar{t}$	½	$+^2/_3$	173,100 ± 600
	bottom	b	$\bar{b}$	½	$-^1/_3$	$4{,}180^{+40}_{-30}$

Table 3. Brief characteristics of detected bosons [17, 78].

Name	Symbol	Antiparticle	Spin	Charge (e)	Mass(GeV/c^2)	Interaction mediated	Observed
Photon	γ	Self	1	0	0	Electromagnetism	Yes
W boson	W^-	W^+	1	−1	80.385 ± 0.015	Weak interaction	Yes
Z boson	Z	Self	1	0	91.1875 ± 0.0021	Weak interaction	Yes
Gluon	g	Self	1	0	0	Strong interaction	Yes
Higgs boson	H^0	Self	0	0	125.09 ± 0.24	Mass	Yes

Table 4. Brief characteristics of undetected boson [17].

Name	Symbol	Antiparticle	Spin	Charge (e)	Mass(GeV/c^2)	Interaction mediated	Observed
Graviton	G	Self	2	0	0	Gravitation	No

The *p*/*n*, mesons, and their antiparticles are classified to a hadron (heavy particle) family, while e^-–ν_e, μ–ν_e, τ–ν_τ and their antiparticles belong to a lepton (light particle) family. Three quarks and three anti-quarks located at three generations are elementary fermion/anti-fermion particles (spin 1/2, charges ±1/3, ±2/3). Likewise, three pairs at three generations of e^- and e^+, μ- and anti-μ, τ- and anti-τ, and the corresponding three pairs of e^-- and e^+-neutrinos, μ- and anti-μ neutrinos, and τ- and anti-τ neutrinos are elementary particles as leptons (fermions, a half-integer spin 1/2, charges 0). Herein, it is important to emphasize that spontaneous decay experiments using metastable neutral particles, like ν, anti-ν, and neutral mesons, are likely to unveil hidden CP- and P-symmetry violations.

In addition to these fermions and anti-fermions, elementary bosons were invented. The gluon, photon, and $Z^0/W^{\pm}$ particles are carriers of SF, EMF, and WF, respectively, that are called gauge bosons with an integer spin 1 (Figure 8 and Table 3). The graviton, which is a gauge boson with spin 2, has not been detected yet, is assumed to be a carrier of gravitational force (Table 4). Moreover, the Higgs particle (H^0) — called a scalar boson with an integer spin 0 — is responsible for the mass of all particles. The $Z^0/W^{\pm}$ and H^0 bosons are massive, but photons and gluons are massless.

The rational design of experimental setups to detect any symmetry violations of these particles — apparatus and instruments located at specific locations and within specific organizations, specifically designed detectors, a collaboration team, funding support, and special health and environmental care toward high-energy sources — is inevitably needed to unveil so-called big science.

1.4 P- and CP-Violation in Sub-Atoms

In the mid-twentieth century, physicists knew that the four fundamental forces — GF, SF, EMF, and WF — govern natures of elementary particles, sub-atoms, atoms, nuclear reactions, and the universe (Figure 1). Most physicists concurred that three discrete symmetries (C, P, and T) and the combinational symmetries (CP, PT, CT, and CPT) are conserved in nuclear reactions, quarks, leptons, sub-atoms, atoms, and molecules. In 1927, Eugene Wigner [11] published the first paper describing P-invariance — mirror symmetry conservation and the scientific community had long accepted the invariances of all C, P, T, CP, PT, CT, and CPT as the core principles in physics. To date, P-invariance — mirror symmetry conservation — is the core principle in modern chemistry.

Several experimental proofs of P- and CP-violations at the levels of subatomic and elementary particles became the core principles in physics, stimulated from an open question raised by Lee and Yang [79] in 1956 of whether the P-symmetry at subatomic and elementary particle levels is precisely conserved. The evidence was confirmed by the fact that P-violations at the subatomic level was proven in the β^--decay reaction by Wu *et al.* in 1957 [80] and in the β^+-decay by Ambler *et al.* and Postma *et al.* in 1957 [81, 82].

$$^{60}\mathrm{Co} \rightarrow {}^{60}\mathrm{Ni} + e^- + \text{anti-}\nu$$

$$^{58}\mathrm{Co} \rightarrow {}^{58}\mathrm{Fe} + e^+ + \nu$$

These pioneering experimental results led to ^{60}Co and ^{58}Co, which are metastable radioisotopes, spontaneously relax to the corresponding ^{60}Ni and ^{58}Fe stable isotopes, respectively. These processes allowed to unveil P-violations via spontaneous $\beta^{\pm}$-decay processes mediated by charged W± bosons [83]. The spontaneous decay from the metastable to the ground states, accompanied by P-violation, essentially arises from the far-from, non-equilibrium excited state. Yet, physicists, including Wigner, concurred with the conservation of CP-invariance within the CPT invariance theorem.

In 1964, Cronin and Fitch discovered CP-violation in the decay process of neutral K-meson (K^0) by the observation of fast and slow decays of K_L and K_S [84, 85]. K^0-meson is composed of d and anti-strange (anti-s) quarks, while anti-K-meson (anti-K^0) is made of anti-d and s. These quarks mix with each other as shown in the equations below.

$$K_L = 1/\sqrt{2}\cdot(K^0 + \text{anti-}K^0)$$

$$K_S = 1/\sqrt{2}\cdot(K^0 - \text{anti-}K^0)$$

Under CP-symmetry, K_L- and K_S-mesons decay to π^0-meson with a long lifetime of 50 ps and π^+-π^- meson pairs with a short lifetime of 70 fs, respectively. Due to a subtle CP-symmetry violation, metastable K_L-meson decayed to π^+-π^- with a small probability of 0.2% relative to events of K_S-meson decays. Thus, K^0 and anti-K^0 were no longer eigenstates of neutral kaons. The idea of the historical experiment is applicable to spontaneous radiation from metastable chiral molecular luminophores in the photoexcited state to the stable chiral luminophores in the ground state. If

K^0 and anti-K^0 are regarded as left-hand (L) and right-hand (R) molecular luminophores, respectively, one can expect that L- and R-enantiomers in the photoexcited state do not emit in an equal probability to relax to the stable ground states. A spontaneous radiation process from the metastable photoexcited state could allow us to unveil hidden P-, T-, and PT-violations, while C-, CP-, and CT-violations are impossible to confirm for molecules made of antimatters. Note that, among the four forces, the weak force is responsible for P- and CP-violations.

1.5 C-, P-, and CP-Violations in Atoms

The historical findings of P-violation in the $\beta^{\pm}$-decay processes stimulated Yakov B. Zel'dovich in 1959 to hypothesize so-called anapole moment, causing a tiny PV effect at the atomic level on the order of 10^{-20} via interactions of e^{-} with nucleon with odd nuclei spin, e.g., hydrogen induced by weak neutral current (WNC) [86] (Figure 9).

In 1974, French physicists M.A. Bouchiat and C. Bouchiat theoretically predicted how the tiny APV is amplifiable to a detectable level [87–91]. In 1988, Hegstrom *et al.* theoretically visualized a handed streamline of closed-loop electron probability around nucleus for chiral hydrogen atom that is led by an admixture of P-odd $2p_{1/2}$ and P-even $2s_{1/2}$ states in nearly energetically degenerate states perturbed by PV-WNC mediated by Z^0 boson [92] (Figure 10). Atoms having an odd-number nuclear spin might be crucial to detect APV.

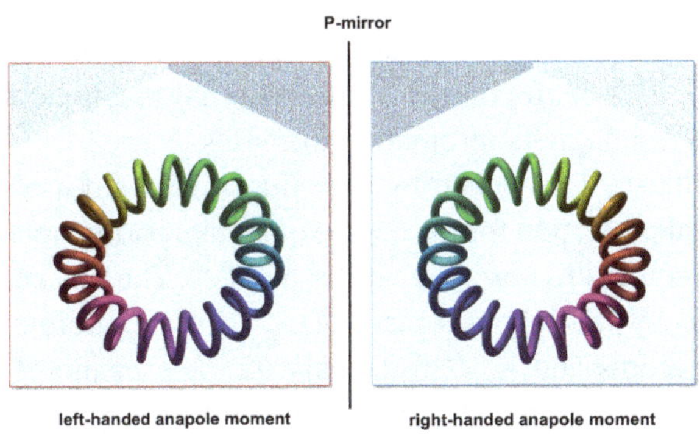

Figure 9. (Left) A handed anapole moment of matter (p, n, and e^{-}) with odd-number nuclear spins in a real world and (right) the opposite hand anapole moment of matter in an unreal world as a consequence of P-violation.

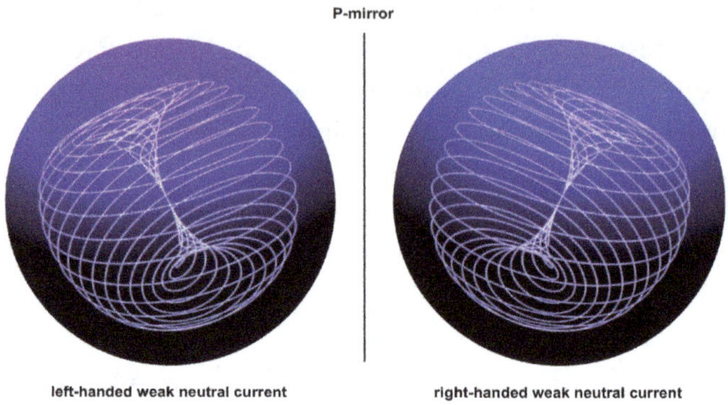

Figure 10. (left) A handed weak neural current (WNC) mediated by Z^0 boson of hydrogen atom (p and e^-) in a real world and (right) the opposite-handed WNC in an unreal world as a consequence of P-violation. Modified by Ref. [92]. In P-violation, hydrogen with the handed WNC is energetically equal to antihydrogen (anti-p and e^+). However, in CP-violation, hydrogen with the handed WNC is no longer energetically equal to antihydrogen.

However, the standard model, or the Glashow–Weinberg–Salam theory, gave rise to a more general prediction that even highly symmetrical neutral atoms comprising p, n, and e^- may be chiral as a consequence of P-violation due to handed WNC mediated by neutral Z^0 bosons. The handed WNC exists in all atoms, including hydrogen. One can, in principle, observe a tiny optical activity deriving from an ultrashort range interaction between e^- and nucleus. In chemistry, WNC is regarded as a persistent torus current flow with handedness in the absence of an external magnetic field. This is in contrast to the external magnetic field-induced ring current of diamagnetic aromatic molecules.

This hypothesis was proven by detecting small values of optical rotation. Theoretical prediction for the degree of atomic parity violation (DAPV) for realistic lighter atoms was in the order of 10^{-15}. The tiny energy inequality is enhanced by the schemes that DAPV $\propto Z^3$ (Z: atomic number) for nonrelativistic atoms and $\propto Z^5$ for relativistic heavier atoms [87, 89, 91, 93–95]. A better enhancement is possible to effectively mix between odd and even parities located at closer energy levels. When a highly forbidden atomic transition is photoexcited, the DAPV reaches a detectable level. In fact, optical rotations as an absorption mode, the Stark effect from vapors of heavier atoms (Bi, Pb, Tl), and a luminescence mode in ^{133}Cs vapor were reported in visible and near-infrared regions. Definitive APV experiments at resonance energy using longitudinally polarized neutron beam as a function of neutron energy were reported to verify the APV hypothesis, as follows:

A cold neutron beam (spin 1/2, <0.01 eV) detected an APV effect from ^{117}Sn with odd-number nuclear spin (p = 50, n = 67, spin 1/2) as positive sign (right-hand) of neutron spin rotation, that was enhanced by nearly two orders of magnitude compared to that of even-number nuclear spin ^{124}Sn (p = 50, n = 74, spin 0) [96, 97]. It is obvious that odd-number nuclear spin is important to detect APV as the solid-state. The low-energy neutron beam is likely to apply to detect the MPV of several molecules consisting of odd-number nuclear spin atoms.

Moreover, in the 1990s and the early 2000s, aiming at pursuing APV in nuclear physics, sophisticated p-wave resonance experiments of isotopically pure metals were designed. Longitudinally polarized neutrons as a function of neutron energy in the range of 1 eV and 1100 eV were transmitted to solid samples [98]. The sign and absolute magnitude in neutron rotation spectrum of ^{117}Sn solid largely depend on neutron resonance energy. The DAPV (%) at four specific resonances among 29 resonances were detected; (+)-0.79% at 1.33 eV, (−)-0.08% at 34.04 eV, (−)-0.30% at 423.3 eV, and (+)-1.07% at 1045.0 eV. The value of DAPV in physics may be half of the g_{NCD} value, known as Kuhn's anisotropy, in NCD spectroscopy in chiroptical chemistry. The positive sign (right-handedness) in physics corresponds to a negative sign (left-handedness) in chemistry. By utilizing the same approach, the enhanced resonance DAPV have been detected for several heavy atoms possessing odd-number nuclear spins; ^{93}Nb (p = 41, n = 52, spin 9/2), ^{103}Rh (p = 45, n = 58, spin 1/2), ^{107,109}Ag (p = 47, n = 60, spin 1/2 (52%) and p = 47, n = 62, spin 1/2 (48%)), ^{113}Cd (p = 48, n = 65, spin 1/2), ^{115}In (p = 49, n = 66, spin 9/2), ^{121,123}Sb (p = 51, n = 70, spin 1/2 (57%) and p = 51, n = 72, spin 7/2 (43%)), ^{127}I (p = 53, n = 74, spin 5/2), and ^{133}Cs (p = 55, n = 78, spin 7/2). In 1985, Khariplovich *et al.* predicted that small intervals between atomic energy levels with the opposite parities in rare-earth atoms may result in the enhancements of P- and T-violations at magnetic-dipole-allowed and/or electrical-dipole-forbidden transitions [99]. The ^{154}Sm and ^{161}Dy are the best candidates to precisely determine the electroweak mixing parameter, $\sin^2\theta_W$ (θ_W, Weinberg's mixing angle), and also to investigate P-odd anapole moments and P-odd/T-odd magnetic quadrupole moments.

In 2009, Budker *et al.* reported that, compared to the APV experiment of odd-number nuclear spin ^{133}Cs, even-number nuclear spin ^{174}Yb with ($p = 70$, $n = 104$, spin 0) revealed APV spectroscopic signals enhanced by ~100 times at a resonance laser line of 408 nm when $6s^2\ ^1S_0$–$5d6s\ ^3D_1$ transition at 24,489 cm^{-1} was excited to mix with $6s6p\ ^1P_1$ (25,068 cm^{-1}) transition under a tiny external magnetic field of ~0.01 T [100]. The resonance effect in the spectral experiment appeared to be crucial to boosting a detectable APV effect.

Finally, the most difficult question remains as to whether matter and antimatter atoms are energetically identical and whether the CPT symmetry invariance is valid. To answer the biggest questions several scientists, for decades, devoted a great deal of effort to stably trap antihydrogen and anti-helium atoms for a fraction of time that are generated without contact with matters, allowing to measure precision spectroscopic datasets. As expected, it is a very difficult task to generate anti-atoms on Earth. In 1978, CERN succeeded in the first storage of several hundred anti-p of 2.1 GeV/c generated from the Proton Synchrotron (PS) accelerator while keeping for a long period of 85 hrs. Following this, in 1995, the Low-Energy Antiproton Ring (LEAR) facility led by Walter Oelert at CERN for the first time created nine anti-H atoms. In 2002, at CERN, ATHENA — an antimatter research project using the Antiproton Decelerator (AD) and Antihydrogen TRAP (ATRAP) collaboration — independently created thousands of cold anti-H [101]. In 2010, at CERN, Antihydrogen Laser Physics Apparatus (ALPHA) and ATRAP independently succeeded in proper amounts of anti-H [102, 103].

In 2016, ASACSA led to the conclusion of the equality between anti-protonic helium, e.g., anti-p^3He^+ and anti-p^4He^+ nucleons and protonic helium by a precise measurement of the antiproton-to-electron mass ratio of 1836.1526734, agreeing with a proton-to-electron experimental value within 8×10^{-10}. In 2020, the ALPHA measured the zero-field fine-structures ($2P_{1/2}$–$2P_{3/2}$ and $2P_{1/2}$–$2P_{1/2}$), of anti-H [104]. Hence, the CPT invariance at a relative precision was validated within a precision of ~10^{-10} [105]; however, the broken CPT invariance may be detectable when an extreme precision of ~10^{-20} is achievable in the far future [106].

In 2011, the STAR collaboration using the Relativistic Heavy Ion Collider (RHIC) at CERN succeeded in the generation of 18 anti-^{4}He among 10^9 Au–Au collision events [107]. In 2015, RIKEN and collaborators confirmed that charge-to-mass ratios between anti-H and H do not reveal any difference in the order of 10^{-12} by comparing their charge-to-mass ratios [108, 109].

On the other hand, several astrophysics teams are searching for the cosmological origin of anti-atoms. In 2018, it was announced that an Alpha Magnetic Spectrometer (AMS-02) on board the International Space Station (ISS) tentatively detected six anti-^{3}He and two anti-^{4}He that were initially attributed to cosmic rays colliding with molecules in the interstellar medium and producing the antimatter in the process. In 2021, Simon Dupourqué, Luigi Tibaldo, and Peter von Ballmoos reported the most probable 14 candidates of anti-stars made of antimatter among nearly 6000 γ-ray sources from the archive γ-ray dataset collected at Fermi Gamma-ray Space Telescope (GLAST) for more than ten years [110, 111]. The General Antiparticle Spectrometer (GAPS) is being designed to detect the cosmic anti-^{3}He nuclei because the astronomers predicted the anti-stars made of antimatter among γ-ray sources. Although the anti-stars are very speculative, the existence may be revealed only by weak γ-ray emission at a specific 70 MeV. If anti-H and anti-He nucleus exist, they are annihilated at 70 MeV with the H and He nucleus. As you may recall, all atoms are inherently chiral with a preference led by PV-WNC mediated with Z^0-boson; an open question is whether a preference in the chirality of anti-atoms has the same or opposite. The annihilation of atom-antiatom may radiate a handed circularly polarized γ-ray if CP-symmetry at the atomic level is violated.

1.6 P-, T, and PT-Violations in Molecules

From the viewpoint of CMSB phenomena, chiral and racemic fields in the realm of chemistry and biochemistry are very common topics relevant to the origin of biomolecule handedness on Earth. For example, an inevitable influence on Earth will be one-way rotation combining with a static magnetic field may be hypothesized as a long-term chiral gravitational force (g^*) [112]. All experimental evidence and discussion of stereochemistry on Earth are employed under these unavoidable conditions that may be classified as the universal cause. Any planet or satellite rotates and revolves in the presence of the cosmic magnetic field. In addition to

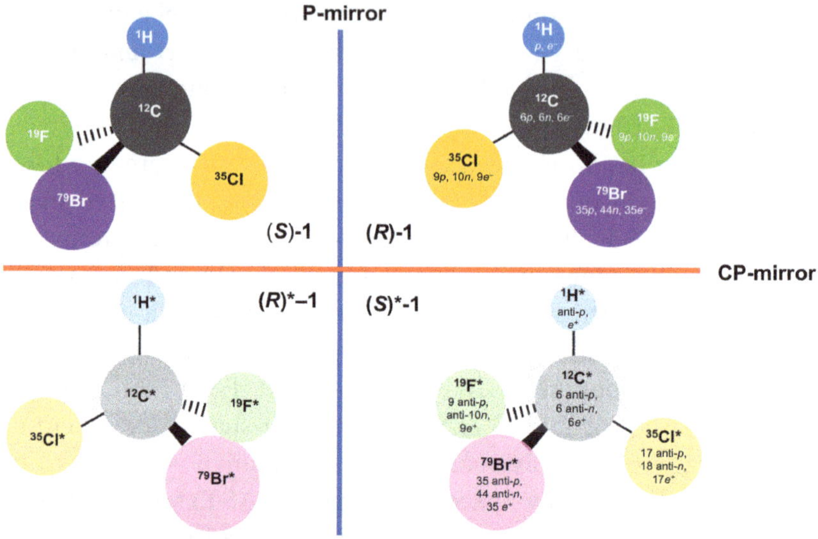

Figure 11. (Top right); (*R*)-^{12}C-^{1}H-^{19}F-^{35}Cl-^{79}Br, **(*R*)–1** (matters) and (top left); **(*S*)–1** (matters) and (bottom right); (*S*)*-^{12}C*-^{1}H*-^{19}F*-^{35}Cl*-^{79}Br*, **(*S*)–1*** (antimatters) and (bottom left) **(*R*)–1*** (antimatters). In the framework of P-violation, **(*R*)–1*** and **(*S*)–1*** are diastereomers and **(*R*)–1** and **(*S*)–1** are as well, while **(*R*)–1*** and **(*S*)–1*** are enantiomers of **(*S*)–1** and **(*R*)–1**, respectively. In the case of CP-violation, **(*R*)–1*** and **(*S*)–1** are no longer enantiomers. Likewise, **(*S*)–1*** and **(*R*)–1** are not enantiomers but diastereomers.

Z^0-origin parity violation energy difference (PVED) and WNC, the chiral environmental influences on Earth may act as chiral triggers on mirror symmetry breaking (MSB) effects observable on Earth. These effects could explain a conformational transition of right-handed DNA to left-handed DNA duplexes, known as *B-Z* DNA transition [112]. In this section, we assume that PVED≫g^*.

Four stereoisomers of (*R*)- and (*S*)-^{12}C-^{1}H-^{19}F-^{35}Cl-^{79}Br as matters, **(*R*)–1** and **(*S*)–1**, and the corresponding antimatter, **(*R*)–1*** and **(*S*)–1***, are illustrated in Figure 11. In the framework of P-symmetry (textbook of orthodox stereochemistry), **(*R*)–1** and **(*S*)–1** in a matter world as well as **(*R*)–1*** and **(*S*)–1*** in an antimatter world are enantiomers, respectively. In the framework of P-violation, a pair of **(*R*)–1** and **(*S*)–1** are no longer enantiomers and should be diastereomer. Likewise, **(*R*)–1*** and **(*S*)–1*** are not enantiomers; therefore, a pair of **(*R*)–1*** and **(*S*)–1**, as well as a pair of (*S*)-1* and (*R*)-1, are enantiomers. On the other hand, when CP-symmetry was violated, **(*R*)–1*** and **(*S*)–1** are no longer enantiomers. Likewise, **(*S*)–1*** and **(*R*)–1** are diastereomers. Therefore, all the four **(*R*)–1**, **(*S*)–1**, **(*R*)–1***, and **(*S*)–1*** are diastereomer because of energy inequality.

To experimentally validate hypotheses of P- and CP-violations in cosmology, subatomic, elementary particle, and atomic physics, well-designed high-energy accelerators, spacecraft, and telescopes, those are equipped with special probes, are inevitably required because special events rarely occur regardless of by-chance and/or predictable by theories. On the other hand, to experimentally test P-, T-, and PT-violation theories for a pair of enantiomers, high-energy accelerators, spacecraft, and special probes are not always needed because special events are detectable by desk-top instruments/apparatuses at any laboratory on Earth, at any time, by anyone with a limitation of anti-chemicals made of anti-*p*, anti-*n*, and e^+ are not feasible.

Chemistry, non-covalent chemistry (supramolecular chemistry), and biochemistry performed on Earth are owed to light atoms with nucleosynthesis and heavier atoms as the result of the supernova explosion. Emerging a supramolecular chirality refers to MSB in the framework of P-conservation law associated with energy equality of left-right enantiomers. Hence, it is safe to state that MSB occurs from achiral and/or racemic multiple molecular systems upon non-covalent interactions through molecular chirality/helicity transfer mechanisms.

1.6.1 The Early History of Stereochemistry — P-Invariance in Organic and Inorganic Molecules

In 1808, French physicist, Étienne-Louis Malus, was aware of the fact that two polarizations of light in the calcite ($CaCO_3$) crystal exist, and in 1811, he published the finding in several papers. The papers stimulated François Arago to observe the colored polarization of a thin plate of mica. In 1815, Jean-Baptiste Biot demonstrated that, when polarized light is passing through natural inorganic crystals such as α-quartz and several natural organic substances, light is rotated clockwise (CW) or counterclockwise

(CCW). Biot was also aware that a set of non-superimposed α-quartz leads polarized light to rotate in the direction of CW or CCW, respectively [24]. This phenomenon is often called natural optical activity (NOA) and natural NOR.

Similar to non-superimposed α-quartz, Pasteur in 1848, who was a student of Biot, found clockwise (CW) and counterclockwise (CCW) NOR by dissolving a set of non-superimposed enantiomorph crystals in water, respectively [1, 3]. Biot verified Pasteur's findings by the repeated cross-checking experiment. The mirror-image crystals were spontaneously generated from NOR inactive, racemic sodium ammonium tartrate. In 1846, Faraday discovered the so-called Faraday effect — magnetically induced optical activity from isotropic glass and media. Faraday prompted Pasteur to test generating optically active substances under a magnetic field but obtained a negative outcome [3]. These personal experiences led Pasteur to give his own view at the lecture in 1860 [2, 3]:

> *Chiral interactions mediate the transformation of the organic and the inorganic world alike, possibly, even the generation of organisms. Life is dominated by these dissymmetric actions of whose enveloping and cosmic existence we have some indication. In consequence of living systems must have access to some universal chiral agency, and dissymmetric forces exist at the moment of the elaboration of natural organic products; forces which are absent or ineffectual in the reaction of our laboratories'. These natural chiral forces might be magnetic or electric force or irradiation with light in rotatory motion.*

In my view, *light in rotatory motion* implies circularly polarized light (CP light). In 1862, Faraday attempted to clarify the effect of external magnetic field for field emission spectrum but failed [113]. His foresight led Peter Zeeman in 1897 to confirm the splitting of an emission line into several lines under a static magnetic field, now called the Zeeman effect [114]. Faraday effect/inverse Zeeman effect and Zeeman effect are the consequence of sub-level energy splitting through CP light–matter interactions under the magnetic field with north-up/south-down or south-up/north-down geometry relative to an incident light beam in the absorption and emission modes, respectively.

Pasteur's assumption led Joseph A. Le Bel [115, 116] and Jacobs H. vant' Hoff [117, 118] independently to theoretically hypothesize an eccentric three-dimensional (3D) stereochemistry as a model in which carbon-centered tetrahedron molecules attach with four different atoms. Before the hypothesis, the two-dimensional (2D) model described in a plain paper was the standard principle in chemistry [116, 120], though nowadays, a wide range of 2D semiconductors such as graphene, transition metal dichalcogenides (MoS_2, WS_2), hexagonal boron nitride (h-BN), and artificial superlattices have become important topics in physics, chemistry, and materials science.

The stereochemical model of organic molecular chirality with tetrahedron geometry led to the successful isolation of an optically active cobalt complex with organic ligand, *cis*-$[Co^{III}(en)_2(NH_3)_2]^{2+}$ X_2 (en: ethylene diamine, X = Cl^-, Br^-, NO_3^-, SO_4^{2-}, $S_2O_6^{2-}$) [121, 122] and purely inorganic tetranuclear cobalt complex $[\{Co^{III}(NH_3)_4(OH^-)_2\}_3Co^{III}]^{6+}(Br^-)_6$ [122, 123], supported by chiroptical NOR and NCD data. However, a naive question remains unanswered whether P-symmetry for pairs of the chiral cobalt enantiomers is rigorously conserved.

1.6.2 Mirror Symmetry Breaking in Stereochemistry

Later, the stereochemistry or 3D chemistry in organic and inorganic chemistry, as well as polymer chemistry, was experimentally and theoretically established as a concrete concept since the early twentieth century up to the present day. However, although we are living in a spatiotemporal (four-dimensional, 4D) world requiring four variables (x, y, z, and t), the stereochemistry does not involve explicitly a time-dependent variable t, e.g., any dynamics, reversibility of time such as oscillating, tunneling in a symmetrical double-well potential, superposition of left-and-right, the arrow of time, negative entropy, non-equilibrium thermodynamics allowing open-flow of energy and substances.

In 1898, Frederic S. Kipping and William J. Pope were possibly the first to use statistical analysis to account for MSB of chiral *d*- or *l*-$NaClO_3$ cubic crystal (space group $P2_13$) obtained from water solutions of achiral $NaClO_3$ in the absence and in the presence of D-glucitol under stagnant conditions as to whether the chance or necessity mechanism is valid regarding the origin of biological chirality [124]; whether MSB is merely as a result of by-chance mechanism, obeying a mirror-symmetrical gaussian function, or a handed MSB (HMSB) is determined by the necessity due to unresolved factors, obeying a non-mirror-symmetrical gaussian function [125, 126]. Note that MSB does not mean P-violation in chemistry, while, possibly, an HMSB means P-violation in chemistry. They concluded the by-chance mechanism [124].

Similar to the use of D-glucitol shown above, in 1919, Perucca succeeded in observing, possibly, the first supramolecular MSB phenomenon by seeding the crystal chirality; Perucca observed NOR triarylmethane dye from NOR inactive one in the presence of *D*-$NaClO_3$ crystal [127]. The MSB belongs to controlled MSB (CMSB). In 1932, Pfeiffer *et al.* were aware of the CMSB phenomenon from labile racemic metal complexes (Zn^{2+}, Cd^{2+}, Ni^{2+}) in the presence of *L*-amino acids [128, 129]. Further reports to confirm P-conservation using *L*-$NaClO_3$ and *D*-amino acids were not conducted in those days. Nowadays, the MSB phenomena are widely recognized in crystallography, artificial systems, and naturally occurring systems. In 2000, Borovkov coined *chirogenesis* to account for emerging NCD active bis(Zn^{II}porphyrin) supramolecules from the corresponding NCD-inactive ones upon non-covalent interaction with D-L chirality of amines and alcohols; the chirogenesis is nowadays popularly known as a generalized concept of chiroptical generation in various supramolecular systems [130–134, 135].

1.6.3 Theories of Molecular Parity Violation

The CMSB plays a key role in designing chiroptical functionalities, such as chirality recognition and chirality sensing. The chirogenesis capability unifying scenarios of attractive/repulsive intermolecular chirality transfers is widely extendable to binary/multiple molecular and polymer hybridized systems upon several non-covalent interactions. However, the non-covalent attractive/repulsive interactions, followed by all chemical reactions/processes, rely on the standard model of static stereochemistry that does not involve spatiotemporal chemistry with an idea of the coherence-and-decoherence.

Physics in the early twentieth century was certainly a ground-breaking era. In 1900, Max Planck proposed a quantum theory ($E = h\nu = hc/\lambda$; h Planck constant, ν frequency, c light speed, λ wavelength) to account for the long-standing issue of blackbody radiation-absorption of light as a function of ν that was formulated by Max Planck in 1900. The radiation-absorption process is a collection of discrete energies. Subatomic, atomic, molecular, and, possibly, supramolecular processes should obey quantum mechanics, and their spectroscopic datasets are analyzed by quantized variables with h or $\hbar$ (= $h/2\pi$) called Dirac's constant or reduced Planck's constant. The newly defined light was named quantum of light (lichtquant in Germany) by Albert Einstein in 1905 and coined photon, playing an essential role in every radiation process, by Gilbert N. Lewis [136, 137]. As mentioned earlier, light is a massless elementary particle carrying an integer spin, ±1 (boson), while an electron carries a half-integer spin, 1/2 (fermion). Nowadays, photon as terminology is popular in physics but perhaps rare in chemistry.

In 1929, Léon Rosenfeld, who is a pioneer in quantum mechanical theory to interpret molecular optical activity [138]. Rosenfeld formulated a second-order time-dependent perturbation theory involving transition electric and magnetic dipole moments. However, in 1927, Friedrich H. Hund posed the fundamental question, widely known as Hund's paradox among molecular physicists and spectroscopists, but rarely recognized among most chiral molecular and supramolecular chemists [78, 139–141]. From the relationship between optical activity and molecular chirality in a symmetrical potential energy double-well, quantum tunneling mechanism allowed to propose that mirror-image molecules exist as superposition of the odd- and even-parity eigenstates, $|L\rangle = 1/\sqrt{2}\ (|+\rangle - |-\rangle)$ and $|R\rangle = 1/\sqrt{2}\ (|+\rangle + |-\rangle)$, or, $|L\rangle = 1/\sqrt{2}\ (|+\rangle + |-\rangle)$ and $|R\rangle = 1/\sqrt{2}\ (|+\rangle - |-\rangle)$. Hund was the first to introduce the concept of tunneling in chemistry. The apparent optical activity and mirror-image molecules were assumed to be observed as a superposition of true mirror-image molecules.

As mentioned earlier, the discovery of P-violation at $\beta^{\pm}$-decay experiments immediately inspired Russian nuclear and atomic theoreticians to arise the P-violation in atoms and molecules. Zel'dovich was the first to propose the anapole moment responsible for APV of atoms with odd nuclei spin ≠ 0 in 1959 [86] and 1961 [141]. A quarter-century later, in 1977, he invoked an inherent P-violation between left-and-right molecules and pointed out the importance of Z^3-dependent spin–orbit coupling (ΔE_{ST}) due to Z-charges of the nucleus (Zel'dovich, 1977). In 1975,

Vladilen Letokhov estimated the degree of P-violation in hypothetical molecules due to weak neutral interaction between electrons and nuclei of molecules [142]. However, the MPV hypothesis is yet not accepted among most of all chemists. Only a few chemists are seriously thinking of very tiny MPV effects.

Due to the *P*-conservation in energetically equivalent mirror-image molecules, the tunneling time between two local minima separated by a potential barrier (E_b) is inversely proportional to the energy splitting $\Delta E_\pm$ between the odd- and even-parity eigenstates. When E_b is sufficiently low, the molecules spontaneously oscillate between the |L> and |R> states, leading to the novel concept of spatiotemporal chirality. In the case of rigid tri-substituted phosphines, P-$R_1R_2R_3$, $E_b \approx 37$ kcal mol^{-1} and the tunneling splitting, $\Delta E_\pm$, is in the order of 10^{-17} eV [78]; the quantum oscillation of optical activity is impossible. On the other hand, when the inversion barrier of non-rigid ammonia molecule due to quantum tunneling is ~5.8 kcal mol^{-1}, leading to $\Delta E_\pm = 0.8$ cm^{-1} ($3.6 \cdot 10^{-6}$ eV or $2.3 \cdot 10^{-3}$ kcal mol^{-1}) [78].

In 1974, a chemical physicist, D. Rein, theoretically pursued P-violation between L-alanine and D-alanine as a realistic model of left-hand and right-hand enantiomer in terms of the origin of biomolecular handedness and chemical evolution on Earth [143, 144]. Rein and coworkers further theoretically investigated the effect of spin–orbit coupling (Z number) and the degree of P-violation when twisted carbon (lighter atom, Z = 6) of ethylene and twisted sulfur (heavier atom, Z = 16) of dialkylsulfide with cholestane framework [145, 146].

In 1978, Harris and Stodolsky assumed an interference between the two quantum oscillations between P-conserved EMF and P-violated EWF-origins for a hypothetical chiral molecule in a double-well [147]. The work may be the first theory to theoretically discuss dynamic spatiotemporal behaviors in conjunction with the MPV hypothesis.

In 1986, Martin Quack discussed PVED of hypothetical enantiomers at the ground state in a double-well and the electronically excited state in a single well [148]; the paper postulated how one could measure a subtle PVED on the order of 10^{-7}–10^{-14} J mol^{-1} or 10^{-8}–10^{-15} cm^{-1} on a spectroscopic scale. For example, isolated molecular beams of chiral molecules consisting of heavier elements, such as under collision-free gas-phase conditions, are suited to avoid the effects of minor uncharacterizable impurities and intermolecular perturbation. This work with comprehensive reviews in 1989, 1992, and 2002 promoted Quack and coworkers to theoretically investigate MPV [149, 150, 151]. A series of chiral molecules made of rigid frameworks and rotamers with non-rigid frameworks were listed as the candidates.

The theory may encourage high-resolution and femtosecond spectroscopists to detect the subtle PVED which are detectable as very small differences in ro-vibrational frequencies in infrared and microwave regions. In 1977, Arimond *et al.* were the first to measure a tiny PVED between *d*- and *l*-camphor, rigid chiral bicyclic framework, but confirmed no difference between their ro-vibrational frequencies within $<10^{-8}$ accuracy [152]. Two decades later, Daussy and coworkers devoted to detecting the PVED between tetrahedral chiral molecules, (*S*)- and (*R*)-CHFClBr in 72% and 56% enantiomeric excess (*ee*), that were composed of mixtures of stable isotopes, e.g., ^{19}F, ^{35}Cl/^{37}Cl/, and ^{79}Br/^{81}Br [153]. They could not obtain a positive conclusion because of an actual $\Delta\nu/\nu << 10^{-13}$ as absorption mode beyond the detection limit even using sophisticated high-resolution laser spectroscopy [154]. The great challenge to conclusively answer MPV theories with the help of the high-resolution laser spectroscopy appears in progress [155].

Alternatively, the MPV in ammonia-like chiral molecules with flip-flop motion as a gas phase may be observable due to an oscillation with time as temperature-independent tunneling in the infrared region. In 2004, MacDermott and Hegstrom theoretically suggested that PVED for ammonia-like chiral pyramidal "cat" molecules, like P*-HDT (P*; stereogenic center) with lone pair and fluxional metal complexes [156, 157]. The ammonia-like molecules allow flip-flop motions associated with tunneling times in the range of 10^{-7}–10^{-2} s. Their PVED is possibly measurable as optical rotations using state-of-the-art polarimeters. In addition to the ammonia-like molecule, in 2000, MacDermott suggested optically active "cat" rotamers analogous to twisted H_2S_2 and twisted biphenyls to detect the PVED [158]. Here, "cat" means a superposition of left- and right-handed molecules. In other words, non-rigid stereogenic bonded rotamers without stereogenic centers and a mixture of Δ-Λ metal complexes with achiral ligands are regarded as candidates of the "cat" molecules and fluxional molecules.

Most theories discussed the degree of PVED values at the electronic, vibrational, and ro-vibrational ground states. Experimentalists attempted to spectroscopically detect the PVED as absorption mode. To my knowledge, Berger in 2003 was the first to suggest that achiral planar fluoroformyl aldehyde in the ground state, HFC*=O, can adopt a chiral pyramidal structure at the pseudo-stereogenic center C* in the photoexcited state, is suited to experimentally test the PVED because a well-defined parity at the photoexcited leads to clearly enhance the PVED [159]. The formyl aldehyde is regarded as an ammonia-like molecule capable of flip-flop motion in the photoexcited state [160, 161]. The formyl aldehyde is the simplest building block of several ribose and sugar structures. MacDermott thought that the EWF unifying P-conserving EMF and P-violating WF is intimately connected to the origin of biomolecular substances on Earth beyond our solar system [156, 162].

Upon excitation of pulsed unpolarized light, the PVED as a spontaneous radiation mode is detectable as optically active emission spectrum through the relaxation process from the photoexcited to the ground states by utilizing the state-of-the-art NCPL spectroscopy in the absence of an external magnetic field and magnetic circularly polarized light (MCPL) in the presence of the external magnetic field spectroscopy. The signal-to-noise ratios of NCPL and MCPL spectrometers commercialized recently are greatly improved. The current NCPL and MCPL spectrometers allow us to measure the photoexcited chirality of achiral and racemic molecular emitters in a homogeneous solution by alternating circularly polarized light as an incident light source. The PVED phenomenon may be detected as non-zero NCPL and non-zero MCPL signals when the photoexcited achiral and racemic molecular emitters are employed by dispersing in fluidic solutions, as suggested previously [164].

In 1967 and 1982, a chiroptical spectroscopist, Stephen F. Mason, described a review and a book based on standard chiral stereochemistry [125, 164]. However, by inspiring those theoretical works of P-violation of molecules and $\beta^{\pm}$ decay processes, in 1983, Mason and Tranter theoretically estimated the degree of P-violation in rotational isomers of L and D-alanine and their secondary structures of α-helix and β-sheet [165]. Soon, in 1983 and 1984, they published the first comprehensive reviews of MPV theories of biological molecules, biopolymers, and other small molecules, that possess rotational freedom along with single bonds such as C–C, C–O, C–N, and O–O [126, 166, 167].

Regarding the origin of biomolecular handedness on Earth and exoplanets, theoretical calculations of left-right preferences between D- and L-sugars, D- and L-amino acids, and left- and right-helical *n*-alkanes are interesting issues. In 1986, Tranter showed P-violation of O-sugars led by WF [168], followed by P-violation for other sugars [169]. In 1989, MacDermott and Tranter calculated the P-violation of twisted and puckered structures of tetrahydrofuran as a simple model of DNA/RNA, pentose, and α-aminoproionitrile [170]. In 1995 and 2000, MacDermott tabulated PVED values with corrected signs in the order of 10^{-14}–10^{-17} kT or 10^{-11}–10^{-14} J mol^{-1} (298K) of bio-related substances such as L-amino acids, D-ribose, DNA, and RNA that stably exist in nature and L-quartz and chiral fullerenes [162]. In 1998, Kikuchi *et al.* theoretically calculated the degree of PVED for L-amino acids and helical *n*-alkanes by taking spin–orbit coupling wavefunctions into account for calculations [171, 172].

1.6.4 Amplification Scenarios of Parity-Violating Energy Difference

The PVED of most organic molecules is on the order of 10^{-11}–10^{-14} J mol^{-1} (298K) and is far from detection limits in any state-of-the-art instruments. To amplify the PVED to the detectable level, four different scenarios were proposed.

The PV theory and $\beta^{\pm}$-decay experiments in 1956–1957 stimulated Yukio Yamagata in 1966 to theoretically consider a possible scenario only with three references, enabling to significantly amplify PVED, named an accumulation principle, later called a linear amplification model [173]. Achiral (or racemic) biomolecules possessing a tiny D-preference of PVED irreversibly get polymerized to yield naturally occurring ultrahigh biopolymers D-DNA over unnatural L-DNA(PVED (natural DNA) $\propto N$ (repeating units) $\times$ PVED (small molecule), when $N > 10^8$ during a chain propagation process). The oversimplified model is somewhat optimistic and was long a matter of debate [174]. Rigorous theoretical calculations concluded that the model is beyond pure speculation. Nevertheless, probably, most people might agree that the first paper greatly contributed to the rigorous theorists of MPV and stimulated experimentalists to design and test several feasible scenarios, enabling to significantly amplify from the tiny to detectable levels of PVED on macroscopic PV phenomena, e.g., as oligomers, crystals, colloids, and bio-/artificial polymers.

In 1982, in an analogy of BCS theory of superconductivity, Abdus Salam theoretically argued a possibility that certain chiral molecular system spontaneously undergoes a handed second-order structural phase transition associated with PVED induced by quantum tunneling at a critical temperature [175]; D–L amino acids, having rotational freedom at multiple single bonds; $T_c \propto <\phi>/10^3 \exp [-2/g_{eff}\, \sigma(1-4 \sin^2\theta_w) \sim 2.5 \times 10^2$ K, ϕ a complex auxiliary Higgs scalar field, θ_w Weinberg angle, $\sin^2\theta_w$ 0.24, g_{eff} an attractive coupling constant between spin-up (↑) and spin-down (↓) electrons, a term $(1-4 \sin^2\theta_w) \sim 1/13$, $\sigma = m_Z{}^2$, so that $g_{eff}\, \sigma(0)$ ~1. Soft crystals, high-molecular-weight polymers, gel, and colloids often cause first-order and second-order phase transitions. Although the high T_c value appears optimistic and largely depends on a subtle difference in coupling constant of electrons-and-nucleon between enantiomers, differential scanning calorimetry, crystallographic data, chiroptical and vibrational spectroscopy, and other physical properties as a function of temperature, down to cryogenic temperatures, is testable.

The most common approach to enhance PVED in line with the MPV hypothesis is known to utilize the Z^3-dependency for lighter atoms whereas Z^{5-6} dependency for heavy atoms [176–178]; PVED $\propto Z^3$ (Z: atomic number) for relativistic lighter atoms and PVED $\propto Z^5$ for significant relativistic heavier atoms. Possibly, in 1974, M. A. Bouchiat and C. Bouchiat were the first in APV physics to show Z^3-dependency of APV enhancement for relativistic heavier atoms with $Z > 50$ [88, 93]. In 1977, Zel'dovich showed Z^5-dependency of the degree of APV [179]. In 1983, Mason in his comprehensive review [166] showed that Z^5-dependency spin–orbit interaction (ξ) is proportional to Z^2 and PVED by electrons and nucleon is proportional to Z^3; thus PVED $\propto \eta Z^5\, 10^{-18}$ eV ~ $\eta Z^5\, 10^{-19}$ J, η is a chirality factor, possibly, equivalent to Kuhn's dissymmetry factors, g_{NCD} (in NCD spectroscopy) and g_{NCPL} (in NCPL spectroscopy). According to recent studies, the absolute value of η, $|\eta|$ can be boosted in the orders of 10^{-1}–10^0.

By developing the concept of dual quantum oscillations (quantum beat) reported by Harris and Stodolsky [147], the most attractive scenario is to effectively tune tunneling splitting and PVED violation in a non-rigid molecular rotamer in a double-well.

In 2004, MacDermott and Hegstrom [157] and in 2012, MacDermott [156] showed that $\Theta_{\pm} \approx \Theta \times 2\varepsilon_{PV}/\delta$ or $\Theta_{\pm}/\Theta \approx 2\varepsilon_{PV}/\delta$, where $\Theta_{\pm}$ is directly connected to PVED and the small optical rotation at the cat state of L- and D-optical isomers, $2\varepsilon_{PV}$ of the PVED, 2δ tunneling splitting, and Θ the magnitude of optical rotation for a resolved L- or D-isomer. If the condition $\varepsilon_{PV} \sim \delta$ is achieved, one could detect PVED by chiroptical spectroscopy, such as NOR, NORD, NCD, NCPL, MCPL, and magnetic circular dichroism (MCD). The value of ε_{PV} is largely dependent on the nature of the fluxional molecular rotamer. On the other hand, the δ is adjustable by the environment, e.g., temperature and barrier height (E_g) for rotational motion along single bonds. In this case, an introduction of proper external stimuli such as inert chemicals and gasses acts as friction for tunneling [180, 181].

The idea of friction in tunneling is practically tunable by varying the degree of collisions with liquified molecules as viscosity at a specific temperature. The authors applied this idea to test the MPV hypothesis for a series of π-conjugated rotamer luminophores using a sophisticated NCPL and NCD spectroscopy, that allowed us to sensitively detect their chiroptical signals as low as ~0.5 mdeg (~10^{-6} radian) in response to a subtle left-right imbalance at the S_1 and S_0 states, rather than insensitive frequency (or wavelength) shifts [182–184], as given in section 6.5.4. The sensitivity of the 10^{-6} radian is almost comparable to apply the APV experiments in the 1980s and 1990s.

1.6.5 Experimental Tests of Molecular Parity Violation Hypothesis

1.6.5.1 *Chiral molecules at the ground state in a gas phase*

To experimentally verify the APV hypothesis, researchers designed an experimental set-up to detect heavier atoms at elevated temperatures as their vapor phase. Likewise, to test the MPV hypothesis, in which ideal pairs of enantiomerically pure chiral molecules at absolute zero Kelvin are calculated, people have attempted to synthesize special chiral molecules including heavier atoms, possibly, permitting to detect their spectral shifts using *achiral* high-resolution vibrational spectroscopy [152–154] To vaporize the chiral molecules to minimize a decomposition and racemization thermally, a maximum molecular weight will be less than 500–700 dalton. At the present stage, several experimental attempts are unsuccessful because of limited spectroscopic resolution.

1.6.5.2 *Crystals*

Most people who are interested in the MPV hypothesis have investigated to compare physicochemical properties, X-ray structural analyses, and statistical analysis between L–D crystals during a spontaneous crystallization from a racemic mixture or achiral molecules without external assistances. It is noted that the statistical analysis of L–D fractions is often biased by uncharacterized impurities [185] and weak parity-conserving chiral physical and mechanical forces such as stirring, swirling, CP light, vortex light, and cosmological origin PV handed neutrinos.

As mentioned earlier, Kipping and Pope in 1898 were the first to apply the statistical analysis of L–D fractions of chiral crystals from water solution of achiral $NaClO_3$ [124] Their analysis of the chiral crystals indicated that the spontaneous resolution relies on a random, by-chance mechanism. A half-century later, in 1954, Havinga reported an interesting finding of a spontaneous asymmetric crystallization from supersaturated chloroform solution of a racemic mixture of methyl-ethyl-allyl-anilinium iodide, capable of auto-racemization, in sealed in a glass tube [186] (Figure 12); among 14 experiments, 12 runs showed (+)-NOR crystals,

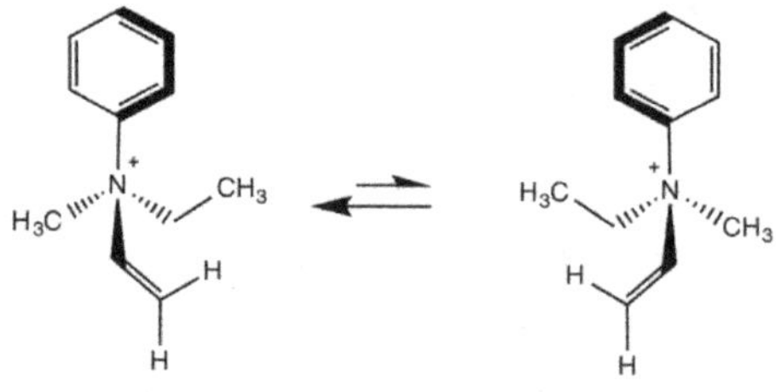

Figure 12. A spontaneous chirogenesis of only (+)-NOR crystals from a handed thermal equilibrium due to unresolved reason between (*S*)- and (*R*)-methyl-ethyl-allyl-anilinium iodide in supersaturated solutions in a sealed glass tube [186].

but did not produce (−)-NOR crystals. In line with PVED $\propto Z^{5-6}$, heavier iodine may affect a handed reorganization between (+)–(−) isomers during the dynamic racemization process. The brief, fact-based work was reported by a chemist just before the historical discovery of P-violation in $\beta^{\pm}$ decay of radioactive ^{60}Co and ^{58}Co elements elucidated by nuclear and particle physicists (1956–1957) [79–82].

In 1999 and 2003, Lajos Keszthelyi reported a preferential crystallization with (−)-NCD at 215 nm, 270 nm, and 490 nm in water crystallized from an L–D mixture of $Na^{+}K^{+}$ tartrate, tris(1,2-ethanediamine)Co^{III}, and tris(1,2-ethanediamine)Ir^{III}; the degree of (−)-NCD signals remarkably altered in the order of $Ir^{III} \gg Co^{III} \gg Na^{+}K^{+}$, that may be interpreted to PVED $\propto Z^{5-6}$ [187, 188]. In 1995, Marian Szurgot based on statistical analysis reported a possibility of MPV crystallization of $NaClO_3$ and $NaBrO_3$ under stagnant conditions and natural convection varying crystallization temperature to avoid any second-nucleation processes [189].

In 2005, Chritobal Viedma found an attrition-enhanced deracemization of $NaClO_3$, called Viedma ripening; a spontaneous MSB occurred in a mixture of solid/liquid enantiomorphous crystals [190]. The finding led him to test the PVED hypothesis based on a population of L-and D-crystals of $NaClO_3$ and $NaBrO_3$; though a marked preference of L-crystals in the four among the six experiments was observed, he concluded the preference led by unknown cryptochiral impurity on a macroscopic level, but not due to PVED [191]. Recently, Pavlov *et al.* thought that the MSB with L-preference reported by Viedma may be ascribed to the PVED. [112]. On the other hand, in 2021, Rubik obtained a preference of the opposite D-crystals from water solutions of $NaClO_3$ in open systems; the D-preference may be impacted by certain subtle external stimuli [192].

Either $NaClO_3$ or $NaBrO_3$ is an achiral molecule, but its crystal form is chiral. In 2000, by inspiring the Salam's handed phase-transition scenario, Wenqing Wang and her coworkers reported detectable differences in several physicochemical properties and solid-state ^{13}C-/^{1}H-NMR/Raman spectroscopy between D–L alanine crystals as a function of temperature down to cryogenic temperatures [193]. However, by repeated purification of L-alanine to remove certain impurities, Rodney Sullivan, Robert Compton, and coworkers partly supported Wang's results by observing certain differences in differential scanning calorimetry (DSC) between D–L alanine and D–L valine crystals [194]. However, Wilson *et al.* did not observe significant differences in crystallographic data between D–L alanine crystals by means of neutron diffraction experiments at 295 K and 60 K, but only realized considerable changes in the N–H covalent distance between the D- and L-crystals [195].

It is a matter of debate whether even purified D–L crystals of amino acids that might be susceptible to certain impurities, enantiopurities, and water, are suited to verify the MPV hypothesis [194, 185]. However, in 2018, based on Wilson's report, Bordallo *et al.* re-examined whether fully deuterated D- and L-alanine as single crystal forms adopt similar structures as crystallographic analysis by means of polarized Raman scattering spectra (514.5 nm, 250 mW) in a temperature range of 20 K to 295 K associated with neutron powder diffraction experiment [196–197]. They confirmed obvious dissimilarity in N–D bond lengths between perdeuterated D- and L-alanine. They were associated with emerging new Raman-active bands in the range of 200 cm^{-1}–300 cm^{-1}, which are unique Raman bands for D-alanine. However, the anomaly observed by Bordallo *et al.* [196–197] is not direct evidence to support or refute Salam's scenarios, predicting the PVED-origin handed second-order phase transition at ~250 K.

On the other hand, noticeable differences between non-natural L- and natural D-RNA oligomers beyond classical chiral chemistry were reported over two decades. Single-strand RNA is classified as flexible macromolecular rotamers with rotational and translational freedom, turning into a semi-flexible macromolecular double-strand with a loss of such freedom. In 2004, Christian Betzel and coworkers were the first to be aware of the dissimilarity between L-RNA and D-RNA based on crystallographic analysis [198]; L-RNA adopts two types of helical duplexes with a head-to-tail packing, in a space group *R*32, with $a = 45.7$, $c = 264.6$ Å unit-cell, whereas D-RNA adopts a Watson–Crick type, end-to-end duplex with wobble-like G–C^+ pair. As one of the explanations, they ascribed this discrepancy to be an inherent nature of D-ribose over L-ribose provided by the PVED theory for D-ribose [168–170]. In 2007, they reported evidence from Raman spectroscopic data that the D-RNA has a different electronic structure compared to the L-RNA [199]. When varying the incident photon energy of the ultraviolet Raman probe across 5 eV, D- and L-RNA duplexes with the same CUGGGCGG sequence show different intensity at two Raman bands of 124.0 meV (1000 cm^{-1}) to 210.8 meV (1700 cm^{-1}). The difference in intensity is related to differences in the electronic levels between D- and L-RNA.

In 2021, Wojciech Rypniewski and coworkers confirmed an asymmetric manner in the crystallization of duplexes from short CUGGGCGG sequences of L-RNA and D-L-RNA, respectively [200]; D- and L-RNA duplexes as crystals had different lattice contacts and have different exposures to water and Zn^{2+} ions in the crystal. Asymmetric structural behaviors between D- and L-RNA may be connected to the origin of homochiral life in the early biosphere. Although these RNA octamers do not include any heavier atoms, the author supposes that a dominant species of surrounding H_2O consisting of *ortho*-water (75%, nuclear spin parallel (↑↑), triplet) *vs.* *para*-water (25%, nuclear spin antiparallel (↑↓), singlet) may affect different pathways in the chirogenesis of D-RNA duplex and L-RNA duplex at ambient temperature if the linear and nonlinear amplification mechanisms (Yamagata [173] and Salam [194]) are applied. The NCD spectra of D- and L-RNA duplexes in the solid states will no longer be mirror-image spectra, and the corresponding UV–Vis spectra also might be considerably different from each other.

Recently, Svetlana Kozlova and coworkers in a series of papers based on the MPV and Salam hypotheses reported MSB phenomena arising from an anomaly in specific heat characteristics and solid-state ^{1}H-NMR spectral width (T_1 spin-lattice relaxation time), including three T_c (~160 K, ~60 K, and ~25 K)) associated with four phases from 300 K down to 8 K [201–204]. MSB of $Zn_2(C_8H_4O_4)_2 \bullet C_6H_{12}N_2$, metal-organic framework (so-called MOF) consisting of 1,4-diazabicyclo[2.2.2]octane (DABCO), that is a roto-symmetric molecule, and Zn^{2+}. NCD and/or NOR measurements at the four phases as a function of temperature remain now awaiting to clarify a preference of left-or-right or of (–)-or-(+)-sign.

Likewise, other non-mirror Cu^{II}-MOF crystals made with D- and L-amino acid-derived ligands were reported. In 2015, Daya Shankar Pandey and coworkers synthesized the homochiral Cu^{II}-MOF, $[Cu_{1.5}(H_2L^{L\text{-}leu})(Ac)H_2O]_n \cdot 3H_2O$ and $[Cu_{1.5}(H_2L^{D\text{-}leu})(Ac)H_2O]_n \cdot 3H_2O$, starting from the corresponding D- and L-leucine-derived ligands in the presence of $Cu(OAc)_2$ [205]. The D-origin and L-origin homochiral Cu^{II}-MOFs revealed structurally distinct crystal packing that revealed significantly different proton conductivity, i.e., L-MOF showed significantly higher conductivity than D-MOF. The D- and L-leucine differently caused *non-mirror chirogenesis* in Cu^{II}-MOFpresumably as a consequence of the cooperative MPV phenomenon on a macroscopic level.

Although Werner was the first to isolate optically active purely inorganic tetranuclear cobalt complexes, (+)- and (−)-$[\{Co^{III}(NH_3)_4(OH^-)_2\}_3 Co^{III}]^{6+}(Br^-)_6 \cdot 2H_2O$, a question remained whether the (+)- and (−)-complexes are energetically equal, therefore, belong to enantiomers [122, 123]. In 1992, Yasui, Ama, and Kauffman described in detail the synthesis of the cobalt complexes with a high enantiopurity and compared NOR and NCD data reported by themselves and previous workers [206].

The (+)-complex: Yasui *et al.* $[\alpha]_{589}$ = +2640° (0.001 M HCl). They noted that additional recrystallization did not increase the optical purity: Kudo and Shimura, $\Delta\varepsilon_{505}$ = +11.3 and $\Delta\varepsilon_{614}$ = −13.5 (0.01 M HCl at 18°C); while, the (−)-complex: Yasui *et al.* $[\alpha]_{589}$ = −4350°, $\Delta\varepsilon_{507}$ = −17.86 and $\Delta\varepsilon_{614}$ = +20.75 (0.01 M HCl at 18°C).

From a comparison, it is obvious that absolute magnitudes in $[\alpha]_{589}$, $\Delta\varepsilon_{507/505}$, and $\Delta\varepsilon_{614}$ between the (+) and (−) complexes differ by ~60%; the absolute magnitudes of chiroptical data tend to shift to (−)-values. Likewise, the absolute $[\alpha]$ values of (+)- and (−)-$[Co^{III}(en)_2(NH_3)_2]^{2+}(X)_2$ tend to shift to (−) values slightly. However, it remains unclear for the author whether the noticeable and subtle discrepancy solely arises from an optical purity or a difference in racemization rates between the (+) and (−) complexes. Recently, a pair of Δ/Λ and Λ/Δ co-complexes comprising tricobalt and di-arsenate, $[\Delta\text{-}Co^{III}(dpa)_4(MeCN)_2](NBu_4)_2[\Lambda\text{-}As_2(tartrate)_2]$ and $[\Lambda\text{-}Co^{III}(dpa)_4(MeCN)_2](NBu_4)_2[\Delta\text{-}As_2(tartrate)_2]$ (dpa: twisted 2,2′-dipyridylamine anion) was isolated [207]. Although the complexes reveal an ideal mirror-image NCD spectra in acetonitrile, though their UV–Vis absorption spectra are slightly different from each other, the D–L enantiomers in the solid-state did not reveal ideal mirror-image in X-ray natural circular dichroism (XNCD) spectra. The considerable discrepancy in the XNCD spectra is likely to connect subtle differences in crystallographic data as $P42_12$ space group; *a*-, *b*-, *c*-axes for the D–L pair appears not exactly identical and differs by 0.1–0.2%. These complexes consisting of polynuclear Co^{III} (*p* 27, *n*20, spin 7/2), As^{III} (*p* 33, *n*42, spin 3/2), with odd-number nuclear spins should be assumed to be a possibility of an enhanced PVED in the solid-state.

1.6.5.3 *Molecules and oligomers at the ground state in liquids*

As shown in a previous section, in my view, to verify the MPV hypothesis based on several scenarios of significantly amplified PVED, crystallographic analysis along with spectroscopic data (e.g., NCD, XNCD, NROA, NVCD), possibly, and NCPL appears reliable rather than statistical analysis of external-and-internal impurity sensitive crystallization as conglomerates from racemic and achiral substances. Alternatively, well-defined chiral molecules and oligomers certified with enantiopurity and/or well-defined roto-symmetrical molecules and oligomers are candidates. Solution-state chiroptical properties and other physicochemical properties by these molecules dissolved homogeneously in achiral solvents are viable.

In 2006, Yosef Scolnik, Meir Shinitzky, and coworkers showed distinct differences in physicochemical and spectroscopic properties of well-defined D- and L-oligoglutamic acids with 24 repeating units dissolved in H_2O and D_2O by comparing their first-order helix–coil phase-transition behaviors (ΔH) associated with their NCD/UV spectra [208]. They ascribed the differences to an idea that only the L-isomer has an affinity to long-lived *ortho*-H_2O, but D-isomer is not; H_2O statistically consists of two nuclear spin isomers, such that triplet (*ortho*) 75% and singlet (*para*) 25% at room temperature [209].

Most chemists are familiar with the concept of triplet and singlet electronic states in the ground and photoexcited states, but rarely know the nuclear spin isomers called isotopomers. However, researchers in astrophysics, astronomy, molecular physics, photophysical materials science, and NMR spectroscopy recognize several isotopomers, *e.g.*, H_2, NH_3, CH_3F, H_2CO, CH_3OH, ethylene, CH_2 in α-amino acids, D_2O, etc. [209–214]. For example, in astronomy and astrophysics, the ratio of nuclear isotopomers plays an essential role in probing the formation conditions in the interstellar medium in the past.

Recent papers have discussed the usefulness of NMR spectroscopy that allows to sensitively detect ^{13}C-chemical splitting in MPV as a result of diastereomeric interactions between racemic 1-phenylethanol as a sensor and L-/D-1-phenylethylamine as chirality inducers [215]; the detection limit with the current NMR technique is $(1.8 \pm 3.4) \times 10^{-3}$ Hz at 20 T. To detect MPV using NMR tools, molecules with only non-relativistic lighter atoms need an accurate resolution in the range of 10^{-6}–10^{-8} Hz and even a molecule having a relativistic ^{207}Pb atom requires at least 10^{-3} Hz resolution.

In 2020, John W. Blanchard and coworkers also showed that an MPV-driven nuclear spin–spin coupling constants as "By"-component, which are distinguishable from P-conserving ones as "Bx" components in appropriate molecules, like ^{1}H^{19}F diatom in liquid and gas states using zero-field NMR technique under an electric field [216]. Experimental proof is awaiting.

1.6.5.4 *Molecules and oligomers at the photoexcited state in liquids*

Radiation modes from the excited states appear to sensitively unveil P-violation at sub-atoms, atoms, and CMB, as exemplified in the $\beta^{\pm}$-decay experiments associated with handed spinning e^{-}/e^{+}, ν and anti-ν, and circularly polarized γ-ray from radioactive atoms and emission modes of ^{133}Cs as APV experiments. Beyond the most MPV theories in the ground states, the MPV theory in the photoexcited state proposed by Berger is attractive [159]. These experimental outcomes and theoretical prediction stimulated the author and coworkers to detect a difference in emission amplitudes as chiroptical signals between L–D molecular states upon excitation of unpolarized light.

The author with his collaborators measured spontaneous radiation processes from the lowest electronic S_1 state of nearly 60 non-rigid achiral and/or racemic luminophores with roto-symmetry in achiral solvents using NCPL spectroscopy [182–184]. The comprehensive experimental results suggested that, without any exception, all the non-rigid racemic rotamers showed (–)-sign NCPL signals in the UV visible region; energetically inequal, non-racemic rotamers are generated upon irradiation of incoherent unpolarized light (Figures 13–16).

For example, oligo(9,9-di-*n*-alkylfluorene)s (FLs) including dimer, trimer, pentamer, and hexamer cannot adopt coplanar structures at the

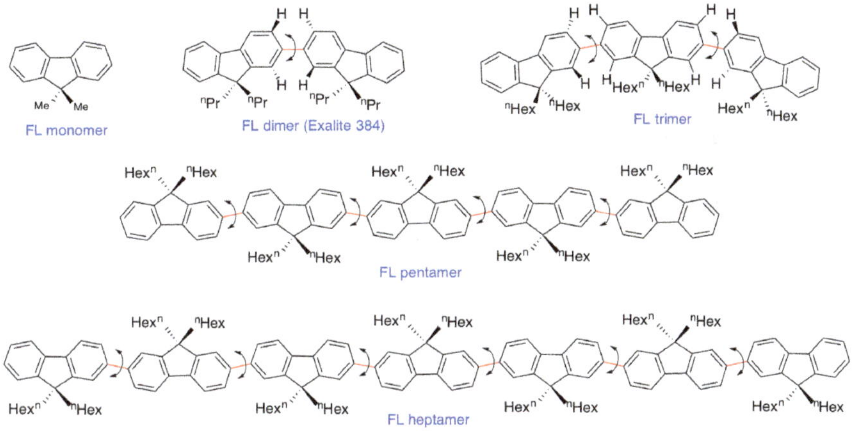

Figure 13. Chemical structures of 9,9-dialkylfluorene (FL) oligomers including dimer, trimer, pentamer, and hexamer, carrying *n*-alkyl groups at 9,9-positions.

ground state, possibly, even at the photoexcited state because of intense CH/HC repulsions [217] (Figure 13). Although the oligomers do not show any NCD spectra at the ground state, they reveal clear (–)-sign NCPL spectra at the photoexcited state. The magnitudes of Kuhn's anisotropy, g_{em}, of the oligomers are progressively shifted to greater (–)-values when the solvent viscosity increases. The absolute g_{em} value at a high viscosity >20 cP approaches 1.5×10^{-3} [182], that is approximately one-sixth compared with $g = 8.7 \pm 3.1 \times 10^{-3}$ evaluated from the hidden CPV [68].

Likewise, unsubstituted *p*-oligophenylenes (**Ph3–Ph6**) and their methyl derivatives (**DMT**, **TMQ**, **Exalite360**, **QUI**, **TBS**) (Figure 14) in the photoexcited state, that are utilized as organic scintillators and emitters of a dye laser, reveal (–)-sign NCPL spectra, while the corresponding NCD signals are zero-signals; *p*-oligophenylenes in the photoexcited state act as main-chain rotors with a handedness, while randomly rotated in the ground state [182]. Multiple bulkier *tert*-butyl groups with three-fold symmetry, compared with compact methyl groups, efficiently work as handed side-chain rotors in the photoexcited state. Similarly, other 6 molecular scintillators, 11 coumarins derivatives, and rhodamine B at the photoexcited state reveal clear (–)-sign NCPL spectra in dilute solutions [183].

Notably, the most striking findings will be that nearly two-fold symmetrical pyrromethene 597 and pyrromethene 546, that are popularly known as BODIPY, show clear (–)-sign NCD spectra in the ground state

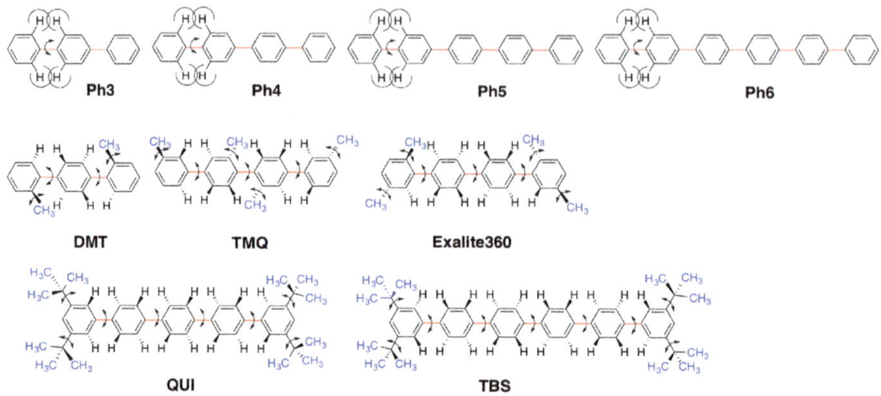

Figure 14. Chemical structures of four unsubstituted oligophenylenes, **Ph3–Ph6**, and their methyl and *tert*-butyl substituted derivatives (**DMT**, **TMQ**, **Exalite360**, **QUI**, and **TBS**) carrying multiple three-fold symmetric rotors. These rotors allow inducing non-planar dynamic structures in the photoexcited and ground states [182].

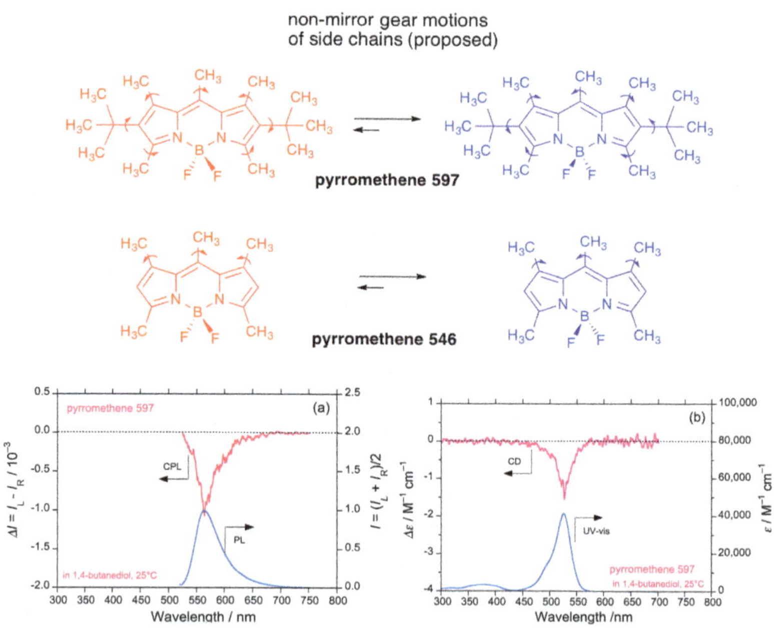

Figure 15. Chemical structures of two BODIPY derivatives, pyrromethene 597 and pyrromethene 546, that carry methyl and *tert*-butyl groups as side-chain rotors, enabling a handed gear motion. (a) The (–)-sign NCPL and PL spectra and (b) (–)-sign NCD and UV–Vis spectra of pyrromethene 597 in 1,4-butanediol. Modified from Ref. [183].

and (–)-sign NCPL spectra at the photoexcited state (Figure 15) [183]. Multiple methyl and *tert*-butyl groups, that possess three-fold symmetry, may be responsible for efficiently acting as a handed gear motion. Pyrromethene 597 and pyrromethene 546 may be no longer achiral or a racemic mixture of rotamers and would exist as a hidden chiral state with the same-handed dynamic motion.

Furthermore, twisted rotamers designed newly with mono- and di-anthryl moieties and two distyryl groups as side-chain rotors, and for comparison, unsubstituted commercial di-anthryl (**BA**) and water-soluble commercial poly-*p*-phenylene (**PP**) reveal clear NCPL spectra at the photoexcited state although any NCD signals are not detectable (Figure 16) [184].

To ensure the above results, the authors confirmed that achiral rigid planar aromatic luminophores (**1**–**4**) generate null-value NCPL and NCD signals, and moreover, rigid chiral D- and L-camphor and (*R*)- and (*S*)-binaphthyl derivatives (**5**) show ideal mirror-image NCPL and NCD spectra (Figure 17) [182–184]. NCPL experiments as probing the photo-excited electronic state from nonrigid molecular rotamers would permit testing of the MPV hypothesis as well as the P-violation from the excited states from the sub-atoms, atoms, and CMB.

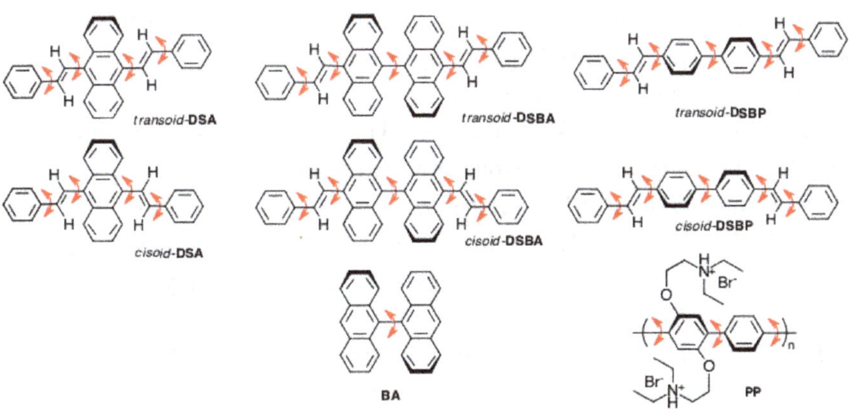

Figure 16. Chemical structures of mono-anthryl and twisted di-anthryl bearing two distyryl groups (**DSA**, **DSBA**, and **DSBP**), unsubstituted bianthryl (**BA**), and water-soluble poly-*p*-phenylene (**PP**), taken from ref [184].

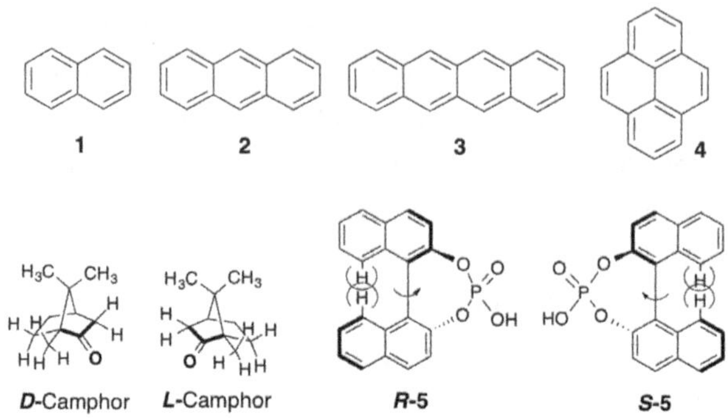

Figure 17. Chemical structures of achiral rigid planar π-conjugated aromatics (**1**–**4**) and two enantiomeric pairs of optically active rigid molecules (D- and L-camphor and ***R*-5**/***S*-5** to test the MPV hypothesis [182].

1.6.5.5 *Colloids and polymers at the ground and photoexcited states in liquids*

Collision-free, gas-phase NCPL/NCD spectroscopy may be a useful tool to provide a definitive answer to the MPV questions. Colloids and polymers under collision conditions might be not suitable for testing the MPV hypotheses because of uncharacterized impurities and no guarantee of 100% enantiopurity when non-natural and non-natural abundant bioresources were employed. However, colloids in a μm size capsule-like world may mimic certain events happening in complex living cells during the chemical evolution of life [218–220].

In 1993, Meir Shinitzky and Rachel Haimovitz reported the considerable differences in the absolute magnitude in NCD spectra that *N*-palmitoyl-D-serine and *N*-stearoyl-D-serine are always greater than those of *N*-palmitoyl-L-serine and *N*-stearoyl-L-serine, although both form micelle above critical micellar concentrations (CMC) [221]. They ascribed this anomaly to the MPV hypothesis.

In 2001, it became recognized that helical Si–Si bond polymers, so-called polysilane or polysilylene, reveal exceptionally very narrow NCD and UV characteristics at 320 nm, endowed with specific (*S*)- and

(*R*)-chiral pendants [222]. In 2001, the author briefly reported his first MPV-related short paper, titled "Experimental tests of parity violation at helical polysilylene level" in line with Salam's scenario for the MPV hypothesis [223]; considerable differences in NCD spectral characteristics, ^{29}Si-NMR chemical shift and linewidth, and viscometric data were recognized for a pair of helix–helix second-order transitioned polysilanes carrying (*S*)- and (*R*)-3,7-dimethyloctyl (96% *ee*) and achiral alkyl groups (**1*S* /1*R*** in Figure 18) in dilute isooctane ranging from −80°C to +80°C. Following, in 2010, based on comprehensive testing using four pairs of helix–helix transition polysilanes (**1*S*/1*R*, 2*S*/2*R*, 3*S*/3*R*, 4*S*/4*R*** in Figure 18) and for comparison, non-helix–helix transition polysilanes (**5*S*/5*R*, 6*S*/6*R*** in Figure 18), a possibility of the MPV on the macroscopic

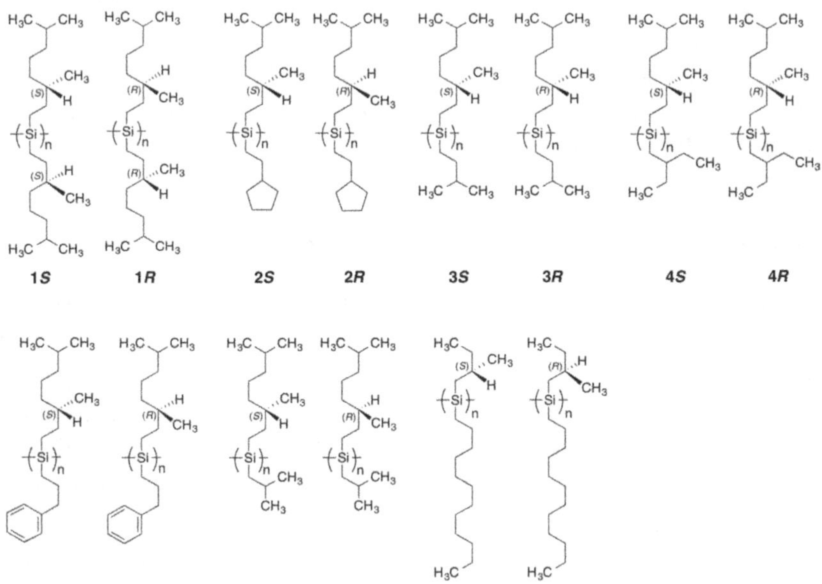

Figure 18. Chemical structures of second-order thermo-driven helix–helix transition polysilanes (**1*S* /1*R*, 2*S*/2*R*, 3*S*/3*R*, 4*S*/4*R***) and non-transition polysilanes (**5*S*/5*R*** and **6*S*/6*R***) [218].

level was published as a regular article [218]. The major critical drawbacks are that chiral pendants are not 100% *ee* and that molecular weights (repeating units) for six pairs of the helical polysilanes are not exactly the same.

Nevertheless, since then, authors and coworkers over the past two decades have been aware of non-ideal mirror-image NCPL and NCD spectroscopic datasets in several colloidal optically active polymers dispersed in optofluidic media with a tuned refractive index [219, 220, 224]. Some readers may be aware of the non-ideal mirror-image NCPL and NCD spectral characteristics in a recent review paper. The authors conjecture yet that the non-ideal mirror-image chiroptical characteristics of the colloidal polymers in the optofluidic medium are connected to optically resonant effects detected as non-mirror chiroptical magnitudes, the same chiroptical sign, and different wavelength, to significantly enhance the tiny PVED effects on μm size macroscopic levels. Particularly, when an idea of optofluidics was applied to the μm size, optically active colloids consisting of π-conjugated polymers and polysilanes, considerable differences in NCD and NCPL spectral characteristics between (*S*)- and (*R*)-derived polymers, presumably, MPV on a macroscopic level can be seen.

1.6.5.6 *Molecules under external magnetic field in liquids*

In 1984, Barron and Vrbancich first introduced the Faraday A-, B-, and C-terms to discuss interactions between the light and chiral matter in an external magnetic field that is parallel (north-up, N-up) or antiparallel (south-up, S-up) to the incident light beam, known as Faraday geometry [225]. Magneto-circular dichroism is popularly called MCD [226, 227] and the concept is developed to magneto-circularly polarized luminescence, MCPL [228, 229].

Precision NCD and NCPL spectroscopy allow researchers to be able to in principle, discuss whether P-symmetry at electronic ground and excited states of chiral molecular chromophores and/or luminophores in the UV–Vis-NIR region is conserved or violated from a prediction by rigorous theoretical calculations.

On the other hand, NCD and NCPL spectroscopy with the north-up (N-up) and the south-up (S-up) Faraday geometries under an external magnetic field, called MCD and MCPL spectroscopy, is possible, respectively. When NCD spectroscopy is applied to achiral diamagnetic molecules, one could anticipate mirror-image MCD spectra between the N-up and S-up geometries: (+)-sign MCD spectrum with N-up (or S-up) will be inverted to (–)-sign MCD spectrum with the S-up (or N-up). T-symmetry at the ground state might be conserved. If non-mirror-image MCD spectra were observed, T-symmetry at the ground state is violated. Likewise, when NCPL spectroscopy is applied to achiral diamagnetic luminophores, one can imagine mirror-image MCPL spectra between the N-up and S-up geometries. When (+)-sign MCPL spectrum with N-up (or S-up) shows an ideal mirror-image of (–)-sign MCPL spectrum with S-up (N-up), T-symmetry at the excited state is conserved. If non-mirror-image MCPL spectra were observed, T-symmetry at the excited state is violated.

In 1981, by assuming the external magnetic field-origin biomolecular handedness during the chemical revolution, Thiemann and Jarzak reported anomalous MCD characteristics, in which achiral D_{6h}-symmetry benzene in cyclohexane reveals non-mirror-symmetric MCD regardless of N-up and S-up geometries under 1.14 T and 0.003 T [230]. As most workers in the past used a heavy-weight electromagnet only with either N-up or S-up Faraday geometry, it was difficult to verify whether mirror-image MCD and MCPL spectra are rigorously inverted by the N-up or S-up Faraday geometry. A recently developed NCPL and NCD spectrometers equipped with a palm-size bipolar permanent magnet able to both N-up and S-up geometries allow us to easily verify MCPL and MCD spectra of several achiral and racemic mixture of achiral, chiral, and helical luminophores and chromophores.

In 2020 and 2021, Imai, Fujiki, and coworkers demonstrated the first mirror-symmetric, mono-signate MCPL spectra at 1.6 T of unsubstituted pyrene and two pyrene derivatives and other achiral aromatic emitters in chloroform solution [231, 232–237]. Likewise, mirror-symmetric, mono-signate MCPL spectra from purely inorganic solids doped with Eu^{III} and Mn^{II} and CdS/ZnS core-shell quantum dots were reported [233, 234].

On the other hand, optically active $Eu^{III}(hfa)_3$ (hfa: hexafluoroacetylacetonate) endowed with three types of (*S*)- and (*R*)-phosphine chirality as ligands showed noticeably non-mirror-symmetric split-shape MCPL and NCPL spectroscopic characteristics at $^5D_0 \rightarrow ^4F_j$ (j = 3 (~650nm),

4 (~690 nm) transitions, implicating an occurrence of PT- and P-violations for the photoexcited Eu^{III}(hfa) [235]. In an analogy of the results in a series of APV experiments using neutron beam for heavier atoms with odd-number nuclear spins, $Eu^{III}(hfa)_3$ with the phosphine ligands comprising odd-number nuclear spin atoms, e.g., $^{151}Eu^{III}$ (5/2, 48%), $^{153}Eu^{III}$ (5/2, 52%), ^{31}P (1/2), ^{19}F (1/2), ^{35}Cl (3/2, 76%) and ^{37}Cl (3/2, 24 %), ^{13}C (1/2, 1.1 %), ^{1}H (1/2, ~100%), was assumed to be responsible for the non-mirror MCPL and NCPL characteristics. Recently, Jean-Pascal Sutter *et al.* reported NCPL characteristics as single crystals of Λ- and Δ-Dy^{III} complex endowed (+)- and (−)-3-(trifluoroacetyl)camphor as ligands, respectively, that contains ^{161}Dy (5/2, 19%) and ^{163}Dy(5/2, 25%)); the Λ- and Δ-Dy^{III} crystals at the photoexcited state provided obvious non-mirror-symmetric (+)- and (−)-sign NCPL profiles and magnitudes, regardless of near mirror-symmetric NCD spectral profiles [236]. To my view, this is implicative that P-symmetry at the photoexcited state of the Dy^{III} complex is violated while P-symmetry at the ground state is conserved.

Likewise, split-shape MCPL spectra at the photoexcited states of racemic [*n*]helicenes (n = 3–5,7) in dilute THF and DMSO solutions were non-mirror-symmetric, inferring PT-violation at the photoexcited state [232]. All the helicenes are substituted with multiple ^{1}H atoms with odd-number nuclear spin. This situation is similar to the anomalous MCD at the ground state of benzene with six ^{1}H atoms.

The most popular hydrogen atom, ^{1}H, at organic molecules has the odd-number nuclear spin. Thus, most of all molecules consisting of ^{1}H at the photoexcited state in the absence and presence of the external magnetic field could bring us to unveil P- and PT-violations at the molecular level utilizing the highly sensitive NCPL and MCPL spectroscopy.

1.7 Conclusion

The present chapter embarks on several paradigm-shifting discoveries that have occurred in cosmology, molecular and biochemistry, and atomic and subatomic particle physics for over a century owing to the intimate interplaying thoughts among theorists and experiments; these are: (i) seven fundamental symmetries that include parity (P), charge conjugation (C), time reversal (T), CP, PT, and CT, and CPT invariance; (ii) four fundamental forces that include gravity, electromagnetic force, strong force, and weak force; (iii) electroweak force unifying weak and electromagnetic forces, so-called the standard model; (iv) detecting four massive bosons (charged $W^{\pm}$, 80.4 GeV, neutral Z^0, 91.2 GeV, Higgs H^0, 125.1 GeV) along with massless photon (γ); and (v) detecting three non-massless electron neutrino (ν_e), muon neutrino (ν_μ), and tau neutrino (ν_τ).

In 2020, Japanese researchers reported two epoch-making papers that were closely connected to the origin of biomolecular handedness. One of the topics was a cosmic parity violation by analyzing asymmetric, curled E- and B-modes in polarized cosmic background radiation at 2.73 K in the Milky Way; the degree of birefringence rotation angle is $\beta = 0.35° \pm 0.14°$ at 68% CL. The value corresponds to Kuhn's g-value of $8.7 \pm 3.1 \times 10^{-3}$. Another hot topic is that the T2K collaboration team demonstrates detection of CP-violation between neutrinos and anti-neutrinos in a lepton family; three neutrinos (0.3%) is one of the most dominant materials (~4%) in the whole universe.

These remarkable discoveries led to theoretical and experimental proof of P-/CP-violations at the elementary particle level, P-/CP-violations at the subatomic level, C-/P-/CP-violations at the atomic level, and P-violation at the cosmological level. Yet, P-/CP-violations at the molecular level, a matter of debate regarding the long-standing question of why nature chose L-amino acids and D-ribose on Earth, remains open. Regrettably, anti-molecules have not been generated yet. Since the time of Pasteur, most chemists have been skeptical of the idea that P-symmetry at the molecular level is violated and/or even whether P-symmetry is violated, with concerns that it cannot be detected owing to the tiny energy inequality between left-hand and right-hand enantiomers predicted by many theories. However, beyond an orthodox P-conserving electromagnetic force-driven stereochemistry established based on the idea of energy equality between left-and-right enantiomers, the present chapter highlighted heresy electroweak force-driven spatiotemporal chemistry.

In my view, among published papers reviewed rigorously, several experimental results may be strongly implicative as positive proofs to support the molecular P-violation hypothesis, that is, chiral crystals, chiral/racemic molecules, and oligomers at the ground and photoexcited states in liquids, optically active colloids and helical polymers in fluidic liquids, achira/racemic molecules by comparing south-up and north-up Faraday geometries under an external magnetic field. To confirm the experimental results, further sophisticated proof is awaiting associated with realistic molecular P-violation theories under highly collision conditions such as liquids, gels, supramolecules, and solids. Several researchers would be aware of the experimental facts revealing non-ideal mirror characteristics; left-and-right enantiomers produced in chirogenesis associated with chiroptical datasets do not rigorously obey P-conserved mirror symmetry derived stereochemistry and chiroptical spectroscopy.

Lastly, I believe that the present chapter throws light on the research gaps still existing between fundamental physics/cosmology, molecular chemistry/biochemistry, and atomic and subatomic particle physics. However, these research areas are no longer segregated but are, in fact, intimately connected with a common denominator of C-, P-, T-, CP-, PT-, and CT-violations of CPT theorem, as briefly illustrated in Figure 2.

1.8 Acknowledgments

Firstly, the author owes a debt of gratitude to Prof. Victor Borovkov (Tallinn University of Technology, Estonia) for giving him the opportunity to contribute to the book chapter. The author expresses special thanks to his coworkers; Prof. Julian R. Koe (International Christian University, Mitaka, Tokyo, Japan), Dr. Puhup Puneet (Department of Chemical Engineering, National Tsing-Hua University, Hsinchu, Taiwan), Prof. Yoshitane Imai (Kindai University, Higashi-Osaka, Osaka, Japan), Dr. Nor Azura Abdul Rahim (Universiti Malaysia Perlis, Perlis, Malaysia), Dr. Abd Jalil Jalilah (Universiti Malaysia Perlis, Jejawi, Malaysia), Prof. Bhanu Nandan (Indian Institute of Technology Delhi, Delhi, India), and his students and coworkers who shared his/her time with him. Finally, Dr. Puneet provided valuable comments to the draft and generated Fig. 2.

References

1. Schurig, V. (1972) On the Origins of Stereochemistry. The Herschel Family — Musicians, Astronomers and Natural Scientists. *Enantiomer*, **2**, 135–142.
2. Haldane, J.B.S. (1960) Pasteur and Cosmic Asymmetry. *Nature,* **185**, 87.

3. Gal, J. (2011) Louis Pasteur, Language, and Molecular Chirality. I. Background and Dissymmetry. *Chirality*, **23**, 1–16.
4. Hegstrom, R.A., and Kondepudi, D.K. (1990) Handedness of the Universe, *Sci. Am.*, **262**, 98–105.
5. Barron, L.D. (1986) Symmetry and Molecular Chirality, *Chem. Soc. Rev.*, **15**, 189–223.
6. Sakharov, A.D. (1967) Violation of CP Invariance, C Asymmetry, and Baryon Asymmetry of the Universe, *JETP Lett.*, **5**, 24–27.
7. Wickramasinghe, N.C. (2014) *Search for Our Cosmic Ancestry.* World Scientific.
8. Globus, N., and Blandford, R.D. (2020) The Chiral Puzzle of Life, *Astrophys. J. Lett.*, **895**, 1, L1.
9. Avnir, D. (2020) Critical Review of Chirality Indicators of Extraterrestrial Life. *New Astron. Rev.*, **92**, 101596 (8 p).
10. Puzzarini, C., and Barone, V. (2020) A Never-Ending Story in the Sky: The secrets of chemical evolution. *Phys. Life Rev.*, **32**, 59–94.
11. Wigner, E.P. (1965) Violation of Symmetry in Physics. *Sci. Am.*, 213(6), 28–36.
12. "Quark." Wikipedia, last edited on 18 July 2021, https://en.wikipedia.org/wiki/Quark
13. National Research Council. (2003) *Connecting Quarks with the Cosmos: Eleven Science Questions for the New Century*, National Academies Press.
14. Pascoli, S., and Turner, J. (2020) Matter–Antimatter Symmetry Violated. *Nature*, **580**, 323–324.
15. T2K collaboration. (2020) Constraint on the Matter-Antimatter Symmetry-Violating Phase in Neutrino Oscillations. *Nature*, **580**, 339–344.
16. Radioactivity.EU.com, "The Neutrino Hypothesis," accessed on 19 September 2021, https://www.radioactivity.eu.com/site/pages/Neutrino_Hypothesis.htm
17. "Particle physics." Wikipedia last edited on 7 August 2021, https://en.wikipedia.org/wiki/Particle_physics
18. "Standard Model," Wikipedia, last edited on 5 September 2021, https://en.wikipedia.org/wiki/Standard_Model
19. "W and Z bosons," Wikipedia, last edited on 9 September 2021, https://en.wikipedia.org/wiki/W_and_Z_bosons
20. "Neutrino," Wikipedia, last edited on 18 September 2021, https://en.wikipedia.org/wiki/Neutrino
21. "Neutrino oscillation," Wikipedia, last edited on 13 September 2021, https://en.wikipedia.org/wiki/Neutrino_oscillation
22. "The History of Neutrino," Fermi National Accelerator Laboratory, accessed on 19 September 2021, https://neutrinos.fnal.gov/history/
23. "Big Bang Expansion and the Fundamental Forces," Astrophysics, HyperPhysics, Department of Physics and Astronomy, Georgia State University, accessed on 19 September 2021, http://hyperphysics.phy-astr.gsu.edu/hbase/Astro/unify.html
24. Leclercq, F. (2011) Biot, Chromatic Polarization, and the Theory of Fits. *Revue d'Histoire des Sciences*, **64**, 121–156.

25. "Electroweak interaction," Wikipedia, last edited on 27 June 2021, https://en.wikipedia.org/wiki/Electroweak_interaction
26. Forbes, N., and Mahan, B. (2014) *Faraday, Maxwell, and the Electromagnetic Field: How Two Men Revolutionized Physics,* Prometheus Books.
27. "Lepton," Wikipedia, last edited on 15 August 2021. https://en.wikipedia.org/wiki/Lepton
28. Tavora, M. (2020) *Why Antiparticles Must Exist in Nature*, issued on 16, April 2020, https://towardsdatascience.com/why-antiparticles-must-exist-in-nature-e9b67ac8983f
29. Zeh, H.D. (2005) Entropy Remarks on the Compatibility of Opposite Arrows of Time. *Entropy,* **7**, 199–207.
30. Zeh, H.D. (2007) *The Physical Basis of the Direction of Time,* 5th Ed., Springer.
31. "Timeline of cosmological theories." Wikipedia, last edited on 13 August 2021, https://en.wikipedia.org/wiki/Timeline_of_cosmological_theories
32. "Big Bang," Wikipedia, last edited on 18 September 2021, https://en.wikipedia.org/wiki/Big_Bang
33. "Microwave radiometer." Wikipedia, last edited on 29 July 2021, https://en.wikipedia.org/wiki/Microwave_radiometer
34. "Cosmic microwave background," Wikipedia, last edited on 6 June 2022, https://en.wikipedia.org/wiki/Cosmic_microwave_background
35. "Wilkinson Microwave Anisotropy Probe," Wikipedia, last edited on 4 September 2021, https://en.wikipedia.org/wiki/Wilkinson_Microwave_Anisotropy_Probe
36. Planck Science Team. (2015) *Planck Image Gallery*, https://www.cosmos.esa.int/web/planck https://www.cosmos.esa.int/web/planck/picture-gallery
37. Planck Science Team. (2018) *Planck Image Gallery*, https://www.cosmos.esa.int/web/planck
38. Eliel, E. L., and Wilen, S.H. (1994) Chiroptical Properties, in *Stereochemistry of Organic Compounds*, Eliel, E. L, Wilen, S.H., and Mander, L.N. (eds), Wiley, pp. 991–1118.
39. Berova, N.K., Nakanishi, K., and Woody, R.W. (eds). (2000) *Circular Dichroism: Principles and Applications*, 2nd Ed. (Wiley-VCH).
40. Berova, N., Polavarapu, P.L., Nakanishi, K., and Woody, R.W. (eds). (2011) *Comprehensive Chiroptical Spectroscopy: Instrumentation, Methodologies, and Theoretical Simulations*, Volume 1. (Wiley).
41. Berova, N., Polavarapu, P.L., Nakanishi, K., and Woody, R.W. (eds). (2012) *Comprehensive Chiroptical Spectroscopy: Instrumentation, Methodologies, and Theoretical Simulations*, Volume 2. (Wiley).
42. "Cosmic Background Explorer," Wikipedia, last edited on 7 August 2021, https://en.wikipedia.org/wiki/Cosmic_Background_Explorer
43. "Planck (spacecraft)." Wikipedia, last edited on 31 May 2021, https://en.wikipedia.org/wiki/Planck_(spacecraft)

44. Perlmutter, S., *et al.* (1998) Measurements of Ω and Λ from 42 High-Redshift Supernovae. *Astrophys. J.*, **517**, 565–586.
45. Rises, A.G., *et al.* (1998) Observational Evidence from Supernovae for an Accelerating Universe and a Cosmological Constant. *Astron. J.*, **116**, 565–586.
46. "Accelerating expansion of the universe," Wikipedia, last edited on 11 September 2021, https://en.wikipedia.org/wiki/Accelerating_expansion_of_the_universe
47. Zwicky, F. (1933) Die Rotverschiebung von Extragalaktischen Nebeln, *Helv. Phys. Acta*, **6**, 110–127.
48. Zwicky, F. (1937) On the Masses of Nebulae and of Clusters of Nebulae, *Astrophys. J.*, **86**, 217–246.
49. Winkler, K. (2014) Fritz Zwicky and the Search for Dark Matter. *Swiss Amer. Hist. Soc. Rev.*, **50**, 3–41.
50. "Galaxy morphological classification," Wikipedia, last edited on 21 July 2021, https://en.wikipedia.org/wiki/Galaxy_morphological_classification
51. Rubin, V.C., and Ford, Jr. W.K. (1970) Rotation of the Andromeda nebula from a spectroscopic survey of emission regions. *Astrophys. J.*, **159**, 379–403.
52. Rubin, V.C. (1983) The Rotation of Spiral Galaxies. *Science,* **220**, 1339–1344.
53. Rubin, V.C. (1983) Dark Matter in Spiral Galaxies. *Sci. Am.*, **248**, 96–108.
54. Minami, Y., and Komatsu, E. (2020) New Extraction of the Cosmic Birefringence from the Planck 2018 Polarization Data. *Phys. Rev. Lett.*, **125**, 221301.
55. "A hint of new physics in polarized radiation from the early Universe," as of 24 November 2020, https://www.kek.jp/en/press-en/pr20201124e/
56. Longo, M.J. (2011) Detection of a Dipole in the Handedness of Spiral Galaxies with Redshifts $z \sim 0.04$. *Phys. Lett. B*, **699**, 224–229.
57. Shamir, L. (2012) Handedness Asymmetry of Spiral Galaxies with $z < 0.3$ Shows Cosmic Parity Violation and a Dipole Axis. *Phys. Lett. B*, **715**, 25–29.
58. Shamir, L. (2016) Asymmetry between Galaxies with Clockwise Handedness and Counterclockwise Handedness. *Astrophys. J.*, **823**, 32.
59. Sachs, R.K., and Wolfe, A.M. (1967) Perturbations of a Cosmological Model and Angular Variations of the Microwave Background. *Astrophys. J.*, **147**, 73–90.
60. Lue, A., Wang, L., and Kamionkowski, M. (1999) Cosmological Signature of New Parity-Violating Interactions. *Phys. Rev. Lett.*, **63**, 1506–1509.
61. The European Space Agency. (2015) Polarised emission from Milky Way dust, released on 5 February 2015, https://www.esa.int/ESA_Multimedia/Images/2015/02/Polarised_emission_from_Milky_Way_dust
62. Planck Collaboration. (2015) Planck intermediate results. XIX. An overview of the polarized thermal emission from Galactic dust. *Astron. Astrophys.*, 576, A104.
63. Planck Collaboration. (2016) Planck intermediate results-XLIX. Parity-violation constraints from polarization data. *Astron. Astrophys.*, 596, A110.
64. Bianchini, F., *et al.* (2020) Searching for Anisotropic Cosmic Birefringence with Polarization Data from SPTpol. *Phys. Rev. D*, **102**, 083504.

65. Tadaki, K.I., Iye, M., Fukumoto, H., Hayashi, M., Rusu, C.E., Shimakawa, R., and Tosaki, T. (2020) Spin Parity of Spiral Galaxies II: A Catalogue of 80k Spiral Galaxies using Big Data from the Subaru Hyper Supreme-Cam Survey and Deep Learning. *Mon. Not. R. Astron. Soc.*, **496**, 4276–4286.
66. The POLARBEAR Collaboration. (2020) A Measurement of the Degree-scale CMB B-mode Angular Power Spectrum with POLARBEAR. *Astrophys. J.*, **897**, 55.
67. Namikawa, T., *et al.* (2020) Atacama Cosmology Telescope: Constraints on Cosmic Birefringence. *Phys. Rev. D*, **101**, 083527.
68. Rubin, V.C. (2006) Seeing Dark Matter in the Andromeda Galaxy. *Phys. Today*, 8–9.
69. Land, K., Slosar, A., Lintott, C., Andreescu, D., Bamford, S., Murray, P., Nichol, R., Raddick, M.J., Schawinski, K., Scala, A., Thomas, D., and Vandenberg, J. (2008) Galaxy Zoo: The Large-Scale Spin Statistics of Spiral Galaxies in the Sloan Digital Sky Survey. *Mon. Not. R. Astron. Soc.,* **388**, 1686–1692.
70. Ghez, A.M., Morris, M., Becklin, E.E., Tanner, A., and Kremenek, T. (2000) The Accelerations of Stars Orbiting the Milky Way's Central Black Hole. *Nature*, **407**, 349–351.
71. The Event Horizon Telescope Collaboration. (2019) First M87 Event Horizon Telescope Results. I. The Shadow of the Supermassive Black Hole. *Astrophys. J. Lett.*, 875, L1.
72. "Supermassive black hole," Wikipedia, last edited on 11 September 2021, https://en.wikipedia.org/wiki/Supermassive_black_hole
73. Pontecorvo, B. (1957) Mesonium and Anti-Mesonium. *J. Exp. Theor. Phys. B*, **6**, 429–431.
74. "Neutrino Properties," created on 17 July /2008, https://pdg.lbl.gov/2008/listings/s066.pdf
75. Carter, A.B., and Sanda, A.I. (1981) CP Violation in B-Meson Decays. *Phys. Rev. D*, **23**, 1567–1579.
76. "Standard Model of Fundamental Particles and Interactions," accessed on 19 September 2021, https://web.archive.org/web/20160304133522/ https://www.pha.jhu.edu/~dfehling/particle.gif
77. "Electron rest mass;" Wikipedia, last edited on 22 January 2021, https://en.wikipedia.org/wiki/Electron_rest_mass
78. Janoschek, R. (1991) Theories on the origin of biomolecular homochirality. In *Chirality–From Weak Bosons to the α-Helix*, Janoschek, R. (ed.). (Springer).
79. Lee, T.D., and Yang, C.N. (1956) Question of Parity Conservation in Weak Interactions. *Phys. Rev.*, **104**, 254–258.
80. Wu, C.S., Ambler, E., Hayward, R.W., Hoppes, D.D., and Hudson, R.P. (1957) Experimental Test on Parity Conservation in Beta Decay. *Phys. Rev.*, **105**, 1413–1415.
81. Ambler, E., Hayward, R.W., Hoppes, D.D., Hudson, R.P., and Wu, C.S. (1957) Further Experiments on β Decay of Polarized Nuclei. *Phys. Rev.*, **106**, 1361–1363.
82. Postma, H., Huiskamp, W.J., Miedema, A.R., Steenland, M.J., Tolhoek, H.A., and Porter, C.J. (1957) Asymmetry of the positron emission by polarized ^{58}Co-nuclei. *Physics*, **23**, 259–260.

83. Lee, T.D., and Wu, C.S. (1965) Weak Interactions. *Annu. Rev. Nucl. Sci.*, **15**, 381–476.
84. "The story of antimatter, 27 07, 1964," CERN, https://timeline.web.cern.ch/taxonomy/term/86?page=1
85. Christenson, J.H., Cronin, J.W., Fitch, V.L., and Turlay, R. (1964) Evidence for the 2π Decay of the K_2^0 Meson, *Phys. Rev. Lett.*, **13**, 138–140.
86. Zel'dovich, Ya. B. (1959) Parity Nonconservation in the First Order in the Weak-Interaction Constant in Electron Scattering and Other Effects. *Soviet Phys. JETP*, **9**, 682–683.
87. Bouchiat, M.A., and Bouchiat, C. (1974) Weak Neutral Current in Atomic Physics. *Phys. Lett. B*, **48**, 111–114.
88. Bouchiat, M.A., and Bouchiat, C. (1974) I. Parity Violation Induced by Weak Neutral Currents in Atomic Physics. *J. Phys.*, **35**, 899–927.
89. Bouchiat, M.A., and Bouchiat, C. (1975) Parity Violation Induced by Weak Neutral Currents in Atomic Physics. Part II. *J. Phys.*, **36**, 493–509.
90. Bouchiat, M.-A., and Pottier, L. (1984) An Atomic Preference between Left and Right. *Sci. Am.*, **250**, 100–111.
91. Bouchiat, M.-A., and Bouchait, C. (1997) Parity Violation in Atoms. *Rep. Prog. Phys.*, **60**, 1351–1396.
92. Hegstrom, R.A., Chamberlain, J.P., Seto, K., and Watson, R.G. (1988) Mapping the Weak Chirality of Atoms. *Am. J. Phys.*, **56**, 1086–1092.
93. Fortson, E.N., and Wilets. L. (1980) Parity Nonconservation in Atoms: Status of Theory and Experiment. *Adv. At. Mol. Phys.*, **16**, 319–373.
94. Frois, B., and Bouchiat, M.-A. (eds). (1999) *Parity Violation in Atoms and Polarized Electron Scattering*. World Scientific.
95. Haxton, W.C., and Wieman, C.E. (2001) Atomic Parity Nonconservation and Nuclear Anapole Moments. *Ann. Rev. Nucl. Part. Sci.*, **51**, 261–293.
96. Forte, M., Heckel, B.R., Ramsey, N.F., Green, K., Greene, G.L., Byrne, J., and Pendlebury, J.M. (1980) First Measurement of Parity-Nonconserving Neutron-Spin Rotation: The Tin Isotopes. *Phys. Rev. Lett.*, **45**, 2088–2092.
97. Stodolsky, L. (1981) Neutron optical activity. *Nature*, **290**, 735–736.
98. Smith, D.A., Bowman, J.D., Crawford, B.E., Grossmann, C.A., Haseyama, T., Johnson, M.B., Masaike, A., Matsuda, Y., Mitchell, G.E., Nazarenko, V.A., Penttila, S.I., Roberson, N.R., Seestrom, S.J., Sharapov, E.I., Smotritsky, L.M., Stephenson, S.L., Tomsovic, S., and Yuan, V.W. (2001) Parity violation in neutron resonances of ^{117}Sn. *Phys. Rev. C*, **64**, 015502.
99. Dzuba, V.A., Flambaum, V.V., and Khariplovich, I.B. (1986) Enhancement of P- and T-Nonconserving Effects in Rare-Earth Atoms. *Z. Phys. D.*, **1**, 243–245.
100. Tsigutkin, K., Dounas-Frazer, D., Family, A., Stalnaker, J.E., Yashchuk, V.V., and Budker. D. (2009) Observation of a Large Atomic Parity Violation Effect in Ytterbium. *Phys. Rev. Lett.*, **103**, 071601.
101. "First storage of antiprotons." CERN, press-released on 18 August 2018, https://timeline.web.cern.ch/first-storage-antiprotons
102. Andresen, G.B., *et al.* (2010) Trapped Antihydrogen. *Nature*, **468**, 673–677.

103. Reich, E.S. (2010) Antimatter Held for Questioning. *Nature*, 468, p. 355.
104. ALPHA Collaboration. (2020) Investigation of the Fine Structure of Antihydrogen. *Nature*, **578**, 375–384.
105. Ahmadi, M., *et al.* (2017) Observation of the 1S–2S Transition in Trapped Antihydrogen. *Nature*, **541**, 506–509.
106. Hori, M., Sótér, A., Barna, D., Dax, A., Hayano, R., Friedreich, S., Juhász, B., Pask, T., Widmann, E., Horváth, D., Venturelli, L., and Zurlo, N. (2011) Two-Photon Laser Spectroscopy of Antiprotonic Helium and the Antiproton-to-Electron Mass Ratio. *Nature*, **475**, 484–488.
107. STAR Collaboration. (2011) Observation of the Antimatter Helium-4 Nucleus. *Nature*, **472**, 353–356.
108. Ulmer, S., Smorral, C., Mooser, A., Franke, K., Nagahama, H., Schneider, G., Higuchi, T., Van Gorp, S., Blaum, K., Matsuda, Y., Quint, W., Walz, J., and Yamazaki. Y. (2015) High-Precision Comparison of the Antiproton-to-Proton Charge-to-Mass Ratio. *Nature*, **524**, 196–199.
109. Jungmann, K.P. (2015) Matter and Antimatter Scrutinized. *Nature*, **524**, 168–169.
110. Dupourque, S., Tibaldo, L., and von Ballmoos, P. (2021) Constraints on the Antistar Fraction in the Solar System Neighborhood from the 10-Year Fermi Large Area Telescope Gamma-Ray Source Catalog. *Phys. Rev. D*, **103**, 083016.
111. PhysicsWorld. (2021) "Are Antimatter Stars Firing Bullets of Antihelium at Earth?" https://physicsworld.com/a/are-antimatter-stars-firing-bullets-of-antihelium-at-earth/
112. Pavlov, V.A., Shushenachev, Y.V., and Zlotin, S.G. (2019) Chiral and Racemic Fields Concept for Understanding of the Homochirality Origin, Asymmetric Catalysis, Chiral Superstructure Formation from Achiral Molecules, and B-Z DNA Conformational Transition. *Symmetry*, **11**, 649 (56 p).
113. Mason, S.F. (1991) *Chemical Evolution — Origin of the Elements, Molecules, and Living Systems*, Oxford University Press.
114. "Zeeman effect." Wikipedia, last edited on 16 September 2021, https://en.wikipedia.org/wiki/Zeeman_effect
115. Le Bel, J.A. (1874) Sur les Relations qui Existent Entre les Formules Atomiques des Corps Organiques et le Pouvoir Rotatoire de Leurs Dissolutions. *Bull. Soc. Chim.*, **22**, 337–347.
116. Wisniak, J. (2002) Joseph Achille Le Bel. His Life and Works. *Revista CENIC. Ciencias Químicas*, **33**, 35–43.
117. vant' Hoff, J.H. (1874) Voorstel tot Uitbreiding der Tegenwoordige in de Scheikunde gebruikte Structuurformules in de Ruimte, benevens een daarmee samenhangende Opmerking omtrent het Verband tusschen Optisch Actief Vermogen en chemische Constitutie van Organische Verbindingen. *Archives Neerlandaises des Sciences Exactes et Naturelles,* **9**, 445–454.
118. Meijer, E.W. (2001) Jacobus Henricus van't Hoff; Hundred Years of Impact on Stereochemistry in the Netherlands. *Angew. Chem. Int. Ed.,* **40**, 3783–3789.

119. Kekule, A. (1858) Ueber die Constitution und die Metamorphosen der Chemischen Verbindungen und über die Chemische Natur des Kohlenstoffs. *Ann. Chem.*, **106**, 129–159.
120. Rocke, A.J. (2015) It Began with a Daydream: The 150th Anniversary of the Kekulé Benzene Structure. *Angew. Chem. Int. Ed.*, **54**, 46–50.
121. Werner, A. (1911) Zur Kenntnis des Asymmetrischen Kobaltatoms. IV. *Ber. Dtsch. Chem. Ges.*, **44**, 3279–3284.
122. Ernst, K.-H., and H. Berke, H. (2010) Optical Activity and Alfred Werner's Coordination Chemistry. *Chirality*, **23**, 187–189.
123. Werner, A. (1914) Zur Kenntnis des asymmetrischen Kobaltatoms XII. Über optische Aktivität bei kohlestofffreien Verbindungen. *Ber. Dtsch. Chem. Ges.*, **47**, 3087–3094.
124. Kipping, F.S., and Pope, W.J. (1898) LXIII. — Enantiomorphism, *J. Chem. Soc. Trans.*, **73**, 606–617.
125. Mason, S.F. (1982) *Molecular Optical Activity and the Chiral Discriminations* (Cambridge University Press).
126. Mason, S.F. (1984) Origins of Biomolecular Handedness. *Nature*, **311**, 19–23.
127. Perucca, E. (1919) Nuove Osservazioni e Misure su Cristalli Otticamente Attivi ($NaClO_3$). *Il Nuovo Cimento,* **18**, 112–154.
128. Pfeiffer, P., and Quehl. K. (1931) Über Einen Neuen Effekt in Lösungen Optischaktiver Substanzen (I. Mitteil). *Chem. Ber.*, **64**, 2667–2671.
129. Pfeiffer, P., and Quehl, K. (1932) Aktivierung von Komplexsalzen in Wäßriger Lösung (II. Mitteil). *Chem. Ber.*, **65**, 560–565.
130. Borovkov, V.V., Lintuluoto, J.M., and Inoue, Y. (2000) Supramolecular Chirogenesis in Bis(zinc porphyrin): An Absolute Configuration Probe Highly Sensitive to Guest Structure. *Org. Lett.*, **2**, 1565–1568.
131. Borovkov, V.V., Lintuluoto, J.M., and Inoue, Y. (2001) Supramolecular Chirogenesis in Zinc Porphyrins: Mechanism, Role of Guest Structure, and Application for the Absolute Configuration Determination. *J. Am. Chem. Soc.*, **123**, 2979–2989.
132. Borovkov, V.V., Hembury, G.A., Yamamoto, N, and Inoue, Y. (2003) Supramolecular Chirogenesis in Zinc Porphyrins: Investigation of Zinc-Freebase Bis-Porphyrin, New Mechanistic Insights, Extension of Sensing Abilities, and Solvent Effect. *J. Phys. Chem. A*, **107**, 8677–8686.
133. Borovkov, V.V., Lintuluoto, J.M., and Inoue, Y. (2004) Origin, Control, and Application of Supramolecular Chirogenesis in Bisporphyrin-Based Systems. *Acc. Chem. Res.*, **37**, 449–459.
134. Borovkov, V.V., Hembury, G.A., and Inoue, Y. (2005) Supramolecular Chirogenesis with Bis-chlorin versus Bis-porphyrin Hosts: Peculiarities of Chirality Induction and Modulation of Optical Activity. *J. Org. Chem.*, **70**, 8743–8754.
135. Borovkov, V. (2014) Supramolecular Chirality in Porphyrin Chemistry. *Symmetry*, **6**, 256–294.
136. Lewis, G.N. (1926) The Conservation of Photons, *Nature*, **118**, 874–875.

137. APS News. (2012) "This Month in Physics History." December 18, 1926: Gilbert Lewis coins "photon" in letter to *Nature*, https://www.aps.org/publications/apsnews/201212/physicshistory.cfm
138. Rosenfeld, L. (1929) Quantenmechanische Theorie der natürlichen optischen Aktivität von Flüssigkeiten und Gasen. *Z. Phys.*, **52**, 161–174.
139. Hund, F. (1927) Symmetriecharaktere von Termen bei Systemen mit gleichen Partikeln in der Quantenmechanik. *Z. Phys.*, **43**, 788–803.
140. Berlin, Y.A., Burin, A.L., and Goldanskii, V.V. (1996) The Hund Paradox and Stabilization of Molecular Chiral States. *Z. Phys.*, **37**, 333–339.
141. Zel'dovich, Y.B., and Perelomov, A.M. (1961) The Effect of Weak Interaction on the Electromagnetic Properties of Particles. *Soviet Phys. JETP*, **12**, 777–784.
142. Letokhov, V.S. (1975) On Difference of Energy Levels of Left and Right Molecules due to Weak Interactions. *Phys. Lett. A.*, **53**, 275–276.
143. Rein, D.W. (1974) Estimation of Energy Differences between Optical Isomers Caused by Parity Non Conceiving Interactions. *Atomki Közlémnyek Suppl.*, **16**, 185–194.
144. Rein, D.W. (1974) Some Remarks on Parity Violating Effects of Intramolecular Interactions. *J. Mol. Evol.*, **4**, 15–22.
145. Rein, D.W., Hegstrom, R.A., and Sandars, P.G.H. (1979) Parity Non-Conserving Energy Difference between Mirror Image Molecules. *Phys. Lett. A*, **71**, 499–502.
146. Hegstrom, R.A., Rein, D.W., and Sandars, P.G.H. (1980) Calculation of the Parity Nonconserving Energy Difference between Mirror-Image Molecules. *J. Chem. Phys.*, **73**, 2329–2341.
147. Harris, R.A., and Stodolsky, L. (1978) Quantum Beats in Optical Activity and Weak Interactions. *Phys. Lett. B*, **78**, 313–317.
148. Quack, M. (1986) On the Measurement of the Parity Violating Energy Difference between Enantiomers. *Chem. Phys. Lett.*, **132**, 147–153.
149. Quack, M. (1989) Structure and Dynamics of Chiral Molecules. *Angew. Chem. Int. Ed. Engl.*, **28**, 571–586.
150. Quack, M, Stohner, J., and Willeke, M. (2008) High-resolution spectroscopic Studies and Theory of Parity Violations in Chiral Molecules. *Annu. Rev. Phys. Chem.*, **59**, 741–769.
151. Quack, M. (2002) How Important is Parity Violation for Molecular and Biomolecular Chirality? *Angew. Chem. Int. Ed.*, **41**, 4618–4630.
152. Arimondo, E., Glorieux, P., and Oka, T. (1977) Observation of Inverted Infrared Lamb Dips in Separated Optical Isomers. *Opt. Commun.*, **23**, 369–372.
153. Daussy, C., Marrel, T., Amy-Klein, A., Nguyen, C.T., Bordé, C.J., and Chardonnet, C. (1999) Limit on the Parity Nonconserving Energy Difference between the Enantiomers of a Chiral Molecule by Laser Spectroscopy. *Phys. Rev. Lett.*, **83**, 1554–1557.

154. Darquié, B., Stoeffler, C., Shelkovnikov, A., Daussy, C., Amy-Klein, A., Chardonnet, C., Trig, S., Guy, L., Crassous, J., Soulard, P., Asselin, P., Huet, T.R., Schwerdtfeger, P., Bast, R., and Saue, T. (2010) Progress toward the First Observation of Parity Violation in Chiral Molecules by High-Resolution Laser Spectroscopy. *Chirality*, **22**, 870–884.
155. Quack, M. (2012) Molecular Parity Violation and Chirality: The Asymmetry of Life and the Symmetry Violations in Physics. In: *Quantum Systems in Chemistry and Physics. Progress in Theoretical Chemistry and Physics*, Nishikawa K., Maruani J., Brändas E., Delgado-Barrio G., and Piecuch P. (eds). Vol 26, Springer.
156. MacDermott, A.J. (2012) Chiroptical Signatures of Life and Fundamental Physics. *Chirality*, **24**, 764–769.
157. MacDermott, A.J., and Hegstrom. R. A. (2004) A Proposed Experiment to Measure the Parity-Violating Energy Difference between Enantiomers from the Optical Rotation of Chiral Ammonia-like "Cat" Molecules. *Chem. Phys.,* **305**, 55–68.
158. MacDermott, A.J. (2000) The Ascent of Parity-Violation: Exochirality in the Solar System and Beyond. *Enantiomer*, **5**, 153–168.
159. Berger, R. (2003) Molecular Parity Violation in Electronically Excited States. *Phys. Chem. Chem. Phys.*, **5**, 12–17.
160. Lin, C.-K., Li, M.-C., Yamaki, M., Hayashi, M., and Lin, S.H. (2010) A theoretical study on the spectroscopy and the radiative and non-radiative relaxation rate constants of the $S_0{}^1A_1$–$S_1{}^1A_2$ vibronic transitions of formaldehyde. *Phys. Chem. Chem. Phys.*, **12**, 11432–11444.
161. Ferro-Costas, D., Pendás, Á.M., González, L., and Mosquera, R.A. (2014) Beyond the Molecular Orbital Conception of Electronically Excited States through the Quantum Theory of Atoms in Molecules. *Phys. Chem. Chem. Phys.*, **16**, 9249–9258.
162. MacDermott, A.J. (1995) Electroweak Enantioselection and the Origin of Life. *Orig. Life Evol. Biosph.,* **25**, 191–199.
163. Bargueño, P., Gonzalo, I., and de Tudela, R.P. (2009) Detection of Parity Violation in Chiral Molecules by External Tuning of Electroweak Optical Activity. *Phys. Rev. A*, **80**, 012110.
164. Mason, S.F. (1967) General Principles. *Proc. R. Soc. Lond. A.*, **297**, 3–15.
165. Mason, S.F., and Tranter, G.E. (1983) Energy Inequivalence of Peptide Enantiomers from Parity Non-Conservation. *J. Chem. Soc., Chem. Comm.*, **1983**, 117–119.
166. Mason, S.F. (1983) Molecular Handedness and the Origins of Chiral Discrimination. *Int. Rev. Phys. Chem.*, **3**, 217–241.
167. Mason, S.F., and Tranter, G.E. (1984) The Parity-Violating Energy Difference between Enantiomeric Molecules. *Mol. Phys.*, **53**, 1091–1111.
168. Tranter, G.E. (1986) Preferential Stabilization of the O-Sugar Series by the Parity-Violating Weak Interaction. *Chem. Commun.*, 60–61.

169. Tranter, G.E., and MacDermott, A.J. (1986) Parity-Violating Energy Differences between Chiral Conformations of Tetrahydrofuran: A Model System for Sugars. *Chem. Phys. Lett.*, **130**, 120–122.
170. MacDermott, A.J., and Tranter, G.E. (1989) The Search for Large Parity-Violating Energy Differences between Enantiomers. *Chem. Phys. Lett.*, **163**, 1–4.
171. Kiyonaga, H., Morihashi, K., and Kikuchi, O. (1998) Calculation of contributions of one- and two-electron spin-orbit coupling terms to the parity-violating energy shifts for amino acids and helical alkanes. *J. Chem. Phys.*, **108**, 2041–2043.
172. Kitayama, T., Kiyonaga, H., Morihashi, K., Takahashi, O., and Kikuchi, O. (2002) Ab initio Spin–Orbit Coupling SCF Calculation of Parity-Violating Energy of Chiral Molecules. *J. Mol. Struct. Theochem.*, **589–590**, 183–193.
173. Yamagata, Y. (1966) A Hypothesis for the Asymmetric Appearance of Biomolecules on Earth. *J. Theor. Biol.*, **11**, 495–498.
174. Wesendrup, R., Laerdahl, J.K., Compton, R.N., and Schwerdtfeger, P. (2003) Biomolecular Homochirality and Electroweak Interactions. I. The Yamagata Hypothesis. *J. Phys. Chem. A*, **107**, 6668–6673.
175. Salam, A. (1991) The Role of Chirality in the Origin of Life. *J. Mol. Evol.*, **33**, 105–113.
176. Laerdahl, J.K., and Schwerdtfeger, P. (1999) Fully Relativistic ab initio Calculations of the Energies of Chiral Molecules including Parity-Violating Weak Interactions. *Phys. Rev. A.*, **60**, 4439–4453.
177. Schwerdtfeger, P., Gierlich, J., and Bollwein. T. (2003) Large Parity-Violation Effects in Heavy-Metal-Containing Chiral Compounds. *Angew. Chem. Int. Ed.*, **42**, 1293–1295.
178. Schwerdtfeger, P., and Bast, R. (2004) Large Parity Violation Effects in the Vibrational Spectrum of Organometallic Compounds. *J. Am. Chem. Soc.*, **126**, 1652–1653.
179. Zel'dovich, Y.B., Saakyan, D.B., and Sobel'man, I.I. (1977) Energy Difference between Right-Hand and Left-Hand Molecules, due to Prity Nonconservation in Weak Interactions of Electrons with Nuclei. *Soviet Phys. JETP Lett.*, **25**, 94–97.
180. Harris, R.A., and Silbey, R. (1983) On the Stabilization of Optical Isomers through Tunneling Friction, *J. Chern. Phys.*, **78**, 7330–7333.
181. Dorta-Urra, A., Peñate-Rodríguez, H.C., Bargueño, P., Rojas-Lorenzo, G., and Miret-Artés, S. (2012) Dissipative Geometric Phase and Decoherence in Parity-Violating Chiral Molecules. *J. Chem. Phys.*, **136**, 174503.
182. Fujiki, M., Koe, J.R., Mori, T., and Kimura, Y. (2018) Questions of Mirror Symmetry at the Photoexcited and Ground States of Non-Rigid Luminophores Raised by Circularly Polarized Luminescence and Circular Dichroism Spectroscopy: Part 1. Oligofluorenes, Oligophenylenes, Binaphthyls and Fused Aromatics. *Molecules*, **23**, 2606.
183. Fujiki, M., Koe, J.R., and Amazumi, S. (2019) Questions of Mirror Symmetry at the Photoexcited and Ground States of Non-Rigid Luminophores Raised by Circularly Polarized Luminescence and Circular Dichroism Spectroscopy. Part 2: Perylenes,

BODIPYs, Molecular Scintillators, Coumarins, Rhodamine B, and DCM. *Symmetry*, **11**, 363.
184. Puneet, P., Singh, S., Fujiki, M., and Nandan, B. (2021) Handed Mirror Symmetry Breaking at the Photo-Excited State of π-Conjugated Rotamers in Solutions. *Symmetry*, **13**, 272.
185. Lahav, M., Weissbuch, I., Shavit, E., Reiner, C., Nicholson, G.J., and Schurig, V. (2006) Parity Violating Energetic Difference and Enantiomorphous Crystals — Caveats; Reinvestigation of Tyrosine Crystallization. *Orig. Life Evol. Biosphe.*, **36**, 151–170.
186. Havinga, E. (1954) Spontaneous Formation of Optically Active Substances. *Biochim. Biophys. Acta*, **13**, 171–174.
187. Szabó-Nagy, A., and Keszthelyi, L. (1999) Demonstration of the Parity-Violating Energy Difference between Enantiomers. *Proc. Nat. Acad. Sci. U.S.A.*, **96**, 4252–4255.
188. Keszthelyi, L. (2003) Parity-Violating Energy Difference between Enantiomers: Recent Developments. *Mendeleev Commun.*, **13**, 129–130.
189. Szurgot, M. (2012) Parity Violation in Unstirred Crystallization from Achiral Solutions, In *Crystallization and Materials Science of Modern Artificial and Natural Crystals*, E. Borisenko, E., and Kolesnikov, N. (eds), InTechOpen, 305–328.
190. Viedma, C. (2005) Chiral Symmetry Breaking During Crystallization: Complete Chiral Purity Induced by Nonlinear Autocatalysis and Recycling. *Phys. Rev. Lett.*, **94**, 065504.
191. Viedma, C. (2007) Selective Chiral Symmetry Breaking during Crystallization: Parity Violation or Cryptochiral Environment in Control? *Cryst. Growth Des.*, **7**, 553–556.
192. Rubik, B. (2021) The Effects of Subtle External Stimuli on Chiral Symmetry Breaking During Crystallization of Sodium Chlorate from Aqueous Solutions. *Water*, **12**, 61–71.
193. Wang, W., Yi, F., Ni, Y., Zhao, Z., Jin, X., and Tang, Y. (2000) Parity Violation of Electroweak Force in Phase Transitions of Single Crystals of D-and L-Alanine and Valine. *J. Biol. Phys.*, **26**, 51–65.
194. Sullivan, R., Pyda, M., Pak, J., Wunderlich, B., Thompson, J.R., Pagni, R., Barnes, C., Schwerdtfeger, P., and Compton, R. (2003) Search for electroweak Interactions in Amino Acid Crystals. II. The Salam Hypothesis, *J. Phys. Chem.* A **107**, 6674–6680.
195. Wilson, C.C., Myles, D., Ghosh, M., Johnson, L.N., and Wang, W. (2005) Neutron Diffraction Investigations of L- and D-Alanine at Different Temperatures: The Search for Structural Evidence for Parity Violation. *New J. Chem.,* **29**, 1318–1322.
196. Belo, E.A., Pereira, J.E., Freire, P.T., Argyriou, D.N., Eckert, J., and Bordallo, H.N. (2018) Hydrogen Bonds in Crystalline D-Alanine: Diffraction and Spectroscopic Evidence for Differences between Enantiomers. *IUCrJ*, **5**, 6–12.

197. Belo, E.A., Pereira, J.E., Freire, P.T., Argyriou, D.N., Eckert, J., and Bordallo, H.N. (2018) Response to Comment on 'Hydrogen Bonds in Crystalline D-Alanine: Diffraction and Spectroscopic Evidence for Differences between Enantiomers'. *IUCrJ*, **5**, 658–659.
198. Vallazza, M., Perbandt, M., Klussmann, S., Rypniewski, W., Einspahr, H.M., Erdmann, V.A., and Betzel, C. (2004) First Lok at RNA in L-Configuration. *Acta Crystallogr. D*, **60**, 1–7.
199. Bolik, S., Rübhausen, M., Binder, S., Schulz, B., Perbandt, M., Genov, N., Erdmann, V., Klussmann, S., and Betzel, C. (2007) First Experimental Evidence for the Preferential Stabilization of the Natural D-Over the Nonnatural L-Configuration in Nucleic Acids. *RNA*, **13**, 1877–1880.
200. Kiliszek, A., Błaszczyk, L., Bejger, M., and Rypniewski, W. (2021) Broken Symmetry between RNA Enantiomers in a Crystal Lattice. *Nucleic Acids Res.*, **49**, 12535–12539.
201. Gabuda, S.P., and Kozlova, S.G. (2014) Chirality-Related Interactions and a Mirror Symmetry Violation in Handed Nano Structures. *J. Chem. Phys.*, **141**, 044701 (5 p).
202. Gabuda, S.P., and Kozlova, S.G. (2015) Abnormal Difference between the Mobilities of Left- and Right-Twisted Conformations of $C_6H_{12}N_2$ Roto-Symmetrical Molecules at Very Low Temperatures. *J. Chem. Phys.*, **142**, 234302.
203. Kozlova, S.G., and Gabuda, S.P. (2017) Thermal Properties of $Zn_2(C_8H_4O_4)_2{\bullet}C_6H_{12}N_2$ Metal-Organic Framework Compound and Mirror Symmetry Violation of DABCO Molecules. *Sci. Rep.*, **7**, 1–5.
204. Kozlova, S., Ryzhikov, M., Pishchur, D., and Mirzaeva, I. (2019) Overview of Low-Temperature Heat Capacity Data for $Zn_2(C_8H_4O_4)_2{\bullet}C_6H_{12}N_2$ and the Salam Hypothesis. *Symmetry*, **11**, 657.
205. Dubey, M., Kumar, A., Dhavale, V.M., Kurungot, S., and Pandey, D.S. (2015) Can Enantiomer Ligands Produce Structurally Distinct Homochiral MOFs? *CrystEngComm*, **17**, 8202–8206.
206. Yasui, T., Ama, T., and Kauffman, G.B. (1992) Resolution of the Dodecaamminehexa-μ-hydroxo-tetracobalt(III) Ion. *Inorg. Synth.*, **29**, 169–174.
207. Srinivasan, A., Cortijo, M., Bulicanu, V., Naim, A., Clérac, R., Sainctavit, P., Rogalev, A., Wilhelm, F., Rosa, P., and Hillard, E.A. (2018) Enantiomeric Resolution and X-ray Optical activity of a Tricobalt Extended Metal Atom Chain. *Chem. Sci.*, **9**, 1136–1143.
208. Scolnik, T., Portnaya, I., Cogan, U., Tal, S., Haimovitz, R., Fridkin, M., Elitzur, A.C., Deamer, D.W., Shinitzky, M. (2006) Subtle Differences in Structural Transitions between Poly-L- and Poly-D-Amino Acids of Equal Length in Water. *Phys. Chem. Chem. Phys.*, **8**, 333–339.
209. Tikhonov, V.I., and Volkov, A.A. (2002) Separation of Water into Its Ortho and Para Isomers. *Science*, **296**, 2363.
210. Sun, Z.D., Takagi, K., and Matsushima, F. (2005) Separation and Conversion Dynamics of Four Nuclear Spin isomers of Ethylene. *Science*, **310**, 1938–1941.

211. Sun, Z.D., Ge, M., and Zheng, Y. (2015) Separation and Conversion Dynamics of Nuclear-Spin Isomers of Gaseous Methanol. *Nat. Commun.*, **6**, 6877.
212. Shinitzky, M., and Elitzur, A.C. (2006) Ortho–Para Spin Isomers of the Protons in the Methylene Group — Possible Implications for Protein Structure. *Chirality*, **18**, 754–756.
213. Turro, N.J., Chen, J.Y.C., Sartori, E., Ruzzi, M., Marti, A., Lawler, R., Jockusch, S., López-Gejo, J. Komatsu, K., and Murata, Y. (2010) The spin chemistry and magnetic resonance of H_2@C60. From the Pauli principle to trapping a long lived nuclear excited spin state inside a buckyball. *Acc. Chem. Res.*, **43**, 335–345.
214. Maity, A., Maithani, S., Pal, A., and Pradhan, M. (2021) Highresolution Spectroscopic Probing of Ortho and Para Nuclear-Spin Isomers of Heavy Water in the Gas Phase. *Chem. Phys.*, **541**, 111041.
215. Eills, J., Blanchard, J.W., Bougas, L., Kozlov, M.G., Pines, A., and Budker, D. (2017) Measuring Molecular Parity Nonconservation using Nuclear-Magnetic-Resonance Spectroscopy. *Phys. Rev. A*, **96**, 042119 (7 p).
216. Blanchard, J.W., King, J.P., Sjolander, T.F., Kozlov, M.G., and Budker, D. (2020) Molecular Parity Nonconservation in Nuclear Spin Couplings. *Phys. Rev. Res.*, **2**, 023258.
217. Jasinski, J.P., Jasinski, J.M., and Crosby, D.J. (2003) Crystal structures of (I) 2-(9,9-dipropylfluorene-2-yl)-9,9-dipropylfluorene and (II) 2-(1,1-dimethylpropyl)-7-{4-[(1,1-dimethylpropyl)-9,9-diethylfluoren-2-yl]phenyl}-9,9-diethylfluorene and (III) 2-(4-ethylphenyl)-7-[7-(ethylphenyl)-9,9-dipropylfluoren-2-yl]-9,9-dipropylfluorene. *J. Chem. Crystllogr.*, **33**, 365–374.
218. Fujiki, M. (2010) Mirror Symmetry Breaking in Helical Polysilanes: Preference between Left and Right of Chemical and Physical Origin. *Symmetry*, **2**, 1625–1652.
219. Fujiki, M. (2021) Resonance in Chirogenesis and Photochirogenesis: Colloidal Polymers Meet Chiral Optofluidics. *Symmetry*, **13**, 199.
220. Fujiki, M., Okazaki, S., Rahim, N.A.A., Yamada, T., and Nomura, K. (2021) Synchronization in Non Mirror-Symmetrical Chirogenesis: Non-Helical π–Conjugated Polymers with Helical Polysilane Copolymers in Co-Colloids. *Symmetry*, **13**, 594 (34 p).
221. Shinitzky, M., and Haimovitz, R. (1993) Chiral Surfaces in Micelles of Enantiomeric N-Palmitoyl- and N-Stearoylserine, *J. Am. Chem. Soc.*, **115**, 12545–12549.
222. Fujiki, M. (2001) Optically Active Polysilylenes: State-of-the-Art Chiroptical Polymers. *Macromol. Rapid Commun.*, **22**, 539–563.
223. Fujiki, M. (2001) Experimental Tests of Parity Violation at Helical Polysilylene Level. *Macromol. Rapid Commun.*, **22**, 669–674.
224. Fujiki, M., Kawagoe, Y., Nakano, Y., and Nakao, A. (2013) Mirror-Symmetry-Breaking in Poly[(9,9-di-*n*-octylfluorenyl-2,7-diyl)-*alt*-biphenyl] (PF8P2) is Susceptible to Terpene Chirality, Achiral Solvents, and Mechanical Stirring. *Molecules*, **18**, 7035–7057.

225. Barron, L.D., and Vrbancich, J. (1984) Magneto-Chiral Birefringence and Dichroism. *Mol. Phys.*, **51**, 715–730.
226. Buckingham, A.D., and Stephens, P.J. (1966) Magnetic Optical Activity. *Ann. Rev. Phys. Chem.*, **17**, 399–432.
227. Stephens, P.J. (1974) Magnetic Circular Dichroism. *Ann. Rev. Phys. Chem.*, **25**, 201–232.
228. Richardson, F.S., and Riehl, J.P. (1977) Circularly Polarized Luminescence Spectroscopy. *Chem. Rev.*, **77**, 773–792.
229. Richardson, F.S., and Brittain, H.G. (1981) A Structural Study of Tris(β-diketonate) europium(III) Complexes in Solution Using Magnetic Circularly Polarized Luminescence Spectroscopy. *J. Am. Chem. Soc.*, **103**, 18–24.
230. Thiemann, W., and Jarzak, U. (1981) A New Idea and Experiment Related to the Possible Interaction between Magnetic Field and Stereoselectivity. *Orig. life*, **11**, 85–92.
231. Toda, H., Hara, N., Fujiki, M., and Imai, Y. (2021) Sign inversion in magnetic circularly polarised luminescence of fused aromatics with 1.6 T N-up/ S-up Faraday geometry. *RSC Adv.*, **11**, 1581–1585.
232. Toda, H., Otake, S., Ito, A., Miyasaka, M., Fujiki, M., and Imai, Y. (2021) Magnetic Circularly Polarized Luminescence in the Photoexcited States of Racemic [*n*] Helicenes (n = 3–5,7) in Tetrahydrofuran and Dimethyl Sulfoxide Solutions. *ChemPhysChem*, **22**, 2058. doi: 10.1002/cphc.202100346
233. Kimoto, T., Mimura, Y., Fujiki, M., and Imai, Y. (2021) Ambidextrous Solid-state Magnetic Circularly Polarized Luminescence (MCPL) from Red-Green-Blue Inorganic Luminophores without Molecular Chirality. *Chem. Lett.*, **50**, 916–919.
234. Mimura, Y., Fujiki, M., and Imai, Y. (2021) Mirror-Symmetric Magnetic Circularly Polarized Luminescence from CdS/ZnS Core-Shell Quantum Dots: Faraday Effect in the Photoexcited State. *Chem. Phys. Lett.*, **767**, 138353.
235. Mishima, K., Kaji, D., Fujiki, M., and Imai, Y. (2021) Remarkable Effects of External Magnetic Field on Circularly Polarized Luminescence of $Eu^{III}(hfa)_3$ with Phosphine Chirality. *ChemPhysChem*, **22**, 1728–1737.
236. El Rez, B., Liu, J., Béreau, V., Duhayon, C., Horino, Y., Suzuki, T., Coolen, L., and Sutter, J.-P. (2020) Concomitant Emergence of Circularly Polarized Luminescence and Single-Molecule Magnet Behavior in Chiral-at-Metal Dy Complex. *Inorg. Chem. Front.*, **7**, 4527–4534.
237. Kaji, D., Okada, H., Hara, N., Kondo, Y., Suzuki, S., Miyasaka, M., Fujiki, M., and Imai, Y. (2020) Non-classically Controlled Sign in a 1.6 Tesla Magnetic Circularly Polarized Luminescence of Three Pyrenes in a Chloroform and a PMMA Film. *Chem. Lett.*, **49**, 674–676.

https://doi.org/10.1142/9789811259227_0002

2 Chirogenesis in Supramolecular Systems

Lukas Ustrnul,* Victor Borovkov† and Riina Aav‡

Department of Chemistry and Biotechnology,
Tallinn University of Technology,
Tallinn, Estonia
*lukas.ustrnul@taltech.ee
†victor.borovkov@taltech.ee
‡riina.aav@taltech.ee

Supramolecular systems have dynamic character, which leads to sensitivity to subtle change of chemical or physical conditions. Therefore, chirality can occur in the system as a consequence of symmetry breaking upon the complex formation or as a result of presence of an asymmetric component. This chapter provides overview on general concepts of supramolecular chemistry by discussing types of the interactions, importance of solvation, and also general types of chirality, including the kinetic point of view related to terms "dynamic chirality" and "atropisomerism." The concept of supramolecular chirality and chirogenesis is outlined with representative examples, while listing the analytical methods in use to determine chirality. Chirality transfer from molecular level to chiral aggregates achieved through self-assembly is briefly discussed in relation to planar chirality. Further, chirality induction and switching in various systems, such as catenanes and rotaxanes, is comprehensively covered. In addition, an overview on supramolecular chiroptical sensors that rely on porphyrins, cucurbiturils, cages, and other supramolecular building blocks is given. Based on 181 references and outlooks presented in this chapter, one can see that chirogenesis is a dynamic process, which has great potential to be employed in future molecular machines, molecular electronics, programmed chemical transformations, and other newly emerging fields.

1.1 Introduction

Supramolecular chemistry is a well-established field, which has twice earned awards for the Nobel Prize in chemistry. The first Nobel was awarded in 1987, "for development and use of molecules with structure-specific interactions of high selectivity" (D.J. Cram, J.-M. Lehn, and C.J. Pedersen) [1] and the second in 2016, "for the design and synthesis of molecular machines" (J.-P. Sauvage, J.F. Stoddart, B.L. Feringa) [2]. The field's beginnings are connected to crown-ethers and their ability to bind cations discovered by Pedersen in late 1960s [3, 4]; this was later followed by Lehn's cryptands [5, 6] and Cram's carcerands [7, 8]. These rather simple cyclic organic molecules are able to bind other smaller molecules or ions to form host–guest systems. The second Nobel Prize-winning project featured induction of controlled motion via host–guest or other molecular systems. In addition, with the development of the design of supramolecular building blocks, the scope of interactions and discoveries now assumed to be related to the field has widened to include works published well before the term "supramolecular chemistry" had been introduced. The work of J.D. van der Waals suggesting the existence of intermolecular forces (1873) [9], A. Werner's works in the field of coordination chemistry (1893) [10], and H. E. Fischer's hypothesis of "lock and key" (1894) in the interactions between enzymes and substrates [11] are the most prominent early cases of supramolecular chemistry. A cartoon view on the formation of supramolecular systems is shown in Figure 1.

The most important question to be answered is: "What exactly is supramolecular chemistry?" The International Union of Pure and Applied Chemistry (IUPAC) "Gold Book" says:

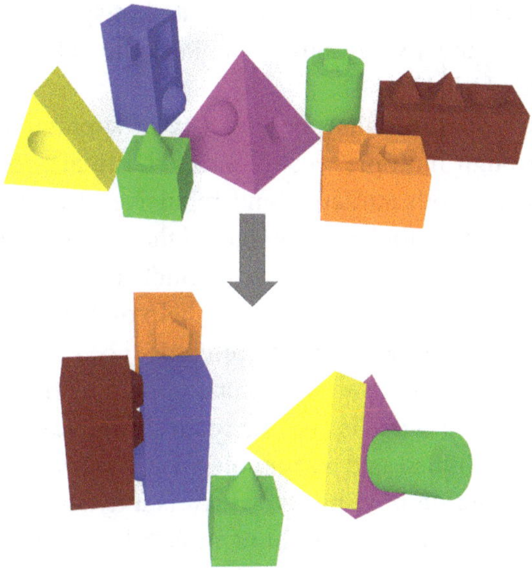

Figure 1. General illustration of the supramolecular concept. Molecules are like building blocks possessing different shapes and capable of interlocking with each other through mutually compatible holes and protuberances.

> Supramolecular chemistry is a field of chemistry related to species of greater complexity than molecules, that are held together and organized by means of intermolecular interactions. The objects of supramolecular chemistry are supermolecules and other polymolecular entities that result from the spontaneous association of a large number of components into a specific phase (membranes, vesicles, micelles, solid state structures etc.) [12].

The first sentence of the IUPAC definition provides an excellent explanation, but the second might lead to confusion, as it describes a subfield of supramolecular chemistry: self-assembled systems. Supramolecular chemistry does not require a "large number of components," as some systems are composed of only two molecules (complexes observed in gas phase). Nevertheless, most systems are studied in solid phase and in solution, which inevitably means the presence and influence of larger numbers of molecules in close proximity. Moreover, to widen the scope of participating parties in supramolecular systems, more supramolecular interactions have emerged, including tetrel bonds, pnictogen bonds, chalcogen bonds, and aerogen bonds — which are related, respectively, to elements from the 14th–16th, and 18th groups of the Periodic Table [13, 14] — and mechanical bonds [15–17].

It is necessary to stress that the majority of supramolecular systems gain stability through multiple weak bonds. The final geometry of complexes is shaped by both attractive and repulsive effects, which are the reason the word "interaction" is better suited than "bond" in the definition of the term "supramolecular chemistry." Such attractive and repulsive forces are easy to follow in solids with the help of X-ray crystallography, and in a solution the geometry can be often deduced using nuclear magnetic resonance (NMR) and various computational methods.

In this chapter, we will discuss the main differences and similarities between covalently linked and supramolecular systems applicable in the generation and induction of chirality, which is a dynamic process termed "chirogenesis" [18]. Typically, a chiral molecule needs to be part of a supramolecular system to achieve chirogenensis; however, many achiral molecules are able to form chiral complexes or achieve a chiral conformation (spatial arrangement) providing racemic mixtures due to lack of a single-handed influence. For example, we can imagine a solution of achiral molecules that self-assemble into helixes with random helicity. Another way to induce chiral arrangement of achiral molecules is via the application of a chiral field, such as circularly polarized light (CPL). We will present some practical examples of systems providing macroscopic enantiomeric excess (exhibiting chiral properties), as well as examples of specific racemic mixtures of supramolecular complexes, which await the development of new approaches achieving an enantiomeric excess and further exploitation.

1.2 Host–Guest Systems

The term "host-guest chemistry" is often used as a synonym for "supramolecular chemistry." However, a conventional point of view is that the "host" is a larger molecule (or molecular assembly) possessing convergent binding sites capable of binding a "guest" that is a smaller species (molecule or ion) possessing divergent binding sites. Complementarity of size, shape, and binding sites (e.g., with host as a hydrogen bond donor and guest as a hydrogen bond acceptor) is often the main factor determining the possibility and strength of the interaction [19]. Considering the wide range of non-covalent interactions and diverse supramolecular systems described to date (including self-assembled systems, molecular cages, mechanically interlocked molecules, and so on), it is more reasonable to consider host–guest chemistry as a specific subfield of supramolecular chemistry.

A conventional approach to the terms "host" and "guest" is to use them more generally, simplifying the identification of binding partners in both written and verbal formats. While the host and guest are larger and smaller molecules, correspondingly, the divergence and convergence of binding sites are inferior.

1.3 Types of Interactions and Their Strengths

Various chemical interactions and their strengths should be outlined (Figure 2). A baseline understanding of the difference between covalent bonds and non-covalent interactions can be reached by inspecting the energy of bonds and their origin. Simply, a covalent bond is based on the sharing of an electron pair between two atoms, where either both atoms provide one electron to form an electron pair, or one atom provides the entire electron pair and the other provides a vacant orbital. In contrast, non-covalent interaction mostly originates from much more subtle differences in the electrostatic fields of covalent molecules or their parts, and possibly ions, leading to stabilization of the whole system. However, a principle of "more electron-rich interacts with less electron-rich" can be unintuitive for some types of non-covalent interactions. For example, in the case of an anion–π interaction, we seemingly have two electron-rich particles forming a stable complex, but this works only with electron-deficient arenes [20]. Similarly, in the case of halogen bonds, we see electron donors (usually acceptors of the hydrogen bond) interacting with electron-rich halogens, because covalently bound halogen atoms have a so-called σ-hole, which is in part positively charged [21–24].

More importantly, to understand supramolecular interactions, we must consider their relatively small binding energies in comparison to covalent bonds. The typical bond dissociation energies (BDE) of organic

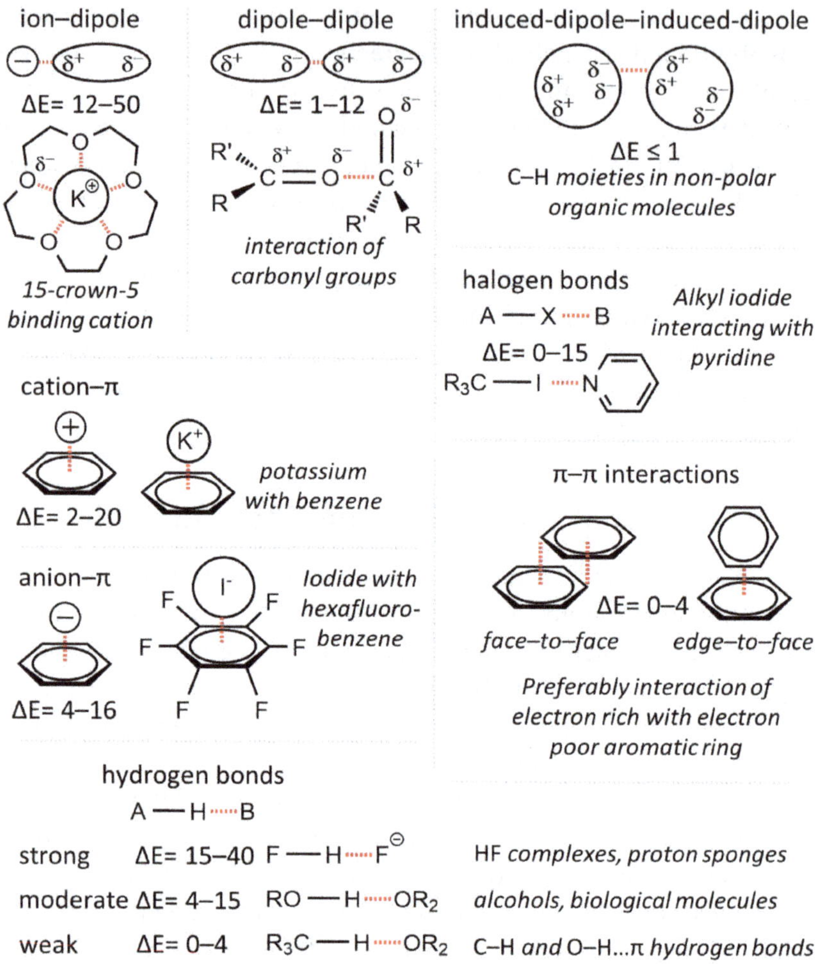

Figure 2. General illustration and representative examples of the most common types of non-covalent interactions [19–21, 25–29]. Values for typical bond energies are given in kcal/mol; notably, the energy can significantly deviate from the presented range for particular supramolecular systems.

molecules are around 100 kcal mol^{-1} for most C–C and C–H bonds, while these become weaker for bonds with good leaving groups (such as halide ions). Peroxides are assumed to have weak and reactive covalent bonds, and their BDE is around 40 kcal mol^{-1} [30]. In the case of non-covalent binding, the BDE values are generally smaller. Although we can find some specific examples of strong enough non-covalent bonds with the energy equal to peroxides (such as ion–dipole interactions), typically the BDE values are considered to be up to 20 kcal mol^{-1} (Figure 2) [19, 25–27]. Most often, the bond energies are only several kcal mol^{-1}; for example, the strength of the interior bridges of proteins dissolved in water is generally assumed to be 0.5–1.5 kcal mol^{-1} [31]. Finally, it is necessary to emphasize that the strength of non-covalent interactions can be strongly dependent upon an environment: for example, a solvent.

1.4 Influence of Solvent on Supramolecular Interactions

Generally, solvation can be considered to be a weak non-covalent binding; the complexes in a solution are strongly affected by the surrounding medium. Thus, solvent molecules are able to exhibit interactions at specific sites of the binding partners, and these interactions can also hinder the complex formation or influence its geometry. In other words, a solvent molecule can behave as a binding partner itself and compete with other hosts or guests in the solution during the complexation. In addition to molecule–molecule interaction, described above, a solvent contributes to equilibria of binding partners, by nonspecific interactions arising from the solvent acting as a dielectric continuum.

A straightforward example of the solvent effect is a 1:1 complex between potassium thiocyanate and cyclic polyether 18-crown-6, which was studied in various solvents [32]. The association constant of the system was the weakest in water, and for other used solvents (including DMSO, DMF, and methanol) correlates well with the energies necessary to transfer the potassium cation from water to the given solvent. This illustrates how the interaction of a solvent with one of the binding partners (in this case, solvation of cation and energy necessary for its desolvation) hinders the complex formation.

Even in cases lacking the specific interactions between solvent and binding sites of dissolved molecules, the solvent molecules in close proximity to the binding partners may reduce opportunities to apply their intermolecular forces as compared to a bulk solvent (Figure 3a) [33]. One can imagine that in order to accommodate solute molecules during the process of dissolution, it is necessary to create cavities in a liquid. This

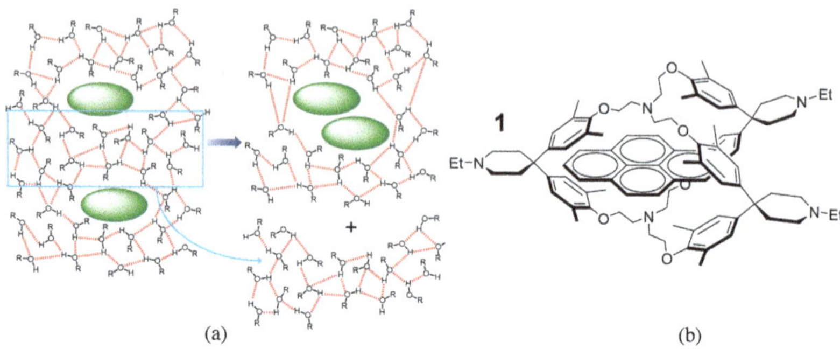

Figure 3. (a) Polar solvent molecules next to a nonpolar solutes can form only limited amount of stabilizing non-covalent interactions in comparison with the bulk solvent. Solvent's intramolecular interactions are restored by minimizing the volume of cavities in the solvent [25]. (b) Cyclophane derivative **1** binding a pyrene molecule in its cavity [34].

may lead to a difference in the energy related to the cohesive pressure of a solvent, which is the energy necessary to break all intermolecular solvent–solvent interactions (energy of vaporization) per a molar volume of solvent. As a result, solvents with a higher cohesive pressure (polar solvents) strongly drive away the other molecules from solution to restore the solvent's intermolecular interactions, causing phase separation. For example, in water, this is the main factor contributing to the hydrophobic effect. The cohesive pressure can assist in forming a complex between molecules, which hold together by an only very weak or nonspecific bond (dispersion forces). One example of this type of system is a molecule of pyrene bound inside a derivative of cyclophane **1** (Figure 3b), which displayed a linear correlation between the standard Gibbs free energy of formation, ΔG^0, and the solvent polarity parameter, $E_T(30)$ [34]. Astonishingly, the corresponding association constant varies from 12 M^{-1} in low-polar benzene to $6{\cdot}10^6$ M^{-1} in highly polar water. Thus, the strength of binding arises entirely from solvation effects. Specifically, benzene has a low cohesive pressure, and the energy necessary for the formation of cavities to accommodate binding partners is low, while water has a large cohesive pressure; this drives pyrene into macrocyclic cyclophane **1** effectively, to minimize the volume of cavities in the water.

In conclusion, supramolecular systems are influenced by solvent in a rather complex way. Therefore, the extent of the solvent dependence or specific solvent role must be analyzed and evaluated carefully for each supramolecular system [33].

1.5 General Types of Chirality

After addressing the general concepts of supramolecular chemistry and interactions, focus can be turned to the appearance of chirality in chemical systems, which can originate both from covalently bound atoms and via supramolecular interactions. The only necessary criterion for chirality of a single molecule or another chemical object (such as a complex) is the absence of the improper rotation S_n axis: i.e., if it cannot be superimposed with its mirror image. This condition is achieved in covalently linked molecules mostly via existence of a stereogenic center, or in other words by configurational chirality. This may, for example, take the form of carbon bearing four different substituents (Figure 4). However, this center can be based on other suitable atoms (nitrogen, phosphorus, sulfur) or a group of atoms (cumulene), in which case the latter is then called a stereogenic unit. Less common is a conformational chirality appearing in systems lacking a stereogenic center from their stable conformations (such as helicenes), which can be usually interconverted via rotations about single bonds. Conformational chirality can be divided into two categories (Figure 4): axial chirality (descriptors: R_a/S_a or *P*/*M*), which arises from the non-planar arrangement of four groups in pairs about a chirality axis, and planar chirality (descriptors: R_p/S_p or *P*/*M*), which originates from the arrangement of out-of-plane groups with respect to a chirality plane. Axial chirality is also called helical chirality, and can be demonstrated on helicenes and enantiomeric gauche conformers of a butane molecule (Figure 4); these are, of course, not possible to separate due to very low rotational barriers, although they provide a useful example of finding asymmetry in a very simple system. A typical example of planar chirality comprises molecules containing two dissymmetric non-coplanar rings, which cannot easily rotate about a chemical bond connecting them, such as substituted paracyclophanes and ferrocene derivatives. However, "planar chirality" is also known as "2D chirality," and can be related to simple planar molecules (such as nitrophenol)

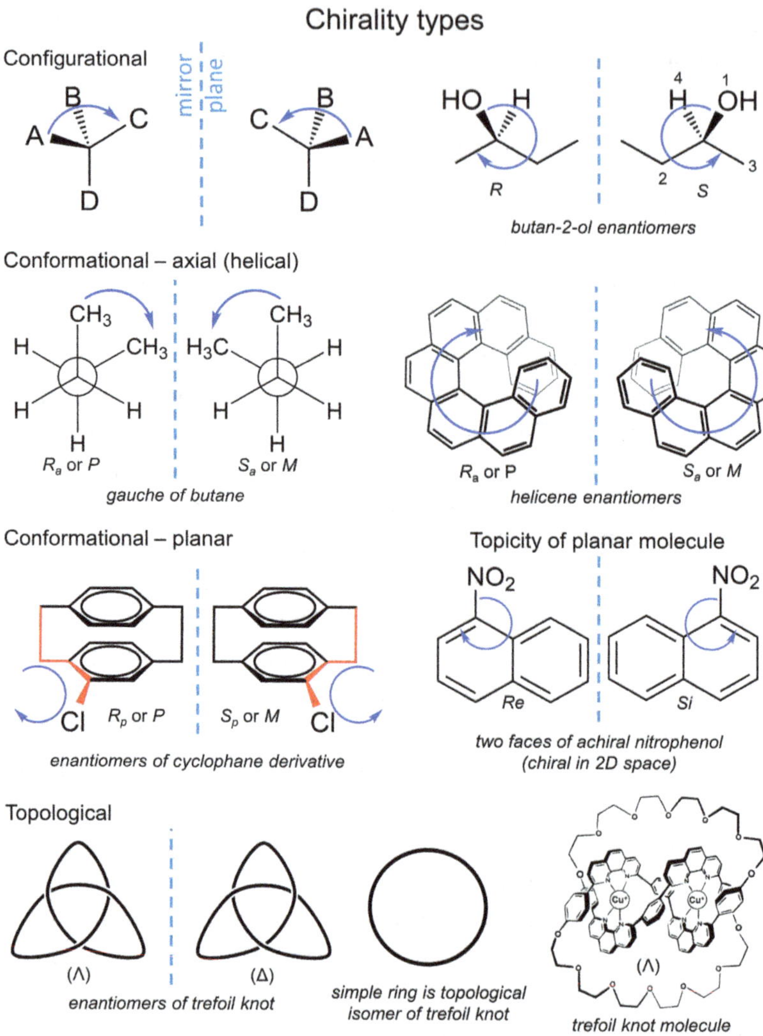

Figure 4. Types of chirality and their examples including stereochemical descriptors [35, 36].

having two different faces (*Re*/*Si* and pro-*R*/pro-*S* for prochiral planar atoms), thus being chiral in two-dimensional (2D) space.

In general, the stereochemistry of these systems which lack a single-atom stereogenic center (such as cumulenes, helicenes, chiral conformations, atropisomers, and so on) can be assigned based on their sense of twist by using the Cahn–Ingold–Prelog priority rules [35] and helical descriptors: *P* (plus) for clockwise twist (right-handed helix), and *M* (minus) for counterclockwise (left-handed helix).

The last, relatively rare, option to obtain covalent molecules lacking the improper rotation S_n axis is a case of cyclic molecules exhibiting topological chirality (Figure 4), when they are able to form various knots [36]. This type of chirality is based on topologically equivalent molecules, which can be transformed into each other by continuous deformations — such as twisting, stretching, and bending — but not by breaking or forming new bonds. It means, for example, that all enantiomers and diastereomers based on an asymmetric carbon are topologically identical. A large amount of topologically chiral molecules can be found within mechanically interlocked molecules, which will be discussed further.

Aside from the strictly spatial perspective of chirality discussed above, pointing out terminology related to the kinetics of structural changes that lead to switching between stereoisomers is necessary. Firstly, there are molecules that have chiral conformations but are traditionally viewed as achiral, due to their fast rotation around single bonds (such as butane; Figure 5), which prevents the direct observation of chiral conformers by conventional experimental techniques. Secondly, the systems exhibiting "dynamic chirality" have a higher, but still surmountable, barrier for interconversion between configurational isomers. For dynamically chiral systems, the barrier is much higher than the free rotation around a single bond in the molecule, like butane, but still low enough to obtain fast interconversion at room temperature; therefore, rotamers can be clearly

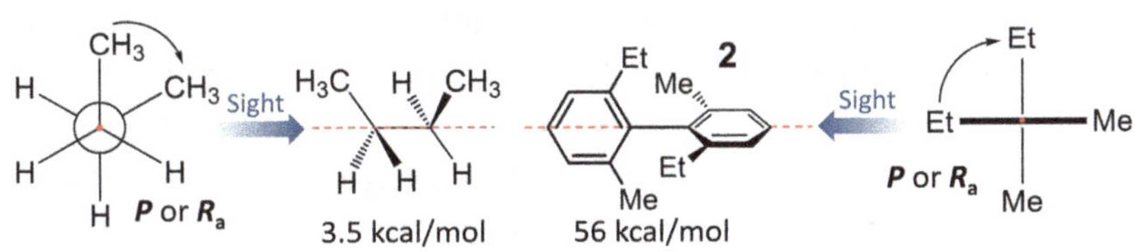

Figure 5. Rotational barrier of interconversion between chiral conformations of butane and atropisomer **2**. The energy necessary for rotation about a single bond is negligible for butane (3.5 kcal mol^{-1}), and high for atropisomer **2** (56 kcal mol^{-1}) [25, 37].

distinguished only at lower temperatures, for example those accessible by NMR. Dynamic chirality is often discussed in relation to mechanically interlocked systems, which will be described further. Finally, it is necessary to mention the phenomenon of atropisomerism (Figure 5), which is based on stereoisomers becoming interconvertible by rotation around a single bond (**2**) with a large rotational barrier causing sufficient stability, hence allowing their separation at room temperature.

1.6 Concept of Supramolecular Chirality and Chirogenesis

In contrast to the conventional chemistry of covalently linked molecules, supramolecular chemistry offers a broader diversity of chirality phenomena, and furthermore opens a new field of induced chirality, termed "chirogenesis." Indeed, non-covalent interactions significantly broaden the scope of approaches to induce chirality in comparison to covalent bonds. Supramolecular chirality can arise both from the geometrical properties of components, and from the way in which they associate [38, 39]. In the simplest case, a system can be chiral if at least one of its components is asymmetric. However, interaction between achiral molecules can also cause dissymmetry, if the symmetrical elements of each component are destroyed upon association. For example, if two simple molecules — each having just one plane of symmetry — form a complex such that these two planes are perpendicular to each other, then the entire complex becomes chiral (Figure 6a). Moreover, a simple achiral ion or small molecule is able to induce a chiral conformation of a flexible achiral host, as can be seen mainly in metal coordination complexes (Figure 6b) [40–43].

Noticeably, the association of both enantiomers can form an achiral complex of the *meso* type (**3**) through interaction with a symmetrical bridging molecule (Figure 7) [38]. However, this type of system can also produce diastereomeric complexes (**4** and its mirror image), if two enantiopure molecules are bridged. Because enantiomers are chemically equal and a bridging molecule is achiral, it then has no basis for enantioselectivity.

It is important to realize that all of the chiral complexes discussed above will generate only racemic mixtures without an external chiral field or influence. This is because the enantiomeric complexes have the same energy and probability of appearance in achiral conditions, which leads to

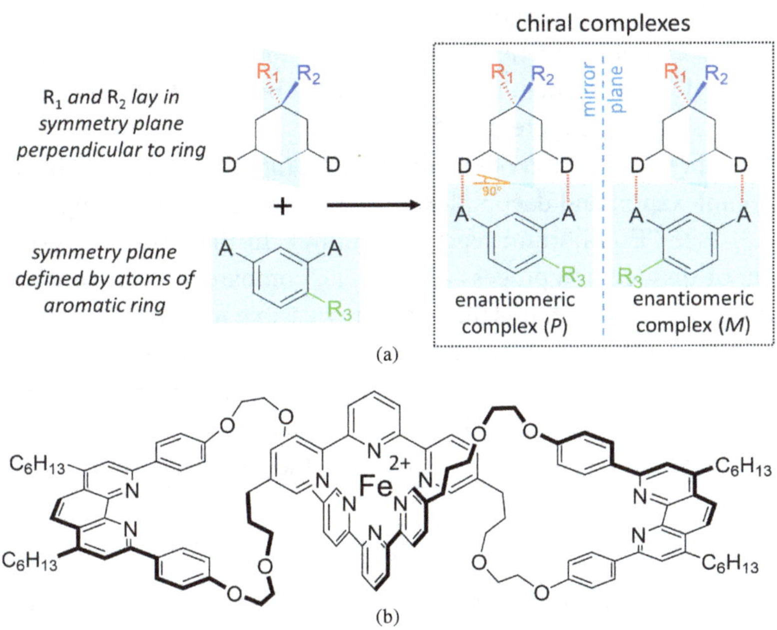

Figure 6. (a) A general example of two achiral molecules forming chiral complexes as a consequence of destroyed symmetry elements of parent molecules; (b) enantiomer of a metal complex of macrocycle, which was obtained as a racemate [42].

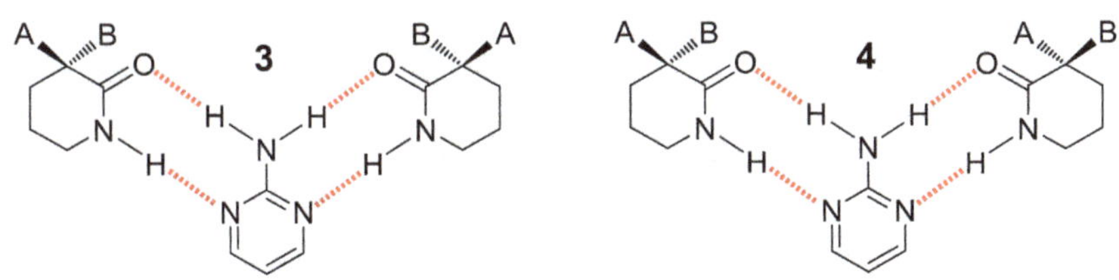

Figure 7. Achiral meso complex **3** formed from both enantiomers and its diastereomeric complex **4** composed of two enantiopure molecules.

a more or less rapidly exchanging racemic mixture. Therefore, some chiral influence (such as a chiral solvent, chiral auxiliary, unidirectional vortex, circularly polarized irradiation, and so on) should be used to increase the preference for one of the enantiomeric forms via non-covalent interactions [44–59]. For example, propeller-shaped molecules **6** derived from triphenylamine and carboxylic acid **5** are able to self-aggregate into helixes with specific handedness, dictated by the rotational direction of CPL. Furthermore, the desired handedness of a supramolecular complex can be permanently locked via photopolymerization between the diacetylene moieties using circularly polarized ultraviolet light (CPUL) irradiation (Figure 8) [55].

Constant racemization of enantiomeric complexes is reached by a constant complexation and decomplexation process between binding partners A and B (see "Equilibrium reaction" below). In the case of a slow rate constant of dissociation process (k_b) (eq. 1), compared to the complex formation rate constant (k_f), it is theoretically possible to separate the enantiomeric complexes from solution before an ongoing racemization eliminates any obtained enantiomeric excess. Practically, such a separation is achieved via spontaneous resolution and crystallization (Figure 9), because in this process racemization is virtually stopped in the solid state [60].

$$\text{Equilibrium reaction: } A + B \underset{k_b}{\overset{k_f}{\rightleftharpoons}} AB$$

$$\text{In equilibrium: } k_f[A][B] = k_b[AB] \qquad \text{(eq. 1)}$$

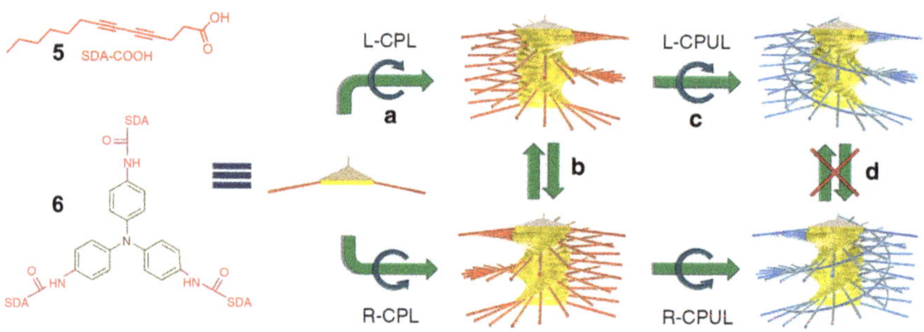

Figure 8. Manipulation of supramolecular chirality entirely by light. The entire process of induction, control, and locking of supramolecular chirality by circularly polarized light (CPL) and circularly polarized ultraviolet light (CPUL) from an achiral compound **6** containing diacetylene **5** (red: non-polymerized, blue: polymerized) moieties is schematically depicted. SDA-COOH states for trideca-4,6-diynoic acid. (a) Irradiation with CPL induces self-assembly of **6** with control over handedness, and (b) the handedness can be reversibly switched by irradiating with the counter CPL. (c) Photopolymerization of the diacetylene moieties by CPUL irradiation knits together the self-assembled structure, to produce a covalently joined nano-object, and (d) the supramolecular chirality is permanently locked, exhibiting no further handedness change upon CPL irradiation. Reprinted with permission from Ref. [55]. Copyright 2015, Springer Nature.

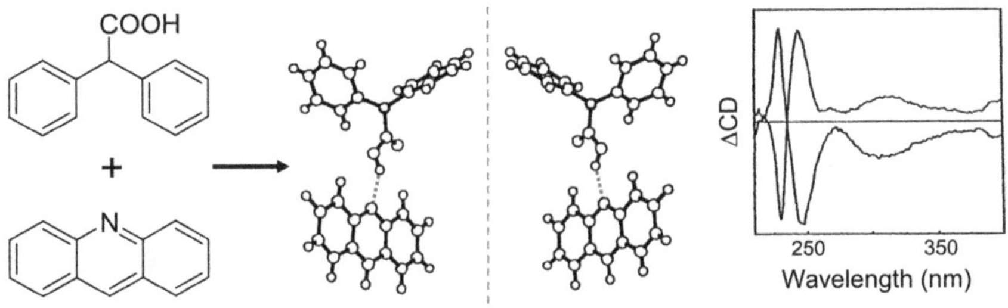

Figure 9. A complex forming enantiomorphic crystals and its solid circular dichroism (CD) spectra. Adapted with permission from Ref. [61]. Copyright 1997, American Chemical Society.

$$\text{Association constant: } K_a = \frac{k_f}{k_b} = \frac{[\mathrm{AB}]}{[\mathrm{A}][\mathrm{B}]} \quad \text{(eq. 2)}$$

Despite the differences between covalent and supramolecular chiral systems, approaches to studying and describing the two types of system are similar. Our capability of recognizing the extent of dissymmetry in a particular system is limited by the availability of suitable experimental techniques — such as electronic circular dichroism (CD), vibrational CD, optical rotatory dispersion, CPL, X-ray diffraction, and chiral shift reagent NMR — and sufficient intensity and resolution of the corresponding signal. For example, it is known that dissymmetric chromophores (e.g., **9** and **10**) have a larger dissymmetry factor g (eq. 3), which is related to stronger CD in comparison to symmetric chromophores with a nearby stereogenic carbon (such as **7** and **8**; Figure 10a) [62, 63]. A similar behavior is also expected in supramolecular systems; for example, naturally achiral chromophores will provide a measurable signal only if the chirogenic process leads to their sufficient dissymmetry. In other words, even when the entire complex is chiral, we may still lack the corresponding expression of chirality (such as a CD signal), because a chirality transfer from a chiral component to an achiral chromophore, through which chirality should be measured, is insufficiently efficient [39, 64–67]. For example, if a chiral molecule is derivatized by introducing a chromophore at a position that is considerably distant from a stereogenic center, there is no (or is a negligibly weak) measurable CD signal in the absorption region of the chromophore; a

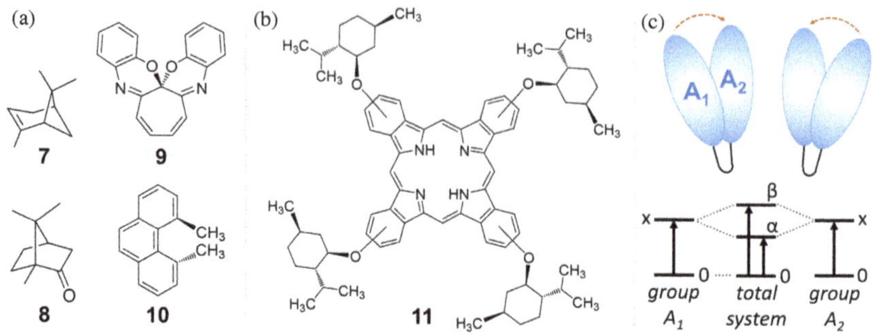

Figure 10. (a) Examples of chiral molecules containing chromophore nearby stereogenic unit (**7, 8**) and dissymmetric chromophore, which involves stereogenic unit (**9, 10**) [63]. (b) A completely circular dichroism (CD)-silent phthalocyanine **11** derivatized by chiral menthol [68]. (c) Transitions and illustration of an exciton coupling interaction between two chromophores (A_1 and A_2).

phthalocyanine chromophore bearing four chiral menthol moieties (**11**) is completely CD-silent (Figure 10b) [68].

$$\text{dissymmetry factor: } g = \frac{2(I_L - I_R)}{I} = \frac{2(I_L - I_R)}{I_L + I_R} \qquad \text{(eq. 3)}$$

I_L and I_R are the absorption intersities for left and right circularly polarized light and I is the absorption intensity for non-polarized light

Generally, dissymmetric chromophores are more sensitive and exhibit notable changes in their CD signal upon complexation or aggregation, compared to achiral chromophores incorporated in a chiral supramolecular environment [69]. The disadvantage of the induced signals being at low intensity can be reduced by introducing two or more chromophores to cause corresponding exciton coupling interaction (Figure 10c), which appears in the system (molecule) where two chromophores (A_1 and A_2) interact with each other, leading the excited state (x) to split into two energy levels (α and β states), while the ground state (0) remains unsplit. Therefore, using achiral chromophores to detect the chirality of asymmetric molecules through supramolecular interactions is beneficial, as they are originally CD-silent and able to follow the chirality via the chirogenic mechanisms [65–67, 70–75].

1.7 Aggregation and Importance of Planar Chirality in Non-covalent Interactions

Apart from rather simple host–guest systems and supramolecular complexes consisting of several components, non-covalent interactions often comprise the main force behind the creation of large assemblies, which can grow up to the macroscopic level. The most striking example are chiral crystals, for which, in connection to chirogenesis, there is a possibility of spontaneous resolution [60]. However, here we shall briefly mention a formation of various aggregates emphasizing the main chirogenic aspects of this process, while the detailed description of chiral aggregation is given in Chapter 3.

In general, the aggregation of molecules can lead to various types of achiral and chiral structures, such as helical ribbons, rods, tubes, 2D sheets, and so on [60, 67, 76–80]. Depending on the size and shape of the structure and polarity of its surface, an aggregate may be either solubilized or solid. From the viewpoint of chirogenesis we can observe the same effects that were originally found in chiral polymers [81]: specifically, the "sergeant and soldiers" effect and the "majority rule" principle (Figure 11) [67, 82, 83]. The former implies the helicity control of a large number of cooperative achiral units (the soldiers) by a few chiral units (the sergeants), whereas in the latter, a slight excess of one enantiomer determines the chirality of the entire system, or at least strongly biases it. Moreover, for supramolecular aggregates, a phenomenon of chiral memory was observed, in which chirality is first induced and then preserved after the source of chirality is removed

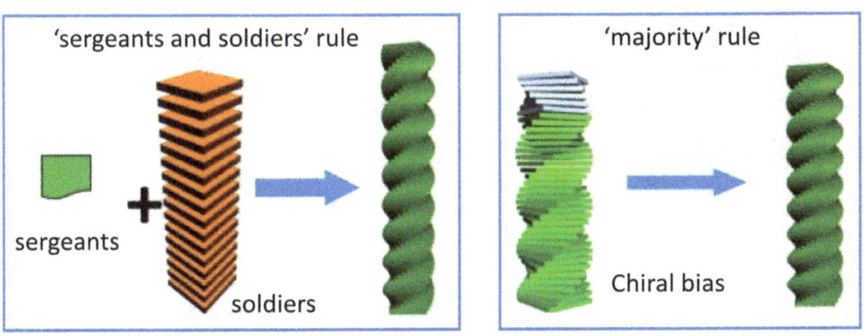

Figure 11. Illustration of the "sergeant and soldiers" effect and the "majority rule" principle for supramolecular aggregates. Adapted with permission from Ref. [67]. Copyright 2015, American Chemical Society.

or replaced by an achiral component. Hence, it can be simply compared to the "sergeant and soldiers" type of chirality induction followed by removal of the "sergeant" [67, 84–89]. Additionally, a preference for the left- or right-handed twist can be enhanced by using CPL [55, 56].

Interestingly, many aggregates are built from rather flat or completely planar molecules or platforms, which are assembled from a distinct number of pre-complexed molecules [67, 79]. These flat building blocks are stacked on the top of each other in the elongation process, with chirality appearing mostly as a consequence of the twist in the entire aggregated architecture. Planar chirality, also called 2D chirality, plays a crucial role in the aggregation. In order to understand this effect, we need to consider that every building block has two faces, which can be non-equivalent; therefore, what is chiral in 2D space can be called "pro chiral" in conventional three-dimensional space. During elongation, the growing flat ends of aggregate represent a 2D space, and the next block is bound with the preference to one of its faces. Apparently, this phenomenon is also important in the case of complexes composed of only several molecules. This means that eventually, we can prepare 2D chiral complexes and aggregates on surfaces, thus making a design of the layers of one molecule thickness [60, 90, 91]. An example of a simple molecule forming 2D chiral clusters and aggregates is 1-nitronaphthalene (Figure 12) [92, 93]. The field of 2D aggregates and

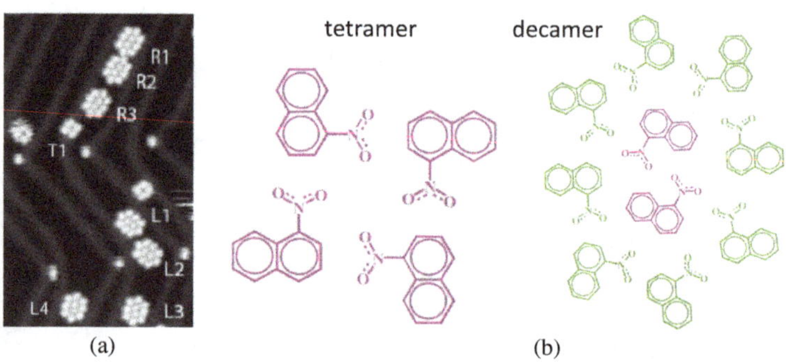

Figure 12. (a) Scanning tunneling microscopy image of 1-nitronaphthalene on Au(111) shows sorted left- (L) and right-handed (R) decamers and homochiral tetramer (T1). (b) A chemical representation of the structures; different coloring of molecules represents their chirality in 2D space. Adapted with permission from Ref. [60]. Copyright 2002, The Royal Society of Chemistry.

advanced 2D materials has undergone rapid development in recent decades, thanks to improvements in and wider availability of atomic force microscopy, scanning tunneling microscopy, scanning electron microscopy, and transmission electron microscopy.

1.8 Supramolecular Systems Exhibiting Chirogenesis and Chirality Switching

While in the previous subchapters we focused on characterizing supramolecular interactions and describing various chirogenic mechanisms that may lead to chirality induction in self-assembled complexes and aggregates, this section will discuss the most representative examples of systems exhibiting chirogenesis. Furthermore, one of the most intriguing properties of chirogenesis as a dynamic process will be discussed: the phenomenon of chirality switching, which can be controlled by various internal and external factors.

A large group of supramolecular systems is based on coordination chemistry, exploiting metal cations as a connecting element for organic ligands and varying remarkably in size and complexity. One of the simplest chiral complexes is a metal cation coordinating three bidentate ligands, such as bipyridines (Figure 13a) [94]. Examples of more sophisticated structures include various metal–organic frameworks (MOF),

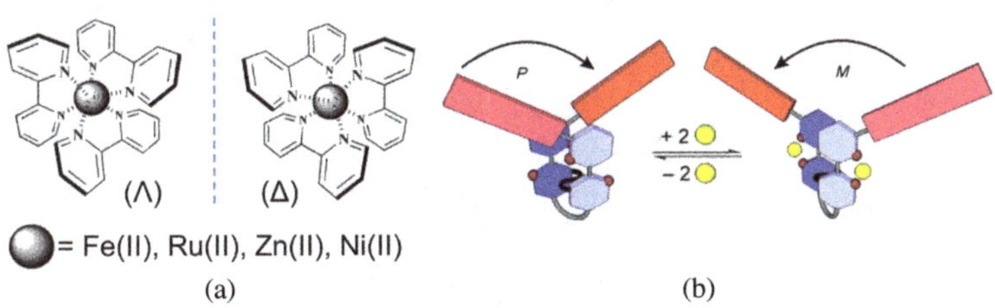

Figure 13. (a) The two chiral forms of metal(II) tris(2,2′-bipyridine) complexes, which form either racemic compounds or conglomerates in their crystals as hexafluorophosphate salts. Adapted with permission from Ref. [60]. Copyright 2002, The Royal Society of Chemistry. (b) The principle of a metal-ion-driven supramolecular chirality pendulum; the red dots are nitrogens of 2,2′-bipyridine ligands and the yellow are metal cations, which can be removed via stronger binding cyclam. Reprinted with permission from Ref. [106]. Copyright 2010, John Wiley & Sons.

molecular ladders, and so on [95–105]. In this section, we will highlight only a few representative examples, especially those which are based not only on coordination bonds, but also on exploiting other types of supramolecular interactions: for example, a system exclusively based on binding a metal cation and its subsequent removal, yielding the effect of a molecular pendulum [106]. In this case, metal cations were employed to achieve a mechanical movement which led to swinging between two diastereomeric conformations of opposite helical chirality (Figure 13b).

The nanorings composed of 6, 7, and 12 units of zinc(II) porphyrins were prepared using relatively rigid templates based on chiral α- (**12**) and β-cyclodextrins [107]. The porphyrin nanorings form a very stable complex with the template due to multiple interactions (Figure 14); this is a useful example of induced chirality and the influence of symmetry breaking on corresponding spectral properties. While macrocycle **13** of the

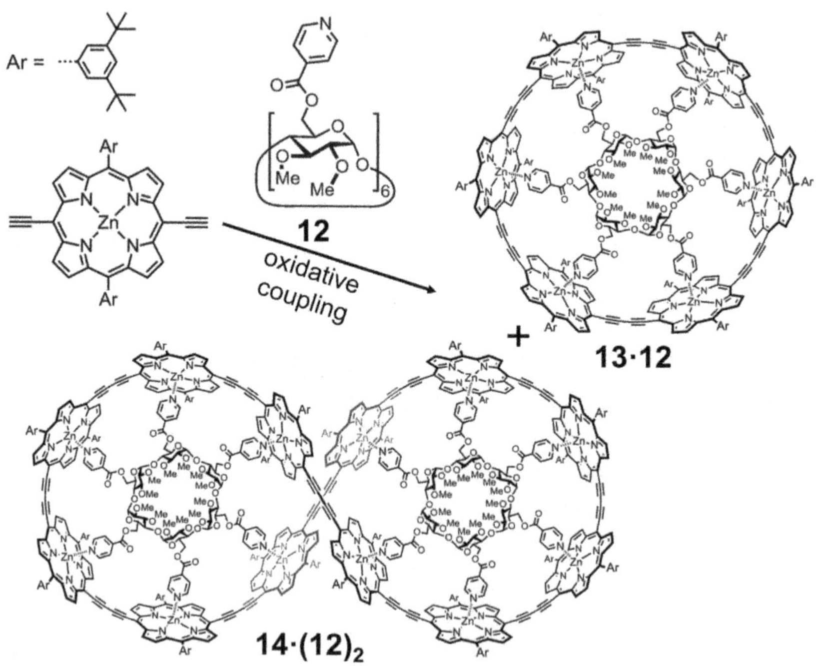

Figure 14. Syntheses of porphyrin macrocycles from a starting monomer employing cyclodextrine-based template **12** and structures of complexes **13·12** and **14·(12)$_2$**. Adapted with permission from Ref. [107]. Copyright 2014, John Wiley & Sons.

complex **13·12** is only slightly distorted from its relaxed D_{6H} geometry, the π-system of 14 has a figure-eight shape, which has inherently chiral D_2 symmetry. The equilibrium between two enantiomeric conformations of the figure-eight structure is biased by the presence of chiral template **12** in the complex. A striking difference in dissymmetry is clearly visible in the CD spectra, where the CD signal of **14·(12)$_2$** is about 20-fold stronger than that of **13·12**.

One of the most interesting MOF systems is based on prototypal MOF-5, which was synthesized in the presence of L-proline or D-proline, leading to its chiral versions Λ-CMOF-5 and Δ-CMOF-5. In turn, these structures are transformed back into achiral MOF-5 when they are immersed in a variety of organic solvents (such as acetone, ethanol, hexane, and so on), but not when immersed in *N*-methyl-2-pyrrolidone (NMP). Surprisingly, however, racemic conglomerate of CMOF was produced when achiral MOF-5 is immersed in NMP (Figure 15a). A computational study explained this fascinating effect as resulting from several supramolecular interactions between NMP and MOF-5, which induce a twist in the MOF-5 structure and force it into the chiral metastable guest-loaded state (Figure 15b) [108, 109].

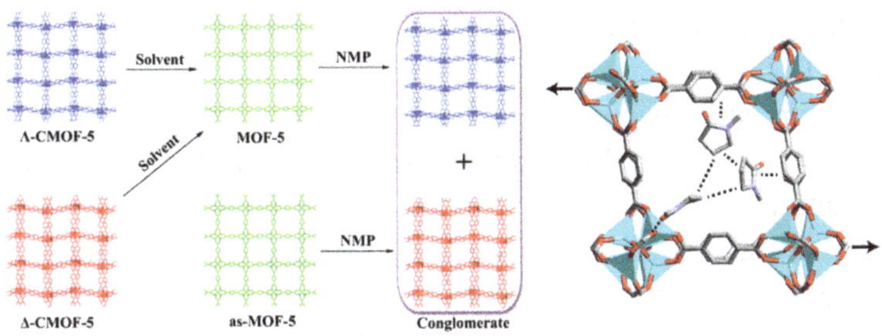

(a) (b)

Figure 15. (a) Most organic solvents converted chiral CMOF-5 into achiral MOF-5, whereas *N*-methyl-2-pyrrolidone (NMP) induced MOF-5 to revert to a racemic conglomerate of CMOF-5. Reprinted with permission from Ref. [108]. Copyright 2015, American Chemical Society. (b) Depicts the distinct local ordering of guest molecules obtained from computational study; in addition, close contacts (dotted lines) are depicted, along with the change in framework, simplified in 2D. Hydrogen atoms have been omitted for clarity. Adapted with permission from Ref. [109]. Copyright 2016, American Chemical Society.

Another large group of supramolecular complexes employing coordination of metals are metal–organic cages (MOCs) [101, 105, 110–112], which are similar to MOFs in that they are composed of a variety of organic ligands connected through metal centers. The difference is that MOFs are polymeric materials, whereas MOCs are objects of a specific size containing usually one cavity, which may bind a guest. Typically, chiral MOCs contain chiral ligands [112]. However, more interesting are racemic mixtures of cages formed from achiral ligands assembled in a helical shape. Enantiomeric excess of specific helicity can be induced by addition of chiral counterion for cages undergoing rapid chirality inversion [113, 114], or suitable chiral resolving agents can be used to resolve enantiomers of stable cages [115, 116]. Further, chirogenesis at a completely symmetrical cage can be achieved via distortion or twisting of achiral ligands, for example, via encapsulation of a chiral guest [117]. Figure 16 shows an unusual approach to obtaining enantiomerically pure MOC composed of initially achiral components [118]. Tritopic trialdehyde component **15** forms a racemic mixture of cage **19** upon reaction with achiral triamine **17** and Fe(II) salt. However, if (*S*)-1-cyclohexylethylamine **16** is used, then an analogous homochiral cage **18** is assembled, and subsequently displaced by achiral triamine to provide **19** while maintaining stereochemistry of the cage. The cage retained its stereochemistry, exhibiting a superb chiral memory effect even after 4 days at 90°C.

Tautomerization is another process that may lead to chirality switching, as was demonstrated on various iminoresorcin[4]arenes **20** derivatized by amino acid moieties that are stabilized in chloroform in an inherently chiral conformation due to intramolecular hydrogen bonds (Figure 17) [119]. The NMR experiments with additions of *N*-methylacetamide **21** to macrocycle (*M,S*)-**20** showed the surprising appearance of new diastereomer (*P,S*)-**20**, which was in slow exchange with (*M,S*)-**20** on the NMR time scale and also binding with **21** in fast exchange. Finally, the authors confirmed that achiral **21** influenced the tautomeric equilibrium and induced a switch to another diastereomer with the opposite sense of chirality.

The above-discussed topological chirality of molecular knots is rather common among supramolecular systems, and often discussed in

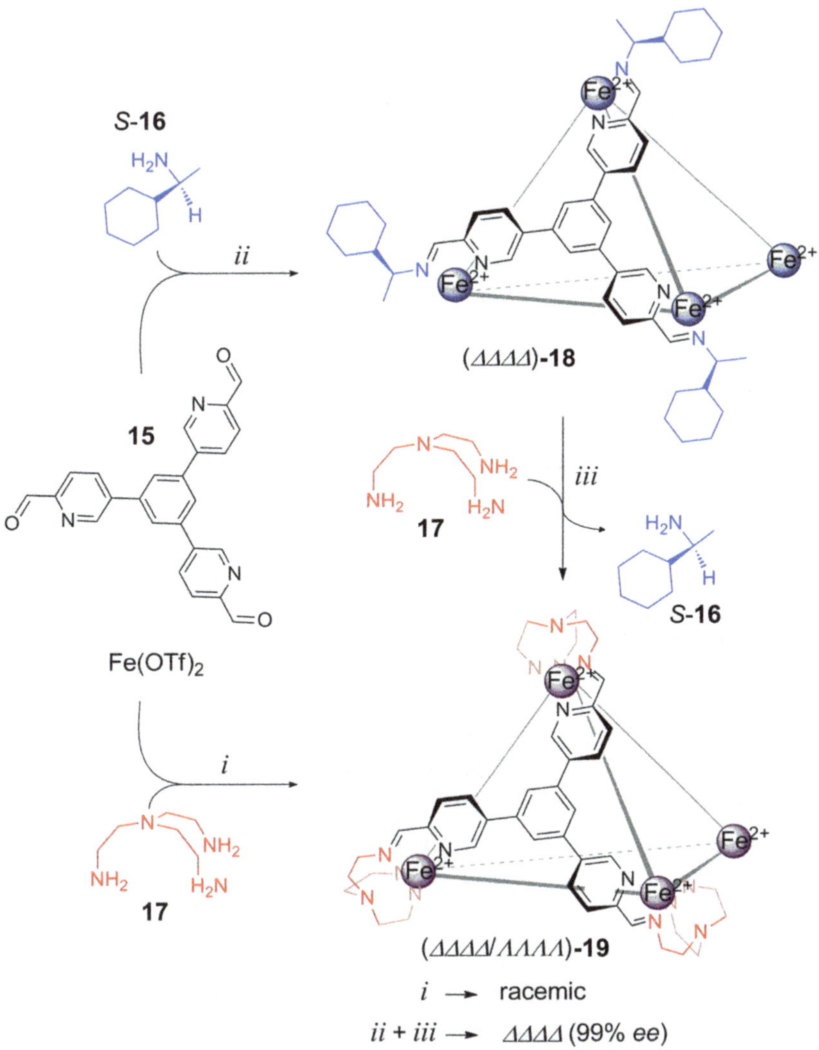

Figure 16. Route *i*: Formation of racemic cage **19** through subcomponent self-assembly. Route *ii* then *iii*: Enantioselective formation of cage **19** through subcomponent substitution. Only one ligand face is drawn per structure, for clarity. Reprinted with permission from Ref. [118]. Copyright 2013, American Chemical Society.

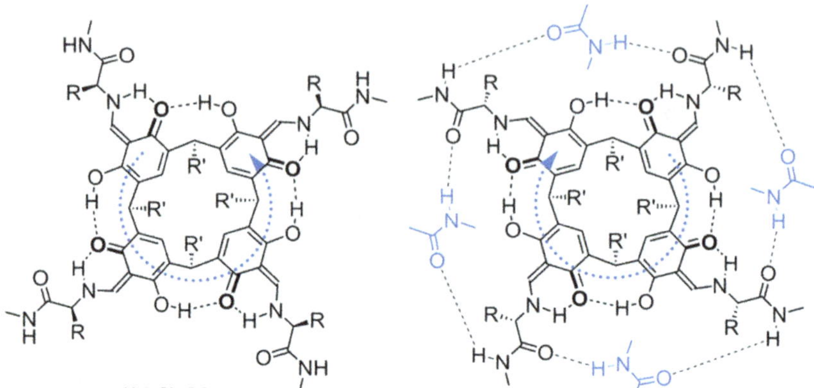

(*M*,*S*)-**20** **21** (*P*,*S*)-**20**

(*S*)-**20a** R=CH_2Ph, R'=$CH(CH_3)_2$
(*S*)-**20b** R=CH_3, R'=$CH_2CH(CH_3)_2$
(*S*)-**20c** R=CH_2Ph, R'=CH_3
(*S*)-**20d** R=$CH(CH_3)_2$, R'=CH_3

Figure 17. Structures of keto-enol tautomers of iminoresorcin[4]arenes **20** with inherent chirality descriptors, and the structure of *N*-methylacetamide **21**. Adapted with permission from Ref. [119]. Copyright 2015, The Royal Society of Chemistry.

connection with catenane structures (Figure 18) [120] and other mechanically interlocked molecules; these are considered to constitute a dim line between covalent and supramolecular chemistry. Their interlocked parts are not covalently bound, which makes them supramolecular, but at the same time they cannot be separated without breaking a covalent bond.

In the presented example, neither of the rings in a catenane **22** is chiral itself, but both have a sense of direction, which leads to topological enantiomers (Figure 18) [121]. Asymmetry arising from such interlocked systems, including rotaxanes, can be named "mechanical chirality." However, rotaxanes (such as **23**) are topologically trivial — non-topological — and the term "mechanically planar chirality" is thus more accurate (Figure 19) [123, 124]. In a similar example described as a "topological rubber glove" (**24**), only one catenane's ring has a sense of direction. However, due to rotation of the 1,5-dioxynaphthalene unit, it readily interconverts between two enantiomeric conformations without being achiral in any moment (Figure 19) [125]. This is a case of "dynamic chirality," a term used for those systems with a barrier for interconversion between its configurational isomers lower than that of atropisomers. Practically, the interconversion is

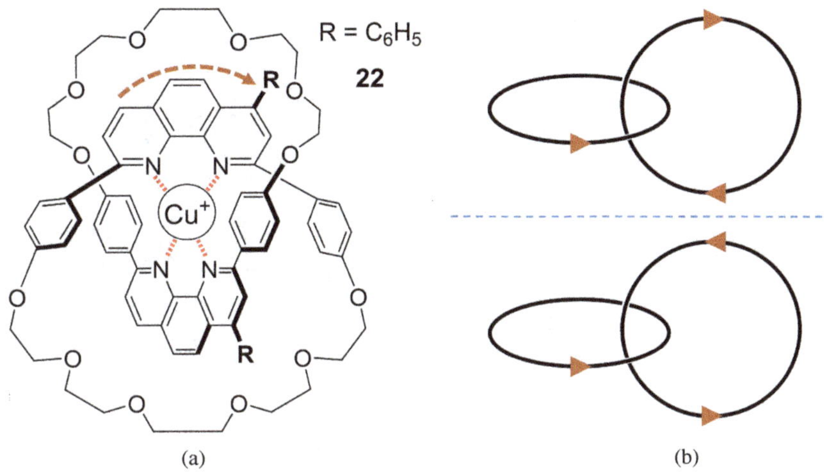

Figure 18. (a) Catenane **22** containing rings with preferred orientation as an example of a topologically chiral system, and (b) a schematic illustration of enantiomeric structures [121, 122].

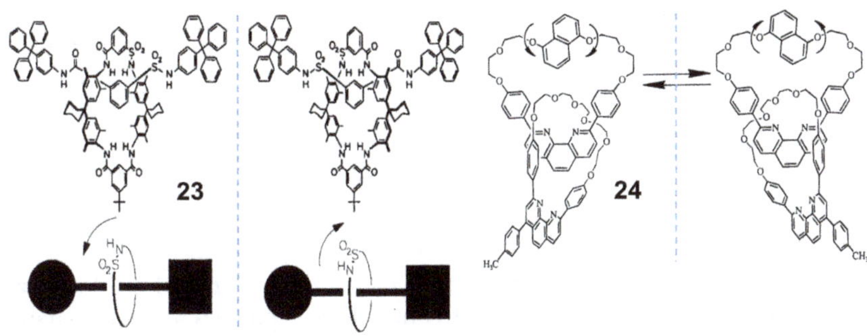

Figure 19. Mechanically but not topologically chiral molecule of rotaxane **23**. Reprinted with permission from Ref. [126]. Copyright 1997, American Chemical Society; and a "topological rubber glove" **24**. Reprinted with permission from Ref. [125]. Copyright 1997, American Chemical Society.

too fast at room temperature to clearly distinguish the rotamers by NMR, but slow enough at lower temperatures to differentiate them.

Generally, dynamic chirality is mostly discussed in connection with mechanically interlocked molecules and systems exhibiting supramolecular interactions [124]; here, we present several representative examples. The first of these is [2]rotaxane **27** with axial dynamic chirality originating from the biphenyl moiety of a crown-ether ring [127]. Twisting of biphenyl

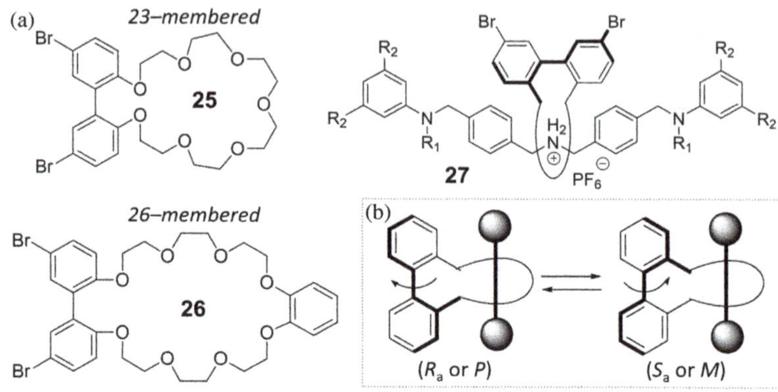

Figure 20. (a) Structural parts of rotaxanes with dynamic chirality and (b) illustration of switching between enantiomers. Reprinted with permission from Ref. [127]. Copyright 2020, John Wiley & Sons.

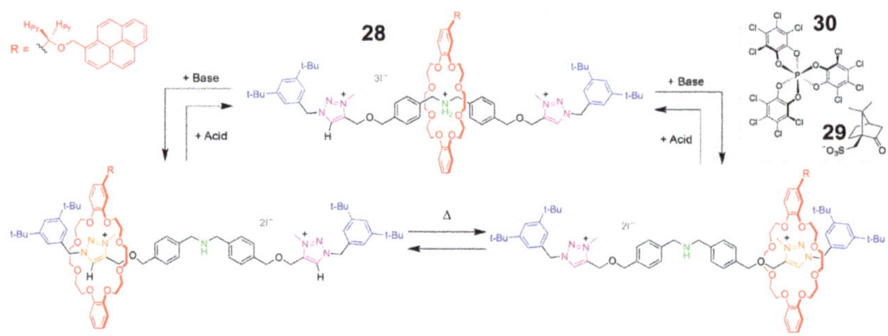

Figure 21. Example of rotaxane (**28**) with dynamic chirality based on shuttling of ring between triazolium cations. Guests **29** and **30** are capable of inducing enantiomeric excess [128].

between *P* and *M* conformation is restricted by a symmetrical thread, and also influenced by the ring size (Figure 20). Separation of enantiomers showed that 26-membered crown-ether **26** racemized at room temperature, but that the conformations of smaller 23-membered ring **25** were more stable and required elevated temperature for racemization. The system can be assumed to be a mechanically interlocked equivalent of atropisomerism.

The second example is [2]rotaxane **28** composed of the symmetrical thread and "directional" ring (C_s symmetry), which can occupy a central position, making the rotaxane achiral (Figure 21) [128]. After deprotonation of central dibenzylammonium, the ring starts to shuttle between the triazolium moieties to form a dynamically chiral pair of enantiomers. Compared to the previous example, in this case chirality emerges only from the ring position, and no part of the system is itself chiral. Moreover, the interaction with chiral monoanions ((1*S*)-(+)-10-camphorsulfonate **29**, tris(tetrachloro-benzenediolato) phosphate(V) **30**) results in an enantiomeric excess, thereby indicating a potential use as a chiral sensor or asymmetric catalyst.

The last example is [2]catenane **31** (Figure 22), which exhibits great complexity, leading to four diastereomeric pairs of enantiomers upon restricted conformational exchange [129]. The first crown-ether-based ring contains one hydroquinone moiety and one 1,5-dioxynaphthalene

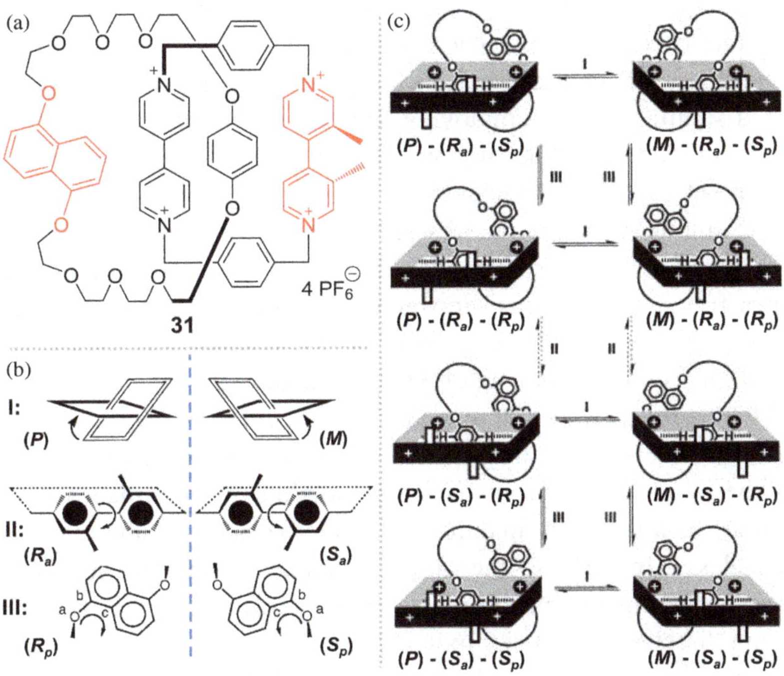

Figure 22. (a) Structure of [2]catenane **31**. (b) Subunits of **31** exhibit three different types of stereochemical interconversions: inversion of helicity (I), axial chirality (II), and planar chirality (III). (c) Therefore, **31** forms four diastereomeric pairs of enantiomers. Not all the possible interconversion pathways for processes I, II, and III are shown. Reprinted with permission from Ref. [129]. Copyright 2003, John Wiley & Sons.

moiety. The second ring is tetracationic cyclophane, composed of the bipyridinium unit, bipicolinium moiety, and two phenylenes on opposing sides of the ring. When the rings are interlocked, the freedom of the rotation of structural moieties is restricted, and the hydroquinone moiety is preferably positioned inside the cavity of cyclophane. Three particular processes play main roles (Figure 22b and c): the first is tilting of the crown-ether ring with respect to the mean plane of cyclophane ring (axial/helical chirality assigned *P*/*M*), giving rise to a rocking motion between two stereoisomers. The second is atropisomerism on the bipicolinium unit (axial chirality), and the third is inversion of 1,5-dioxynaphthalene between two of the faces (planar chirality). It is crucial to stress that these processes vary in their activation barriers; in particular, the bipicolinium barrier prevents a direct conversion between the enantiomers in solution at room temperature. However, the system is nevertheless highly dynamic as the other two activation barriers are surmountable.

The range of approaches for chirogenesis in molecular systems is enlarged significantly by employing supramolecular interactions. Recent progress in understanding supramolecular interactions has led to the development of simple machines at a molecular level, often with the desire to mimic natural processes [120, 130–133]. Molecular rotors are perfect examples of systems with the other type of chirality, which emerges from the directional movement of one part of the complex or molecule with respect to other parts [134]. We can imagine a simple symmetrical rotor composed of stator **32** and a rotating shaft **33**, connected by axle **34** (Figure 23) frozen in time [135–139], which in any possible position of the shaft is indistinguishable from its mirror image. However, when the system in motion, the corresponding mirror images become dissimilar, with the shaft spinning either clockwise or counterclockwise. Nevertheless, similarly to the above examples generating a racemate, there is no dominant spin direction; moreover, there is a possibility of changing directionality at any moment.

To date, researchers had described only a limited number of systems capable of unidirectional rotation; the most developed of these is based on sterically crowded chiral covalent compounds where the rotation on a double bond is induced by light [140]. Therefore, it is fairly different from the supramolecular symmetrical rotors described above (Figure 23).

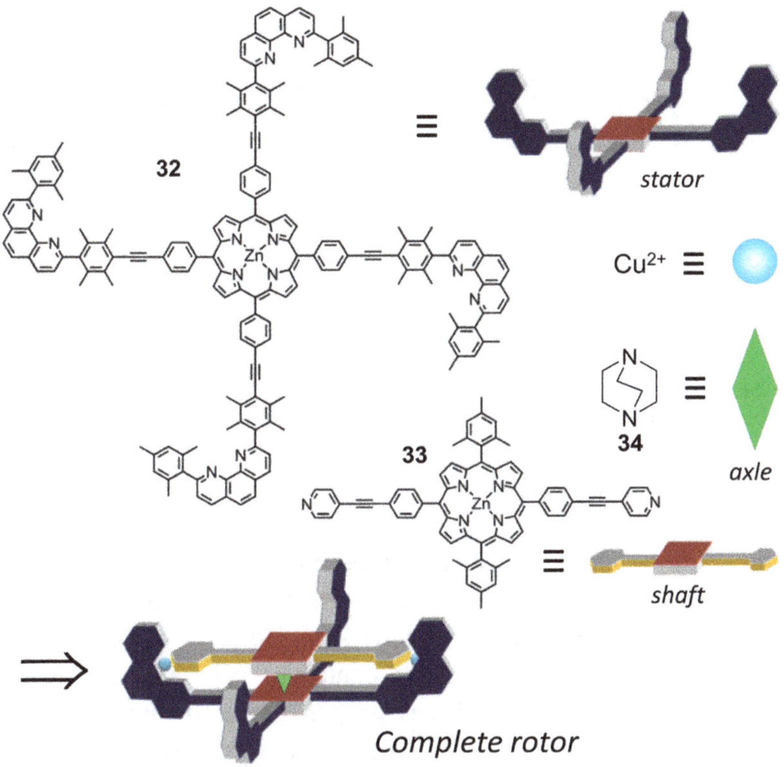

Figure 23. Chemical structures and cartoon representation of a molecular nanorotor. The authors have shown various versions of such rotors including their application as a catalyst; they discovered how to influence spin frequency and its effects on catalytic function. Adapted with permission from Ref. [136]. Copyright 2013, American Chemical Society.

Interestingly, if we were able to force a symmetrical system to spin preferably in one direction, it would resemble Barron's spinning cone model [141], bringing a question regarding whether such system is really chiral. In particular, Barron proposed a new definition of chirality, which would be satisfactory for all scientific fields: "True chirality is exhibited by systems that exists in two distinct enantiomeric states that are interconverted by space inversion, but not by time reversal combined with any proper spatial rotation." Barron explained the consequences of this definition on a cone spinning about its axis of symmetry. A stationary spinning

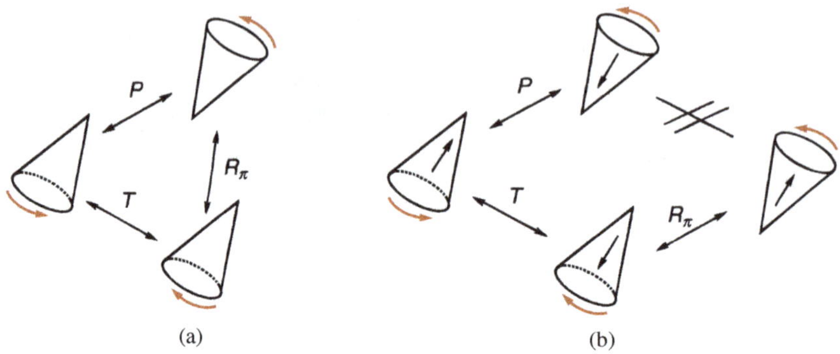

Figure 24. The effects of parity, time reversal, and spatial rotation on (a) a stationary and (b) a translating spinning cone. Used operations: *P* is space inversion, *T* is time inversion, and R_π is proper rotation axis. Adapted with permission from Ref. [44]. Copyright 1998, American Chemical Society.

cone seems to be chiral, but can be superimposed with its mirror image (space inversion *P*) by applying the time reversal (*T*) followed by rotation (R_π); therefore, it is falsely chiral (Figure 24a). However, the spinning cone translating along the axis of rotation cannot be superimposed using the time reversal (*T*) combined with any proper spatial rotation (R_π); thus, it exhibits true chirality (Figure 24b). As one can see, chirality of an object is motion-dependent (translation) in this example. Later, Mislow argued that the stationary spinning cone, when exemplified by an achiral molecule in a pure rotational state, is chiral because it belongs to the chiral point group [142]. This old theoretical controversy remains ongoing [143, 144], and the details of this topic are far beyond the scope of this chapter.

1.9 Supramolecular Chiroptical Sensors Based on Chirogenesis

1.9.1 Porphyrin-based Sensor Molecules

Chiroptical sensor application is one of the most important properties of supramolecular chirogenesis. For example, typical cases of chirogenic implementation are related to the hosts capable of sensing the chiral sense of an asymmetric guest. In this case, the hosts are originally CD-silent, mainly composed of flexible parts and achiral chromophores near binding sites. However, a racemic mixture of sensitive and selective chiral hosts can also be used. As we have pointed out previously, it is beneficial to use more than one chromophore in the host structure, to increase the intensity of induced CD signal by utilizing exciton coupling interaction [69].

Apparently, one of the most developed and researched types of host molecules utilizing chirogenesis for sensing chirality are derivatives of bis-porphyrins [65, 70–72, 75, 145–158]. Generally, a sensor is composed of two metalloporphyrin cores connected by one covalent link, which can be rigid as well as long and flexible (Figure 25). Another option is the presence of two or more covalent links, leading to a macrocyclic structure [159]. Indeed, bis-porphyrins are highly versatile molecules, and their selectivity and sensitivity can be finely tuned in multiple different ways. The most conventional metal coordinated in a porphyrin's center is zinc(II) cation, but many other metal ions, such as magnesium, copper, cobalt, nickel, and so on, are also used, including a combination of two different metals in one bis-porphyrin [160]. Moreover, the core can be metal-free. The dominance of zinc(II) bis-porphyrins comes from a facile preparation procedure, and moderate to strong binding with a variety of nitrogen- and oxygen-containing guests. The link between the porphyrin cores can be a simple ethane bridge (**35**) ensuring sufficient flexibility, or other structural moieties providing additional functionalities to a bis-porphyrin. Namely, the linkage can be chiral (**36**), based on biologically related units (such as peptides), affording additional binding sites (**37**),

Figure 25. Examples of various bis-porphyrin derivatives: simple ethane bridged **35** with good flexibility [150]; **36** linked by chiral rigid moiety [156]; **37** containing a linker providing additional binding sites [75]. Figure of structure **37** reprinted with permission from Ref. [75]. Copyright 2021, American Chemical Society.

rigidity, and other functions [75, 156–158]. Finally, the properties of bis-porphyrins can be influenced by substitution at the *beta* and *meso* positions of a porphyrin core.

A guest-free bis-porphyrin host is, in non-coordinating solvents, typically in the *syn* face-to-face conformation, due to strong π-π interactions between the chromophores. A general sensing mechanism is based on the complexation with a chiral guest yielding a conformationally rigid, twisted system. In turn, this arrangement gives rise to exciton coupling between the helically oriented electric transition dipole moments of covalently linked chromophores [65, 66, 75, 145, 161–164]. The result is induced circular dichroism (ICD) spectra, exhibiting bisignate Cotton effects, which reflect the absolute configuration of a guest (Figure 26).

Bidentate guests include the formation of specific "tweezer-like" conformations of the corresponding 1:1 complexes. In excess of the bidentate guest, or upon interaction with monodentate guests, the bis-porphyrin structure turns to extended *anti* conformation (if a linker is flexible), with each porphyrin moiety binding one guest. Importantly, the unidirectional helical shape and ICD can be observed even in the case of *anti* form; however, exciton coupling is diminished [65, 145].

The adopted helicity often correlates to the steric demands imposed by the asymmetric carbon center of the bound guest. The porphyrin core closest to the chiral center adopts a specific orientation to avoid steric repulsion with the largest group at the stereogenic center [66, 75, 145,

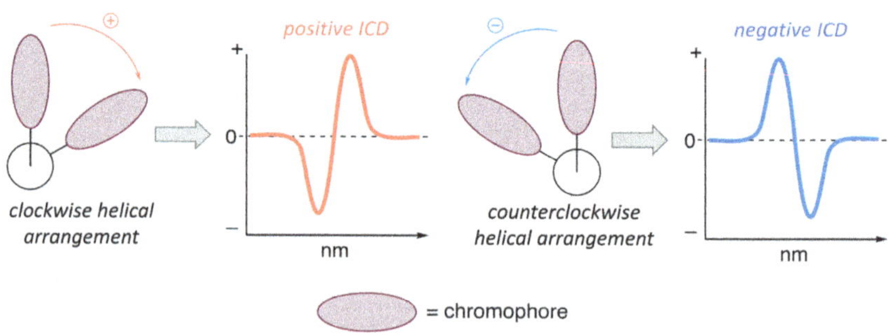

Figure 26. A general representation of an exciton-coupled induced circular dichroism (ICD) active complex. Adapted with permission from Ref. [75]. Copyright 2021, American Chemical Society.

162]. Detailed studies conducted mainly with ethane-bridged bis(zinc(II) octaethylporphyrin) (**35**) explored the specific effects of guests' substituents bulkiness (Figure 27), as well as the influence of solvent on ICD (Figure 28) [65, 70, 145–148].

In particular, a monodentate guest with a bulkier substituent causes stronger steric hindrance with the bis-porphyrin's (**35**) ethyl groups, leading to a larger change in the dihedral angles between the respective B band electronic transitions (Figure 27). This results in a more intense ICD signal in comparison to an analogous guest bearing a smaller substituent [65, 70, 145–148].

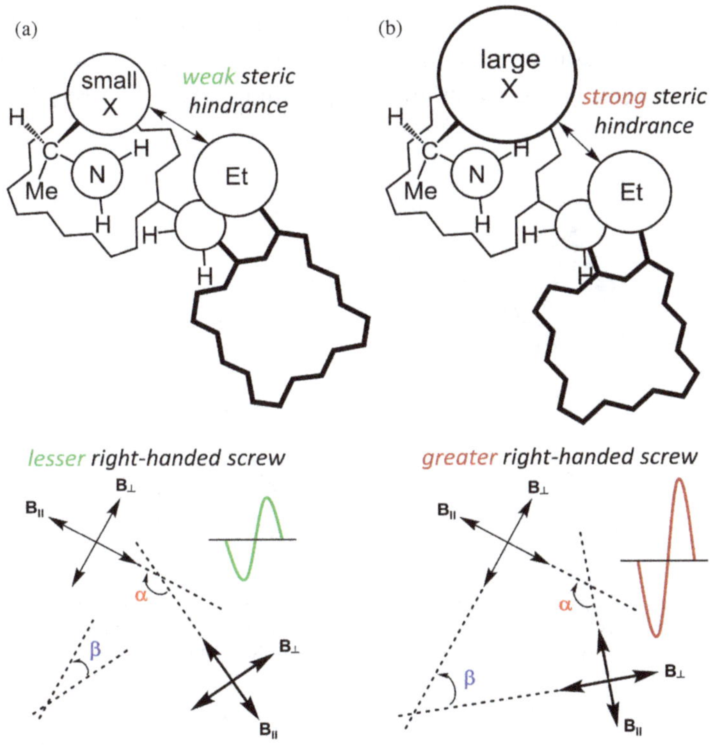

Figure 27. Bis-porphyrin **35** binding monodentate (*S*)-guest with (a) small and (b) large substituent X. Intensity of the induced signal is related to the size of a ligand's substituent as a larger group increases steric repulsion, which leads to change of the angle between Soret band (B band) electronic transitions, and to larger induced circular dichroism (ICD). Adapted with permission from Ref. [145]. Copyright 2004, American Chemical Society.

Notably, the importance of the solvent on the chirogenic processes is often overlooked as a significant contributing factor. However, it has been discovered that the change of solvent polarity can influence the intensity of ICD signal, and even cause the chirality inversion of **35**. Such solvent effects on the bis-porphyrin distortion are related to steric hindrance and bulkiness of the substituents, as has been described above. Simply, polar substituents of the guest can electrostatically interact with solvent molecules. A more polar solvent results in stronger interaction, which subsequently leads to a larger solvation shell, in turn increasing the effective size of polar substituents (Figure 28) [65, 145, 148, 151]. The best examples are derivatives of L-amino acids: specifically, methyl esters of alanine, valine, and isoleucine. The amino group is coordinated to zinc(II) cation, and comparably large sidechains will drive the sense of induced chirality. A slightly bigger nonpolar sidechain repels the porphyrin's ethyl

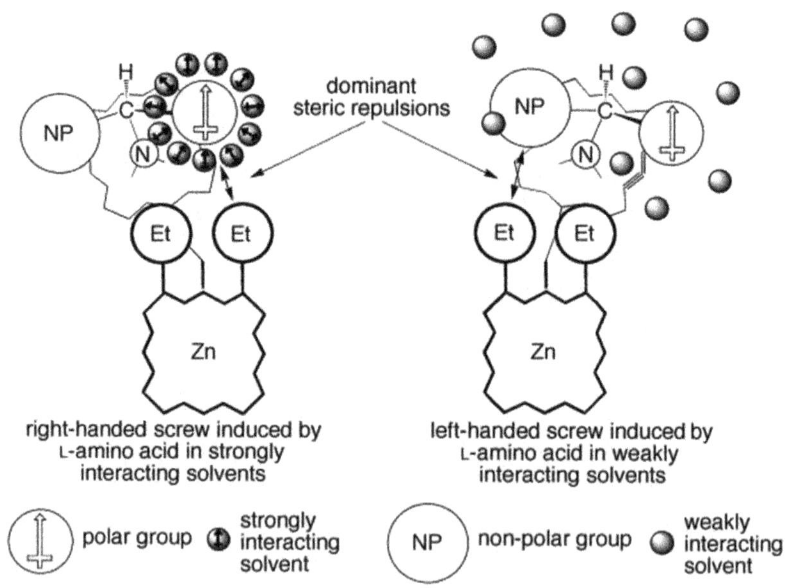

Figure 28. Solvent effect on the mechanism of supramolecular chirality induction in bis-porphyrin **35** in strongly and weakly interacting solvents, leading to inversion of induced chirality (the ligand coordinated to the upper porphyrin has been omitted for clarity). Reprinted with permission from Ref. [148]. Copyright 2003, John Wiley & Sons.

substituent and promotes the corresponding left-handed screw in a nonpolar solvent (e.g., hexane or cyclohexane). However, the situation is entirely different in a more polar solvent (e.g., chloroform or ethyl acetate): in this case, the effective size of the polar ester group becomes bigger than that of the nonpolar sidechain, due to the interaction with solvent molecules resulting in chirality switching to the right-handed screw. Apparently, different guests bearing larger polar than nonpolar groups would give the same sign of ICD couplet in all solvent, with the intensity of signal enhancing upon increased solvent polarity.

While the majority of bis-porphyrins are used to sense the chirality of bidentate guests due to significantly stronger ICD as a result of more effective exciton coupling in the tweezer conformation [75, 145], there are a few examples of bis-porphyrins designed to efficiently sense the chirality of monodentate guests. The first of these is **38**, based on the short –CH_2COOCH_2– ester linker, which brings both porphyrin cores close to each other, and results in binding mono-alcohols by simultaneous double coordination of the hydroxyl group to two central metal ions of the porphyrin subunits (Figure 29) [152]. The second is a more versatile host, **37**, with the porphyrin units linked by biphenol, which is atropisomeric with a relatively low rotational barrier and may act as an additional binding site (Figure 25) [75]. The chirality of monoamines can be simply recognized via interaction with the biphenol linkage of metal-free bis-porphyrin. After the introduction of a zinc(II) ion, the host can efficiently sense the chirality of cyanohydrins, sulfoxides, and phosphorus oxides.

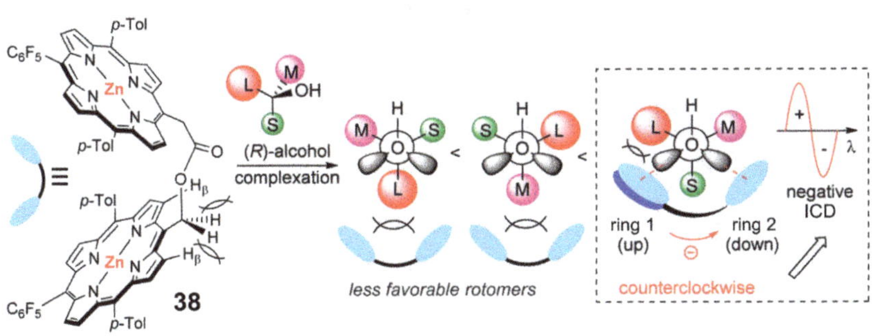

Figure 29. Structure of bis-porphyrin **38**, capable of forming complex with mono-alcohol and recognizing its absolute configuration. Adapted with permission from Ref. [152]. Copyright 2015, The Royal Society of Chemistry.

A pronounced exciton coupling effect of the bis-porphyrins is a great advantage of such host systems. However, decent ICD and chirality sensing can be reached via other metalated or metal-free porphyrinoids (such as phtahalocyanines and chlorines) and their derivatives [65, 156, 165, 166]. Typically, it is necessary to derivatize a host with suitable substituents providing an additional binding site. In general, a guest is then conventionally coordinated to a central metal ion; at the same time, the guest interacts with the host's substituents of **39** (Figure 30), which can include both attractive and repulsive intermolecular forces [166]. Some guests can induce a moderate CD signal even in the case of simple conventional porphyrins, as was recently reported with chiral cyclohexanohemicucurbit[*n*]urils binding to zinc(II)tetraphenyl porphyrin (**40**): the main attractive interaction occurs between zinc ion and carbonyl oxygen of urea moieties.

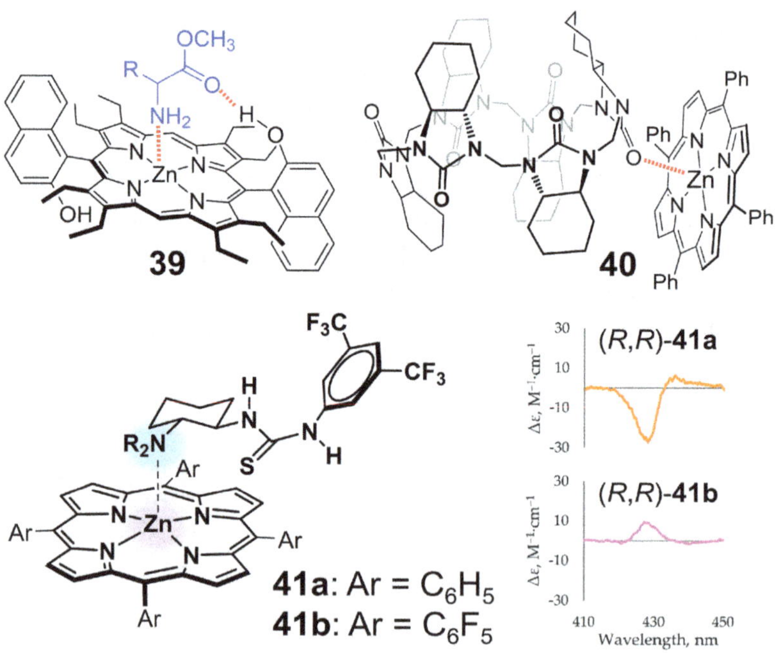

Figure 30. Examples of single porphyrins exhibiting induced circular dichroism (ICD) upon binding chiral guest: porphyrin **39** binding amino acids [166] and complexes of zinc(II)tetraphenyl porphyrin derivatives with hemicucurbituril macrocycle (**40**) [167], and thioureas (**41**) [169].

However, additional interactions between porphyrin's aromatic system and polarized C–H groups of macrocycle participate in the overall stability of the complex, and are described as "outer surface interactions" leading to ICD of opposite sign for both macrocycle enantiomers [167]. Similarly, zinc(II)tetraphenyl porphyrin derivatives binds thiourea-based organocatalysts (**41**) as emerging pollutants via coordination of the primary amino group [168, 169]. Interestingly, the ICD of the complex is dependent on the geometry of secondary π–π interaction between aryl groups of host and guest; therefore, significantly different CD spectra are obtained for derivatives substituted at aryl groups. Specific CD patterns were rationalized in the publication by the additional data from computational modeling and crystal structure analysis.

1.9.2 Polymeric and Macrocyclic Sensors

Apart from the porphyrin and bis-porphyrin-based hosts, various polymeric structures can also effectively be used for chirality sensing. For example, the system based on poly(*m*-ethynylpyridine) binds saccharides through the hydrogen bonding between the saccharide's OH groups and nitrogen atoms of the polymer's pyridine moieties (Figure 31) [170, 171]. The polymer derivative **42** bearing the octaethylene glycol groups forms spontaneously helical structures in protic solvents, through the intramolecular solvophobic interactions. A saccharide molecule is bound to a cavity inside the helix, and the helicity of the host polymer can adapt to the structure of the guest, which leads to induction of a strong CD signal. Surprisingly, the ICD of the complex with D-glucose was rather weak. Further experiments proved a correlation of helix inversion with the equilibrium of mutarotation between α-D-glucose and β-D-glucose in solution. In other words, each anomer of glucose caused the polymer to bias a single-handed helix in the opposite sense. Thus, the ICD effects are almost cancelled out as the equilibrium for α/β-glucose is nearly 1:1.

The polymer derivative **43** bearing a nonpolar substituent was studied in less polar aprotic solvents, where pure **43** adapts dipole-driven unfolded conformation. The binding of saccharide drives the folding of disordered **43** into a helical shape [170]. The chiral sense of the helixes was guided by bound saccharides and caused ICD, as in the previous example in

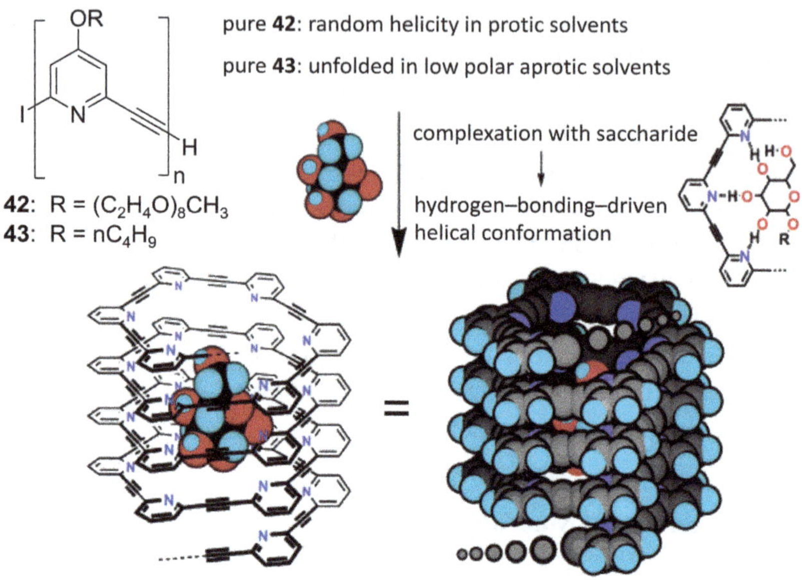

Figure 31. Structures of poly(*m*-ethynylpyridine)s and conformation change driven by the complexation with saccharide resulting in a helix with strong induced circular dichroism (ICD). Adapted with permission from Ref. [170]. Copyright 2004, American Chemical Society.

protic solvents. Importantly, it was possible to distinguish glucosides and even natural glucose from other monosaccharides and their derivatives based on the shape of the ICD spectra. Furthermore, one turn of the helical shape necessitates six ethynylpyridine units, and the whole helix must include at least three turns to produce a binding cavity, which surrounds a saccharide and allows ICD.

Macrocyclic compounds are another class of supramolecular hosts that are efficiently used for chirogenesis. To date, a large number of different macrocyclic cavitands, such as calixarenes, cyclophanes, cucurbiturils, cyclodextrines, and pillararenes capable of binding guest molecules in their cavities have been synthesized. Cyclodextrines are composed of saccharides, making them naturally chiral, while all other macrocycles can be altered to obtain corresponding chiral derivatives. On examples of chiral cyclophanes and cyclodextrines, a transfer of chirality has been demonstrated from cavitand to achiral chromophoric guest [66, 162, 172, 173]. The appearance of ICD in the absorption region of a small aromatic guest is especially evident in the case of calixarenes, as they are practically CD-silent above 200 nm. However, efficient induction of chirality on a guest is related to dissymmetry of the macrocycle cavity itself, rather than to a distant chiral substituent on a fairly symmetrical cavitand.

Phenylacetylene macrocycles (PAMs) constitute a good example of dissymmetry induction; these have been used as achiral building blocks (**44**) and connected by two-fold bridges from terephthalamide [174]. The authors obtained multiple chiral molecules varying significantly in their shapes: simply stacked [6]PAM rings (**45**, **46**), enlarged [12]PAM helically folded rings (**47**, **48**), and mechanically interlocked [6]PAM rings (**49–52**) (Figure 32a). It is of note that structures **45–48** are dynamically

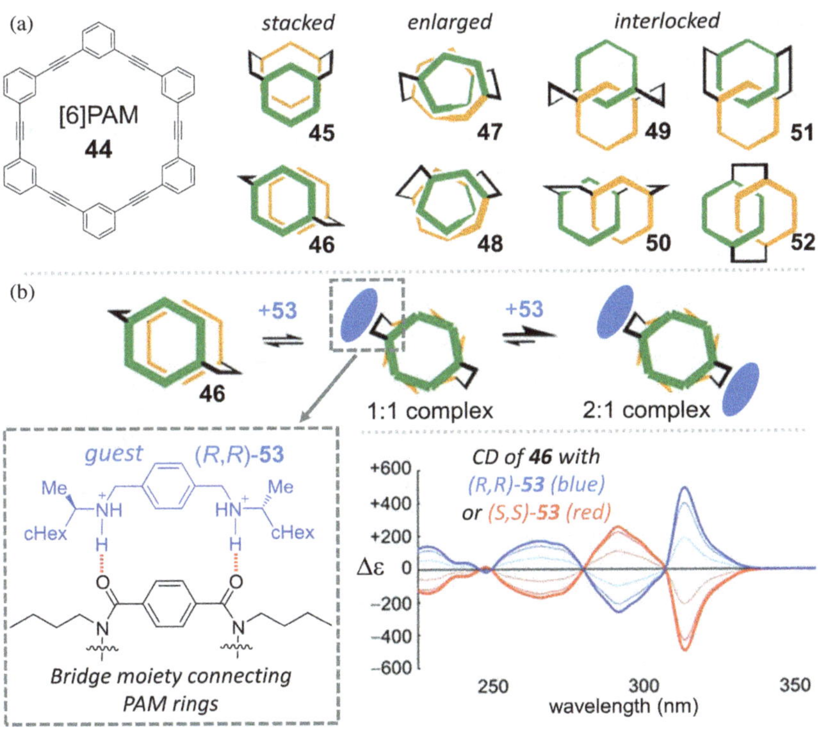

Figure 32. (a) Structure of [6]PAM **44** and schematic structures of obtained [6]PAM and [12]PAM two-fold bridged derivatives **45–52**. (b) Illustration of 2:1 complex formation with guest **53**, which takes place at the macrocycles bridges (inset) and results in induced circular dichroism (ICD) for dynamically chiral derivatives **45–48**. The example of circular dichroism (CD) for **46** in the presence of 1, 2, and 3 equivalents of **53** is shown. Adapted with permission from Ref. [174]. Copyright 2019, The Royal Society of Chemistry.

chiral. Upon addition of the xylylenediamine-based chiral guest **53** interacting with the ring's bridges, a stability of the chiral conformation is enhanced, and an ICD signal is observed (Figure 32b). The intensity of induced Cotton effects was remarkably large in the case of host **46**. With only two-molecule stacking, it was comparable to the Cotton effects of the helical columnar assemblies of [6]PAMs.

A novel approach for recognition of enantiomers by a macrocyclic host employed rigid cucurbit[8]uril (CB[8]). A large cavity of CB[8] works as a confined space, and it is known for its ability to bind two guests at the same time, therefore forming ternary complexes (Figure 33) [175]. It was shown that a synchronous binding of both a dye (**54–56**) and a guest (amino acid derivatives in this study) imprinted and induced the sense of chirality to the dye. This results in the ICD spectra, which can be measured at the dye's absorption range. The main benefits were a high sensitivity (in the range of μM) due to strong association constants, broad scope of compatible analytes, and capability to distinguish different analytes based on their ICD, including mirror-image ICD for corresponding enantiomers.

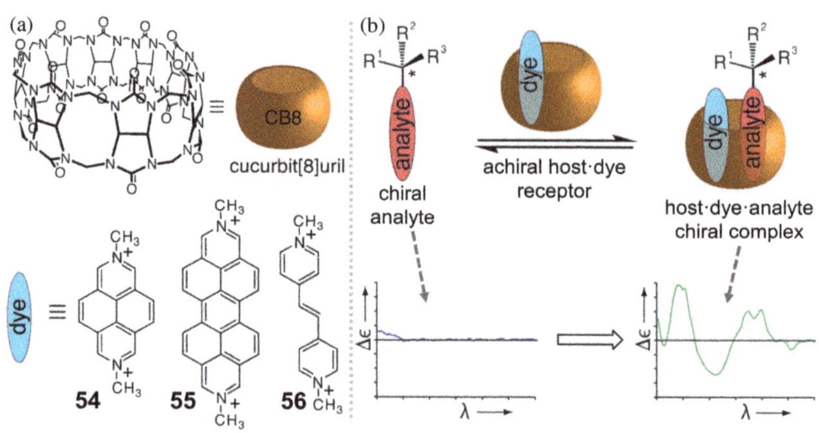

Figure 33. (a) Chemical structures and schematic representations of the macrocyclic host cucurbit[8]uril (CB8) and the dyes **54–56**. (b) The complexation of chiral analytes by a self-assembled, achiral host·dye receptor converts a weak intrinsic circular dichroism (CD) signal in the far-ultraviolet (UV) region into a strong induced circular dichroism (ICD) effect in the near-UV and visible regions. Reprinted with permission from Ref. [175]. Copyright 2014, John Wiley & Sons.

Indeed, various molecular cages are a more advanced class of macrocyclic hosts and are usually built from ligands coordinating metal cations. Nevertheless, in recent years, a new strategy using anion-coordination by multiple hydrogen bonds to construct similar well-defined architectures has been developed (Figure 34) [176–180]. Compared to more

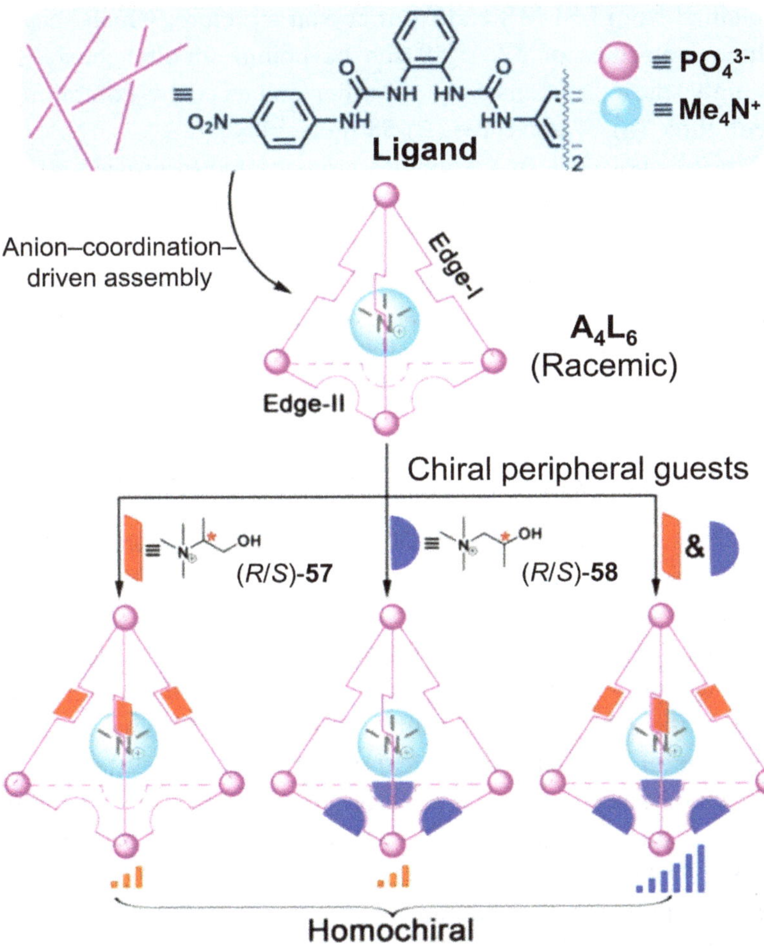

Figure 34. Illustration of the site-selective chiral induction behavior of the guests (*R/S*)-**57** and (*R/S*)-**58** (cations used as $CF_3SO_3^-$ salts for solubility reasons) bound in "peripheral" sites of the A_4L_6 anion tetrahedron. The signals below the tetrahedra indicate the strength of the Cotton effect. Reprinted with permission from Ref. [180]. Copyright 2020, American Chemical Society.

conventional MOCs, this type of assembly is more flexible, thus being more easily adaptable toward various guests; it also allows a significant template effect of the peripheral guests. For example, binding of a chiral cationic guest — α- (**57**) and β-methylcholine (**58**) — induces a CD signal, the intensity of which grows, up to the addition of three equivalents of a particular guest [180]. This observation perfectly corroborates the data obtained from ESI-MS and from crystal structure, which show that only three molecules of **57** or **58** can be bound simultaneously. More interestingly, the CD intensity was further enhanced by combining the cage with three (*S*)-**57** and three (*R*)-**58** molecules.

All of the above selected examples are expected to provide a general overview, showing the variability and versatility of supramolecular systems that can be used for chirality sensing based on the chirogenic approach. Clearly, this chapter's purpose is not to list all such supramolecular systems; thus, many other host molecules capable of chirogenesis (such as arylacetylene or helicene derivatives) are out of the scope of this review, and can be found in the literature [39, 65–67, 69, 72–74, 162, 173, 181].

1.10 Conclusion and Further Perspectives

Non-covalent interactions have a robust dynamic character due to their mostly low energy of binding per individual bond, and their utilization is a cornerstone of supramolecular chemistry. The fast pace of formation and destruction of individual interactions in a supramolecular system leads naturally to quick stabilization in the state with minimum energy. Importantly, the achieved equilibrium can be easily reorganized by even a subtle change of chemical or physical conditions in the system (such as the addition of a new compound, dilution, irradiation, change in temperature, polarity, pressure, and so on). Nevertheless, supramolecular systems composed of a large amount of interactions (e.g., aggregates, folded proteins, DNA, lipid membranes, and other self-assembly systems) can be significantly more stable, and their reorganization may require more time, or considerable changes in conditions. Thus, these general properties of supramolecular systems provide strong foundations for effective chirogenesis and the possible preservation or modulation of chirality.

In this chapter, we have shown various types of dissymmetry in supramolecular systems and mechanisms, and discussed how dissymmetry can be induced or altered. While some examples provide symmetry breaking leading to racemic mixtures as a natural consequence of complex formation, the other examples exhibit chiral properties due to the induced enantiomeric excess of a particular complex, as a consequence of chiral information's presence or transfer in the system. Undoubtedly, the supramolecular approach offers fascinating, and often hardly imaginable, scenarios of chirogenesis and its application. However, the presented examples indicate wide room for future development, as only a few systems can be used for real practical purposes, mainly in the sensing and recognition of chiral compounds. Other appealing applications could include production of chiral compounds via asymmetric synthesis and catalysis, enantioselective separation, chiral memory devices, and molecular machinery.

One could imagine an initially achiral supramolecular catalyst, which could be tuned to produce a specific enantiomer of desired compound by the presence of "chiral instruction", such as a simple chiral compound. Later, the "chiral instruction" could be switched to produce an opposite enantiomer or be removed, while the catalyst could be used again by employing a different "chiral instruction" to produce another chiral compound with the desired stereoselectivity. This hypothetical example illustrates how the supramolecular approach could eventually lead to cheaper, more versatile chirogenic systems, useful in asymmetric catalysis and in other fields of modern chemistry.

Moreover, in the current state of molecular machinery, such systems are able to demonstrate only basic concepts of this field and perform isolated tasks without practical use. However, this is a remarkably promising field since living organisms are essentially an assembly of well-cooperating molecular machines performing specific biological actions. To understand the relation with chirogenesis, we must realize that when we are determining the sense of chirality, our answer is related to the corresponding handedness or clock movement, which implies that chirality is related to a sort of directionality. Therefore, chirogenesis itself as a dynamic process has great potential to be employed in future molecular machines based on supramolecular interactions, to fulfill real machines' purposes of performing useful work that requires directionality, such as drug delivery, molecular electronics, programmed chemical transformations, and other newly emerging fields.

1.11 Acknowledgments

We acknowledge the support of this work by the Estonian Research Council through grant PRG399, the H2020-FETOPEN 828779 INITIO project, and LU offers thanks for the personal postdoctoral grant MOBJD592. We would also like to thank our families for their patience and understanding of our passion for science.

References

1. The Nobel Prize in Chemistry 1987. *NobelPrize.org*.
2. The Nobel Prize in Chemistry 2016. *NobelPrize.org*.
3. Pedersen, C.J. (1967) Cyclic polyethers and their complexes with metal salts. *J. Am. Chem. Soc.*, **89** (10), 2495–2496.
4. Pedersen, C.J. (1967) Cyclic polyethers and their complexes with metal salts. *J. Am. Chem. Soc.*, **89** (26), 7017–7036.
5. Lehn, J.M. (1977) Cryptates: macropolycyclic inclusion complexes. *Pure Appl. Chem.*, **49** (6), 857–870.
6. Lehn, J.M. (1978) Cryptates: inclusion complexes of macropolycyclic receptor molecules. *Pure Appl. Chem.*, **50** (9–10), 871–892.
7. Cram, D.J. (1983) Cavitands: organic hosts with enforced cavities. *Science*, **219** (4589), 1177–1183.
8. Cram, D.J., Karbach, S., Kim, Y.H., Baczynskyj, L., and Kallemeyn, G.W. (1985) Shell closure of two cavitands forms carcerand complexes with components of the medium as permanent guests. *J. Am. Chem. Soc.*, **107** (8), 2575–2576.
9. van der Waals, J.D. (1873) Over de continuïteit van den gas- en vloeistoftoestand.
10. Werner, A. (1893) Beitrag zur Konstitution anorganischer Verbindungen. *Z. Anorg. Chem.*, **3** (1), 267–330.
11. Fischer, E. (1894) Einfluss der Configuration auf die Wirkung der Enzyme. *Ber. Dtsch. Chem. Ges.*, **27** (3), 2985–2993.
12. Chalk, S.J., McNaught, A.D., and Wilkinson, A. (1997) *IUPAC. Compendium of Chemical Terminology, 2nd ed. (the "Gold Book")*, Blackwell Scientific Publications, Oxford.
13. Brammer, L. (2017) Halogen bonding, chalcogen bonding, pnictogen bonding, tetrel bonding: origins, current status and discussion. *Faraday Discuss.*, **203**, 485–507.
14. Alkorta, I., Elguero, J., and Frontera, A. (2020) Not only hydrogen bonds: other noncovalent interactions. *Crystals*, **10** (3), 180.
15. Fraser Stoddart, J. (2009) The chemistry of the mechanical bond. *Chem. Soc. Rev.*, **38** (6), 1802–1820.
16. Fang, L., Olson, M.A., Benítez, D., Tkatchouk, E., Iii, W.A.G., and Stoddart, J.F. (2009) Mechanically bonded macromolecules. *Chem. Soc. Rev.*, **39** (1), 17–29.

17. Bruns, C.J. and Stoddart, J.F. (2016) An introduction to the mechanical bond, in *The Nature of the Mechanical Bond*, John Wiley & Sons, Ltd., pp. 1–54.
18. Sun, Y., Aav, R., Tsuda, A., Miyake, H., Hirose, K., and Borovkov, V. (2021) Editorial: supramolecular chirogenesis in chemical and related sciences. *Front. Chem.* **9**, 679332.
19. Steed, J.W. and Atwood, J.L. (2009) *Supramolecular Chemistry*, John Wiley & Sons, Inc.
20. Rather, I.A., Wagay, S.A., and Ali, R. (2020) Emergence of anion-π interactions: the land of opportunity in supramolecular chemistry and beyond. *Coord. Chem. Rev.*, **415**, 213327.
21. Cavallo, G., Metrangolo, P., Milani, R., Pilati, T., Priimagi, A., Resnati, G., and Terraneo, G. (2016) The halogen bond. *Chem. Rev.*, **116** (4), 2478–2601.
22. Wolters, L.P., Schyman, P., Pavan, M.J., Jorgensen, W.L., Bickelhaupt, F.M., and Kozuch, S. (2014) The many faces of halogen bonding: a review of theoretical models and methods. *WIREs Comp. Mol. Sci.*, **4** (6), 523–540.
23. Wang, H., Wang, W., and Jin, W.J. (2016) σ-Hole bond vs. π-hole bond: a comparison based on halogen bond. *Chem. Rev.*, **116** (9), 5072–5104.
24. Riel, A.M.S., Rowe, R.K., Ho, E.N., Carlsson, A.-C.C., Rappé, A.K., Berryman, O.B., and Ho, P.S. (2019) Hydrogen bond enhanced halogen bonds: a synergistic interaction in chemistry and biochemistry. *Acc. Chem. Res.*, **52** (10), 2870–2880.
25. Anslyn, E.V. and Dougherty, D.A. (2006) *Modern Physical Organic Chemistry*, University Science Books.
26. Takahashi, O., Kohno, Y., and Nishio, M. (2010) Relevance of weak hydrogen bonds in the conformation of organic compounds and bioconjugates: evidence from recent experimental data and high-level *ab initio* MO calculations. *Chem. Rev.*, **110** (10), 6049–6076.
27. Biedermann, F. and Schneider, H.-J. (2016) Experimental binding energies in supramolecular complexes. *Chem. Rev.*, **116** (9), 5216–5300.
28. Dougherty, D.A. (2013) The cation–π interaction. *Acc. Chem. Res.*, **46** (4), 885–893.
29. Liptrot, D.J. and Power, P.P. (2017) London dispersion forces in sterically crowded inorganic and organometallic molecules. *Nat. Rev. Chem.*, **1** (1), 1–12.
30. Blanksby, S.J. and Ellison, G.B. (2003) Bond dissociation energies of organic molecules. *Acc. Chem. Res.*, **36** (4), 255–263.
31. Fersht, A.R., Shi, J.-P., Knill-Jones, J., Lowe, D.M., Wilkinson, A.J., Blow, D.M., Brick, P., Carter, P., Waye, M.M.Y., and Winter, G. (1985) Hydrogen bonding and biological specificity analysed by protein engineering. *Nature*, **314** (6008), 235–238.
32. Solov'ev, V.P., Strakhova, N.N., Raevsky, O.A., Rüdiger, V., and Schneider, H.-J. (1996) Solvent effects on crown ether complexations. *J. Org. Chem.*, **61** (16), 5221–5226.
33. Reichardt, C. (2002) Solvent effects on the position of homogeneous chemical equilibria, in *Solvents and Solvent Effects in Organic Chemistry*, John Wiley & Sons, Ltd, pp. 93–145.

34. Smithrud, D.B. and Diederich, F. (1990) Strength of molecular complexation of apolar solutes in water and in organic solvents is predictable by linear free energy relationships: a general model for solvation effects on apolar binding. *J. Am. Chem. Soc.*, **112** (1), 339–343.
35. Cahn, R.S., Ingold, C., and Prelog, V. (1966) Specification of molecular chirality. *Angew. Chem. Int. Ed. Engl.*, **5** (4), 385–415.
36. Lukin, O. and Vögtle, F. (2005) Knotting and threading of molecules: chemistry and chirality of molecular knots and their assemblies. *Angew. Chem. In. Ed.*, **44** (10), 1456–1477.
37. Zimmerman, H.E. and Crumrine, D.S. (1972) Duality of mechanism in photoracemization of optically active biphenyl. Mechanistic and exploratory organic photochemistry. LXV. *J. Am. Chem. Soc.*, **94** (2), 498–506.
38. Lehn, J.M. (1995) Self-processes — programmed supramolecular systems, in *Supramolecular Chemistry*, John Wiley & Sons, Ltd., pp. 139–197.
39. Mateos-Timoneda, M.A., Crego-Calama, M., and Reinhoudt, D.N. (2004) Supramolecular chirality of self-assembled systems in solution. *Chem. Soc. Rev.*, **33** (6), 363–372.
40. Hofmeier, H. and Schubert, U.S. (2004) Recent developments in the supramolecular chemistry of terpyridine–metal complexes. *Chem. Soc. Rev.*, **33** (6), 373–399.
41. Bazzicalupi, C., Bianchi, A., Biver, T., Giorgi, C., Santarelli, S., and Savastano, M. (2014) Formation of double-strand dimetallic helicates with a terpyridine-based macrocycle. *Inorg. Chem.*, **53** (22), 12215–12224.
42. Niess, F., Duplan, V., Diercks, C.S., and Sauvage, J.-P. (2015) Contractile and extensible molecular figures-of-eight. *Chem. Eur. J.*, **21** (41), 14393–14400.
43. Brandl, T., Hoffmann, V., Pannwitz, A., Häussinger, D., Neuburger, M., Fuhr, O., Bernhard, S., Wenger, O.S., and Mayor, M. (2018) Chiral macrocyclic terpyridine complexes. *Chem. Sci.*, **9** (15), 3837–3843.
44. Avalos, M., Babiano, R., Cintas, P., Jiménez, J.L., Palacios, J.C., and Barron, L.D. (1998) Absolute asymmetric synthesis under physical fields: facts and fictions. *Chem. Rev.*, **98** (7), 2391–2404.
45. Feringa, B.L. and van Delden, R.A. (1999) Absolute asymmetric synthesis: the origin, control, and amplification of chirality. *Angew. Chem. Int. Ed.*, **38** (23), 3418–3438.
46. Feringa, B.L., van Delden, R.A., Koumura, N., and Geertsema, E.M. (2000) Chiroptical molecular switches. *Chem. Rev.*, **100** (5), 1789–1816.
47. Ribó, J.M., Crusats, J., Sagués, F., Claret, J., and Rubires, R. (2001) Chiral sign induction by vortices during the formation of mesophases in stirred solutions. *Science*, **292** (5524), 2063–2066.
48. Cintas, P. (2002) Chirality of living systems: a helping hand from crystals and oligopeptides. *Angew. Chem. Int. Ed.*, **41** (7), 1139–1145.
49. Tsuda, A., Alam, M.A., Harada, T., Yamaguchi, T., Ishii, N., and Aida, T. (2007) Spectroscopic visualization of vortex flows using dye-containing nanofibers. *Angew. Chem. Int. Ed.*, **46** (43), 8198–8202.

50. Wolffs, M., George, S.J., Tomović, Ž., Meskers, S.C.J., Schenning, A.P.H.J., and Meijer, E.W. (2007) Macroscopic origin of circular dichroism effects by alignment of self-assembled fibers in solution. *Angew. Chem. Int. Ed.*, **46** (43), 8203–8205.
51. D'Urso, A., Randazzo, R., Lo Faro, L., and Purrello, R. (2010) Vortexes and nanoscale chirality. *Angew. Chem. Int. Ed.*, **49** (1), 108–112.
52. Petit-Garrido, N., Claret, J., Ignés-Mullol, J., and Sagués, F. (2012) Stirring competes with chemical induction in chiral selection of soft matter aggregates. *Nat. Commun.*, **3** (1), 1001.
53. Fujiki, M. (2014) Supramolecular chirality: solvent chirality transfer in molecular chemistry and polymer chemistry. *Symmetry*, **6** (3), 677–703.
54. Kalachyova, Y., Lyutakov, O., Goncharova, I., and Svorcik, V. (2015) "Artificial" chirality induced in doped polymer by irradiation with circularly polarized excimer laser light. *Opt. Mater. Express,* **5** (12), 2761–2767.
55. Kim, J., Lee, J., Kim, W.Y., Kim, H., Lee, S., Lee, H.C., Lee, Y.S., Seo, M., and Kim, S.Y. (2015) Induction and control of supramolecular chirality by light in self-assembled helical nanostructures. *Nat. Commun.*, **6** (1), 6959.
56. Bisoyi, H.K. and Li, Q. (2016) Light-directed dynamic chirality inversion in functional self-organized helical superstructures. *Angew. Chem. Int. Ed.*, **55** (9), 2994–3010.
57. Wang, L., Yin, L., Zhang, W., Zhu, X., and Fujiki, M. (2017) Circularly polarized light with sense and wavelengths to regulate azobenzene supramolecular chirality in optofluidic medium. *J. Am. Chem. Soc.*, **139** (37), 13218–13226.
58. Sun, J., Li, Y., Yan, F., Liu, C., Sang, Y., Tian, F., Feng, Q., Duan, P., Zhang, L., Shi, X., Ding, B., and Liu, M. (2018) Control over the emerging chirality in supramolecular gels and solutions by chiral microvortices in milliseconds. *Nat. Commun.*, **9** (1), 2599.
59. Yang, G., Zhang, S., Hu, J., Fujiki, M., and Zou, G. (2019) The chirality induction and modulation of polymers by circularly polarized light. *Symmetry*, **11** (4), 474.
60. Pérez-García, L. and Amabilino, D.B. (2002) Spontaneous resolution under supramolecular control. *Chem. Soc. Rev.*, **31** (6), 342–356.
61. Koshima, H. (2000) Generation of chirality in two-component molecular crystals from achiral molecules. *J. Mol. Struct.*, **552** (1), 111–116.
62. Kuball, H.-G. (2013) Chiroptical analysis, in *Encyclopedia of Analytical Science, 3rd ed.* (eds. Worsfold, P., Poole, C., Townshend, A., and Miró, M.), Academic Press, Oxford, pp. 1–21.
63. Diedrich, C. and Grimme, S. (2003) Systematic investigation of modern quantum chemical methods to predict electronic circular dichroism spectra. *J. Phys. Chem. A*, **107** (14), 2524–2539.
64. Nicu, V.P., Johannes Neugebauer, and Baerends, E.J. (2008) Effects of complex formation on vibrational circular dichroism spectra. *J. Phys. Chem. A*, **112** (30), 6978–6991.

65. Hembury, G.A., Borovkov, V.V., and Inoue, Y. (2008) Chirality-sensing supramolecular systems. *Chem. Rev.*, **108** (1), 1–73.
66. Pescitelli, G., Bari, L.D., and Berova, N. (2014) Application of electronic circular dichroism in the study of supramolecular systems. *Chem. Soc. Rev.*, **43** (15), 5211–5233.
67. Liu, M., Zhang, L., and Wang, T. (2015) Supramolecular chirality in self-assembled systems. *Chem. Rev.*, **115** (15), 7304–7397.
68. Lv, W., Zhu, P., Bian, Y., Ma, C., Zhang, X., and Jiang, J. (2010) Optically active homoleptic bis(phthalocyaninato) rare earth double-decker complexes bearing peripheral chiral menthol moieties: effect of π–π interaction on the chiral information transfer at the molecular level. *Inorg. Chem.*, **49** (14), 6628–6635.
69. Yang, C. and Inoue, Y. (2012) Electronic circular dichroism of supramolecular systems, in *Comprehensive Chiroptical Spectroscopy*, John Wiley & Sons, Ltd., pp. 317–353.
70. Borovkov, V.V., Lintuluoto, J.M., Sugeta, H., Fujiki, M., Arakawa, R., and Inoue, Y. (2002) Supramolecular chirogenesis in zinc porphyrins: equilibria, binding properties, and thermodynamics. *J. Am. Chem. Soc.*, **124** (12), 2993–3006.
71. Borovkov, V.V. and Inoue, Y. (2006) Supramolecular chirogenesis in host–guest systems containing porphyrinoids, in *Supramolecular Chirality* (eds. Crego-Calama, M. and Reinhoudt, D.N.), Springer Berlin Heidelberg, Berlin, Heidelberg, pp. 89–146.
72. Wolf, C. and Bentley, K.W. (2013) Chirality sensing using stereodynamic probes with distinct electronic circular dichroism output. *Chem. Soc. Rev.*, **42** (12), 5408–5424.
73. Chen, Z., Wang, Q., Wu, X., Li, Z., and Jiang, Y.-B. (2015) Optical chirality sensing using macrocycles, synthetic and supramolecular oligomers/polymers, and nanoparticle based sensors. *Chem. Soc. Rev.*, **44** (13), 4249–4263.
74. Hasan, M. and Borovkov, V. (2018) Helicene-based chiral auxiliaries and chirogenesis. *Symmetry*, **10** (1), 10.
75. Gholami, H., Chakraborty, D., Zhang, J., and Borhan, B. (2021) Absolute stereochemical determination of organic molecules through induction of helicity in host–guest complexes. *Acc. Chem. Res.*, **54** (3), 654–667
76. Yang, W., Chai, X., Chi, L., Liu, X., Cao, Y., Lu, R., Jiang, Y., Tang, X., Fuchs, H., and Li, T. (1999) From achiral molecular components to chiral supermolecules and supercoil self-assembly. *Chem. Eur. J.*, **5** (4), 1144–1149.
77. Arias, S., Freire, F., Quiñoá, E., and Riguera, R. (2014) Nanospheres, nanotubes, toroids, and gels with controlled macroscopic chirality. *Angew. Chem. Int. Ed.*, **53** (50), 13720–13724.
78. Xing, P. and Zhao, Y. (2018) Controlling supramolecular chirality in multicomponent self-assembled systems. *Acc. Chem. Res.*, **51** (9), 2324–2334.

79. Dorca, Y., Greciano, E.E., Valera, J.S., Gómez, R., and Sánchez, L. (2019) Hierarchy of asymmetry in chiral supramolecular polymers: toward functional, helical supramolecular structures. *Chem. Eur. J.*, **25** (23), 5848–5864.
80. Yan, X., Wang, Q., Chen, X., and Jiang, Y.-B. (2020) Supramolecular chiral aggregates exhibiting nonlinear CD–ee dependence. *Adv. Mater.*, **32** (41), 1905667.
81. Lifson, S., Green, M.M., Andreola, C., and Peterson, N.C. (1989) Macromolecular stereochemistry: helical sense preference in optically active polyisocyanates. Amplification of a conformational equilibrium deuterium isotope effect. *J. Am. Chem. Soc.*, **111** (24), 8850–8858.
82. Palmans, A.R.A. and Meijer, E.W. (2007) Amplification of chirality in dynamic supramolecular aggregates. *Angew. Chem. Int. Ed.*, **46** (47), 8948–8968.
83. Smulders, M.M.J., Stals, P.J.M., Mes, T., Paffen, T.F.E., Schenning, A.P.H.J., Palmans, A.R.A., and Meijer, E.W. (2010) Probing the limits of the majority-rules principle in a dynamic supramolecular polymer. *J. Am. Chem. Soc.*, **132** (2), 620–626.
84. Purrello, R., Raudino, A., Monsù Scolaro, L., Loisi, A., Bellacchio, E., and Lauceri, R. (2000) Ternary porphyrin aggregates and their chiral memory. *J. Phys. Chem. B*, **104** (46), 10900–10908.
85. Mammana, A., D'Urso, A., Lauceri, R., and Purrello, R. (2007) Switching off and on the supramolecular chiral memory in porphyrin assemblies. *J. Am. Chem. Soc.*, **129** (26), 8062–8063.
86. Rosaria, L., D'urso, A., Mammana, A., and Purrello, R. (2008) Chiral memory: induction, amplification, and switching in porphyrin assemblies. *Chirality*, **20** (3–4), 411–419.
87. Yashima, E., Maeda, K., Iida, H., Furusho, Y., and Nagai, K. (2009) Helical polymers: synthesis, structures, and functions. *Chem. Rev.*, **109** (11), 6102–6211.
88. Helmich, F., Lee, C.C., Schenning, A.P.H.J., and Meijer, E.W. (2010) Chiral memory via chiral amplification and selective depolymerization of porphyrin aggregates. *J. Am. Chem. Soc.*, **132** (47), 16753–16755.
89. Yang, D., Zhao, Y., Lv, K., Wang, X., Zhang, W., Zhang, L., and Liu, M. (2016) A strategy for tuning achiral main-chain polymers into helical assemblies and chiral memory systems. *Soft Matter*, **12** (4), 1170–1175.
90. Feyter, S.D. and Schryver, F.C.D. (2003) Two-dimensional supramolecular self-assembly probed by scanning tunneling microscopy. *Chem. Soc. Rev.*, **32** (3), 139–150.
91. Shen, B., Kim, Y., and Lee, M. (2020) Supramolecular chiral 2D materials and emerging functions. *Adv. Mater.*, **32** (41), 1905669.
92. Böhringer, M., Morgenstern, K., Schneider, W.-D., and Berndt, R. (1999) Separation of a racemic mixture of two-dimensional molecular clusters by scanning tunneling microscopy. *Angew. Chem. Int. Ed.*, **38** (6), 821–823.

93. Böhringer, M., Morgenstern, K., Schneider, W.-D., Berndt, R., Mauri, F., De Vita, A., and Car, R. (1999) Two-dimensional self-assembly of supramolecular clusters and chains. *Phys. Rev. Lett.*, **83** (2), 324–327.
94. Breu, J., Domel, H., and Stoll, A. (2000) Racemic compound formation versus conglomerate formation with [M(bpy)3](PF6)2 (M = Ni, Zn, Ru); molecular and crystal structures. *Eur. J. Inorg. Chem.*, **2000** (11), 2401–2408.
95. Serrano, J.L. and Sierra, T. (2003) Helical supramolecular organizations from metal-organic liquid crystals. *Coord. Chem. Rev.*, **242** (1), 73–85.
96. Seeber, G., Tiedemann, B.E.F., and Raymond, K.N. (2006) Supramolecular chirality in coordination chemistry, in *Supramolecular Chirality* (eds. Crego-Calama, M. and Reinhoudt, D.N.), Springer, Berlin, Heidelberg, pp. 147–183.
97. Maeda, C., Kamada, T., Aratani, N., and Osuka, A. (2007) Chiral self-discriminative self-assembling of meso–meso linked diporphyrins. *Coord. Chem. Rev.*, **251** (21), 2743–2752.
98. Crassous, J. (2009) Chiral transfer in coordination complexes: towards molecular materials. *Chem. Soc. Rev.*, **38** (3), 830–845.
99. Du, D.-Y., Yan, L.-K., Su, Z.-M., Li, S.-L., Lan, Y.-Q., and Wang, E.-B. (2013) Chiral polyoxometalate-based materials: from design syntheses to functional applications. *Coord. Chem. Rev.*, **257** (3), 702–717.
100. Constable, E.C. (2013) Stereogenic metal centres — from Werner to supramolecular chemistry. *Chem. Soc. Rev.*, **42** (4), 1637–1651.
101. Castilla, A.M., Ramsay, W.J., and Nitschke, J.R. (2014) Stereochemistry in subcomponent self-assembly. *Acc. Chem. Res.*, **47** (7), 2063–2073.
102. Gu, Z.-G., Zhan, C., Zhang, J., and Bu, X. (2016) Chiral chemistry of metal–camphorate frameworks. *Chem. Soc. Rev.*, **45** (11), 3122–3144.
103. Cui, Y., Li, B., He, H., Zhou, W., Chen, B., and Qian, G. (2016) Metal–organic frameworks as platforms for functional materials. *Acc. Chem. Res.*, **49** (3), 483–493.
104. Ariga, K., Mori, T., Kitao, T., and Uemura, T. (2020) Supramolecular chiral nanoarchitectonics. *Adv. Mater.*, **32** (41), 1905657.
105. Dong, J., Liu, Y., and Cui, Y. (2021) Supramolecular chirality in metal–organic complexes. *Acc. Chem. Res.*, **54** (1), 194–206.
106. Haberhauer, G. (2010) A Metal-ion-driven supramolecular chirality pendulum. *Angew. Chem. Int. Ed.*, **49** (48), 9286–9289.
107. Liu, P., Neuhaus, P., Kondratuk, D.V., Balaban, T.S., and Anderson, H.L. (2014) Cyclodextrin-templated porphyrin nanorings. *Angew. Chem. Int. Ed.*, **53** (30), 7770–7773.
108. Zhang, S.-Y., Li, D., Guo, D., Zhang, H., Shi, W., Cheng, P., Wojtas, L., and Zaworotko, M.J. (2015) Synthesis of a chiral crystal form of MOF-5, CMOF-5, by chiral induction. *J. Am. Chem. Soc.*, **137** (49), 15406–15409.
109. Evans, J.D. and Coudert, F.-X. (2016) Microscopic mechanism of chiral induction in a metal–organic framework. *J. Am. Chem. Soc.*, **138** (19), 6131–6134.

110. Chakrabarty, R., Mukherjee, P.S., and Stang, P.J. (2011) Supramolecular coordination: self-assembly of finite two- and three-dimensional ensembles. *Chem. Rev.*, **111** (11), 6810–6918.
111. Cook, T.R. and Stang, P.J. (2015) Recent developments in the preparation and chemistry of metallacycles and metallacages via coordination. *Chem. Rev.*, **115** (15), 7001–7045.
112. Chen, L.-J., Yang, H.-B., and Shionoya, M. (2017) Chiral metallosupramolecular architectures. *Chem. Soc. Rev.*, **46** (9), 2555–2576.
113. Yeh, R.M., Ziegler, M., Johnson, D.W., Terpin, A.J., and Raymond, K.N. (2001) Imposition of chirality in a dinuclear triple-stranded helicate by ion pair formation1. *Inorg. Chem.*, **40** (10), 2216–2217.
114. Yeh, R.M. and Raymond, K.N. (2006) Supramolecular asymmetric induction in dinuclear triple-stranded helicates1. *Inorg. Chem.*, **45** (3), 1130–1139.
115. Jodry, J.J. and Lacour, J. (2000) Efficient resolution of a dinuclear triple helicate by asymmetric extraction/precipitation with TRISPHAT anions as resolving agents. *Chem. Eur. J.*, **6** (23), 4297–4304.
116. Wan, S., Lin, L.-R., Zeng, L., Lin, Y., and Zhang, H. (2014) Efficient optical resolution of water-soluble self-assembled tetrahedral M4L6 cages with 1,1′-bi-2-naphthol. *Chem. Commun.*, **50** (97), 15301–15304.
117. Hiraoka, S. and Fujita, M. (1999) Guest-selected formation of Pd(II)-linked cages from a prototypical dynamic library. *J. Am. Chem. Soc.*, **121** (43), 10239–10240.
118. Castilla, A.M., Ousaka, N., Bilbeisi, R.A., Valeri, E., Ronson, T.K., and Nitschke, J.R. (2013) High-fidelity stereochemical memory in a FeII4L4 tetrahedral capsule. *J. Am. Chem. Soc.*, **135** (47), 17999–18006.
119. Jędrzejewska, H., Kwit, M., and Szumna, A. (2015) Switching of inherent chirality driven by self-assembly. *Chem. Commun.*, **51** (72), 13799–13801.
120. Erbas-Cakmak, S., Leigh, D.A., McTernan, C.T., and Nussbaumer, A.L. (2015) Artificial molecular machines. *Chem. Rev.*, **115** (18), 10081–10206.
121. Mitchell, D.K. and Sauvage, J.-P. (1988) A topologically chiral [2]catenand. *Angew. Chem. Int. Ed.*, **27** (7), 930–931.
122. Chambron, J.C., Mitchell, D.K., and Sauvage, J.P. (1992) Synthesis, characterization, and a proton NMR study of topologically chiral copper(I) [2]-catenates and achiral analogs. *J. Am. Chem. Soc.*, **114** (12), 4625–4631.
123. Evans, N.H. (2018) Chiral catenanes and rotaxanes: fundamentals and emerging applications. *Chem. Eur. J.*, **24** (13), 3101–3112.
124. Jamieson, E.M.G, Modicom, F., and Goldup, S.M. (2018) Chirality in rotaxanes and catenanes. *Chem. Soc. Rev.*, **47** (14), 5266–5311.
125. Chambron, J.-C., Sauvage, J.-P., and Mislow, K. (1997) A chemically achiral molecule with no rigidly achiral presentations. *J. Am. Chem. Soc.*, **119** (40), 9558–9559.
126. Yamamoto, C., Okamoto, Y., Schmidt, T., Jäger, R., and Vögtle, F. (1997) Enantiomeric resolution of cycloenantiomeric rotaxane, topologically chiral cat-

enane, and pretzel-shaped molecules: observation of pronounced circular dichroism. *J. Am. Chem. Soc.*, **119** (43), 10547–10548.
127. Kimura, T., Miyagawa, S., Takaya, H., Naito, M., and Tokunaga, Y. (2020) Locking the dynamic axial chirality of biphenyl crown ethers through threading. *Chem. Eur. J.*, **15** (22), 3897–3903.
128. Corra, S., de Vet, C., Groppi, J., La Rosa, M., Silvi, S., Baroncini, M., and Credi, A. (2019) Chemical on/off switching of mechanically planar chirality and chiral anion recognition in a [2]rotaxane molecular shuttle. *J. Am. Chem. Soc.*, **141** (23), 9129–9133.
129. Tseng, H.-R., Vignon, S.A., Celestre, P.C., Stoddart, J.F., White, A.J.P., and Williams, D.J. (2003) Dynamic chirality: keen selection in the face of stereochemical diversity in mechanically bonded compounds. *Chem. Eur. J.*, **9** (2), 543–556.
130. Schalley, C.A., Beizai, K., and Vögtle, F. (2001) On the way to rotaxane-based molecular motors: studies in molecular mobility and topological chirality. *Acc. Chem. Res.*, **34** (6), 465–476.
131. Aprahamian, I. (2020) The future of molecular machines. *ACS Cent. Sci.*, **6** (3), 347–358.
132. Lancia, F., Ryabchun, A., and Katsonis, N. (2019) Life-like motion driven by artificial molecular machines. *Nat Rev Chem*, **3** (9), 536–551.
133. Goswami, A., Saha, S., Biswas, P.K., and Schmittel, M. (2020) (Nano)mechanical motion triggered by metal coordination: from functional devices to networked multicomponent catalytic machinery. *Chem. Rev.*, **120** (1), 125–199.
134. Kottas, G.S., Clarke, L.I., Horinek, D., and Michl, J. (2005) Artificial molecular rotors. *Chem. Rev.*, **105** (4), 1281–1376.
135. Samanta, S.K., Samanta, D., Bats, J.W., and Schmittel, M. (2011) DABCO as a dynamic hinge between cofacial porphyrin panels and its tumbling inside a supramolecular cavity. *J. Org. Chem.*, **76** (18), 7466–7473.
136. Samanta, S.K. and Schmittel, M. (2013) Four-component supramolecular nanorotors. *J. Am. Chem. Soc.*, **135** (50), 18794–18797.
137. Samanta, S.K., Bats, J.W., and Schmittel, M. (2014) A five-component nanorotor with speed regulation. *Chem. Commun.*, **50** (18), 2364–2366.
138. Samanta, S.K., Rana, A., and Schmittel, M. (2016) Conformational slippage determines rotational frequency in five-component nanorotors. *Angew. Chem. Int. Ed.*, **55** (6), 2267–2272.
139. Biswas, P.K., Saha, S., Paululat, T., and Schmittel, M. (2018) Rotating catalysts are superior: suppressing product inhibition by anchimeric assistance in four-component catalytic machinery. *J. Am. Chem. Soc.*, **140** (29), 9038–9041.
140. Roke, D., Wezenberg, S.J., and Feringa, B.L. (2018) Molecular rotary motors: unidirectional motion around double bonds. *PNAS*, **115** (38), 9423–9431.
141. Barron, L.D. (1986) Symmetry and molecular chirality. *Chem. Soc. Rev.*, **15** (2), 189–223.

142. Mislow, K. (1999) Molecular chirality, in *Topics in Stereochemistry*, John Wiley & Sons, Ltd., pp. 1–82.
143. Petitjean, M. (2020) Molecular chirality in classical spacetime: solving the controversy about the spinning cone model of rotating molecules. *Chem. Eur. J.*, **26** (47), 10648–10652.
144. Barron, L.D. and Cintas, P. (2021) Comments on "molecular chirality in classical spacetime: solving the controversy about the spinning cone model of rotating molecules." *Chem. Eur. J.*, **27** (4), 1476–1477.
145. Borovkov, V.V., Hembury, G.A., and Inoue, Y. (2004) Origin, control, and application of supramolecular chirogenesis in bisporphyrin-based systems. *Acc. Chem. Res.*, **37** (7), 449–459.
146. Borovkov, V.V., Lintuluoto, J.M., and Inoue, Y. (2000) Supramolecular chirogenesis in bis(zinc porphyrin): an absolute configuration probe highly sensitive to guest structure. *Org. Lett.*, **2** (11), 1565–1568.
147. Borovkov, V.V., Yamamoto, N., Lintuluoto, J.M., Tanaka, T., and Inoue, Y. (2001) Supramolecular chirality induction in bis(zinc porphyrin) by amino acid derivatives: rationalization and applications of the ligand bulkiness effect. *Chirality*, **13** (6), 329–335.
148. Borovkov, V.V., Hembury, G.A., and Inoue, Y. (2003) The origin of solvent-controlled supramolecular chirality switching in a bis(zinc porphyrin) system. *Angew. Chem. Int. Ed.*, **42** (43), 5310–5314.
149. Borovkov, V.V., Fujii, I., Muranaka, A., Hembury, G.A., Tanaka, T., Ceulemans, A., Kobayashi, N., and Inoue, Y. (2004) Rationalization of supramolecular chirality in a bisporphyrin system. *Angew. Chem. Int. Ed.*, **43** (41), 5481–5485.
150. Borovkov, V. and Inoue, Y. (2009) A versatile bisporphyrinoid motif for supramolecular chirogenesis. *Eur. J. Org. Chem.*, **2009** (2), 189–197.
151. Borovkov, V. (2010) Effective supramolecular chirogenesis in ethane-bridged bisporphyrinoids. *Symmetry*, **2** (1), 184–200.
152. Hayashi, S., Yotsukura, M., Noji, M., and Takanami, T. (2015) Bis(zinc porphyrin) as a CD-sensitive bidentate host molecule: direct determination of absolute configuration of mono-alcohols. *Chem. Commun.*, **51** (55), 11068–11071.
153. Tanasova, M., Anyika, M., and Borhan, B. (2015) Sensing remote chirality: stereochemical determination of β-, γ-, and δ-chiral carboxylic acids. *Angew. Chem. Int. Ed.*, **54** (14), 4274–4278.
154. Dhamija, A., Saha, B., and Rath, S.P. (2017) Metal-center-driven supramolecular chirogenesis in tweezer amino alcohol complexes: structural, spectroscopic, and theoretical investigations. *Inorg. Chem.*, **56** (24), 15203–15215.
155. Saha, B., Ikbal, S.A., Petrovic, A.G., Berova, N., and Rath, S.P. (2017) Complexation of chiral zinc-porphyrin tweezer with achiral diamines: induction and two-step inversion of interporphyrin helicity monitored by ECD. *Inorg. Chem.*, **56** (7), 3849–3860.

156. Lu, H. and Kobayashi, N. (2016) Optically active porphyrin and phthalocyanine systems. *Chem. Rev.*, **116** (10), 6184–6261.
157. Lu, W., Yang, H., Li, X., Wang, C., Zhan, X., Qi, D., Bian, Y., and Jiang, J. (2017) Chiral discrimination of diamines by a binaphthalene-bridged porphyrin dimer. *Inorg. Chem.*, **56** (14), 8223–8231.
158. Saha, B., Petrovic, A.G., Dhamija, A., Berova, N., and Rath, S.P. (2019) complexation of chiral zinc(II) porphyrin tweezer with achiral aliphatic diamines revisited: molecular dynamics, electronic CD, and ^{1}H NMR analysis. *Inorg. Chem.*, **58** (17), 11420–11438.
159. Mondal, P. and Rath, S.P. (2020) Cyclic metalloporphyrin dimers: conformational flexibility, applications and future prospects. *Coord. Chem. Rev.*, **405**, 213117.
160. Miyake, Y., López-Moreno, A., Yang, J., Xu, H.-J., Desbois, N., Gros, C.P., and Komatsu, N. (2018) Synthesis of flexible nanotweezers with various metals and their application in carbon nanotube extraction. *New J. Chem.*, **42** (10), 7592–7594.
161. Huang, X., Rickman, B.H., Borhan, B., Berova, N., and Nakanishi, K. (1998) Zinc porphyrin tweezer in host–guest complexation: determination of absolute configurations of diamines, amino acids, and amino alcohols by circular dichroism. *J. Am. Chem. Soc.*, **120** (24), 6185–6186.
162. Berova, N., Bari, L.D., and Pescitelli, G. (2007) Application of electronic circular dichroism in configurational and conformational analysis of organic compounds. *Chem. Soc. Rev.*, **36** (6), 914–931.
163. Matile, S., Berova, N., Nakanishi, K., Novkova, S., Philipova, I., and Blagoev, B. (1995) Porphyrins: powerful chromophores for structural studies by exciton-coupled circular dichroism. *J. Am. Chem. Soc.*, **117** (26), 7021–7022.
164. Harada, N., Nakanishi, K., and Berova, N. (2012) Electronic CD exciton chirality method: principles and applications, in *Comprehensive Chiroptical Spectroscopy*, John Wiley & Sons, Ltd., pp. 115–166.
165. Labuta, J., Hill, J.P., Ishihara, S., Hanyková, L., and Ariga, K. (2015) Chiral sensing by nonchiral tetrapyrroles. *Acc. Chem. Res.*, **48** (3), 521–529.
166. Ogoshi, H. and Mizutani, T. (1998) Multifunctional and chiral porphyrins: model receptors for chiral recognition. *Acc. Chem. Res.*, **31** (2), 81–89.
167. Ustrnul, L., Kaabel, S., Burankova, T., Martõnova, J., Adamson, J., Konrad, N., Burk, P., Borovkov, V., and Aav, R. (2019) Supramolecular chirogenesis in zinc porphyrins by enantiopure hemicucurbit[n]urils (n = 6, 8). *Chem. Commun.*, **55** (96), 14434–14437.
168. Konrad, N., Meniailava, D., Osadchuk, I., Adamson, J., Hasan, M., Clot, E., Aav, R., Borovkov, V., and Kananovich, D. (2019) Supramolecular chirogenesis in zinc porphyrins: complexation with enantiopure thiourea derivatives, binding studies and chirality transfer mechanism. *J. Porphyrins Phthalocyanines*, **24** (05n07), 840–849.
169. Konrad, N., Horetski, M., Sihtmäe, M., Truong, K.-N., Osadchuk, I., Burankova, T., Kielmann, M., Adamson, J., Kahru, A., Rissanen, K., Senge, M.O., Borovkov, V., Aav, R., and Kananovich, D. (2021) Thiourea organocatalysts as emerging chiral

pollutants: en route to porphyrin-based (chir)optical sensing. *Chemosensors*, **9** (10), 278.

170. Inouye, M., Waki, M., and Abe, H. (2004) Saccharide-dependent induction of chiral helicity in achiral synthetic hydrogen-bonding oligomers. *J. Am. Chem. Soc.*, **126** (7), 2022–2027.
171. Waki, M., Abe, H., and Inouye, M. (2007) Translation of mutarotation into induced circular dichroism signals through helix inversion of host polymers. *Angew. Chem. Int. Ed.*, **46** (17), 3059–3061.
172. Forman, J.E., Barrans, R.E., and Dougherty, D.A. (1995) Circular dichroism studies of molecular recognition with cyclophane hosts in aqueous media. *J. Am. Chem. Soc.*, **117** (36), 9213–9228.
173. Allenmark, S. (2003) Induced circular dichroism by chiral molecular interaction. *Chirality*, **15** (5), 409–422.
174. Katoono, R., Kusaka, K., Saito, Y., Sakamoto, K., and Suzuki, T. (2019) Chiral diversification through the assembly of achiral phenylacetylene macrocycles with a two-fold bridge. *Chem. Sci.*, **10** (18), 4782–4791.
175. Biedermann, F. and Nau, W.M. (2014) Noncovalent chirality sensing ensembles for the detection and reaction monitoring of amino acids, peptides, proteins, and aromatic drugs. *Angew. Chem. Int. Ed.*, **53** (22), 5694–5699.
176. Li, S., Jia, C., Wu, B., Luo, Q., Huang, X., Yang, Z., Li, Q.-S., and Yang, X.-J. (2011) A triple anion helicate assembled from a bis(biurea) ligand and phosphate ions. *Angew. Chem. Int. Ed.*, **50** (25), 5721–5724.
177. Wu, B., Cui, F., Lei, Y., Li, S., Amadeu, N. de S., Janiak, C., Lin, Y.-J., Weng, L.-H., Wang, Y.-Y., and Yang, X.-J. (2013) Tetrahedral anion cage: self-assembly of a (PO4)4L4 complex from a tris(bisurea) ligand. *Angew. Chem. Int. Ed.*, **52** (19), 5096–5100.
178. Bai, X., Jia, C., Zhao, Y., Yang, D., Wang, S.-C., Li, A., Chan, Y.-T., Wang, Y.-Y., Yang, X.-J., and Wu, B. (2018) Peripheral templation-modulated interconversion between an A4L6 tetrahedral anion cage and A2L3 triple helicate with guest capture/release. *Angew. Chem. Int. Ed.*, **57** (7), 1851–1855.
179. Fan, X., Zhang, D., Jiang, S., Wang, H., Lin, L.-T., Zheng, B., Xu, W.-H., Zhao, Y., Hay, B.P., Chan, Y.-T., Yang, X.-J., Li, X., and Wu, B. (2019) Construction and interconversion of anion-coordination-based ('aniono') grids and double helicates modulated by counter-cations. *Chem. Sci.*, **10** (25), 6278–6284.
180. Li, B., Zheng, B., Zhang, W., Zhang, D., Yang, X.-J., and Wu, B. (2020) Site-selective binding of peripheral chiral guests induces stereospecificity in A4L6 tetrahedral anion cages. *J. Am. Chem. Soc.*, **142** (13), 6304–6311.
181. Jo, H.H., Lin, C.-Y., and Anslyn, E.V. (2014) Rapid optical methods for enantiomeric excess analysis: from enantioselective indicator displacement assays to exciton-coupled circular dichroism. *Acc. Chem. Res.*, **47** (7), 2212–2221.

https://doi.org/10.1142/9789811259227_0003

3 Chirogenesis in Molecular Aggregates

Massimiliano Gaeta,* Alessandro D'Urso† and Roberto Purrello‡

Dipartimento di Scienze Chimiche,
Università degli Studi di Catania, Catania, Italy
*mgaeta@unict.it
†adurso@unict.it
‡rpurrello@unict.it

Self-assembly process appears as a powerful and attractive strategy for constructing complex supramolecules by spontaneous organization of appropriate building blocks. Molecular recognition, which allows building ordered architectures starting from small messy units, is the basic process behind self-assembly. Therefore, exploiting these phenomena one can convey asymmetry to supramolecular level. The accurate insight of the chirogenic phenomenon via weak intermolecular forces in supramolecular systems may elucidate the origin of homochirality in Nature and Life. For this reason, several strategies have been implemented to investigate how molecular aggregates become chiral and which processes govern the bias for a certain enantiomeric assembly rather than another. This chapter first intends to address the basic principles regulating the self-assembly phenomena and their chirogenesis aspects from single units to multicomponent systems with particular regard to porphyrin-based assemblages. Moreover, the fundamental strategies for designing and building chirally oriented supramolecular aggregates will be taken into account.

1.1 Introduction

The importance of molecular aggregates is observed in Nature, where most of the systems are based on several aggregates — or their own derivatives — and arranged in a well-organized structure in order to define electronic or catalytic properties. Undoubtedly, the light-harvesting complex, exploited by cyanobacteria and green plants for their photosynthesis, depicts the major example of chromophoric array [1–5] as well as *cytochrome c3* hemoprotein isolated from the sulfate-reducing bacteria of the genus *Desulfovibrio*, where four hemes of the system are arranged in a nonparallel fashion to foster electron transport chains in anaerobically respiring microorganisms [6].

With the attempt to caught inspiration from the biological systems to design supramolecular functional architectures, scientists developed several experimental methods. However, the conventional synthetic strategies to build molecular arrays have generally provided quite limits: numerous steps followed by laborious separation of the reaction mixture, resulting in a low product yield [7]. On the other hand, the building of large multicomponent architectures via *self-assembly* has been emerging as a feasible alternative to covalent synthesis [8].

Nowadays, the supramolecular strategy can be intended as a multidisciplinary field that engages different areas, from the traditional organic and inorganic chemistry, from synthesizing the precursors for a supermolecule, to physical chemistry and computational chemistry, needed to understand the properties and the behavior of complex supramolecular systems [9]. Therefore, the main motivation for continuing the investigations in this field is the promise of useful molecular devices in future.

In principle, a super(*supra*)molecule refers to two large groups: (a) host–guest chemistry, and (b) self-assembly. The key difference between these two categories can be described in terms of size and shape:

if a large molecule can enclose around it "something," thus it is labeled as *host* and the smaller molecule is its *guest* entrapped within the host [10]. On the contrary, whether there are no species acting as a host for another, more species may form supramolecular assemblies spontaneously, in a process termed *self-assembly*. In principle, *self-assembly* is an equilibrium between two or more species — i.e., building blocks — leading to molecular aggregates whose contained information strictly depends on the chemical monomers [11]. In this chapter, we will analyze just some examples of molecular aggregates obtained by self-assembly of the selected building blocks.

Noteworthy, self-assembly is usually a spontaneous process but could be influenced by solvents [12] or the presence of template, e.g., small molecules [13] or large hosts [14–16], amino acids [17, 18], polymers [19, 20], inorganic materials and NPs [21–24], polyelectrolytes and polypeptides [25–27], polynucleotides [28–30], and so on. In both cases, host–guest interactions and self-assembly, some concepts of *complementarity*, *preorganization*, and *cooperativity* of the binding sites play a key role in the deep comprehension of the supramolecular phenomena [31–33].

In other words, when a host — or a template — reveals a preferential binding toward a specific guest — or family of guests — is stated to prove *selectivity*. As such, the selectivity depends on several factors, including the *complementary*, i.e., when the host and guest are shown to possess both mutual spatially and electronically matching of its own binding sites [10]. The idea of *pre-organization*, first proposed by D.Cram [34], involves that if the binding sites of a host resemble those of a guest molecule, minimal conformational changes are expected. As a consequence, the entropic costs — in terms of a loss of degrees of freedom — are minimized, thus making favorable the overall free energy of the host–guest complex formation. Another remarkable aspect that can reduce the entropic costs is the abovementioned *cooperativity*. Indeed, if two or more binding sites — acting in a concerted fashion — produce an interaction stronger than when the binding sites act independently, we refer to cooperativity effects [35, 36]. For instance, we observe a *positive* cooperativity when the presence of the first species increases the receptor's affinity for the second species and arises in various biological processes and contributes to explain the allosteric effect in enzymes. On the contrary, *negative cooperativity* is the reverse of positive cooperativity and there are only few examples in Nature.

Furthermore, a supramolecular synthesis follows rigid hierarchical rules. *Hierarchy* is a kinetic time-dependent process in which the self-organization of simple elements, following a specific sequence of complexation events, leads to multipart supramolecular architectures. This phenomenon is also called *sergeant–soldier* principle [37–39]: the sequence of the individual molecular components in the final product derives from the sequence of chemical events.

The principles and phenomena described above are the basic concepts of supramolecular chemistry; the union of these phenomena can lead to realize multicomponent supramolecular architectures with desired functionality.

One of the most intriguing properties of the matter found in nature at all levels is chirality. Many biological mechanisms, that are essential to life, involve the use of chiral molecules [40]. In this respect, the chiral recognition of chiral molecules assumes a key role in biotechnology and biochemistry [41, 42]. In addition, chirality has applications in numerous scientific fields from medicinal chemistry to material science [43–45], and sensing [46, 47]. For these reasons, the introduction of chirality into the supramolecular species has been widely studied [48].

Chirality, often termed *asymmetry*, arises in any respect levels, being not only ascribable to single or covalently bound molecules with clearly outlined configuration, but also to weak-noncovalently linked systems where supramolecular forces (i.e., electrostatic and van der Waals forces, steric, solvophobic, hydrogen bonding, π–π stacking, and metal coordination interactions) drive the generation of chiral *super*-structures through a self-assembly process. Further, the usage of weak interactions with opportune reversibility allows for a "*self-error-correcting*" route resulting in thermodynamically stable species during which the asymmetric information is efficaciously transferred/amplified to the molecular assembly [49–52]. This strategy, generally known as *Supramolecular Chirogenesis*, is an astonishing and multidisciplinary area of the chemical sciences whichsinks its roots into the chiral communication among and beyond the molecules [39, 48, 53, 54].

The chirogenesis is a unique process originally involved in many living systems and acting in several biological activities [40, 55]. However, in the last two decades, numerous efforts have been conducted to expand the chirogenesis concepts toward non-biological systems including wide areas of material sciences [56–58] and nanotechnology [59, 60] such as chiral separation and recognition [61–63], sensing [64], photonics [65, 66], catalysis [67, 68], nonlinear optics [69], and nanoscale asymmetric reaction chambers [70–72].

Due to different degree of preorganization and synthetic versatility, large cyclic molecules or macrocycles represent important groups of compounds employed in supramolecular chemistry. Porphyrins, phthalocyanines, and related porphyrinoids constitute significant examples of molecular building blocks to design and realize chiral supramolecular systems [73].

Therefore, the present chapter intends to describe the chirogenesis examples into the porphyrinoid-based self-assembled systems in the last 20 years.

1.2 Porphyrinoids as Ideal Molecular Building Blocks

The term porphyrin derives from the Greek word *πορφύρα* (porphura, meaning purple) and comprises an interesting class of macrocycle rings involved in a wide variety of biological processes ranging from oxygen transport to photosynthesis and catalysis [74–77]. It is not hyperbole to assert that no life can exist on Earth without porphyrins.

In chemical terms, porphyrins consist of four pyrrole subunits linked via methine bridges at the pyrroles' α carbon atoms, as illustrated in Figure 1. According to Huckel's rule, this macrocycle is a planar and aromatic ring containing 22-π electrons, of which only 18 are delocalized ($4n$+2 delocalized π-electrons, where $n = 4$). As a result of this extensive conjugation, porphyrins are intensely colored and strong absorption in the visible zone of the ultraviolet–visible (UV–Vis) spectrum are expected. The porphyrin core is a tetradentate ligand with a diameter of 3.7 Å approximately, suitable for binding most of the transition metals (e.g., Fe, Cu, Zn, Sn, Ni, Mn, etc.) and the stability constant, for some metals, are very high [78].

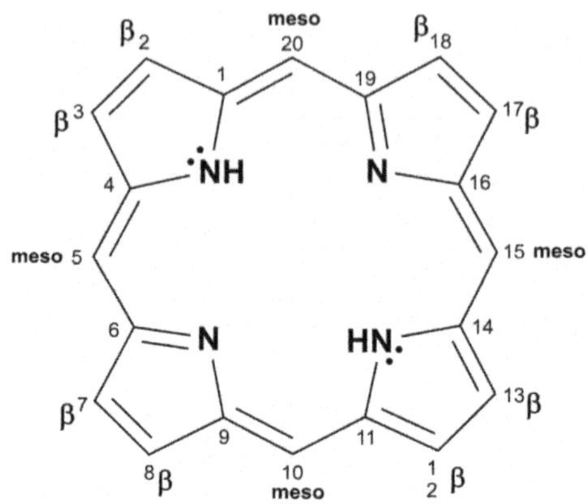

Figure 1. Illustration of a generic porphyrin's ring.

However, the design of tailored porphyrin building blocks is essential in order to achieve fascinating features useful in functional supramolecular aggregates. For instance, (a) hydrophilic or charged substituents can be inserted to increase solubility in water; (b) porphyrins for photovoltaic applications require donor and acceptor moieties at exact positions; (c) in electron transfer based on porphyrin aggregates appropriate redox potential is necessary.

In particular, the *meso-functionalization* is easy, and the corresponding derivatives give rise to the *meso-porphyrin* group. On the other hand, *meso*-substituents can be further functionalized with specific groups or charged moieties in order to confer unique features in the whole macrocycle. Moreover, as abovementioned, metals can be inserted into the porphyrin core leading to the formation of the metallo-porphyrins [79].

It is undeniable that the spectroscopic behaviors are highly dependent on the presence of substituents and/or metals in the porphyrin ring. The large aromatic nature of porphyrins and their derivatives is responsive for the intense absorption properties in the UV–Vis region, mainly owing to π–π^* (HOMO–LUMO) electronic transitions [80].

A typical UV–Vis porphyrin spectrum (Figure 2) exhibits absorptions in two regions, labeled as *Soret* or *B-band* (~380–450 nm) and *Q-bands* (~500–800 nm). The strong *Soret* band is due to the second excited transition state ($\mathbf{S_2 \leftarrow S_0}$), whilst the weak *Q*-(*quasi*-allowed) bands are

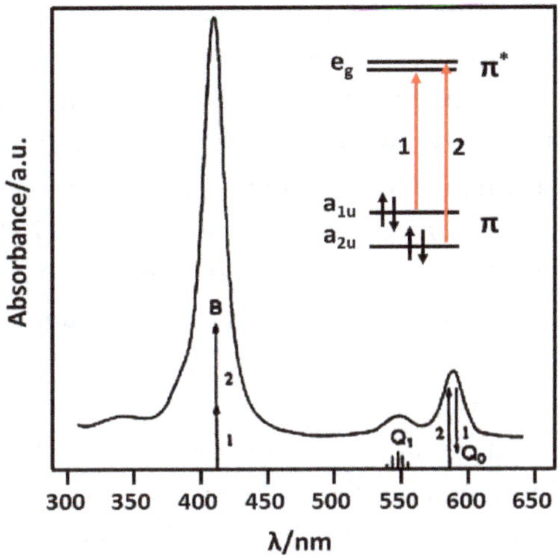

Figure 2. Generic ultraviolet–visible spectrum of a metallo-porphyrin with D4h symmetry. In inset is shown in detail the electronic transition responsible for the absorption bands.

ascribable to the first excited state ($\mathbf{S_1 \leftarrow S_0}$) [80]. Theoretical analysis of the *Soret* band and *Q*-bands have been developed by Martin Gouterman in the 1960s who proposed his semiquantitative *Four Orbital Model* to explain the spectra of porphyrins, a mixture of *Hückel* and *Configuration Interaction* (CI) theory [81]. According to his theory, the main frontier molecular orbitals are four molecular orbitals, two highest energy occupied (HOMOs) and two lowest energy unoccupied molecular orbitals (LUMOs), respectively. In a metallo-porphyrin, with symmetry group *D4h*, the two HOMOs transform as $\mathbf{a_{1u}}$ and $\mathbf{a_{2u}}$, instead, the two LUMOs transform both as $\mathbf{e_g}$ (Figure 2, inset). In accordance with Gouterman's model: (a) the $\mathbf{e_g}$ molecular orbitals are strictly degenerate, whereas (b) the $\mathbf{a_{1u}}$ and $\mathbf{a_{2u}}$ molecular orbitals are *quasi*-degenerate. One may envision that this would lead to two absorption bands of very similar energy due to the $\mathbf{e_g \leftarrow a_{1u}}$ and $\mathbf{e_g \leftarrow a_{2u}}$ electronic transitions. However, by analyzing the UV–Vis spectrum of a hypothetic metallo-porphyrin (Figure 2) we observe an intense band about 400 nm (i.e., the *Soret* or *B*-band) and two weak bands above 500 nm (i.e., Q_0 and Q_1-bands). Gouterman explained this spectroscopic evidence in terms of *configuration interaction* (*CI*) between the two near-degenerate singlet excited states. In particular, when the transition dipoles of the two configurations are constructive, the resulting resonance yields the intense Soret or $B(0, 0)$ band. On the contrary, if the transition dipoles of the two configurations interact deconstructively, the corresponding resonance leads to the weak $Q(0, 0)$ band — namely Q_0 in the spectrum shown in Figure 2. Noteworthy, the Q_0 band is actually forbidden; however, it can "borrow" the intensity by the Soret band through a vibronic coupling, decreasing the Soret band's intensity about 10%. Besides, the higher energy Q-band, namely Q_1, is its vibrational overtone, also denoted as $Q(1, 0)$.

The difference in the positions and the intensities of the bands observed in the various porphyrin and porphyrin-like derivatives, including the metal complexes, is explained by the perturbation induced in the relative energies of the four orbitals by the substituent on the porphyrin macrocycle and/or by the interaction with the metal orbitals. In fact, the inclusion of a metal atom in the porphyrin macrocycle can generate strong electrostatic charge transfer (CT) and/or metal–ligand CT interactions [82].

However, a key difference between the spectra of metallo-porphyrins and their free-base counterparts is the number of Q-bands: a free-porphyrin owns the typical B-band and further *four* Q-bands. This results from the reduced symmetry of the free-base (D2h) so that the two axes defined by opposite pyrrole nitrogen atoms are no longer equivalent affecting further splitting of the electron transitions. In other words, in free-base porphyrin, the $\mathbf{e_g}$ orbitals are further *quasi*-degenerate and more electronic transitions are expected. Consequently, the CI between the new transition dipoles generates four Q-bands.

The intriguing electronic and spectroscopic features discussed so far make porphyrins appealing molecular synthons for building functional multi-porphyrin assembly. In the next paragraphs, these arguments and related topics will be treated.

1.3 Fundamental Theoretical Aspects of the Origin of Chirality in Molecular Aggregates

Although an aggregation process may lead to various types of assemblies, in a more wide-ranging view, of particular interest are those aggregates in which the molecular arrangement is highly ordered, and unique electronic properties are expected. In this sense, pure H- or J-aggregates appear as striking models for understanding molecular interactions in aggregation processes and for technological applications in molecular devices [25, 83]. In J-aggregates ("J-" from E.E. Jelley, the first discoverer [84]), monomers are organized in such a way as to ensure an *edge-to-edge* (or side-by-side) arrangement in one dimension. As such, the monomers' electric dipole moments are parallel and, in the ideal case, the angle between the line connecting the molecular centers and the transition moment is zero [85]. Conversely, H-aggregates ("H-" for hypsochromic, see below) are a one-dimensional *face-to-face* arrangement of strongly coupled monomers. Here, the monomers' electric dipole moments are perpendicular and the angle between the transition moment and the line across molecular centers is 90° (ideal case). These aggregates exhibit remarkable spectroscopic differences in the absorption band compared to the monomeric species [83]. Perhaps, the first significant explanation concerning their photophysical properties dates back to five decades ago, thanks to A.S. Davydov — in molecular crystals [86] — and resumed by M. Kasha in non-crystal molecular aggregate [87].

In comparison with the single monomer, Kasha showed that molecular dimers arranged *face-to-face* — i.e., H-aggregates — exhibit a blue-shifted absorption maximum (hypsochromic effect), whereas dimers stacked *edge-to-edge* exhibit a red-shifted absorption maximum. To explain these shifts, M. Kasha applies the "Molecular Exciton Theory" to molecular dimers resulting in the "Exciton Coupling Model."

The *Exciton Coupling Model* lends itself a quasi-classical vector model by considering a monomer as a chromophore characterized by an electronic HOMO–LUMO transition with energy $\boldsymbol{\Delta E}$, and transition dipole moment $\vec{\boldsymbol{\mu}}$, both of which are independent of the chromophore's position. In a dimer, the single chromophores' transition moments can interact through long-range Coulombic interactions and/or by short-range exchange (i.e., Dexter's electron overlap) [88]. However, the *Exciton Coupling Model* retains only the electrostatic interactions between the two dipole moments in the chromophores' coupling. Therefore, the electrostatic coupling provokes a split in the exciton states, also labeled as exciton splitting, by generating allowed and forbidden electron transitions

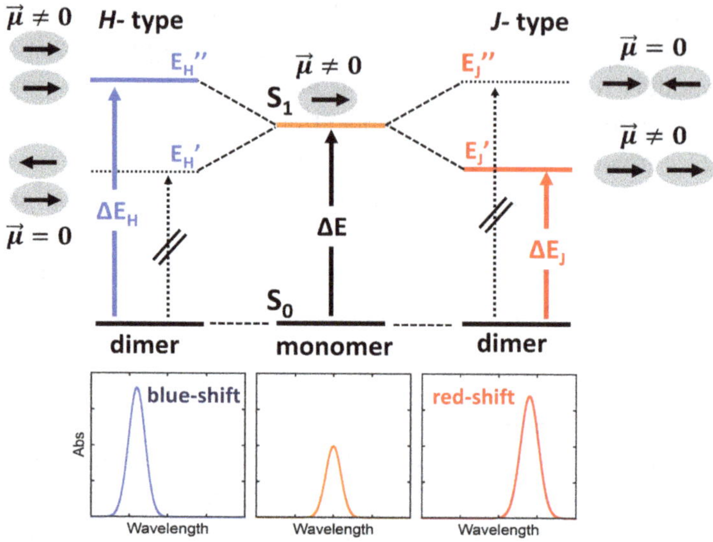

Figure 3. Schematic illustration of the *"Exciton splitting model."* Detailed description is provided in the text.

which depend on geometric arrangements, J- or H-stacking [87]. Figure 3 shows the energy level diagrams of the exciton splitting as a result of the interactions of two identical monomers, in two boundary geometries: (a) parallel arrangement, namely H-type, and (b) inline arrangement, i.e., J-type.

In the case of H-type, when the electric dipoles interact out-of-phase the coupling is electrostatically favored and a lower energy level $\boldsymbol{E_H'}$ is obtained. Otherwise, the in-phase coupling gives repulsion so that the energy level $\boldsymbol{E_H''}$ is shifted upward compared to $\boldsymbol{E_H'}$. Bearing in mind that the transition dipole moment $\vec{\mu}$ is given by the vector sum of the individual transition dipole moments in the molecular aggregates, thus the transitions from the ground state to exciton state $\boldsymbol{E_H'}$ are forbidden ($\vec{\mu} = 0$), whilst transitions from the ground state to higher exciton state $\boldsymbol{E_H''}$ are allowed ($\vec{\mu} \neq 0$) [87]. Consequently, the $\boldsymbol{\Delta E_H}$ transition in H-aggregate is more energetic and a blue-shifted absorption is expected, compared to the electronic transition $\boldsymbol{\Delta E}$ ($\mathbf{S_1} \leftarrow \mathbf{S_0}$) of the single monomer.

Concerning the J-type, from the diagram in Figure 3, it is clearly figured out how the in-phase arrangement of transition dipoles promotes an electrostatic attraction, producing the lower excited state $\boldsymbol{E_J'}$, whereas the unfavorable out-of-phase arrangement leads to repulsion, generating the higher state $\boldsymbol{E_J''}$. Without doubt, the transition moments are finite for the electronic transitions from the ground state to level $\boldsymbol{E_J'}$ ($\vec{\mu} \neq 0$), and not allowed to the state $\boldsymbol{E_J''}$ ($\vec{\mu} = 0$). As a consequence, it will be undeniable that the transitions in J-aggregate will be less energetic and a strong spectral red-shift is observed in respect to single monomer absorption [87].

The exciton splitting energy (often called also "Davydov splitting") corresponds to the separation $\Delta\xi = \boldsymbol{E''} - \boldsymbol{E'}$ as described accurately in Refs. [87, 89]. A characteristic feature of exciton theory is that (a) the exciton splitting energy is directly related to the square of the transition moment for the monomers: thus, greater the intensity of light absorption in the single monomer, greater is the exciton band splitting; (b) the exciton splitting depends on the inverse cube of the intermolecular distance, suggesting major electronic splits in case of closer coupling; (c) the geometrical parameters enter in the manner characteristic of the structure of the molecular aggregates.

Now, if a single chromophore or a molecular aggregate are absolute — or inherently — chiral, they will interact differently with left-hand or right-hand circularly polarized light. The difference absorption in light-handedness is very specific for characterizing both chiral small molecule and supramolecular structures, and it can be revealed by circular dichroism (CD) spectroscopy only in correspondence to chromophores' absorption bands [90].

Nevertheless, when two — or more — asymmetric chromophores are located near in space and have a proper chiral mutual orientation, the electronic coupling occurs (see Kasha's model above), causing in turn, a strong magnetic coupling. As above reported, the splitting of excited states reflects in a split or broadened absorption band, centered around the wavelength transition λ_0 of the isolated chromophore. If the two transition moments are not coplanar, the magnetic moment generated by the oscillating dipole 1 at the end of the vector $\overrightarrow{\boldsymbol{r}_{12}}$ will be nonorthogonal to dipole 2, and vice-versa [90]. Therefore, a bisignate CD couplet is generated around λ_0 as illustrated in Figure 4. Furthermore, the sign of exciton chirality can be easily assessed as follows: for instance, a negative sign is well defined when an anticlockwise rotation from the

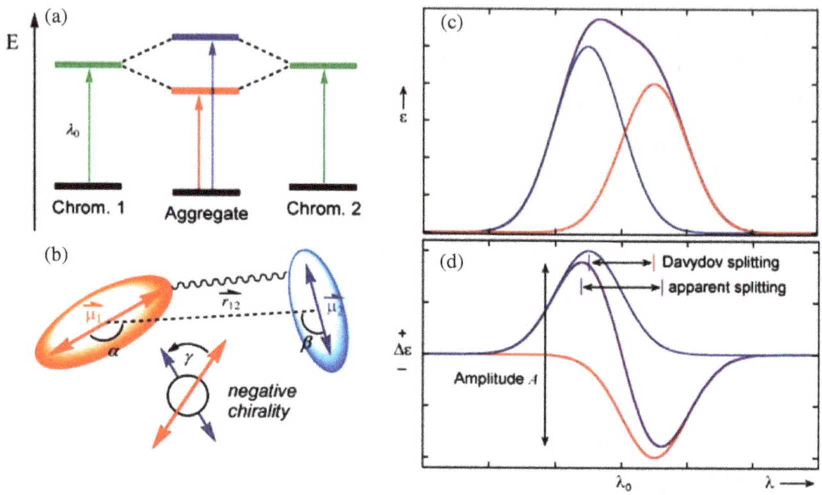

Figure 4. (a) Splitting of the excited states of two degenerate exciton-coupled asymmetric; (b) definition of geometrical parameters necessary for predicting circular dichroism (CD) sign; (c) expected absorption; (d) CD spectra in case of exciton splitting. Modified from Ref. [90].

dipole centers by an acute angle drives the dipole in overlapping (Figure 4b). Besides, it is worth mentioning how the exciton chirality rule states that a positive chirality corresponds to a positive CD couplet and vice versa [90].

Beyond such basic and fundamental theoretical aspects, a clear comprehension of the facts underlying the origin of chirality — i.e., *chirogenesis* — in single molecules and/or multi-assembly structures may allow for the design of systems with appropriate functionalities and applications. For instance, artificial bio-inspired systems borrow their conceptualization from the examination on chiral natural self-assembly such as light-harvesting complex, enzymes, the α-helical structure of proteins, DNA double helix, and so forth. In this sense, artificial chiral super(supra)structures tend to mimic the efficacy of natural chirogenic systems. As such, the success of chirogenicity in a supramolecule depends largely on corresponding building blocks, on what sort of electronic communication occurs and how the chiral information is transferred.

1.4 En Route to Chiral Supramolecular Aggregates: Strategies and Methods

Over the years numerous attempts and strategies have been fruitfully addressed to convey chirality in molecules or into discrete multicomponent systems [48].

Because of the vastness of the subject, it may be advantageous to narrow the field to self-assembly (supramolecular) systems, as regards two classes selected on the basis of the origin of chirality (Figure 5): route (**a**) intrinsically chiral supramolecular aggregates, or route (**b**) induced chiral supramolecular systems.

In the former case, the asymmetric route of the self-assembly process is driven by the presence of chiral groups/moieties/substituents directly bound to monomers [91, 92]. In the latter instance, a final not-symmetric arrangement is realized via hierarchical self-assembly of achiral building blocks by interaction with (a) chiral templates, i.e., polymeric substrates [52, 93–95], surfactants [96–98], chiral ligands [53, 99, 100] and small molecules [101–104], or (b) some external physical factors, i.e., vortices or magnetic fields [105–108] and hydrodynamic directional forces/flows [109–114].

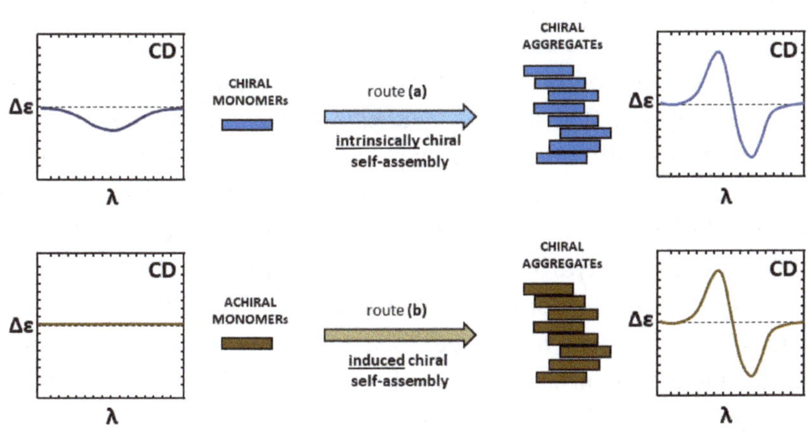

Figure 5. General scheme that displays the two common self-assembly *routes* to realize chiral aggregates from chiral — route (**a**) — or achiral — route (**b**) monomers.

1.4.1 Self-assembly of Intrinsically Chiral Systems

It is commonly recognised the fact that molecules having stereogenic center may assemble into asymmetric molecular aggregates giving rise to so-called intrinsically chiral self-assembly systems. In this sense, CD spectroscopy is a valuable technique to inspect the supramolecular chirality of the aggregates with regard to feasible arrangement adopted by chromophores. However, such transfer of chirality depends on several aspects, and as a result, it is not always quite so straightforward. The relative force of the noncovalent interactions involved or the mutual distance of chiral centers into the aggregate are just some examples to name. Furthermore, the chirality transfer, from single chiral molecules to supramolecular arrangements, will also depend largely on the structure of the building block considering the requirement to reach an asymmetric spatial that minimizes the total energy of the system.

As reported in the introduction among numerous chromophores, porphyrinoid compounds constitute the most highly investigated systems, owing to their exceptional ability to self-assemble. In this respect, we are well aware of the fact that the literature is remarkably extensive. Therefore, herein we intend to report only some relevant works based on porphyrinoids which demonstrate how to build asymmetric final supramolecular structures starting from chiral porphyrin monomer.

Most likely the first case of chiral supramolecular structures generated by the aggregation of chiral dyes was reported in 2003 by Balaban and coworkers [115]. The authors synthesized two enantiomers based on Zn(II)-porphyrin with different groups in the β positions (acetyl and hydroxyethyl) and in the 5–15-*meso* positions (3, 5-di-*tert*-butylphenyl groups) as illustrated in Figure 6a. Due to an extensive hydrogen-bonding network between the hydroxyl group, axially coordinated to Zn-core, and the carbonyl group of another porphyrin ring, self-aggregation at room temperature occurred in *n*-heptane solvent for both enantiomers. As a result, an intense dichroic signal in the porphyrins' absorption band is revealed (Figure 6b). Noteworthy, the different chirality (R or S) of the hydroxyethyl substituents, on the β position, induces mirror-conformation in the supramolecular helical assemblies. Moreover, the addition of methanol to preformed aggregates in *n*-heptane competes for the zinc ligation,

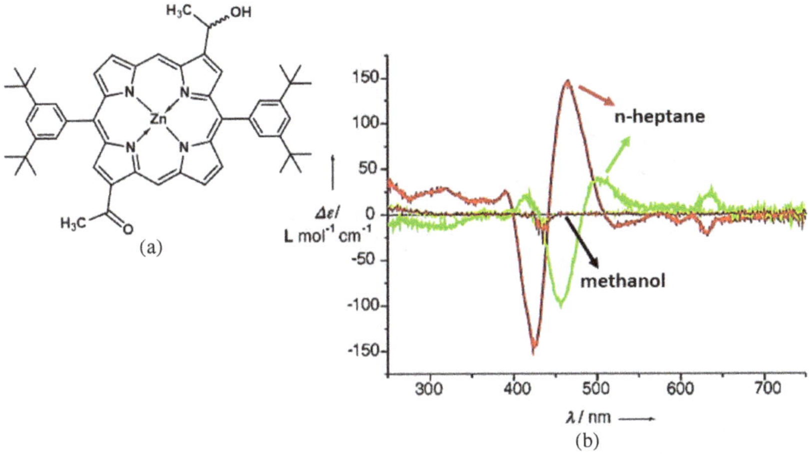

Figure 6. (a) Molecular structure of Zn(II)-porphyrin used in Balaban's work. (b) Circular dichroism spectra of the enantiomerically enriched (R)- and (S)- Zn(II)-porphyrin after self-assembly in *n*-heptane (thick red and green lines) and after addition of methanol which disrupts the assemblies (thin red and green lines overlapping the 0 line). Modified from Ref. [115].

leading to a disaggregation as demonstrated by the loss of both enantiomers' dichroic signals.

The lack of well-defined CD signals for the monomeric porphyrin forms clearly highlights the following facts: (a) the self-assembly of single asymmetric monomers can lead to enhanced chiral amplification resulting in a sort of supramolecular chirality; (b) the handedness of specific supramolecular aggregates is independent by the stereoisomerism of its monomer.

Could the initial molecular structures of monomers affect the corresponding aggregate's final features, including chirality? This effect was investigated by Meijer *et al.* which proposed the employment of chiral oligo(*p*-phenylenevinylene)-appended porphyrins containing *trans*-vinylene or amide connections (see red-marked linkage in Figures 7a and 8b, respectively) [116].

The comparative study of the self-aggregation process demonstrated that the chirality and the arrangement of the assembly strongly depend on the nature of the covalent linkage. In particular, the conjugated pendants

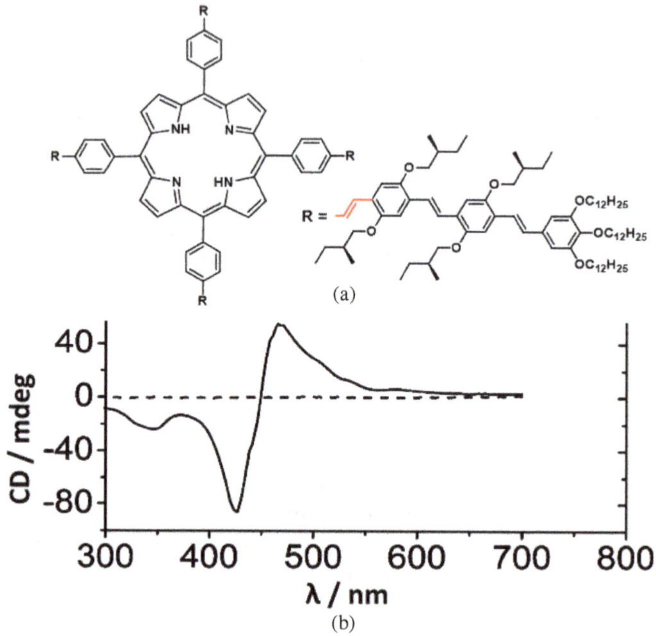

Figure 7. (a) Molecular structure of chiral oligo(*p*-phenylenevinylene)-appended porphyrin conjugated with *trans*-vinylene moieties; (b) circular dichroism spectrum of its self-aggregate in methylcyclohexane solution (solid black line). Modified from Ref. [116].

for the *trans*-vinylene porphyrin derivative (Figure 7a) tend to stack by π–π interactions in methylcyclohexane environment, resulting in a J-type arrangement with a typical bisignate Cotton effect (Figure 7b).

Conversely, in the system with amide-linked pendant (Figure 8a), the favorable intermolecular hydrogen bonding toward the piling direction suggest an H-type arrangement of higher stability with a characteristic blue-shifted exciton splitting dichroic signal (Figure 8b).

Monti and coworkers reported an elegant investigation about the solvodichroic effect on chiral porphyrin aggregation [117]. The system object of study was a chiral functionalized porphyrin having an l-prolinium group in 5-meso position (Figure 9a). Between 50% and 100% of the ethanol/water mixture ration (v/v), the dye remains into monomeric form; however, below the 50% v/v the self-aggregation occurs as a result of the increase of water. The authors noted the chiral self-assembly process strictly depends

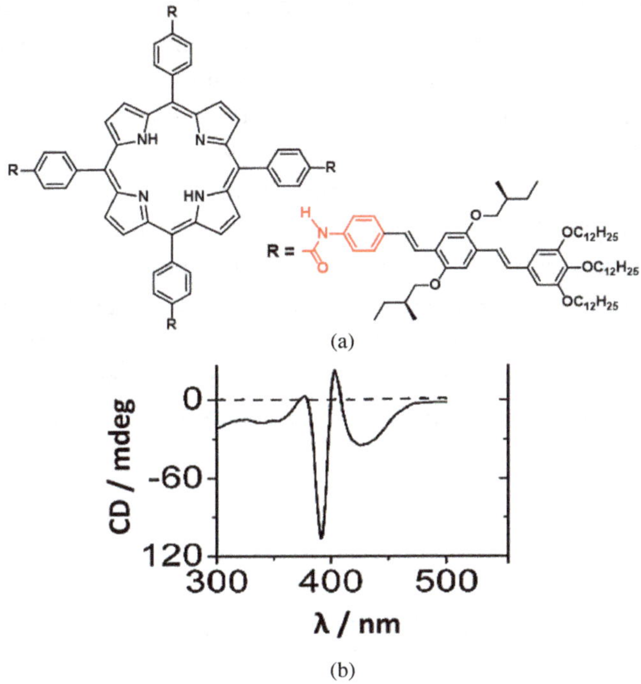

Figure 8. (a) Molecular structure of chiral oligo(*p*-phenylenevinylene)-appended porphyrin conjugated with amide moieties; (b) circular dichroism (CD) spectrum of its self-aggregate in methylcyclohexane solution (solid black line). Modified from Ref. [116].

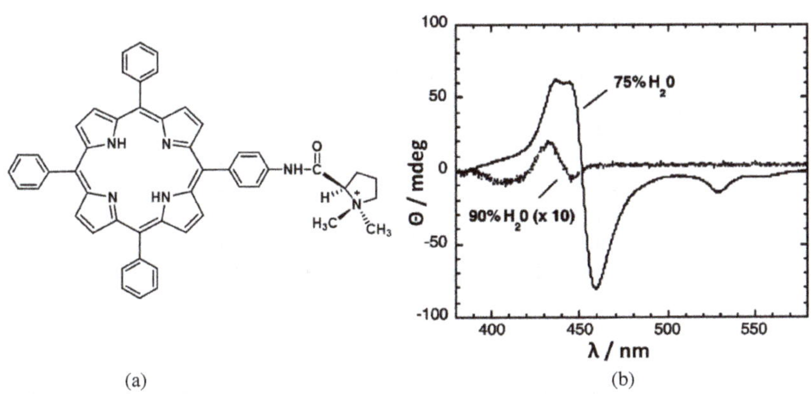

Figure 9. (a) Molecular structure of porphyrin used in the Monti *et al.*'s work; (b) circular dichroism spectra of the system at different water/ethanol mixture compositions (90/10 and 75/25 v/v). Modified from Ref. [117].

on the composition of mixture solvent as revealed by CD spectra of the corresponding aggregates. In fact, at higher water composition (>90%) only a small signal is observed, which, however, becomes two magnitudes larger if the water composition is reduced to 75% (Figure 9b).

An explanation of this phenomenon was due to kinetic studies [117]: the self-assembly process evolves quickly at higher presence of water resulting in aggregates less structured. Conversely, a major amount of ethanol leads to slow down the aggregation kinetic entailing the formation of more organized J-type aggregates with an intense bisignate Cotton effect.

Another example of solvent effect inducing aggregation of chiral systems was reported by Zhang *et al.* [118]. They used an optically active Zn(II)-porphyrin conjugated with a pentapeptide (Figure 10a). In a

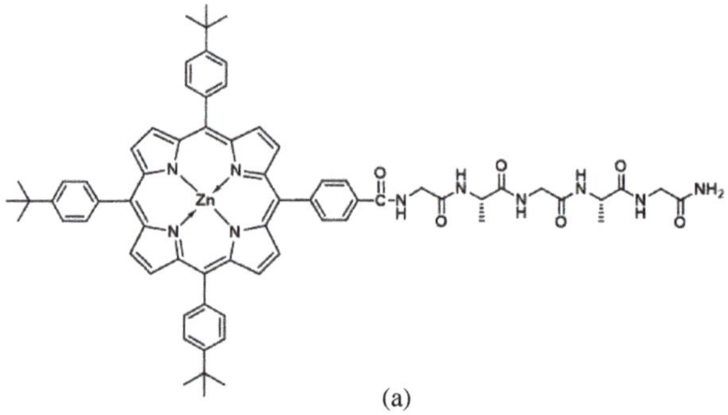

(a)

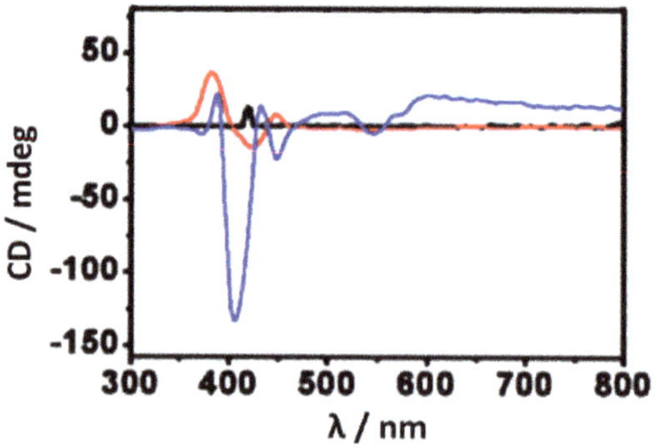

(b)

Figure 10. (a) Molecular structure of Zn(II)porphyrin bearing pentapeptide; (b) circular dichroism (CD) spectra of the monomeric porphyrin (black line) and in the aggregated system at different mixture solvents: (a) THF/*n*-hexane 1:3 (red line) and (b) THF/water 1:3 (blue line). Modified from Ref. [118].

nonpolar mixture (THF/*n*-hexane 1:3), the system spontaneously aggregates in a right-handed helix with a typical positive bisignate dichroic signal. On the contrary, the same porphyrin in a polar mixture (THF/water 1:3) self-aggregates with a left-handed helical arrangement resulting in two negative bisignate Cotton effects (Figure 10b). The choice to employ a peptide-conjugated porphyrin comes from basic evidence: peptides self-organize in α-helix in nonpolar solvent and antiparallel β-sheet secondary conformations in a polar mixture. As a consequence, in the THF/*n*-hexane mixture, the self-assembly of hydrophilic peptide chains in the right-handed helix drives the organization of the corresponding linked hydrophobic porphyrin macrocycles into the same conformation. On the contrary, in polar THF/water mixture the peptide chains aggregate in β-sheet conformations resulting in dimeric structures which, afterward, self-assemble into left-handedness arrangement.

As final example, it is worth mentioning the chiral self-assembly reported from Monsù Scolaro *et al.* [119] of two enantiomeric forms of 5, 10, 15, 20-mesotetrakis-(2, 6-dimethylheptyl)porphyrin, i.e., a *meso*-substituted porphyrin for chiral-hydrogenated citronellal groups (Figure 11a). Although the chiral citronella units are directly linked to macrocycle, in dichloromethane solvent no asymmetrical perturbation are transferred to the porphyrin core as clearly by the absence of induced CD in porphyrins' Soret region. In other words, in a hydrophobic environment, both enantiomers preserve their monomeric form, further confirmed by their monoexponential trend with a long-living lifetime in emission decay measurements, characteristic for not-aggregated forms [119, 120]. However, in the acetone/water mixture (40/60 v/v) the presence of aggregated species was first indicated by spectroscopic investigations, resulting in the splitting of the Soret band — ascribable to H- and J-type stacking — arrangements, including DLS and emission decay experiments too (herein data not shown).

This latter behavior can be elucidated with regard to the hydrophobic of the alkyl citronella chains which encourage the self-assembly process in a more hydrophilic environment. Remarkably, the phenomenon is chirally driven since the CD spectra obtained in the same conditions (acetone/water, 40/60 v/v) exhibit a negative and positive bisignate Cotton effect for S and R enantiomers, respectively (Figure 11b).

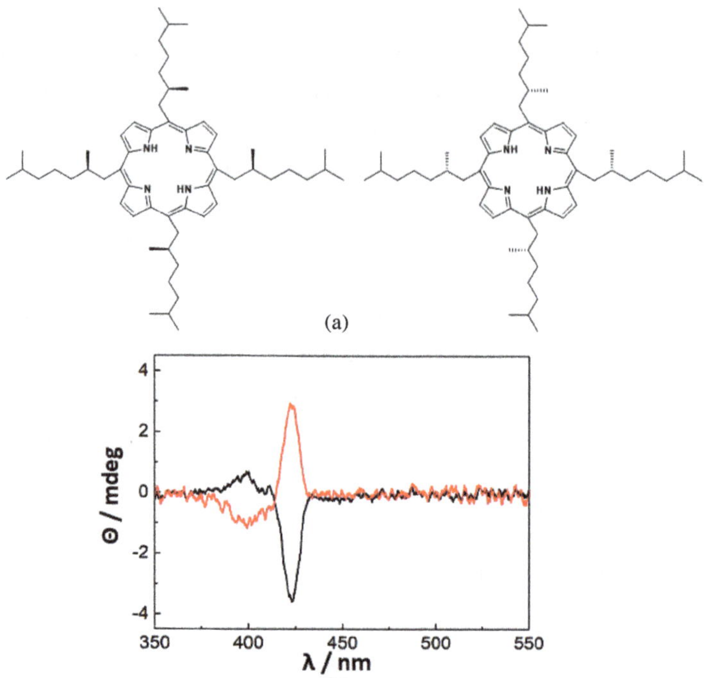

Figure 11. (a) Molecular structure of R and S enantiomers used in the work; (b) circular dichroism spectra for R (red line) and S (black line) enantiomers in the mixture acetone/water (40/60 v/v). Modified from Ref. [119].

Another captivating self-assembly of intrinsically chiral systems was reported by Amabilino *et al.* [92] by designing chiral *meso*-substituted Zn(II)-porphyrins (Figure 12). In principle, both the ability of zinc to coordinate with axial ligands containing nitrogen or oxygen and an extensive network of hydrogen bonds allow building well-organized self-superstructures (Figure 12a). The self-assembly of a Zn(II)-porphyrin (Figure 12b) was investigated by UV–Vis and CD spectroscopies at different temperatures in order to obtain information about the electronic transitions' coupling and, as a consequence, the adopted orientation of the chromophores into the self-assembly. For instance, the temperature dependence of the CD spectra of the compound (Figure 12c) clearly shows two distinct states of self-assembly. A weak Cotton effect arises at the position of the band associated with the pyridyl-Zn(II)-porphyrin

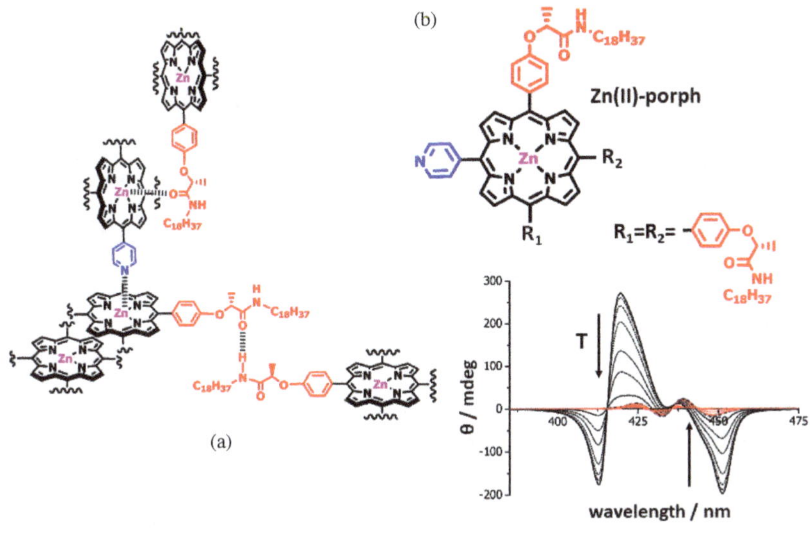

Figure 12. (a) Illustration of some potential porphyrin aggregates by an extensive noncovalent network used in the cited work; (b) molecular structure of the chiral porphyrin from Amabilino *et al.*'s work; (c) circular dichroism spectra at different temperatures (from 338 K to 263 K) of the chiral *meso*-substituted Zn(II)-porphyrin (5 μM) in methylcyclohexane/3% $CHCl_3$ solvent. The temperature of each spectrum increases toward the direction of the arrow. Modified from Ref. [92].

complex, suggesting a perpendicular disposition of the two metalloporphyrins and a thermodynamically stable aggregate in the range of temperatures from 300.5 K to 338 K (Figure 12b, red curves). When the temperature is decreased, a highly intense positive Cotton effect and a huge negative CD signal indicate extension of the interchromophore interactions; thus, the evolution from aggregate to a larger superstructure is expected (Figure 12b).

Owing to high self-assembling tendencies of both enantiomers of diphenylalanine peptides, namely F_DF_D and F_LF_L, into tubular or spherical forms [121, 122], recently Balaban and coworkers reported the synthesis of two enantiomers of *meso*-tetraphenylporphyrin (TPP) dipeptide conjugates by covalently linking to the C-terminus of diphenylalanine (Figure 13a). Such chiral porphyrinoid systems self-assemble from fibrils to quasi-spherical agglomerates depending on the solvent composition, in

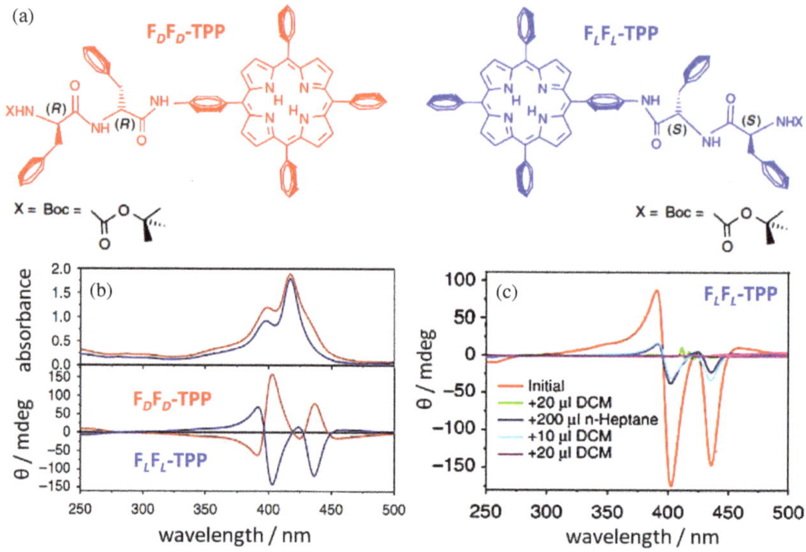

Figure 13. (a) Molecular structures of the two porphyrin-diphenylalanine conjugate enantiomers (FF-TPP); (b) ultraviolet–visible (Top) and circular dichroism (CD) spectra (Bottom) of the self-assemblies obtained by injecting FF-TPP enantiomers from a stock solution in pure dichloromethane into dry *n*-heptane; (c) CD spectra of the self-assemble system from F_LF_L-TPP into *n*-heptane (red curve) by adding small amounts of dichloromethane.

terms of mixture ratio of dichloromethane to *n*-heptane [123]. That behaviour can lead to the transfer of chirality from molecular to the supramolecular level, switching as well, the chiral information just changing the polarity of the solvent. Indeed, in dichloromethane, the porphyrin conjugates arise in monomeric form; however, after "injecting" into a nonpolar *n*-heptane solvent, they observed broadened, red-shifted, and split Soret bands typical of porphyrinic J-aggregates (Figure 13b, top). After self-assembly, a very intense and multiple Cotton effects prove that the porphyrin chromophores are excitonically coupled with a strong dichroic signal centered at 403 and 437 nm (Figure 13b, bottom). Upon addition of pure dichloromethane to the solvated self-assemblies into *n*-heptane, the aggregates get destroyed gradually and a quasi-silent ECD spectrum is obtained (Figure 13c). Noteworthy, a restoration of exciton couplets is accomplished right after the further addition of *n*-heptane, providing an efficient switchable self-assembling chiral system.

1.4.2 Chiral Supramolecular Aggregates by Achiral Building Blocks

So far, we discussed the supramolecular chirality acquired as a result of chiral amplification of self-arrangement of building blocks having their own asymmetry, with particular regard to porphyrinoid systems.

However, a self-assembled aggregate can exhibit supramolecular chirality even though exclusively made up of achiral components, by a phenomenon well-known as "induction of chirality." In fact, induced chirality refers to chiral supramolecular systems in which chirality is transferred to achiral monomers via external interactions with either chiral molecular templates or asymmetric factors [48].

Notable, in induced chiral systems, an aspect strictly correlated with the transfer of chirality is the so-called "chiral memory." Basically, the memory of chirality describes a supramolecular arrangement in which the chiral information is induced, stored, and maintained also when the pristine chiral source is either removed or suppressed or replaced. Thus, the success to convey chirality in a supramolecular system depends on how much the resulting aggregate is able to preserve the induced asymmetry over time. In fact, the noncovalent assemblies are susceptible to disaggregating — losing the *imprinted* chiral information — by changing those factors and conditions (i.e., solvent, pH, ionic strength, temperature, etc.) which can move the intermolecular forces, amongst the molecular blocks, from the minimum of energy reached [48, 124].

In recent years, transfer and memory of chirality have been becoming a cutting-edge field as the synthetic steps to introduce chiral groups to realize a given asymmetric superstructure may be overcome by an easy "mix and shake" of appropriate achiral monomer units with more accessible chiral factors, whether they are molecules or chiral external forces. In order to successfully build a so intriguing system some features have to be taken into account: (a) make sure to have efficient and available chiral inductors; (b) the induced chiral nanostructure should be stable enough (thermodynamically and/or kinetically); (c) the memorized chiral information should be maintained after chiral inducer's removal.

In this respect, strong J- or H-type aggregates from self-assembly of achiral porphyrinoids meet all those properties required and discussed so far, and for that reason, now we intend to report some relevant works.

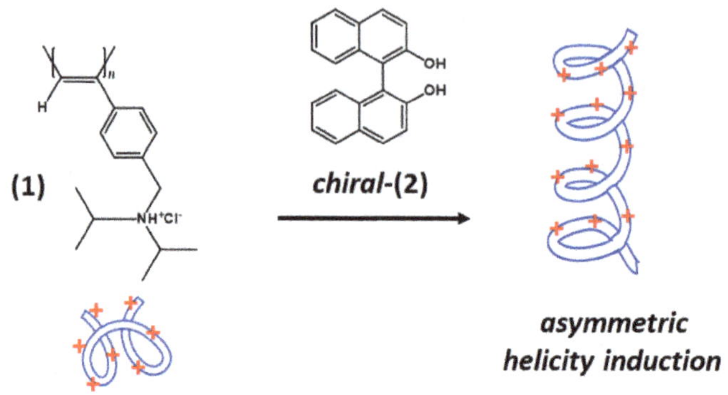

Figure 14. Asymmetric helicity induction in optical inactive **1** (positively charged poly-phenylacetylene) by interaction with chiral guest **2** (R- or S-1, 1′-bi-2-naphthol).

Yashima and coworkers synthesized a stereoregular and positively charged poly(phenylacetylene), namely poly(4-(N, N-diisopropylaminomethyl)phenylacetylene) hydrochloride (Figure 14, structure **1**), soluble in water but optically inactive. However, in aqueous medium the interaction with several oppositely charged amino acids or chiral acids induces an asymmetric helical winding into the polymer backbone, arranging the cationic pendants outward [125]. Afterward, they discovered that the polymer can also entrap hydrophobic chiral guest, such as chiral atropisomeric 1, 1′-bi-2-naphthol (Figure 14, structure **2**), resulting in an intense induced CD (Figure 14) [93].

The asymmetric induced helix can act as a template for further supramolecular assembly by cationic pendants arranged along the outside of the polymer. Indeed, in acidic condition the achiral tetra-anionic *meso*-tetrakis(4-sulfonatophenyl)porphyrin, H_4TPPS^{2-}, aggregates onto the poly(phenylacetylene) surface giving rise to homo J-aggregates characterized by an induced dichroic signal in the Soret region (Figure 15) [93]. The binary complex, poly(phenylacetylene)/H_4TPPS^{2-}, is also able to "store" the chiral imprinting even after the loss of chirality for the polymer (e.g., by introducing an excess of opposite enantiomer **2** of Figure 14). This case represents the first example of chiral memory in a binary complex from porphyrin homoaggregates as compared to ternary ones — based on porphyrin heteroaggregates — reported by Purrello *et al.* [126–128].

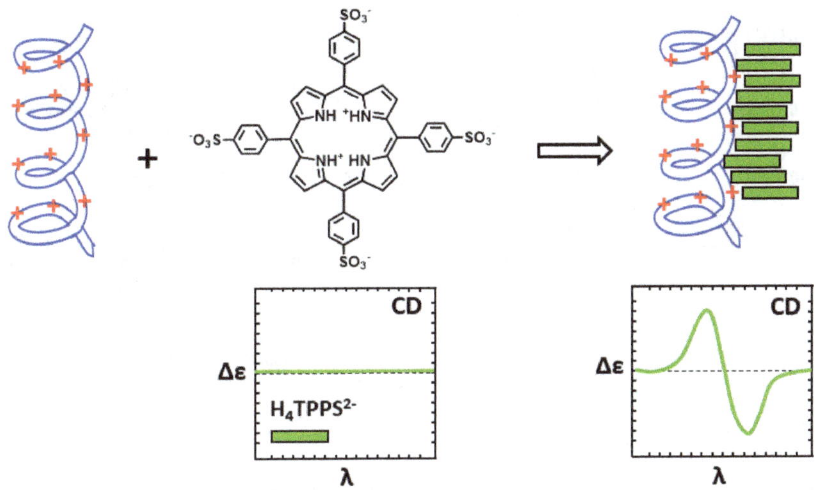

Figure 15. The supramolecular interaction of achiral H_4TPPS^{2-} with asymmetric helical poly-phenylacetylene induces optical activity in porphyrin Soret region.

Another very interesting examples concerning the aggregation of the tetra-anionic H_2TPPS^{4-} onto polymeric templates were published by Professors L. Zhang and M. Liu in 2009 [129]. They investigated the interaction of the protonated, zwitterionic form of H_2TPPS^{4-} (H_4TPPS^{2-} with L- and D-poly-lysine (PLL and PDL, respectively), observing spontaneous porphyrin self-assembly along the groove of PLL (or PDL) into the H- and J-type aggregates.

Interestingly, the induced exciton couplet for H-band always keeps a negative coupling — according to the chirality of PLL (or PDL), whereas the sign of the exciton couplet in the J-band ($\lambda \approx 490$ nm) can have the same or opposite signs to that of the H-band ($\lambda \approx 420$ nm), depending on the mixing sequence and also the ratio of PLL (or PDL) to H_4TPPS^{2-} (P/T ratio). Thus, that suggests a different arrangement of J-aggregate with a distinct transition moment's direction resulting in its inversion of chirality. In particular, if PLL is added into H_2TPPS^{4-} water solution (pH = 3.1) a negative couplet will be for the J-band (Figure 16a, black curve) with an identical sign for H-band; conversely, if the mixing order is inverted (H_2TPPS^{4-} to PLL water solution), a positive couplet will be obtained, having opposite sign to that H-band (Figure 16a, red curve).

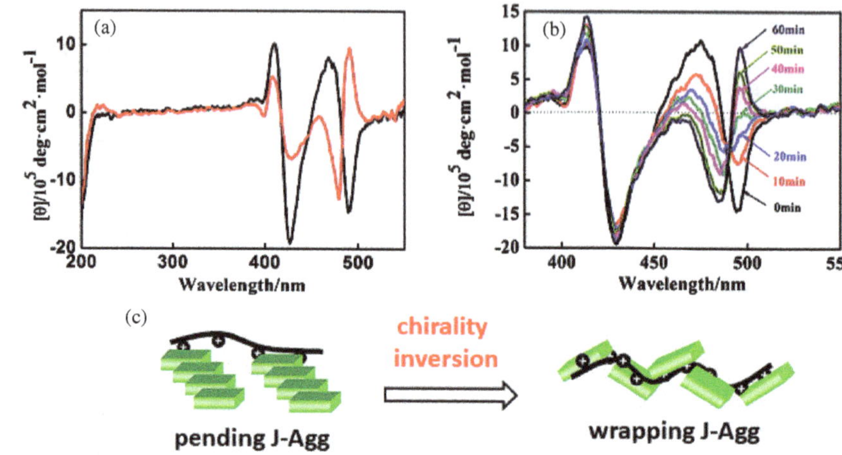

Figure 16. (a) Induced circular dichroism (CD) of H_2TPPS^{4-} in presence of PLL with different mixing order (PLL-to-H_2TPPS^{4-}, black curve; H_2TPPS^{4-}-to-PLL, red curve); (b) Time-dependent CD spectra upon heating of PLL-to-H_2TPPS^{4-} system; (c) cartoon illustration of the pending-type J-aggregates, the wrapping-type J-aggregates, and their chiral inversion under heating. Modified from Ref. [129].

The authors investigated the thermodynamic stability of J-aggregate by elevating the temperature from room condition to 35°C in order to explain such hierarchical order. When H_2TPPS^{4-} is added into PLL no changes in the CD spectra are observed. On the contrary, if PLL is added to H_2TPPS^{4-}, significant variations in J-band are displayed over time by heating to 35°C (Figure 14b). Noteworthy, after 60 min a complete chiral inversion in J-band is reached further kept when the temperature was cooled down to room condition. Authors clarified the phenomenon in terms of major thermodynamic stability for J-aggregate obtained from H_2TPPS^4-to-PLL mixing order (and not *vice versa*). They invoked the self-assembly model illustrated in Figure 16c: (a) in the PLL-to-H_2TPPS^{4-} mixing sequence, due to locally overconcentrated condition, the porphyrin tends to assemble on several limited binding sites of PLL, giving the "pending" J-agg, in which one site of porphyrin binds on the PLL, while the other units stacked on the first in a head-to-tail manner; (b) in the H_2TPPS^{4-}-to-PLL mixing order, PLL is in large excess and as a consequence, porphyrins prefer to bind at least two sites along the chain, resulting in "wrapping" J-agg. Notable, these latter J-agg are thermodynamically more stable and for this reason, upon heating, we observe a rearrangement of porphyrin units from pending-type to wrapping-type aggregates entailing a chirality inversion (Figure 16c).

The effect of both preorganization of the porphyrin units and template chain-length in inducing supramolecular chirality was further investigated by D'Urso and coworkers a few years later [130]. For the investigations, they chose the anionic zinc(II) derivative of the *meso*-tetrakis(4-sulfonatophenyl)porphyrin, ZnTPPS, to interact with cationic PLL at neutral condition in aqueous solution. The ZnTPPS porphyrin is a metallo-porphyrin which undergoes a demetalation/protonation process, in water at acidic conditions, toward the aforementioned zwitterionic form, H_4TPPS^{2-} [131, 132]. However, as such, the supramolecular interaction ZnTPPS/PLL slows down the demetallation rate of ZnTPPS, which starts only at pH below 1.5 [131].

Moreover, throughout demetalation/protonation stage, PLL "catalyzes" the aggregation of zwitterionic porphyrin along the groove of the polypeptide, giving the so-called H- and J-aggregates [131]. Nonetheless, by supporting PLL chains serve as skeleton to self-organization, authors supposed that different degrees of polymerization (dp) may be affected the final arrangement of J-aggregate. In order to confirm their hypothesis, a water solution containing system ZnTPPS/PLL with different poly-L-lysine chain-length, dp 36 and 2060, respectively, was incubated for 3 h at strongly acid values (pH = 1.5). As a consequence, the resulting J-aggregate templated onto long PLLs chain-length display an enhanced exciton couplet compared to the shorter ones, confirming as well a structural evolution of the supramolecular aggregates into extensive and more ordered structures (Figure 17) [130].

Deep spectroscopic investigations allow authors for speculating these findings in terms of spatial preorganization of ZnTPPS units along the polypeptide chains and their kinetic evolution, then decreasing the pH value [130]. In other words, short chains promote the self-assembly of ordered dimers [130, 133], shifting toward diffused random aggregates upon increasing the PLL chain-length. Hence, owing to their "labile" nature, ZnTPPS dimers are involved in a fast demetallation/protonation process, yielding rapid families of less-organized J-aggregates; otherwise, random aggregates (templated onto longer PLL chain-length) are

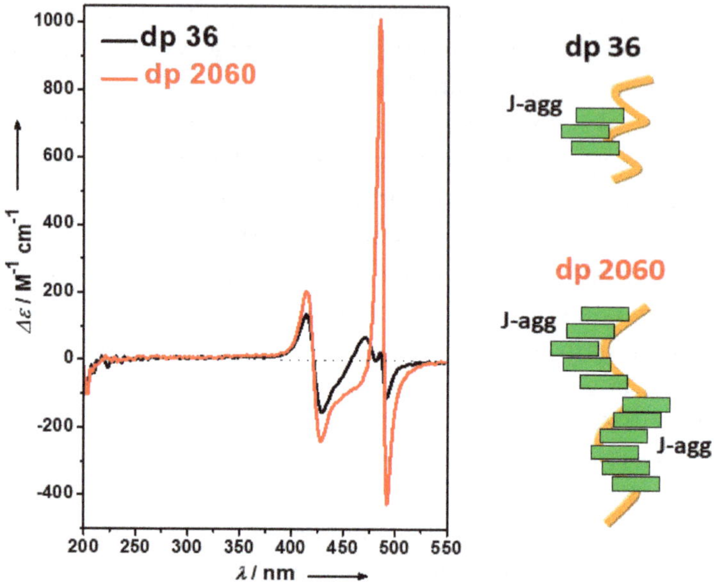

Figure 17. Induced circular dichroism spectra of ZnTPPS in presence of different PLL chain-length (dp = 36, black curve; dp = 2060, red curve) incubated in water solution at pH = 1.5 for 3 h. The cartoon on the right shows the supposed organization of J-aggregates onto PLL dp 36 and 2060, respectively. Modified from Ref. [130].

susceptible of slower kinetic evolution giving rise to long and more-ordered J-aggregates (see scheme in Figure 18).

These latter works clearly evidence the polymeric matrices as templates are able to guarantee a fine control into the chirality of a self-assembled system. However, the property in inducing supramolecular chirality might further be due to small molecules having point chirality, such as amino acids. In this respect, L. Zeng *et al.* reported chiral homoaggregates from H_2TPPS^{4-} induced by interactions with D- and L-alanine (Ala) in acidic water [134]. The mechanism of supramolecular chirogenesis in such a system entails an electrostatic binding between protonated amino acids (cationic at low pH) and negatively charged benzenesulfonate pendants of H_2TPPS^{4-}. Nevertheless, Ala may appear as an efficient chiral inducer owing to its small methyl side group which does not hinder the coulombic interactions by hydrophobic steric hindrance.

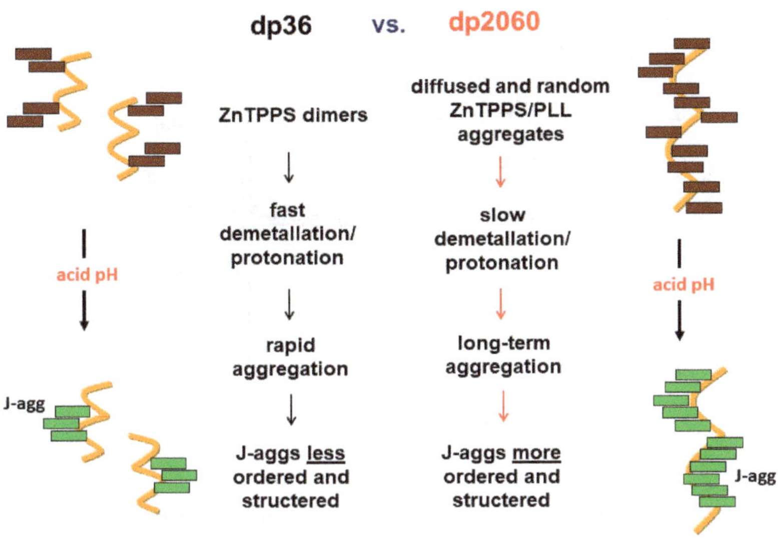

Figure 18. Figurative illustration about the different preorganization and kinetic evolution of ZnTPPS self-assembly in presence of various PLL chain-lengths, dp 36 and 2060, respectively.

As a result, in acidic aqueous solution, J-aggregate assembled from the zwitterionic form of H_2TPPS^{4-} exhibit an intense induced chirality when the chiral inducer (D- or L-alanine) is first introduced into the system. Notable, the transition moment coupling of porphyrin units rises with alanine concentrations, highlighting as well, how Ala may be a versatile and sensitive inducer of chirality for supramolecular systems (Figure 19) [134]. Furthermore, the authors uncovered a negligible J-agg dichroic signal when the Ala concentration is below 0.5 mM. Therefore, a certain minimum number of inducers are essential to drive the chirality of the entire process. Thus, the supramolecular chirogenesis here follows a typical "sergeant and soldiers mechanism" [52, 135, 136].

Positively (or negatively) charged amino acids do not self-assemble in water because their ionic repulsions prevent that. Nevertheless, aromatic amino acids can exist as large aggregates having an intrinsical chirality; for instance, the phenylalanine (Phe) generates aggregates (≈60 nm) up to 30°C [137].

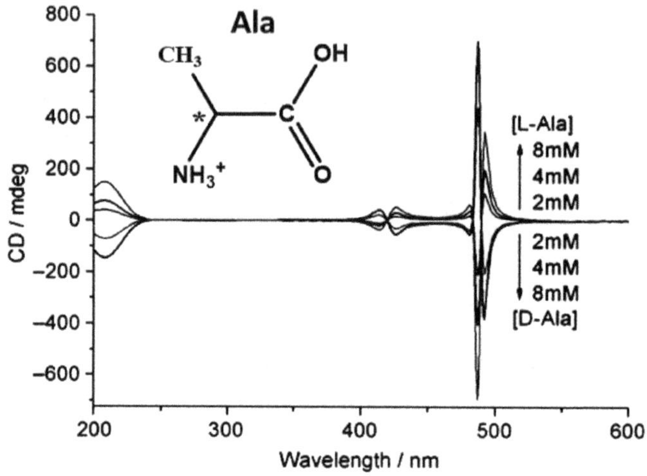

Figure 19. Circular dichroism spectra at pH = 1.1 of J-aggregate from H_2TPPS^{4-} (10 μM) in presence of increasing amount of D- (negative exciton-coupled) and L-alanine (positive exciton-coupled). Inset: molecular structure of protonated alanine. The star indicates its stereogenic center. Modified from Ref. [134].

Hence, may amino acids chiral clusters be used as templates to convey chirality in molecular aggregates? Based on their previous studies about porphyrin heteroaggregation in water [126, 128], Purrello *et al.* proposed a model for imprinting chirality to H_2TPPS/CuT4 aggregates in presence of L- (or D) phenylalanine cluster (Figure 20a) [137]. Notably, the authors exploited the well-known kinetic inertia of porphyrin heteroaggregates to "memorize" the chiral information after phenylalanine removal: the induced CD signal remain almost unaltered to that imprinted aggregates (Figure 20b).

Some years later, in 2007, they take advantage of the remarkable stability of heteroaggregates to design a supramolecular system in which the chiral information can alternatively be induced, memorized, erased, and restored again more and more times, by realizing a fascinating supramolecular memory system [138].

The issue of supramolecular chirogenesis in a self-assembly system and its transmission of chirality across the length, from nano- to mesoscopic scale, has been addressed by Monsù Scolaro's group for several years [139–141]. For instance, they reported the usage of ternary

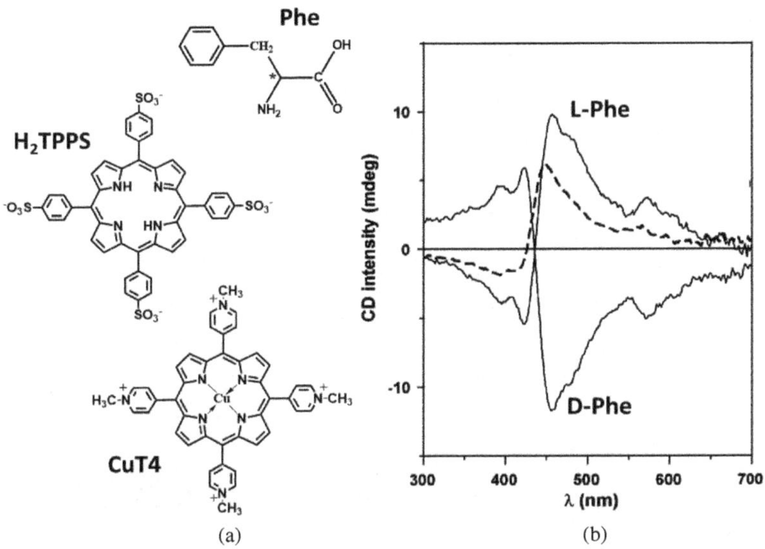

Figure 20. (a) Molecular structures of Phe, H_2TPPS, and CuT4; (b) induced circular dichroism (CD) spectra in water of H_2TPPS/CuT4 porphyrin heteroaggregates in presence of L- and D-phenylalanine. The dashed black plot refers to chiral imprinted heteroaggregates after the removal of L-Phe. Modified from Ref. [137].

reverse microemulsions, based on sodium bis(2-ethylhexyl)sulfosuccinate (AOT)/decane/water, as nanochamber to control the growth of J-aggregates from protonated H_2TPPS^{4-} by changing the ratio W_0 = $[H_2O]/[AOT]$ and under slightly acidic conditions (see illustration in Figure 19) [142]. They also demonstrated a direct matching between the hydrodynamic radius of micelles, coherence length in J-aggregates, and photophysical properties. Therefore, they suppose to "scale" the supramolecular chirality with the size of aggregates, from a few up to hundreds of nanometers, by introducing a chiral inducer throughout the process [143]. As such, the addition of D- (or L-) tartrate reveals the appearance of the two typical CD exciton splits ascribable to J-band around 490 nm and H-band at 420 nm (Figure 21a). The plot of the dissymmetry g-factor ($g = \Delta\varepsilon/\varepsilon$) versus hydrodynamic radii strengthened the hypothesis to scale the chirality, by observing a well-evident linear scaling law even coherent with J-aggregates obtained in bulk (Figure 21b) [143].

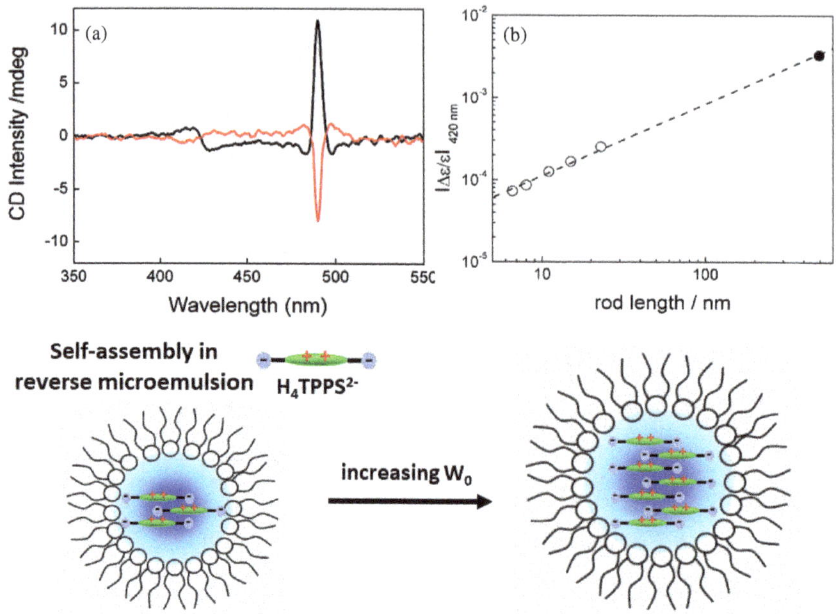

Figure 21. (a) Circular dichroism (CD) spectra (pH = 2.7) of J-aggregates from H_2TPPS^{4-} within reverse microemulsions in presence of D- (red line) or L-tartrate (black line); (b) double logarithmic plot of the absolute value for the dissymmetry *g*-factor ($\lambda = 420$ nm) versus J-aggregates size inside the micelles (empty circles), or in bulk (filled circle). A figurative example of self-assembly in reverse microemulsions is illustrated in the lower part. Modified from Ref. [143].

More simply, the supramolecular chirality in molecular aggregates can be successfully induced, propagated, and fine-tuned, from nano- up to mesoscopic level, by confined environments.

The aggregative process so far reported represent a compromise between thermodynamic and kinetic factors: the intermolecular forces push monomers to self-assemble for minimizing the system's free energy in a certain kinetic instant. In other words, the minimum of energy may not be the absolute one, so that some thermodynamic parameters (solvent, temperature, pH, ionic strength, mixing order, etc.) shift the aggregative kinetics toward a new relative (kinetic) minimum and so forth. Hence, being the supramolecular chirality's expression an asymmetric exciton coupling, its chirogenesis is strictly related to fair "equilibrium" between thermodynamic aspects and kinetic ones. A systematic study regarding kinetics/mechanism of the aggregative process was tackled by Romeo *et al.* in 2014. They demonstrated which the rapid nucleation of few and small oligomers (dimers or trimers) drives the growth of ordered structures, pointing out the fundamental role of kinetic parameters in the expression and transmission of chirality in molecular aggregates [102].

1.5 Concluding Remarks

The chapter intended to give an overall overview on self-assembly phenomena and the conditions underlying the occurrence of chirogenesis in aggregative processes. The fundamentals of supramolecular chemistry have been widely addressed with particular regard to porphyrinoid systems. The building of chirogenic superstructures obeys rigid hierarchical rules in which several conditions (templates, solvents, pH, temperature, ionic strength, mixing order, etc.) can affect the origin or the transmission of chirality, from a single molecule to molecular aggregate, and its amplification at large scale. The asymmetric self-assembly mainly arises in two routes: (a) aggregation of chiral monomeric units ("intrinsically chiral self-assembly"), or (b) asymmetry induced by external factors during the self-aggregation of achiral units ("induced chiral self-assembly"). In this sense, some relevant works, and examples of chiral supramolecular systems based on porphyrin dyes have been reported in order to elucidate the different strategies for designing complex noncovalent structures.

Chiral molecular aggregates represent promising systems for a broad range of potential applications, and in perspective, there are still remarkable tasks to construct innovative supramolecular chirogenic systems for cutting-edge technologies.

1.6 Acknowledgment

We acknowledge the support of this work by programma ricerca di ateneo UNICT 2016-18 linea 2 and 1; and the Ministero dell'Università e della Ricerca (MUR) PRIN Prot. 2017YJMPZN-005, The Department of Chemical Science (finanziamento giovani ricercatori 2020), programma ricerca di ateneo UNICT 2020–22 linea 2.

References

1. Bahatyrova, S., Frese, R.N., Siebert, C.A., Olsen, J.D., Van Der Werf, K.O., Van Grondelle, R., Nlederman, R.A., Bullough, P.A., Otto, C., and Hunter, C.N. (2004) The native architecture of a photosynthetic membrane. *Nature*, **430**, 1058–1062.
2. Zouni, A., Witt, H.T., Kern, J., Fromme, P., Krauss, N., Saenger, W., and Orth, P. (2001) Crystal structure of photosystem II from *Synechococcus elongatus* at 3.8 Å resolution. *Nature*, **409**, 739–743.
3. Jordan, P., Fromme, P., Witt, H.T., Klukas, O., Saenger, W., and Krauß, N. (2001) Three-dimensional structure of cyanobacterial photosystem I at 2.5 Å resolution. *Nature*, **411**, 909–917.
4. Balaban, T.S., Tamiaki, H., and Holzwarth, A.R. (2005) in *Supermolecular Dye Chemistry*, (ed. Würthner, F.), Topics in Current Chemistry. Vol .258. Springer, Berlin, Heidelberg, pp. 1–38.
5. Drain, C.M., Varotto, A., and Radivojevic, I. (2009) Self-organized porphyrinic materials. *Chem. Rev.*, **109**, 1630–1658.
6. Haser, R., Pierrot, M., Frey, M., Payan, F., Astier, J.P., Bruschi, M., and Gall, J.L. (1979) Structure and sequence of the multihaem cytochrome C3. *Nature*, **282**, 806–810.
7. Burrell, A.K., Officer, D.L., Plieger, P.G., and Reid, D.C.W. (2001) Synthetic routes to multiporphyrin arrays. *Chem. Rev.*, **101**, 2751–2796.
8. Philp, D., and Stoddart, J.F. (1996) Self-assembly in natural and unnatural systems. *Angew. Chem. Int. Ed. English*, **35**, 1154–1196.
9. Cragg, P.J. (2005) *A Practical Guide to Supramolecular Chemistry*. John Wiley & Sons, Ltd, Chichester, UK.
10. Jonathan, W., Steed, J.L.A. *Supramolecular Chemistry*. 2nd ed., John Wiley & Sons, Ltd [Accessed January 21, 2021] Available from: https://www.wiley.com/en-us/Supramolecular+Chemistry%2C+2nd+Edition-p-9780470512340.
11. Lindoy, L.F., Atkinson, I.M., and Stoddart, J.F. (2000) *Self-Assembly in Supramolecular Systems*. Royal Society of Chemistry, Cambridge, UK.
12. Korevaar, P.A., Schaefer, C., De Greef, T.F.A., and Meijer, E.W. (2012) Controlling chemical self-assembly by solvent-dependent dynamics. *J. Am. Chem. Soc.*, **134**, 13482–13491.
13. Toyofuku, K., Alam, M.A., Tsuda, A., Fujita, N., Sakamoto, S., Yamaguchi, K., and Aida, T. (2007) Amplified chiral transformation through helical assembly. *Angew. Chem. Int. Ed.*, **46**, 6476–6480.
14. Bonaccorso, C., Brancatelli, G., Forte, G., Arena, G., Geremia, S., Sciotto, D., and Sgarlata, C. (2014) Factors driving the self-assembly of water-soluble calix[4]arene and gemini guests: A combined solution, computational and solid-state study. *RSC Adv.*, **4**, 53575–53587.
15. Gaeta, M., Rodolico, E., Fragalà, M.E., Pappalardo, A., Pisagatti, I., Gattuso, G., Notti, A., Parisi, M.F., Purrello, R., and D'Urso, A. (2021) Self-assembly of discrete

porphyrin/calix[4]tube complexes promoted by potassium ion encapsulation. *Molecules*, **26**, 704.

16. Durso, A., Brancatelli, G., Hickey, N., Farnetti, E., De Zorzi, R., Bonaccorso, C., Purrello, R., and Geremia, S. (2016) Interactions of a water-soluble calix[4]arene with spermine: Solution and solid-state characterisation. *Supramol. Chem.*, **28**, 499–505.
17. Randazzo, R., Gaeta, M., Gangemi, C.M.A., Fragalà, M.E., Purrello, R., and D'Urso, A. (2019) Chiral recognition of L- and D-amino acid by porphyrin supramolecular aggregates. *Molecules*, **24**, 84.
18. Gaeta, M., Randazzo, R., Villari, V., Micali, N., Pezzella, A., Purrello, R., D'Ischia, M., and D'Urso, A. (2020) En route to a chiral melanin: The dynamic "from-imprinted-to-template" supramolecular role of porphyrin hetero-aggregates during the oxidative polymerization of L-DOPA. *Front. Chem.*, **8**, 616961.
19. Carbone, A., Gaeta, M., Romeo, A., Portale, G., Pedicini, R., Gatto, I., and Castriciano, M.A. (2018) Porphyrin/SPEEK membranes with improved conductivity and durability for PEFC technology. *ACS Appl. Energy Mater.*, **1**, 1664–1673.
20. Castriciano, M.A., Carbone, A., Saccà, A., Donato, M.G., Micali, N., Romeo, A., De Luca, G., and Scolaro, L.M. (2010) Optical and sensing features of TPPS4 J-aggregates embedded in Nafion® membranes: Influence of casting solvents. *J. Mater. Chem.*, **20**, 2882–2886.
21. Hasobe, T., Imahori, H., Kamat, P.V., Tae, K.A., Seong, K.K., Kim, D., Fujimoto, A., Hirakawa, T., and Fukuzumi, S. (2005) Photovoltaic cells using composite nanoclusters of porphyrins and fullerenes with gold nanoparticles. *J. Am. Chem. Soc.*, **127**, 1216–1228.
22. Zhang, L., Li, X., and Mu, J. (2007) Self-assembly of porphyrin-based supramolecules and their characteristics on gold nanoparticles. *Colloids Surf. A Physicochem. Eng. Asp.*, **302**, 219–224.
23. Gaeta, M., Sanfilippo, G., Fraix, A., Sortino, G., Barcellona, M., Conti, G.O., Fragalà, M.E., Ferrante, M., Purrello, R. and D'urso, A. (2020) Photodegradation of antibiotics by noncovalent porphyrin-functionalized Tio2 in water for the bacterial antibiotic resistance risk management. *Int. J. Mol. Sci.*, **21**, 3775.
24. Greco, E., Ciliberto, E., Verdura, P.D., Lo Giudice, E., and Navarra, G. (2016) Nanoparticle-based concretes for the restoration of historical and contemporary buildings: A new way for CO_2 reduction in architecture. *Appl. Phys. A Mater. Sci. Process.*, **122**, 1–9.
25. Koti, A.S.R., and Periasamy, N. (2003) Self-assembly of template-directed J-aggregates of porphyrin. *Chem. Mater.*, **15**, 369–371.
26. Ruthard, C., Maskos, M., Kolb, U., and Gröhn, F. (2011) Polystyrene sulfonate-porphyrin assemblies: Influence of polyelectrolyte and porphyrin structure. *J. Phys. Chem. B*, **115**, 5716–5729.
27. Frühbeißer, S., and Gröhn, F. (2017) Porphyrin–polyelectrolyte nanoassemblies: The role of charge and building block architecture in self-assembly. *Macromol. Chem. Phys.*, **218**, 1600526.

28. Gaeta, M., Farini, S., Gangemi, C.M.A., Purrello, R., and D'Urso, A. (2020) Interactions of mono spermine porphyrin derivative with DNAs. *Chirality*, **32**, 1243–1249.
29. D'Urso, A., Mammana, A., Balaz, M., Holmes, A.E., Berova, N., Lauceri, R., and Purrello, R. (2009) Interactions of a tetraanionic porphyrin with DNA: from a Z-DNA sensor to a versatile supramolecular device. *J. Am. Chem. Soc.*, **131**, 2046–2047.
30. Ensslen, P., and Wagenknecht, H.A. (2015) One-dimensional multichromophor arrays based on DNA: from self-assembly to light-harvesting. *Acc. Chem. Res.*, **48**, 2724–2733.
31. Badjić, J.D., Nelson, A., Cantrill, S.J., Turnbull, W.B., and Stoddart, J.F. (2005) Multivalency and cooperativity in supramolecular chemistry. *Acc. Chem. Res.*, **38**, 723–732.
32. Lehn, J.M. (2009) Towards complex matter: Supramolecular chemistry and self-organization. *Eur. Rev.*, **17**, 263–280.
33. Whitty, A. (2008) Cooperativity and biological complexity. *Nat. Chem. Biol.*, **4**, 435–439.
34. Cram, D.J. (1986) Preorganization — From Solvents to Spherands. *Angew. Chem. Int. Ed. English*, 25, 1039–1057.
35. Peacocke, A.R., and Skerrett, J.N.H. (1956) The interaction of aminoacridines with nucleic acids. *Trans. Faraday Soc.*, **52**, 261–279.
36. McGhee, J.D., and von Hippel, P.H. (1974) Theoretical aspects of DNA-protein interactions: Co-operative and non-co-operative binding of large ligands to a one-dimensional homogeneous lattice. *J. Mol. Biol.*, **86**, 469–489.
37. Smulders, M.M.J., Schenning, A.P.H.J., and Meijer, E.W. (2008) Insight into the mechanisms of cooperative self-assembly: The 'sergeants-and-soldiers' principle of chiral and achiral C 3-symmetrical discotic triamides. *J. Am. Chem. Soc.*, **130**, 606–611.
38. Prince, R.B., Moore, J.S., Brunsveld, L., and Meijer, E.W. (2001) Cooperativity in the folding of helical M-phenylene ethynylene oligomers based upon the 'sergeants-and-soldiers' principle. *Chem. A Eur. J.*, **7**, 4150–4154.
39. Morrow, S.M., Bissette, A.J., and Fletcher, S.P. (2017) Transmission of chirality through space and across length scales. *Nat. Nanotechnol.*, **12**, 410–419.
40. Barron, L.D., In Botta, O., Bada, J.L., Gomez-Elvira, J., Javaux, E., Selsis, F.S.R., (eds.) (2008) *Strategies of Life Detection.* Space Sciences Series of ISSI, vol. 25. Springer, Boston, USA, pp. 187–201.
41. Brückner, H., Becker, D., and Lüpke, M. (1993) Chirality of amino acids of micro-organisms used in food biotechnology. *Chirality*, **5**, 385–392.
42. Corradini, R., Sforza, S., Tedeschi, T., and Marchelli, R. (2007) Chirality as a tool in nucleic acid recognition: Principles and relevance in biotechnology and in medicinal chemistry. *Chirality*, **19**, 269–294.
43. Liu, C., Yang, D., Jin, Q., Zhang, L., and Liu, M. (2016) A chiroptical logic circuit based on self-assembled soft materials containing amphiphilic spiropyran. *Adv. Mater.*, **28**, 1644–1649.

44. Green, D.W., Lee, J.-M., Kim, E.-J., Lee, D.-J., and Jung H.-S. (2016) Chiral biomaterials: From molecular design to regenerative medicine. *Adv. Mater. Interfaces*, **3**, 1500411.
45. Caricato, M., Delforge, A., Bonifazi, D., Dondi, D., Mazzanti, A., and Pasini, D. (2015) Chiral nanostructuring of multivalent macrocycles in solution and on surfaces. *Org. Biomol. Chem.*, **13**, 3593–3601.
46. Caricato, M., Leza, N.J., Roy, K., Dondi, D., Gattuso, G., Shimizu, L.S., Vander Griend, D.A., and Pasini, D. (2013) A chiroptical probe for sensing metal ions in water. *Eur. J. Org. Chem.*, **2013**, 6078–6083.
47. Agnes, M., Nitti, A., Vander Griend, D.A., Dondi, D., Merli, D., and Pasini, D. (2016) A chiroptical molecular sensor for ferrocene. *Chem. Commun.*, **52**, 11492–11495.
48. Liu, M., Zhang, L., and Wang, T. (2015) Supramolecular chirality in self-assembled systems. *Chem. Rev.*, **115**, 7304–7397.
49. Rowan, S.J., Cantrill, S.J., Cousins, G.R.L., Sanders, J.K.M., and Stoddart, J.F. (2002) Dynamic covalent chemistry. *Angew. Chem. Int. Ed.*, **41**, 898–952.
50. Elemans, J.A.A.W., Rowan, A.E., and Nolte, R.J.M. (2003) Mastering molecular matter. Supramolecular architectures by hierarchical self-assembly. *J. Mater. Chem.*, **13**, 2661–2670.
51. Lehn, J. (1995) *Supramolecular Chemistry: Concepts and Perspectives. 1st ed.* VCH, Weinheim Germany.
52. Palmans, A.R.A., and Meijer, E.W. (2007) Amplification of chirality in dynamic supramolecular aggregates. *Angew. Chem. Int. Ed.*, **46**, 8948–8968.
53. Borovkov, V.V., and Inoue, Y. (2006) In (Crego-Calama, M., and Reinhoudt D.N., eds.) *Supramolecular Chirality*. Topics in Current Chemistry. vol. 265. Springer-Verlag, Berlin/Heidelberg, pp. 89–146.
54. Mateos-Timoneda, M.A., Crego-Calama, M., and Reinhoudt, D.N. (2004) Supramolecular chirality of self-assembled systems in solution. *Chem. Soc. Rev.*, **33**, 363–372.
55. Cintas, P. (2002) Chirality of living systems: A helping hand from crystals and oligopeptides. *Angew. Chem. Int. Ed.*, **41**, 1139–1145.
56. Zhang, L., Qin, L., Wang, X., Cao, H., and Liu, M. (2014) Supramolecular chirality in self-assembled soft materials: Regulation of chiral nanostructures and chiral functions. *Adv. Mater.*, **26**, 6959–6964.
57. Liu, K.S., Tian, D.L., and Jiang, L. (2017) In (Xu, R., and Xu, Y., eds.) *Modern Inorganic Synthetic Chemistry. 2nd ed.* Elsevier B.V., 2017; 687–721.
58. Marx, S., and Avnir, D. (2007) The induction of chirality in sol-gel materials. *Acc. Chem. Res.*, **40**, 768–776.
59. Amabilino, D.B. (2009) *Chirality at the Nanoscale*. (Amabilino, D.B., ed.). Wiley.
60. Zhang, J., Albelda, M.T., Liu, Y., and Canary J.W. (2005) Chiral nanotechnology. *Chirality*, **17**, 404–420.
61. Berthod, A. (2010) *Chiral Recognition in Separation Methods*. (Berthod, A., ed.). Springer, Berlin Heidelberg.

62. Zhang, L., Jin, Q., and Liu, M. (2016) Enantioselective recognition by chiral supramolecular gels. *Chem. An Asian J.*, **11**, 2642–2649.
63. Rekharsky, M.V., Yamamura, H., Inoue, C., Kawai, M., Osaka, I., Arakawa, R., Shiba, K., Sato, A., Young, H.K., Selvapalam, N., Kim, K., and Inoue, Y. (2006) Chiral recognition in cucurbituril cavities. *J. Am. Chem. Soc.*, **128**, 14871–14880.
64. Hembury, G.A., Borovkov, V.V., and Inoue, Y. (2008) Chirality-sensing supramolecular systems. *Chem. Rev.*, **108**, 1–73.
65. Nowacki, B., Oh, H., Zanlorenzi, C., Jee, H., Baev, A., Prasad, P.N., and Akcelrud, L. (2013) Design and synthesis of polymers for chiral photonics. *Macromolecules*, **46**, 7158–7165.
66. Esposito, M., Tasco, V., Todisco, F., Cuscunà, M., Benedetti, A., Sanvitto, D., and Passaseo, A. (2015) Triple-helical nanowires by tomographic rotatory growth for chiral photonics. *Nat. Commun.*, **6**, 1–7.
67. Taylor, M.S., and Jacobsen, E.N. (2006) Asymmetric catalysis by chiral hydrogen-bond donors. *Angew. Chem. Int. Ed.*, **45**, 1520–1543.
68. Yang, Q., Han, D., Yang, H., and Li, C. (2008) Asymmetric catalysis with metal complexes in nanoreactors. *Chem. An Asian J.*, **3**, 1214–1229.
69. Haupert, L.M., and Simpson, G.J. (2009) Chirality in nonlinear optics. *Annu. Rev. Phys. Chem.*, **60**, 345–365.
70. Deraedt, C., and Astruc, D. (2016) *Coord. Chem. Rev.*, **324**, 106–122.
71. Zhao, B., Deng, J., and Deng, J. (2017) Optically active helical polyacetylene self-assembled into chiral micelles used as nanoreactor for helix-sense-selective polymerization. *ACS Macro Lett.*, **6**, 6–10.
72. Chen, X., Wang, Y., Wang, H., Kim, Y., and Lee, M. (2017) α-helical peptide vesicles with chiral membranes as enantioselective nanoreactors. *Chem. Commun.*, **53**, 10958–10961.
73. Elemans, J.A.A.W., Van Hameren, R., Nolte, R.J.M., and Rowan, A.E. (2006) Molecular materials by self-assembly of porphyrins, phthalocyanines, and perylenes. *Adv. Mater.*, **18**, 1251–1266.
74. Li, L.L., and Diau, E.W.G. (2013) Porphyrin-sensitized solar cells. *Chem. Soc. Rev.*, **42**, 291–304.
75. Biesaga, M., Pyrzyńska, K., and Trojanowicz, M. (2000) Porphyrins in analytical chemistry. A review. *Talanta*, **51**, 209–224.
76. Tsolekile, N., Nelana, S., and Oluwafemi, O.S. (2019) Porphyrin as diagnostic and therapeutic agent. *Molecules*, **24**, 2669.
77. Senge, M., Ryan, A., Letchford, K., MacGowan, S., and Mielke, T. (2014) Chlorophylls, symmetry, chirality, and photosynthesis. *Symmetry (Basel),* **6**, 781–843.
78. Kadish, K., Smith, K.M., and Guilard, R., (eds.) (2000) *The Porphyrin Handbook.* Vol. 3. Academic Press, San Diego, CA, USA.
79. Vicente, M., and Smith, K. (2014) Syntheses and functionalizations of porphyrin macrocycles. *Curr. Org. Synth.*, **11**, 3–28.

80. Karolczak, J., Kowalska, D., Lukaszewicz, A., Maciejewski, A., and Steer, R.P. (2004) Photophysical studies of porphyrins and metalloporphyrins: Accurate measurements of fluorescence spectra and fluorescence quantum yields for soret band excitation of zinc tetraphenylporphyrin. *J. Phys. Chem. A*, **108**, 4570–4575.
81. Gouterman, M., Wagnière, G.H., and Snyder, L.C. (1963) Spectra of porphyrins. Part II. Four orbital model. *J. Mol. Spectrosc.*, **11**, 108–127.
82. Kadish, K.M., Smith, K.M., and Guilard, R., (eds.) (2000) *The Porphyrin Handbook*. Vol. 6. 1st ed. Academic Press, USA.
83. Maiti, N.C., Mazumdar, S., and Periasamy, N. (1998) J- and H-aggregates of porphyrin — surfactant complexes: Time-resolved fluorescence and other spectroscopic studies. *J. Phys. Chem. B*, **102**, 1528–1538.
84. Jelley, E.E. (1936) Spectral Absorption and Fluorescence of Dyes in the Molecular State. *Nature*, 138, 1009–1010.
85. Ohno, O., Kaizu, Y., and Kobayashi, H. (1993) J-aggregate formation of a water-soluble porphyrin in acidic aqueous media. *J. Chem. Phys.*, **99**, 4128–4139.
86. Davydov, A.S. (1964) The theory of molecular excitons. *Sov. Phys. Uspekhi*, **7**, 145–178.
87. Kasha, M., Rawls, H.R., and El-Bayoumi M.A. (1965) The exciton model in molecular spectroscopy. *Pure Appl. Chem.*, **11**, 371–392.
88. Dexter, D.L. (1953) A theory of sensitized luminescence in solids. *J. Chem. Phys.*, **21**, 836–850.
89. Hestand, N.J., and Spano, F.C. (2017) Molecular aggregate photophysics beyond the kasha model: Novel design principles for organic materials. *Acc. Chem. Res.*, **50**, 341–350.
90. Berova, N., Di Bari, L., and Pescitelli, G. (2007) Application of electronic circular dichroism in configurational and conformational analysis of organic compounds. *Chem. Soc. Rev.*, **36**, 914–931.
91. van der Weegen, R., Teunissen, A.J.P., and Meijer, E.W. (2017) Directing the self-assembly behaviour of porphyrin-based supramolecular systems. *Chem. A Eur. J.*, **23**, 3773–3783.
92. Oliveras-González, C., Di Meo, F., González-Campo, A., Beljonne, D., Norman, P., Simón-Sorbed, M., Linares, M., and Amabilino, D.B. (2015) Bottom-up hierarchical self-assembly of chiral porphyrins through coordination and hydrogen bonds. *J. Am. Chem. Soc.*, **137**, 15795–15808.
93. Onouchi, H., Miyagawa, T., Morino, K., and Yashima, E. (2006) Assisted formation of chiral porphyrin homoaggregates by an induced helical poly(phenylacetylene) template and their chiral memory. *Angew. Chem.*, **118**, 2441–2444.
94. Yashima, E., Maeda, K., and Nishimura, T. (2004) Detection and amplification of chirality by helical polymers. *Chem. A Eur. J.*, **10**, 42–51.
95. D'Urso, A., Fragalà, M.E., and Purrello, R. (2012) From self-assembly to noncovalent synthesis of programmable porphyrins' arrays in aqueous solution. *Chem. Commun.*, **48**, 8165–8176.

96. Franke, D., Vos, M., Antonietti, M., Sommerdijk, N.A.J.M., and Faul, C.F.J. (2006) Induced supramolecular chirality in nanostructured materials: Ionic self-assembly of perylene-chiral surfactant complexes. *Chem. Mater.*, **18**, 1839–1847.
97. Huang, Y., Yan, Y., Smarsly, B.M., Wei, Z., and Faul, C.F.J. (2009) Helical supramolecular aggregates, mesoscopic organisation and nanofibers of a perylenebisimide-chiral surfactant complex via ionic self-assembly. *J. Mater. Chem.*, **19**, 2356–2362.
98. El-Hachemi, Z., Mancini, G., Ribó, J.M., and Sorrenti, A. Role of the hydrophobic effect in the transfer of chirality from molecules to complex systems: From chiral surfactants to porphyrin/surfactant aggregates. *J. Am. Chem. Soc.*, **130**, 15176–15184.
99. Borovkov, V., and Inoue, Y. (2009) A versatile bisporphyrinoid motif for supramolecular chirogenesis. *Eur. J. Org. Chem.*, **2009**, 189–197.
100. Borovkov, V.V., Hembury, G.A., and Inoue, Y. (2005) Supramolecular chirogenesis with bis-chlorin versus bis-porphyrin hosts: Peculiarities of chirality induction and modulation of optical activity. *J. Org. Chem.*, **70**, 8743–8754.
101. Castriciano, M.A., Romeo, A., Zagami, R., Micali, N., and Scolaro L.M. (2012) Kinetic effects of tartaric acid on the growth of chiral J-aggregates of tetrakis(4-sulfonatophenyl)porphyrin. *Chem. Commun.*, **48**, 4872–4874.
102. Romeo, A., Castriciano, M.A., Occhiuto, I., Zagami, R., Pasternack, R.F., and Scolaro, L.M. (2014) Kinetic control of chirality in porphyrin J-aggregates. *J. Am. Chem. Soc.*, **136**, 40–43.
103. van Dijken, D.J., Beierle, J.M., Stuart, M.C.A., Szymański, W., Browne, W.R. and Feringa B.L. (2014) Autoamplification of molecular chirality through the induction of supramolecular chirality. *Angew. Chem.*, **126**, 5173–5177.
104. Gaeta, M., Oliveri, I.P., Fragalà, M.E., Failla, S., D'Urso, A., Di Bella, S., and Purrello, R. (2016) Chirality of self-assembled achiral porphyrins induced by chiral Zn(II) Schiff-base complexes and maintained after spontaneous dissociation of the templates: A new case of chiral memory. *Chem. Commun.*, **52**, 8518–8521.
105. Sang, Y., Yang, D., Duan, P., and Liu, M. (2019) Towards homochiral supramolecular entities from achiral molecules by vortex mixing-accompanied self-assembly. *Chem. Sci.*, **10**, 2718–2724.
106. Micali, N., Engelkamp, H., Van Rhee, P.G., Christianen, P.C.M., Scolaro, L.M., and Maan, J.C. (2012) Selection of supramolecular chirality by application of rotational and magnetic forces. *Nat. Chem.*, **4**, 201–207.
107. D'Urso, A., Randazzo, R., Lo Faro, L., and Purrello, R. (2010) Vortexes and nanoscale chirality. *Angew. Chem. Int. Ed.*, **49**, 108–112.
108. Arteaga, O., El-Hachemi, Z., Canillas, A., Crusats, J., Rovira, M., and Ribó, J.M. (2016) Reversible and irreversible emergence of chiroptical signals in J-aggregates of achiral 4-sulfonatophenyl substituted porphyrins: Intrinsic chirality: vs. chiral ordering in the solution. *Chem. Commun.*, **52**, 10874–10877.
109. Ribò, J.M., El-Hachemi, Z., Arteaga, O., Canillas, A., and Crusats, J. (2017) Hydrodynamic effects in soft-matter self-assembly: The case of J-aggregates of amphiphilic porphyrins. *Chem. Rec*, **17**, 713–724.

110. El-Hachemi, Z., Balaban, T.S., Campos, J.L., Cespedes, S., Crusats, J., Escudero, C., Kamma-Lorger, C.S., Llorens, J., Malfois, M., Mitchell, G.R., Tojeira, A.P., and Ribó, J.M. (2016) Effect of hydrodynamic forces on meso-(4-sulfonatophenyl)-substituted porphyrin J-aggregate nanoparticles: Elasticity, plasticity and breaking. *Chem. A Eur. J.*, **22**: 9740–9749.
111. Arteaga, O., Canillas, A., Crusats, J., El-Hachemi, Z., Llorens, J., Sacristan, E., and Ribò, J.M. (2010) Emergence of supramolecular chirality by flows. *ChemPhysChem*, **11**, 3511–3516.
112. Crusats, J., El-Hachemi, Z., and Ribó, J.M. (2010) Hydrodynamic effects on chiral induction. *Chem. Soc. Rev.*, **39**, 569–577.
113. Escudero, C., Crusats, J., Díez-Pérez, I., El-Hachemi, Z., and Ribó, J.M. Folding and hydrodynamic forces in J-aggregates of 5-phenyl-10, 15, 20-tris- (4-sulfophenyl)porphyrin. *Angew. Chem. Int. Ed.* 2006, **45**, 8032–8035.
114. Mineo, P., Villari, V., Scamporrino, E., and Micali, N. (2014) Supramolecular chirality induced by a weak thermal force. *Soft Matter*, **10**, 44–47.
115. Balaban, T.S., Bhise, A.D., Fischer, M., Linke-Schaetzel, M., Roussel, C., and Vanthuyne, N. (2003) Controlling chirality and optical properties of artificial antenna systems with self-assembling porphyrins. *Angew. Chem.*, **115**, 2190–2194.
116. Hoeben, F.J.M., Wolffs, M., Zhang, J., De Feyter, S., Leclère, P., Schenning, A.P.H.J. and Meijer, E.W. (2007) Influence of supramolecular organization on energy transfer properties in chiral oligo(p-phenylene vinylene) porphyrin assemblies. *J. Am. Chem. Soc.*, **129**, 9819–9828.
117. Monti, D., Venanzi, M., Mancini, G., Di Natale, C., and Paolesse, R. (2005) Supramolecular chirality Control by Solvent Changes. Solvodichroic Effect on Chiral Porphyrin Aggregation. *Chem. Commun.*, 2471–2473.
118. Wang, Q., Chen, Y., Ma, P., Lu, J., Zhang, X., and Jiang. J. (2011) Morphology and chirality controlled self-assembled nanostructures of porphyrin–pentapeptide conjugate: Effect of the peptide secondary conformation. *J. Mater. Chem.*, **21**, 8057.
119. Castriciano, M.A., Zagami, R., Trapani, M., Romeo, A., Patanè, S., and Monsù Scolaro, L. (2015) Investigation of the aggregation properties of a chiral porphyrin bearing citronellal meso substituent groups. *Chirality*, **27**, 900–906.
120. Mcrae, E.G., and Kasha, M. (1958) Enhancement of Phosphorescence Ability upon Aggregation of Dye Molecules. *J. Chem. Phys.,* 28, 721–722.
121. Reches, M., and Gazit, E. (2003) Casting metal nanowires within discrete self-assembled peptide nanotubes. *Science,* **300**, 625–627.
122. Song, Y., Challa, S.R., Medforth, C.J., Qiu, Y., Watt, R.K., Peña, D., Miller, J.E., Van Swol, F., and Shelnutt, J.A. (2004) Synthesis of peptide-nanotube platinum-nanoparticle composites. *Chem. Commun.*, **4**, 1044–1045.
123. Charalambidis, G., Georgilis, E., Panda, M.K., Anson, C.E., Powell, A.K., Doyle, S., Moss, D., Jochum, T., Horton, P.N., Coles, S.J., Linares, M., Beljonne, D., Naubron, J.V., Conradt, J., Kalt, H., Mitraki, A., Coutsolelos, A.G. and Balaban, T.S.

(2016) A switchable self-assembling and disassembling chiral system based on a porphyrin-substituted phenylalanine-phenylalanine motif. *Nat. Commun.*, **7**, 1–11.
124. Borovkov, V. (2014) Supramolecular Chirality in Porphyrin Chemistry. *Symmetry (Basel),* 6, 256–294.
125. Nagai, K., Maeda, K., Takeyama, Y., Sakajiri, K., and Yashima, E. (2005) Helicity induction and chiral amplification in a poly(phenylacetylene) bearing N, N-diisopropylaminomethyl groups with chiral acids in water. *Macromolecules*, **38**, 5444–5451.
126. Bellacchio, E., Lauceri, R., Gurrieri, S., Scolaro, L.M., Romeo, A., and Purrello, R. (1998) Template-imprinted chiral porphyrin aggregates. *J. Am. Chem. Soc.*, **120**, 12353–12354.
127. Purrello, R., Monsù Scolaro, L., Bellacchio, E., Gurrieri, S., and Romeo, A. (1998) Chiral H- and J-type aggregates of meso-tetrakis(4-sulfonatophenyl)porphine on α-helical polyglutamic acid induced by cationic porphyrins. *Inorg. Chem.*, **37**, 3647–3648.
128. Purrello, R., Raudino, A., Monsù Scolaro, L., Loisi, A., Bellacchio, E., and Lauceri, R. (2000) Ternary porphyrin aggregates and their chiral memory. *J. Phys. Chem. B*, **104**, 10900–10908.
129. Zhang, L., and Liu, M. (2009) Supramolecular chirality and chiral inversion of tetraphenylsulfonato porphyrin assemblies on optically active polylysine. *J. Phys. Chem. B*, **113**, 14015–14020.
130. Gaeta, M., Raciti, D., Randazzo, R., Gangemi, C.M.A., Raudino, A., D'Urso, A., Fragalà, M.E., and Purrello, R. (2018) Chirality enhancement of porphyrin supramolecular assembly driven by a template preorganization effect. *Angew. Chem. Int. Ed.*, **57**, 10656–10660.
131. Gaeta, M., Randazzo, R., Cristaldi, D.A., D'Urso, A., Purrello, R., and Fragalà, M.E. (2017) ZnTPPS demetalation: Role of polyelectrolytes on aggregation after protonation in acid. *J. Porphyr. Phthalocyanines*, **21**, 426–430.
132. Occhiuto, I.G., Castriciano, M.A., Trapani, M., Zagami, R., Romeo, A., Pasternack, R.F. and Scolaro, L.M. (2020) Controlling J-aggregates formation and chirality induction through demetallation of a zinc(Ii) water soluble porphyrin. *Int. J. Mol. Sci.*, **21**, 4001.
133. Purrello, R., Bellacchio, E., Gurrieri, S., Lauceri, R., Raudino, A., Scolaro, L.M. and Santoro, A.M. (1998) PH modulation of porphyrins self-assembly onto polylysine. *J. Phys. Chem. B*, **102**, 8852–8857.
134. Zeng, L., He, Y., Dai, Z., Wang, J., Cao, Q., and Zhang, Y. (2009) Chiral induction, memory, and amplification in porphyrin homoaggregates based on electrostatic interactions. *ChemPhysChem*, **10**, 954–962.
135. Green, M.M., Reidy, M.P., Johnson, R.J., Darling, G., O'Leary, D.J., and Willson, G. (1989) Macromolecular stereochemistry: The out-of-proportion influence of optically active comonomers on the conformational characteristics of polyisocyanates. The sergeants and soldiers experiment. *J. Am. Chem. Soc.*, **111**, 6452–6454.

136. Prins, L.J., Timmerman, P., and Reinhoudt, D.N. (2001) Amplification of chirality: The 'sergeants and soldiers' principle applied to dynamic hydrogen-bonded assemblies. *J. Am. Chem. Soc.*, **123**, 10153–10163.
137. Lauceri, R., Raudino, A., Scolaro, L.M., Micali, N., and Purrello. R. (2002) From achiral porphyrins to template-imprinted chiral aggregates and further. Self-replication of chiral memory from scratch. *J. Am. Chem. Soc.*, **124**, 894–895.
138. Mammana, A., D'Urso, A., Lauceri, R., and Purrello, R. (2007) Switching off and on the supramolecular chiral memory in porphyrin assemblies. *J. Am. Chem. Soc.*, **129**, 8062–8063.
139. Micali, N., Mallamace, F., Romeo, A., Purrello, R., and Scolaro, L.M. (2000) Mesoscopic structure of meso-tetrakis(4-sulfonatophenyl)porphine J-aggregates. *J. Phys. Chem. B* 2000, **104**, 5897–5904.
140. Castriciano, M.A., Romeo, A., Villari, V., Micali, N., and Scolaro, L.M. (2003) Structural rearrangements in 5, 10, 15, 20-tetrakis(4-sulfonatophenyl)porphyrin J-aggregates under strongly acidic conditions. *J. Phys. Chem. B*, **107**, 8765–8771.
141. Micali, N., Villari, V., Castriciano, M.A., Romeo, A., and Scolaro, L.M. (2006) From fractal to nanorod porphyrin J-aggregates. Concentration-induced tuning of the aggregate size. *J. Phys. Chem. B*, **110**, 8289–8295.
142. Castriciano, M.A., Romeo, A., Villari, V., Micali, N., and Scolaro, L.M. (2004) Nanosized porphyrin J-aggregates in water/AOT/decane microemulsions. *J. Phys. Chem. B*, **108**, 9054–9059.
143. Castriciano, M.A., Romeo, A., De Luca, G., Villari, V., Scolaro, L.M., and Micali, N. (2011) Scaling the chirality in porphyrin J-nanoaggregates. *J. Am. Chem. Soc.*, **133**, 765–767.

https://doi.org/10.1142/9789811259227_0004

4 Chirogenesis in Asymmetric Synthesis and Catalysis

Dzmitry G. Kananovich* and Margus Lopp[†]

Department of Chemistry and Biotechnology,
School of Science, Tallinn University of Technology,
Tallinn, Estonia
*dzmitry.kananovich@taltech.ee
[†]margus.lopp@taltech.ee

Methods of asymmetric synthesis and catalysis are briefly reviewed and illustrated with selected examples, together with a short historical sketch. Particular attention is paid to the genesis of chirality in organic reactions, i.e., the mechanisms of asymmetric inductions and the origin of enantioselectivity. The significance of mechanistic studies for the rational design of efficient catalysts and engineering of highly enantioselective catalytic processes is emphasized. Green chemistry aspects of enantioselective catalysis are discussed. Finally, the authors provide an overview of their own research work devoted to titanium-mediated asymmetric transformations, such as the oxidation of 1,2-cyclopentanediones and enantioselective Kulinkovich reactions.

1.1 Introduction

Asymmetric synthesis is a chemical reaction, or a reaction sequence, in which one or more new elements of chirality are formed in a substrate molecule and which produces the stereoisomeric (enantiomeric or diastereoisomeric) products in unequal amounts [1]. In the event of asymmetric synthesis, violation of the symmetry occurs by virtue of the influence of a chiral feature present in the substrate, reagent, catalyst, or environment. This phenomenon is called *asymmetric induction* [1]. The term *chirogenesis* was first introduced to describe the chirality transfer phenomena in supramolecular chemistry [2–4]. Subsequently, the term was suggested as a general concept which deals with various aspects of asymmetry induction, chirality generation, modulation, transfer, and amplification in various fields of chemical science (see the Preface). Asymmetric induction is undoubtedly the principal chirogenic phenomenon which accounts for chirogenesis in chemical reactions. However, chirogenic phenomena occurring in asymmetric synthesis and catalysis are more diverse and not limited to asymmetric induction only. In this respect, the term chirogenesis has somewhat broader meaning and is used here in that sense or as a synonym of asymmetric induction.

Chiral influence was identified as a key factor responsible for chirogenesis in asymmetric reactions as long ago as the second half of the 19th century, in strong connection with the development of the theory of vitalism [5]. Indeed, the first discovered chiral substances and asymmetric transformations were of biological origin. Even now, the majority of chiral chemicals are still derived from the natural chiral pool, such as amino and hydroxy acids, carbohydrates, alkaloids, terpenoids, etc. As we currently know, highly stereoselective enzymatic processes occurring in the living cell are responsible for chirogenesis in the biosynthesis of natural products. However, the origin of prebiotic chirality and the chain of evolutionary events which engendered the enzymes themselves chiral still remain obscure [6, 7].

In organic synthesis, the field of asymmetric transformations flourished in the second half of the 20th century and continues its rapid development at present. State-of-the-art approaches offer a plethora of chiral reagents and catalysts, which are indispensable in many industrial fields requiring preparation of enantiopure materials. For example, as more drugs appear on the market as single enantiomers, asymmetric transformations become more frequently employed in the pharmaceutical industry [8, 9]. Enantioselective synthesis, especially in a catalytic mode, is a more attractive approach for production than the resolution of racemates, as it delivers higher yields, produces less waste, and requires only a subtle catalytic amount of a precious chiral inducer. Nevertheless, despite the numerous outstanding achievements in the area of asymmetric synthesis and captivating simplicity of its basic principles, attaining high levels of enantiocontrol in organic reactions still represents an arduous task. Many of the synthetically useful transformations are lacking asymmetric variants or are not efficient enough for practical adaptations. The rational development of asymmetric reactions, especially catalytic processes, commonly requires deep mechanistic knowledge and involves judicious analysis of the intrinsic chirogenic processes leading to emergence of chirality or, vice versa, to erosion of stereoselectivity.

In this chapter, the authors intended to provide a short overview of this vast field, its historical development and its current state, emphasizing the mechanisms of chirogenesis in organic reactions. First, we aimed to familiarize the reader with the basic principles, definitions, and concepts, by providing their concise description (Section 1.2), followed by a historical sketch (Section 1.3). The most common types and techniques of asymmetric induction in organic reactions are presented afterward. The methods which utilize a stoichiometric source of chirality are introduced first (Section 1.4). This is the case of *diastereoselective synthesis*, in which asymmetric induction takes place due to an existing element of chirality in the starting material, or introduced with the help of a chiral auxiliary. The *enantioselective reactions* of prochiral substrates with stoichiometric chiral reagents are covered in the same section. Section 1.5 deals entirely with *enantioselective catalysis*, in which chiral molecules are produced with an aid of catalytic amount of a chiral inducer. Besides a few representative examples of venerable enantioselective catalytic reactions, we tried to shed some light on new directions together with green chemistry aspects of various catalytic approaches. Finally, in the concluding Section 1.6 the authors provide a brief account of their own research experience in the area, which illustrates the importance of mechanistic knowledge in understanding the chirality transfer phenomena as a prerequisite for the development of an efficient asymmetric process.

Due to the brief character of this chapter, it is impossible to discuss all aspects of chirogenic phenomena taking place in organic synthesis and catalysis. Therefore, we would like to note that our vision presented here is in no way comprehensive, especially in respect to the topics well covered previously in the scientific literature and textbooks. The selection of presented examples is rather subjective. It was invoked by our pursuit to emphasize the key points of the presented concepts in a concise and unambiguous manner. References cited in the chapter are recommended for further reading and provide much deeper insight to the topics.

1.2 Basic Principles, Definitions, and Concepts

Chiral molecules have a spatial arrangement of atoms which makes them non-superimposable with their mirror images. As a consequence, chiral compounds exist as a pair of *stereoisomers* — *enantiomers* (the older term *optical isomers*). In terms of symmetry operations, chiral molecules do not contain any mirror plane ($\sigma = S_1$), center of inversion ($i = S_2$) and, in general, reflection axis (S_n). Such asymmetry is usually caused by at least one of the *chirality elements* present in a molecule, such as a chiral center (which, most commonly, is represented by a carbon atom with four different substituents), a chiral axis, a chiral plane, or a helix (see Figure 1).

Asymmetric synthesis creates chiral molecules from achiral precursors (*enantioselective synthesis*), or installs a new additional element of chirality into a chiral starting material (*diastereoselective synthesis*). It is crucial that the stereoisomeric products must be formed in unequal amounts.

Let us consider how a new chiral element (e.g., a chiral center) can form in a chemical reaction. It means generally that a *prochiral* precursor must be desymmetrized during the reaction. For example, asymmetry can arise from the transformation of *prostereogenic* (or *prochiral*) *structural elements* present in an achiral starting material. Thus, planar sp^2-carbon (e.g., in a carbonyl group) can be transformed into a chiral stereocenter by means of addition reactions (Figure 2). The addition of hydrogen cyanide to the carbonyl group of benzaldehyde can occur from two distinct

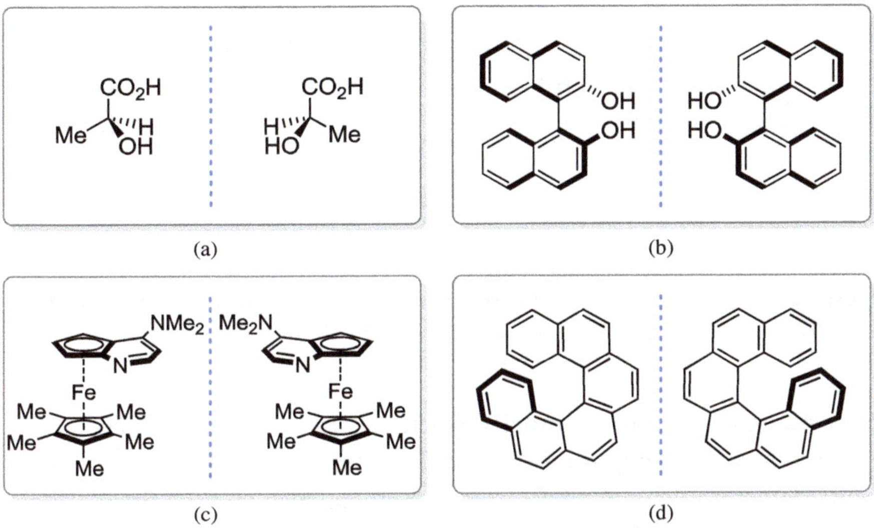

Figure 1. Examples of chiral molecules using a pair of enantiomers with (a) chiral center (lactic acid); (b) axial chirality (BINOL ligand); (c) planar chirality (a ferrocene derivative); (d) chiral helix ([6]helicene). Blue dashed lines represent a mirror plane.

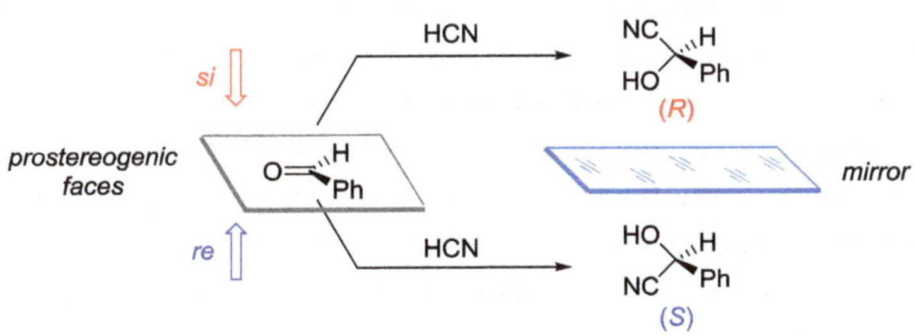

Figure 2. Prostereogenic (enantiotopic) faces of benzaldehyde and their differentiation in a hydrogen cyanide addition reaction.

prostereogenic (enantiotopic) faces of the latter, labeled as *re* and *si* (in accordance with Cahn–Ingold–Prelog priority rules). The addition from the opposite face produces opposite enantiomers of the cyanohydrin product. To discriminate the faces, the presence of an asymmetric inducer is obligatory.

Prostereogenic groups and atoms (e.g., two identical substituents attached to a sp^3-carbon) represent another common structural fragment.

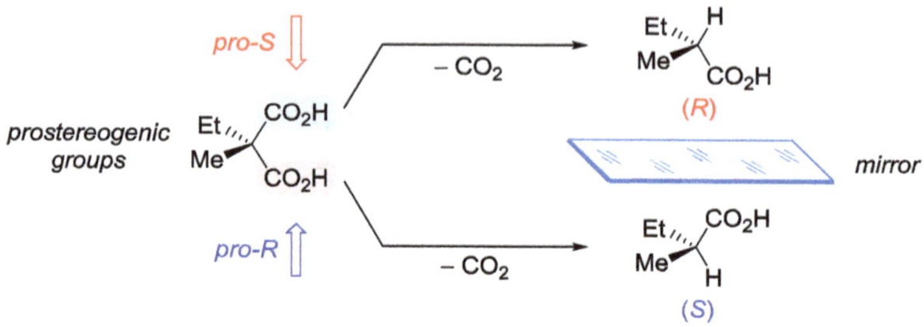

Figure 3. Prostereogenic (enantiotopic) carboxylic groups of methylethyl malonic acid and their differentiation in decarboxylation reaction.

Prostereogenic groups are usually labeled as *pro-S* and *pro-R*, according to Cahn–Ingold–Prelog priority rules. Donations *Re* and *Si* (starting from capital letters to distinguish from *re* and *si*) can also be used for the same purpose. For example, any of two prostereogenic carboxylic groups in methylethyl malonic acid can be removed by means of decarboxylation reaction, which results in formation of enantiomeric products (Figure 3).

Other elements of chirality can be constructed similarly. For example, allenes with axial chirality can be prepared by a palladium-catalyzed 1,4-addition of a hydroborane to prochiral enynes (Figure 4a) [10].

In addition to modification of prochiral groups and faces, a new chirality element can emerge in the event of assembly of several achiral molecules. Thus, chiral binaphthols are synthesized via an iron-mediated oxidative coupling of achiral 2-naphthols (Figure 4b), while a ferrocene derivative with planar chirality can be assembled from planar and achiral cyclopentadienyl precursors (Figure 4c) [11].

Although the new elements of chirality are generated in the reactions mentioned above, these transformations yet cannot be referred as asymmetric since the enantiomers are formed *in equal amounts*, i.e., the reactions generate *racemic mixtures*. Unequal amounts of enantiomers are commonly produced in the presence of a chiral inducer, which can be a chiral catalyst, a chiral reagent, a chiral auxiliary, an internal chiral structural element, or a physical aid like circularly polarized light, or even stirring in one direction [12, 13]. For example, the reactions **A** and **B** shown on the Figure 4 become enantioselective only in the presence of chiral palladium [14] and iron [15] catalysts. In a kinetically controlled reaction,

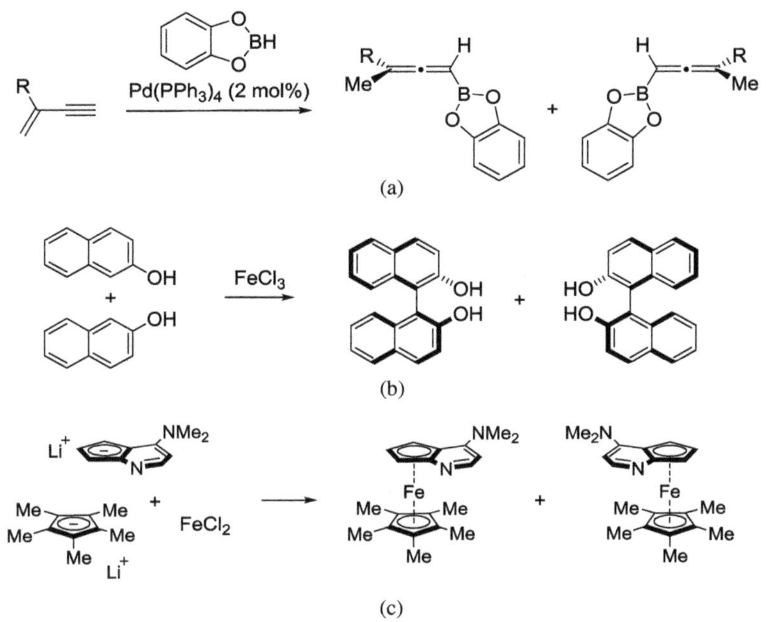

Figure 4. Generation of the molecules with axial and planar chirality from the achiral precursors.

thus equal amounts of enantiomeric products are formed as a consequence of *equienergetic transition states* leading to these products. Such situation usually takes place in the absence of an external chiral force field. Chiral inducers (e.g., a chiral catalyst) eliminate the energetic degeneracy between the transition states due to a distinct set of interactions (or distinct values of their respective energies) occurring between a prochiral substrate and the inducer ("matched" and "mismatched" transition states on the Figure 5). In this case, the equation for *enantiomeric ratio* (*er*) is derived from the Arrhenius equation and shows that enantioselectivity is determined by the difference in activation energies ($\Delta\Delta G^{\neq}$) and temperature:

$$er = \frac{[S]}{[R]} = \frac{k_S}{k_R} = e^{\frac{\Delta\Delta G^{\neq}}{RT}} \tag{1}$$

According to this equation, high enantiomeric ratios, which are suitable for practical applications (er > 98:2), require the difference in

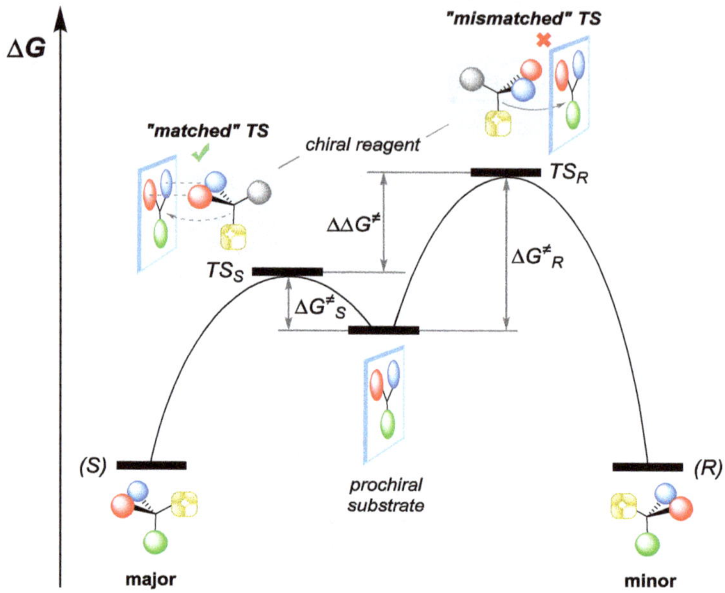

Figure 5. Gibbs free energy plot for the reaction of a prochiral starting material with one enantiomer of a chiral reagent. TS = transition state.

activation energies above 2.3 kcal·mol^{-1} at ambient conditions (T = 293 K). This subtle difference is just slightly lower than the rotation barrier in ethane (3 kcal·mol^{-1}) and is a typical value for a weak hydrogen bond [16]. Nevertheless, even such a little energy difference cannot be securely guaranteed in any given stereoselective chemical reaction. Another conclusion from equation 1 is that a reaction at lower temperature affords a better enantiomeric ratio. However, it is worth noting that such a temperature dependence is strictly valid only for the case of the simplest one-step reaction. Mechanistically complex multistep processes might not necessarily obey the same rule (e.g., Rh-catalyzed asymmetric hydrogenation discussed in the Section 1.5 below).

It also becomes clear from the same energy diagram (Figure 5) that opposite enantiomers must react with an enantiopure chiral reagent with different rates. This makes possible *kinetic resolution* of racemic mixtures [17], a process which is especially efficient if rapid interconversion of enantiomers to each other takes place (Figure 6) [18].

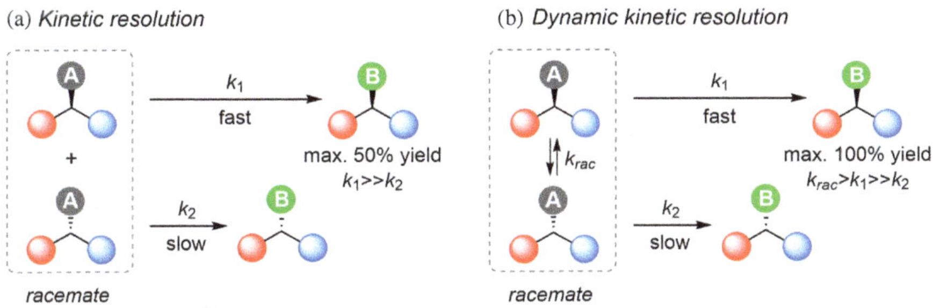

Figure 6. Basic principle of kinetic (a) and dynamic kinetic resolution (b).

Deracemization may also occur under thermodynamic control. In this case, the energy discrimination usually arises from the formation of transient equilibrating intermediates, formed in the reaction of a racemic starting material with a chiral auxiliary [19, 20].

It is worth noting that in rare cases enantioselective transformations may occur even in the absence of any external chiral stimuli. Such a sporadic symmetry breaking is termed *absolute asymmetric synthesis*. The reported cases of absolute asymmetric synthesis [13, 21, 22] had a stochastic nature and occurred due to autocatalytic amplification of the enantiomeric ratio as a result of small fluctuations in the composition of racemic mixtures.

The efficiency of asymmetric transformations is measured by the ratio of the produced enantiomers. Enantiomeric ratio (*er*) correlates with the reaction kinetics (as shown above) and can be determined, for example, by performing HPLC analysis on a chiral stationary phase, probably the most direct and reliable method (Figure 7). The separation of enantiomers is an outcome of their dissimilar absorption energy with a chiral stationary material.

Another frequently used quantity for expressing the ratio of enantiomers is *enantiomeric excess* (*ee*), which is defined as the excess of one enantiomer over the other:

$$ee(\%) = \frac{R - S}{R + S} \times 100, \tag{2}$$

where R and S stand for relative quantities of opposite enantiomers and the R-isomer is dominant. For an enantiopure compound ee = 100%,

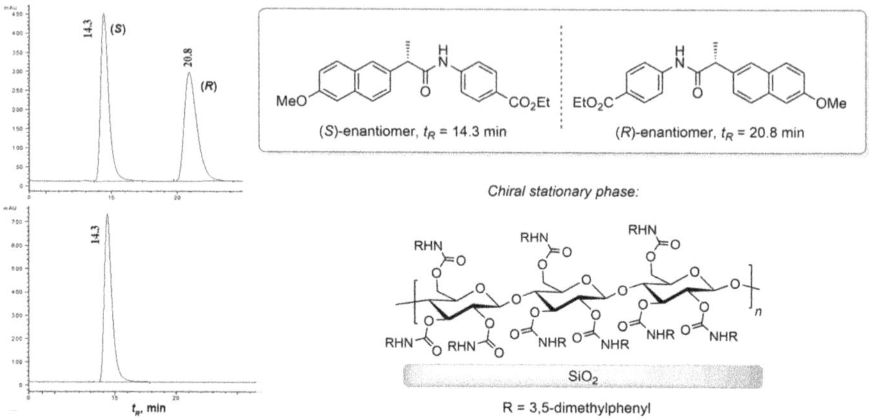

Figure 7. HPLC chromatograms of racemic (upper) and enantiopure (lower) samples of a naproxen derivative [23]. The analysis was performed on a Chiralcel® OD-H column with cellulose tris(3,5-dimethylphenylcarbamate)-coated silica gel as a chiral stationary phase.

while for a racemic mixture the *ee* is zero. Historically, *ee* was widely adopted because it links enantiomeric composition with *optical purity* (*op*):

$$op(\%) = \frac{[\alpha]_{obs}}{[\alpha]_{max}} \times 100, \tag{3}$$

where $[\alpha]_{obs}$ is the observed specific rotation of a sample, and $[\alpha]_{max}$ is the maximum specific rotation (observed for enantiopure material). Optical purity can be determined by polarimetry, which was the only available method for the determination of enantiomeric composition for a long time. In most of cases, *ee* = *op*. However, it was found that optical purity and enantiomeric excess are not necessarily linearly related (the Horeau effect), a phenomenon caused by mutual interaction of enantiomers in solution [24, 25]. Furthermore, even achiral impurities could impact the measured value of specific rotation due to intermolecular interactions. For example, the specific rotation of 1-phenylethanol is increased in the presence of acetophenone [26]. This means that the use of optical purity as a measure of enantiomeric composition is currently inappropriate. Moreover, it was suggested that even the widely adopted *ee* should be avoided and replaced by that of *er*, since modern analytic techniques allow the direct measurement of the latter [27].

$$er = \frac{R}{S} \tag{4}$$

A stereoselective reaction in which a new element of chirality is formed within a chiral starting material usually (but not necessarily) produces unequal amounts of stereoisomers which are not related as mirror images and called *diastereoisomers*. The ratio of products in this case is indicated as a *diastereoisomeric ratio* (*dr*) or, less frequently, as *diastereomeric excess* (*de*), by analogy with the *er* and *ee*. Diastereomers have physical and chemical properties distinguishable without a chiral aid. Therefore, *dr* values can be more easily determined, e.g., by integration of NMR spectra or by chromatography on a non-chiral stationary phase. Derivatization of a sample of unknown optical purity with an optically pure chiral reagent produces two diastereomeric products, with *dr* = *er* of the starting sample. Such a technique can be used as an alternative for *er* measurement. For example, derivatization with α-methoxy-α-trifluoromethyl phenylacetic acid (MTPA, Mosher's acid) is widely used for determination of the enantiomeric purity of alcohols and amines. However, one must ensure that a reaction with a chiral reagent is complete and no kinetic resolution takes place.

1.3 Brief Historical Overview

The dissimilar behavior of enantiomers in biochemical processes was discovered by Louis Pasteur in 1858. He noticed that a mold *Penicillium glaucum* consumed the L-(+)-enantiomer of racemic ammonium tartrate faster than the D-(−)-enantiomer. This discovery represents the first known example of enzymatic kinetic resolution. As a summary of his research in stereochemistry, Pasteur pointed out that "dissymmetric forces" and chiral aid are necessary to create enantiopure chiral compounds [28].

It is extraordinary that the Pasteur's insight happened well before the nature of chirality in organic compounds was rationalized by van't Hoff and Le Bell in their fundamental works on stereochemistry published in 1874. The same scientists also suggested that the presence of optically active

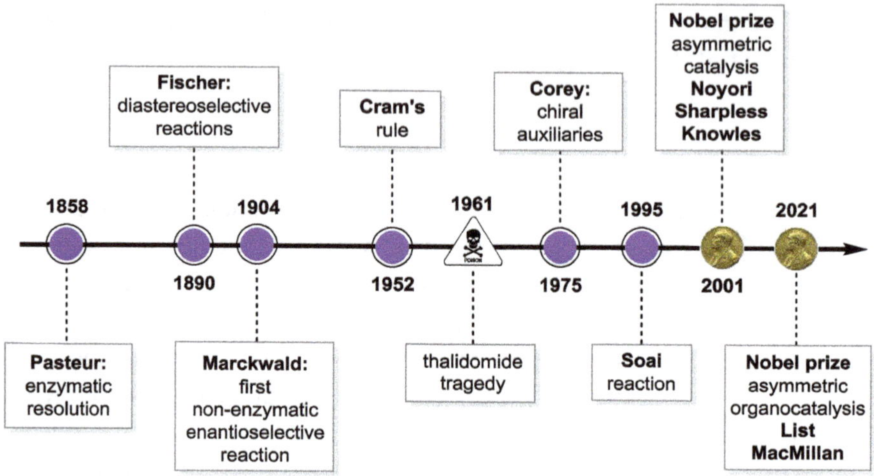

Figure 8. Some notable achievement and pivotal events in the area of asymmetric synthesis and catalysis.

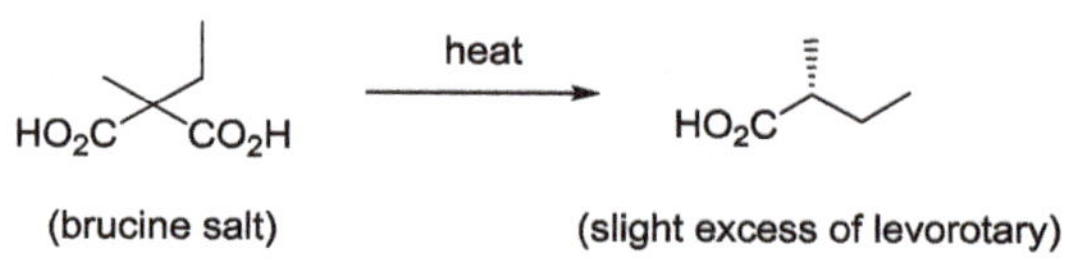

Figure 9. Enantioselective decarboxylation described by Marckwald in 1904.

compounds or irradiation with circularly polarized light were two methods of asymmetric induction [28]. The asymmetric reactions induced by circularly polarized light were discovered subsequently [29] and are suspected to be involved in the prebiotic generation of optically active compounds [30].

In his classical studies on carbohydrate chemistry published in the 1890s, Emil Fischer presented early examples of diastereoselective asymmetric transformations (Kiliani-Fischer synthesis) [28]. He also attributed the stereoselective synthesis of D-glucose by plants to the chiral influence of optically active substances within chlorophyll [31] and proposed the "lock and key" model to explain the high specificity of enzymes [32].

The first examples of non-enzymatic enantioselective transformations and asymmetric catalysis are usually attributed [5, 28] to Marckwald (1904), who found that decarboxylation of the monobrucine salt of methylethyl malonic acid produced 2-methylbutyric acid with a slight excess of the levorotary form (Figure 9), and Bredig and Fiske (1913), who reported

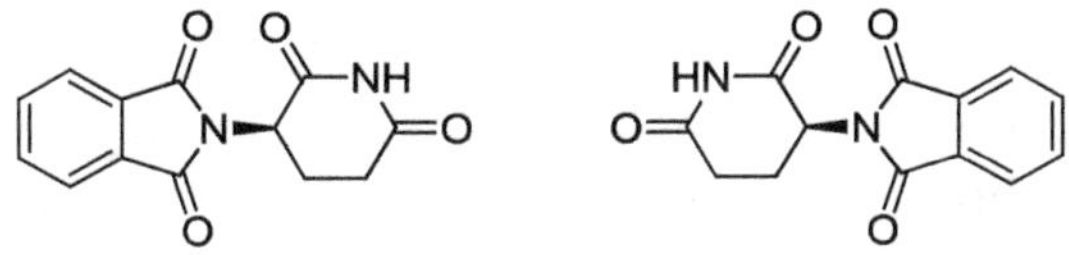

Figure 10. Enantiomers of thalidomide.

the synthesis of slightly enantioenriched (about 8% *ee*) mandelonitrile by the asymmetric addition of HCN to benzaldehyde in the presence of quinine or quinidine as chiral catalysts. Marckwald is also notable by his definition of asymmetric synthesis, as "reactions that produce optically active substances from symmetrically constituted compounds with the intermediate use of optically active materials but with the exclusion of all analytical processes" [28, 33].

Despite these benchmarking early advances, progress in asymmetric synthesis was slow until the mid of the 20th century, when the modern era commenced.

In 1952, Donald Cram formulated his famous rule which linked the stereochemical outcome of carbonyl addition reactions to the configuration of the stereocenter adjacent to the carbonyl [34]. At nearly the same time, Vladimir Prelog developed a similar concept [35]. Cram's rule, together with the descendant chirogenic models, was based on consideration of steric hindrances as a source of asymmetric induction (see Section 1.4 below).

In the beginning of 1960s, the "thalidomide disaster" happened. A racemic drug thalidomide (Figure 10) was widely prescribed until its teratogenic effect was revealed in 1961. The adverse effect was supposedly caused by (*S*)-enantiomer. The racemic drug heavily affected around 10,000 pregnancies and raised concerns about the safety of racemic pharmaceuticals. With thalidomide the problem was even more complex: the "non-toxic" enantiomer racemizes in body giving rise again the "toxic" enantiomer. The tragedy accelerated research in asymmetric synthesis because of increased demand for enantiopure medicines, and also for enhanced the study of the physiological properties of enantiomers.

Although the first use of a chiral auxiliary as an aid in asymmetric synthesis could be credited to McKenzie who studied in 1904 the reduction and Grignard addition reactions of menthyl benzoylformate [36], the concept was revived by Corey and Ensley, who employed (−)-8-phenylmenthol in the synthesis of an optically active prostaglandin intermediate [37]. Further development of the concept allowed high enantiomeric purity of products in numerous other transformations to be obtained and still represents a cornerstone approach in asymmetric synthesis [38–40].

However, the idea of using the catalytic source of chirality was very appealing and economically attractive. The dawn of asymmetric catalysis with chiral metal complexes started from the works of William Knowles and Ryōji Noyori in the 1960s [41, 42]. Despite disappointingly low enantioselectivities attained in the early works, further research resulted in the design of highly enantioselective catalytic hydrogenation processes mediated by rhodium and ruthenium complexes with chiral phosphine ligands as asymmetric inducers [43, 44]. The development also resulted in the first industrial-scale non-enzymatic asymmetric synthesis, i.e., the production of L-DOPA [45], an anti-Parkinson's drug (Figure 11). As another prominent achievement, Barry Sharpless and co-workers reported in 1980 the first practical method for asymmetric epoxidation of allylic alcohols in the presence of a chiral titanium catalyst formed *in situ* from titanium(IV)

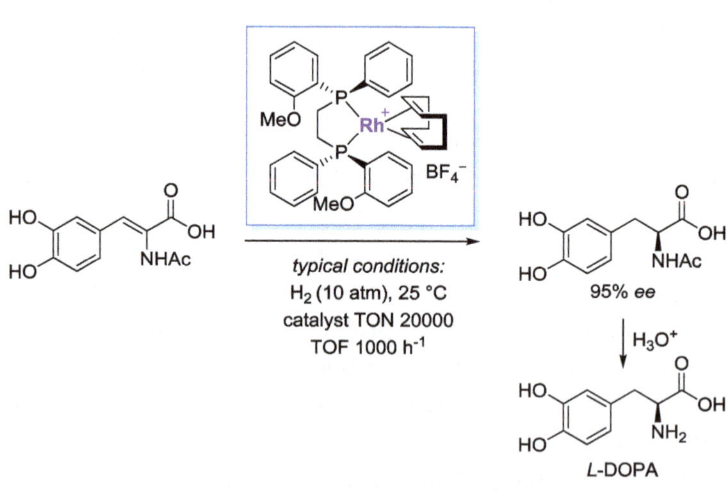

Figure 11. Monsanto's industrial synthesis of L-DOPA. TON = turnover number; TOF = turnover frequency.

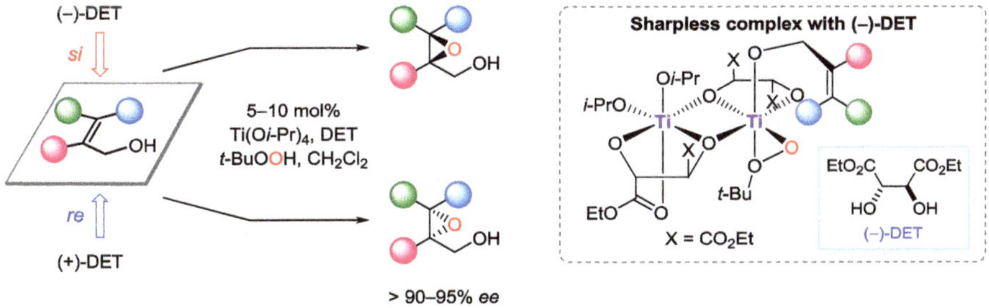

Figure 12. Sharpless asymmetric epoxidation of allylic alcohols [46, 47].

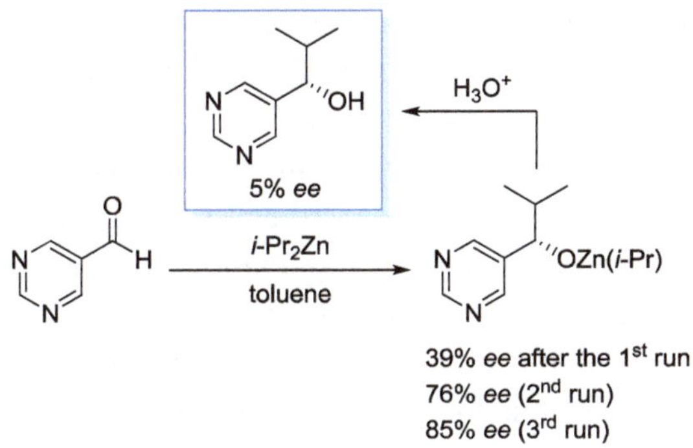

Figure 13. Amplification of chirality in the Soai autocatalytic reaction [49].

isopropoxide and diethyl tartrate (Figure 12) [46, 47]. The pioneering contributions of Knowles, Noyori, and Sharpless on the asymmetric catalytic reactions were acknowledged by the Nobel Prize in Chemistry in 2001.

The Sharpless asymmetric epoxidation reaction was among the first transformations for which a nonlinear relation between enantiomeric purity of the product and the chiral catalyst was observed by Kagan and co-workers [48], i.e., enantiomerically impure catalyst gave a product with an enantiomeric excess higher than that of the catalyst. In a nutshell, nonlinear effects (NLE) are observed due to the formation of several equilibrating homochiral and heterochiral complexes with different catalytic activity, which give the products in dissimilar rates. NLE play an important role in mechanistic investigations but also became a clue in understanding the plausible origin of biological homochirality. The Soai reaction [49] discovered in 1995 (Figure 13) represents a notable case of an asymmetric

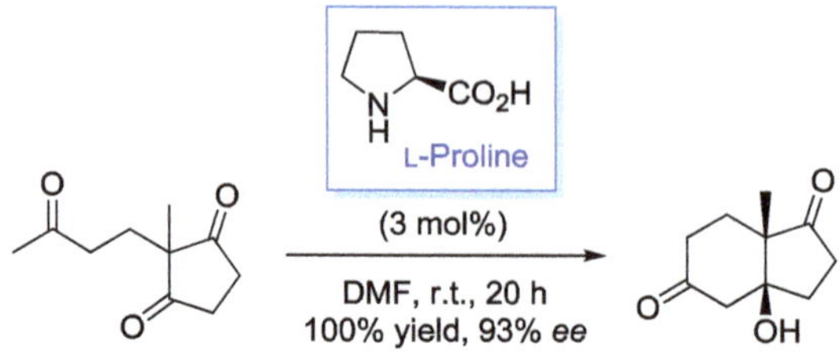

Figure 14. Hajos–Parrish–Eder–Sauer–Wiechert reaction.

autocatalytic transformation with an extremely pronounced positive NLE. It was shown, for example, that asymmetric autocatalysis in the Soai reaction results in a spectacular amplification of chirality (>99.5% *ee*) from an extremely low enantiomeric excess produced by circularly polarized light [50] and other triggers, including spontaneous symmetry breaking and a rare example of absolute asymmetric synthesis [21, 51]. The mechanism of the Soai reaction and its connection to the abiotic origin of homochirality have been the subjects of the intensive studies [7, 52, 53].

Until beginning of the 2000s, the triumph of asymmetric transition metal catalysis had overshadowed another prominent direction, i.e., the catalysis with small chiral organic molecules, currently known as *asymmetric organocatalysis* [54]. Although, historically the first enantioselective organocatalytic transformations could be credited to Marckwald, Bredig, and Fiske (as mentioned above), the pivotal discoveries in the modern era were made in the 1960–1980s [55]. Perhaps, the first notable event was the discovery of the asymmetric aldol reaction (Hajos–Parrish–Eder–Sauer–Wiechert reaction, see Figure 14), in which a natural amino acid (L-proline) acted as a chiral secondary amine catalyst to perform activation of the carbonyl through the formation of enamine [56].

A plethora of other processes have been developed within this concept, such as chiral phase transfer catalysis [57], hydrogen bonding catalysis [58], etc. An especially phenomenal growth has occurred after the revival of organocatalysis in the 2000s. Now, this approach truly stands as one of three major pillars of asymmetric catalysis, together with metal catalysis and biocatalysis. In 2021, two prominent pioneers of asymmetric organocatalysis, Benjamin List and David MacMillan, were awarded the Nobel Prize in Chemistry.

Currently, asymmetric synthesis continues its rapid advance, including intriguing innovative developments at the interface of its well-established three major catalytic pillars [59–61] and in synergy with new emerging directions, such as photo- and electrocatalysis [62]. The underlying mechanisms leading to the genesis of chirality in the different branches of asymmetric synthesis and catalysis are briefly presented in the following subsections.

1.4 Chirogenesis in Substrate- and Reagent-Controlled Asymmetric Reactions

As was already mentioned in the preceding section, the first report on the substrate-controlled asymmetric reaction can be credited to Emil Fisher. Thus, he discovered [63] that treatment of L-arabinose with aqueous hydrocyanic acid followed by hydrolysis of cyanohydrin intermediates (Kiliani reaction) produced L-mannonic acid as the main product and only a minor amount of L-gluconic acid (Figure 15). Hence, the stereochemical configuration of the carbohydrate backbone dictated the outcome of the carbonyl addition reaction in a way that nucleophilic attack occurred predominantly at the *si*-face of the carbonyl group.

Evidently, the diastereoselective nature of the Kiliani–Fisher synthesis is a result of the nonuniform steric environment at the opposite faces of the carbonyl group. However, a mechanistic rationale which allows

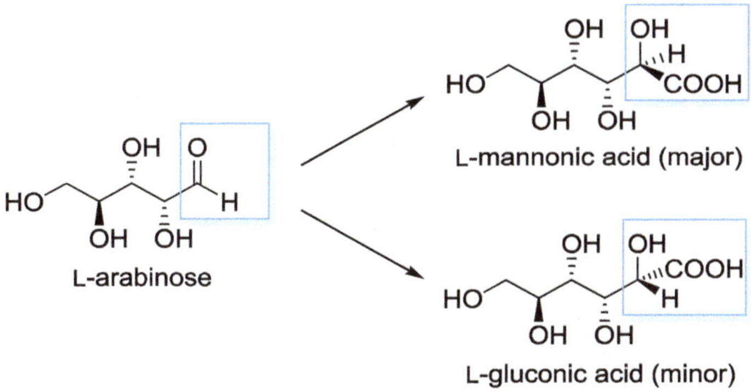

Figure 15. Synthesis of diastereomeric hexonic acids from L-arabinose by Emil Fischer.

prediction of the stereochemical configuration for a newly formed stereocenter was missing at Fischer's time. The problem of the influence of an existing chiral center on a prochiral reaction center attracted a great deal of attention afterward. The fundamental discoveries addressing the problem were mostly made around the carbonyl addition chemistry because of its cornerstone significance for organic synthesis.

Donald Cram suggested the first rule with a predictive power (Figure 16a). The rule was formulated for acyclic carbonyl compounds with an α-stereocenter (the case of 1,2-asymmetric induction) and stated "that diastereomer will predominate which would be formed by the approach of the entering group from the least hindered side of the double bond when the rotational conformation of the C–C bond is such that the double bond is flanked by the two least bulky groups attached to the adjacent asymmetric center" [34]. In other words, the nucleophile prefers approaching from the side of the smallest group (e.g., a hydrogen atom) attached to the nearby asymmetric carbon atom.

Cram's rule was based on simple steric arguments and provided the first rationale for asymmetric induction in carbonyl addition reactions. Nevertheless, the purely empiric Cram's model suffered from several pitfalls

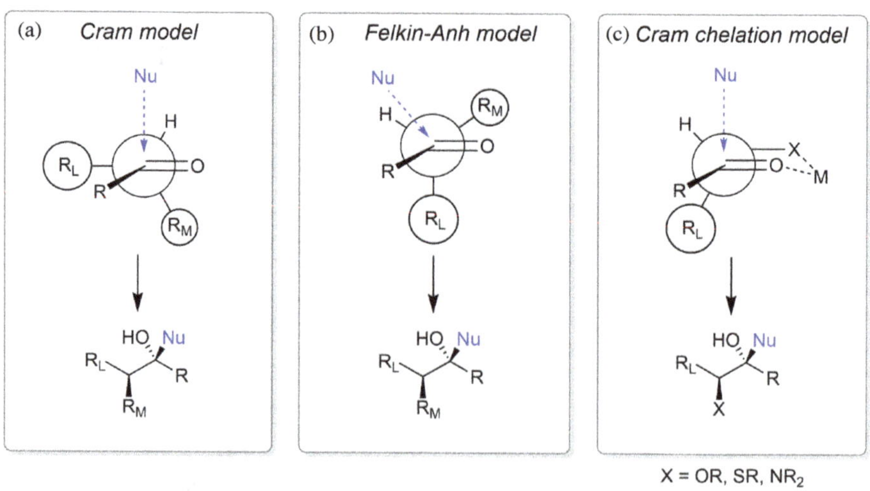

Figure 16. Representative models of 1,2-asymmetric induction in carbonyl addition reactions. R_L and R_M represent large and medium size substituents, respectively. M = metal atom, Nu = nucleophile.

which resulted in incorrect predictions. These limitations encouraged the development of more advanced models, which eliminated the weaknesses of Cram's rule via consideration of the governing steric and electronic effects in more accurately depicted ground and/or transition state geometries. The detailed overview of these models has been comprehensively described [64, 65]. In a nutshell, all of these models attempt to recognize the most sterically accessible face of a carbonyl moiety, although the input assumptions might be distinct. Here we would like to highlight only the Felkin–Anh model (Figure 16b) which predicts the same sense of asymmetric induction as the Cram model but in a more precise manner. The accuracy arises from consideration of the transition state geometry which assumes approach of a nucleophile along the Bürgi–Dunitz trajectory (Nu···C=O angle ~ 107°) in a staggered conformation and takes into account also orbital effects. The model's assumptions were supported by *ab initio* calculations [66]. Another notable case is when the carbonyl compound forms a chelate, typically leading to "anti-Cram" or "anti-Felkin" products (Figure 16c) [67]. It is worth noting that there are no rules without exceptions, so the fundamental problem of asymmetric induction in carbonyl additions remains open for further improvement. For example, addition reactions of highly reactive allylmagnesium reagents often cannot be straightforwardly interpreted in the framework of the Felkin–Ahn or chelation-control models [68].

Prelog's rule (Figure 17) was formulated at nearly the same time as Cram's rule and explains asymmetric induction in the event of Grignard

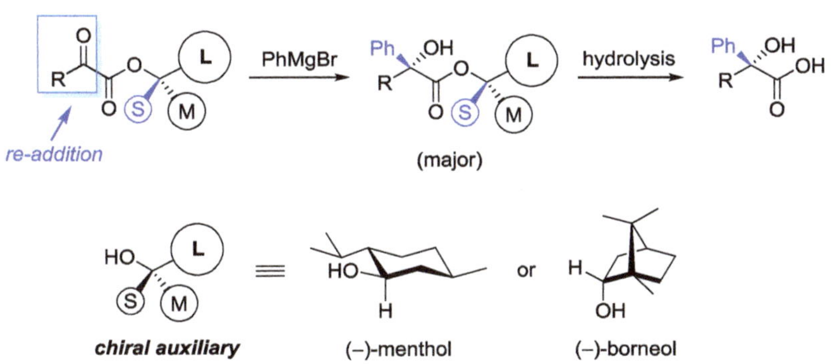

Figure 17. Prelog's rule and asymmetric induction with a chiral auxiliary. L, M, and S stand for large, medium, and small substituents, respectively.

addition to chiral pyruvate esters [35]. As in Cram's case, nucleophile (PhMgBr) prefers to attack from the least hindered side of the smallest substituent. The hydrolysis of produced ester released enantiomerically enriched α-hydroxy acids. This example is important for our discussion since it demonstrates asymmetric induction with a *chiral auxiliary*, i.e., a covalently linked chiral motif which controls the stereochemical course of the reaction and is removed afterward. Natural chiral alcohols such as (−)-menthol and (−)-borneol served that purpose in the Prelog's synthesis. Currently, numerous chiral auxiliaries are known, many of them being derived from the natural building blocks (a few examples are shown on the Figure 18). The chiral auxiliary approach is a classic technique of asymmetric synthesis [38–40] which often offers excellent enantiomeric ratios. The stereochemistry is commonly governed by steric factors, i.e., the incoming group is introduced from the least hindered side. For example, alkylation of enolates with Evans' oxazolidinone auxiliary predominantly occurs from the side opposite to the bulky isopropyl substituent of the auxiliary (Figure 18) [69].

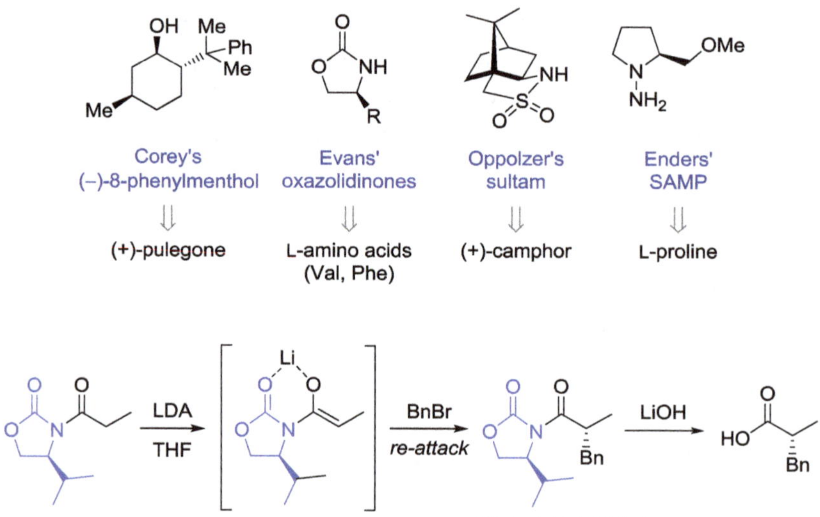

Figure 18. Classic chiral auxiliary groups and their natural precursors. Use of Evans' oxazolidinone auxiliary group for stereoselective enolate alkylation [69].

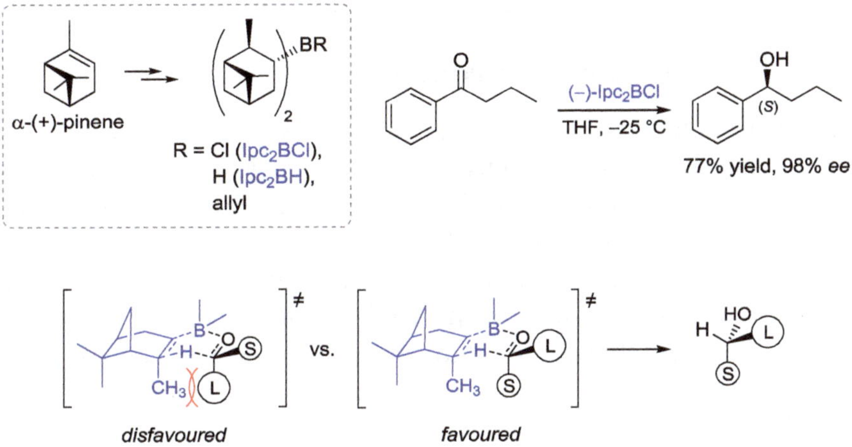

Figure 19. Examples of chiral organoboron reagents for asymmetric synthesis. Asymmetric reduction with (–)-Ipc_2BCl and proposed model of asymmetric induction [74].

Chiral reagents provide an option to perform enantioselective synthesis directly with an achiral starting material. As in the case of chiral auxiliaries, the first examples of chiral reagents were also derived from the easily available natural compounds. A family of diisopinocampheylborane reagents (Figure 19) derived from α-pinene (available in both enantiomeric forms) is a good example illustrating this concept. For example, Ipc_2BCl and the sibling organoboron reagents perform highly enantioselective carbonyl reductions [70, 71], and can be used for other applications, e.g., asymmetric allylations (for R = allyl). The absolute configuration of the reduction product can be predicted by using a six-membered ring transition state model, in which the largest group of the carbonyl substrate occupies the less sterically hindered equatorial-like position (Figure 19). Multiple other chiral reagents have been developed [72]. Metal reagents commonly contain chiral ligands (e.g., BINAL-H for the reduction of ketones) [73], which can be recovered after the reaction and reused.

Concluding this section, we would like to mention that asymmetric reactions can also be triggered by chiral media (solvent) [75, 76] and circularly polarized light [29, 30]. However, these options are omitted from the discussion here. Photochemical reactions are covered in Chapter 7 of this book.

1.5 Enantioselective Catalysis

1.5.1 Introduction and Basic Principles

In contrast to the stoichiometric approaches presented above, enantioselective catalysis is advantageous because it employs only a catalytic amount of a chiral inducer. Principal stages of an enantioselective catalytic reaction and its free energy diagram are schematically depicted in Figure 20. First, chiral catalyst **C*** must form a complex (**Int1**) with reagent **A** and/or prochiral substrate **B**. Assembly of the complex can occur via either covalent linkage or through nonbonding interactions. Such complexation should also provide activation of the reactants, so that the activation energy of the catalytic process is lower than that of non-catalytic reaction leading to racemic products. The reaction between **A** and **B** occurs within the catalytic complex and is commonly a rate-determining and a chirogenic step, on which asymmetric induction takes place. The recovery stage releases the catalyst back from its complex **Int2** with the formed chiral product **AB**. Importantly, the latter must not block ("poison") the catalyst and therefore must be flawlessly expelled from **Int2** as a prerequisite of high catalyst activity.

The catalyst productivity and activity can be given as turnover number (TON) and turnover frequency (TOF). Turnover number shows the number of moles of substrate that a mole of catalyst can convert before

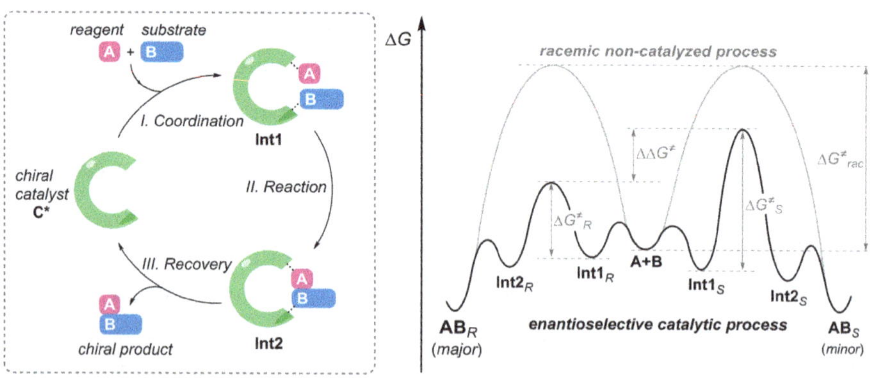

Figure 20. General principle of enantioselective catalysis and the free energy plot for a hypothetical three-step catalytic reaction.

becoming inactivated (higher number is better), while TOF shows the turnover per unit of time:

$$TON = \frac{n_{product}}{n_{cat}} \tag{5}$$

$$TOF = \frac{TON}{t} \tag{6}$$

For industrial applications, the TON should be at least 1000 for a small-scale production [77].

What ensures efficacy of enantioselective catalytic processes delivering products with high optical purity at low catalyst loadings? Briefly, the following criteria must be fulfilled:

— The chiral catalyst must provide sufficient differentiation of enantiomeric transition states with free energy difference $\Delta\Delta G^{\neq}$ above the ~2 kcal·mol^{-1} threshold mentioned above (see Section 1.2), in addition to overall activation of reactants to eliminate a chance for non-catalyzed racemic process to occur.
— A number of catalytic cycle-related issues must be taken into account, such as kinetics and equilibria of the individual steps.
— The chiral catalyst must be recyclable in an unchanged form after each turnover; it must not degrade in the process and must not give other catalytically active species which are less efficient in terms of enantioselectivity but more reactive.
— Especially important is the absence of achiral catalytic impurities which promote a parasitic racemic reaction.

The detailed discussion of these important issues are given below and illustrated by notable examples.

Basics for enantiodiscrimination and catalyst design. Stereodifferentiation lies at the heart of any enantioselective process, including catalytic transformations. Therefore, distinct modes of substrate interaction with a chiral catalyst must take place within the competitive diastereomeric transition states, with adequate high energy difference and, consequently, distinct reaction rates within the catalytic cycle. The chirogenic mechanisms leading to enantiodiscrimination within each branch of enantioselective catalyst will be presented in the forthcoming Subsections 1.5.2–1.5.4. Here we would like to focus on the general point concerning the selection and design of appropriate chiral catalysts for developing new asymmetric transformations.

The traditional approach relies on trial-and-error discovery, which could be serendipitous, led by analogy, or involve high-throughput screening. Such methodology depends on luck, intuition, and is time-consuming. Although the preliminary screening allows promising candidates to be identified for further improvement and is inevitable at the current technology level, rational design should be applied for the next lead structure optimization. First of all, speculative models considering steric and electronic factors in the respective transitions states could be imagined and provide a good starting point for structural optimization [78]. State-of-the-art computer technologies allow the involvement of quantum chemistry calculations [79, 80] and chemoinformatic approaches, such as quantitative structure–selectivity relationships [81–83].

The copper-catalyzed enantioselective Markovnikov hydroboration of aliphatic terminal alkenes developed by Ito and co-workers [84] is a good representative example illustrating the power of computer-assisted chiral ligand design (Figure 21). In this work, the design guidelines were obtained from density functional theory (DFT) calculations of the respective transition states, which allowed the identification of the key structural parameters of the disphosphine ligand responsible for regio- and enantiocontrol. Structural optimization and increasing the steric bulkiness in the appropriate places of the first-generation ligand resulted in finding the optimized structure which delivered chiral alkylboronates with high regioselectivity and excellent enantioselectivity (up to 99% *ee*).

There is no need to say that such catalyst design provides valid and useful results only if the reaction mechanism is known and if the chirogenic step is firmly identified, including the structures of the key catalytic species (which are not necessarily the same as the initially loaded "catalyst"). Moreover, several other issues discussed below should not impede the catalyst performance.

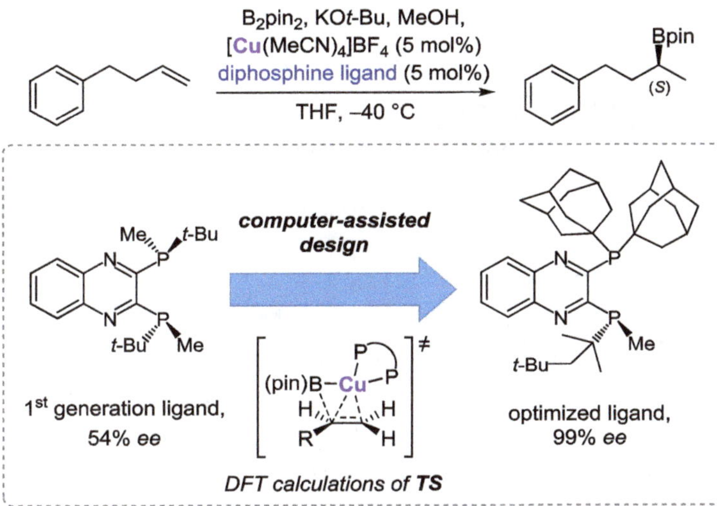

Figure 21. Computational design of a high-performance diphosphine ligand for enantioselective hydroboration [84]. DFT = density functional theory, pin = pinacolato ligand.

The catalytic cycle. The knowledge of the whole mechanistic cycle including the structures of intermediates and the respective kinetic parameters is vital for the rational process design. These data could be helpful in the elimination of problematic hot spots and essential for the identification of the rate-determining step which commonly determines the stereochemical outcome as well. It is worth mentioning that if a pair of diasteromeric intermediates is formed during the complexation step and their rapid mutual interconversion takes place, the distribution of products derived from them reflects the difference in activation energies of the respective rate-limiting steps but not necessarily the equilibrium distribution of the two intermediates (the Curtin–Hammett principle). In other words, *the major diastereomeric intermediate does not necessarily give rise to the major enantiomer.*

The already-mentioned rhodium-catalyzed asymmetric hydrogenation of prochiral olefins like methyl (*Z*)-α-acetamidocinnamate **1** (Figure 22) can serve as a brilliant illustration [43, 45]. The reaction mechanism has been comprehensively investigated [85, 86] with the help of kinetic

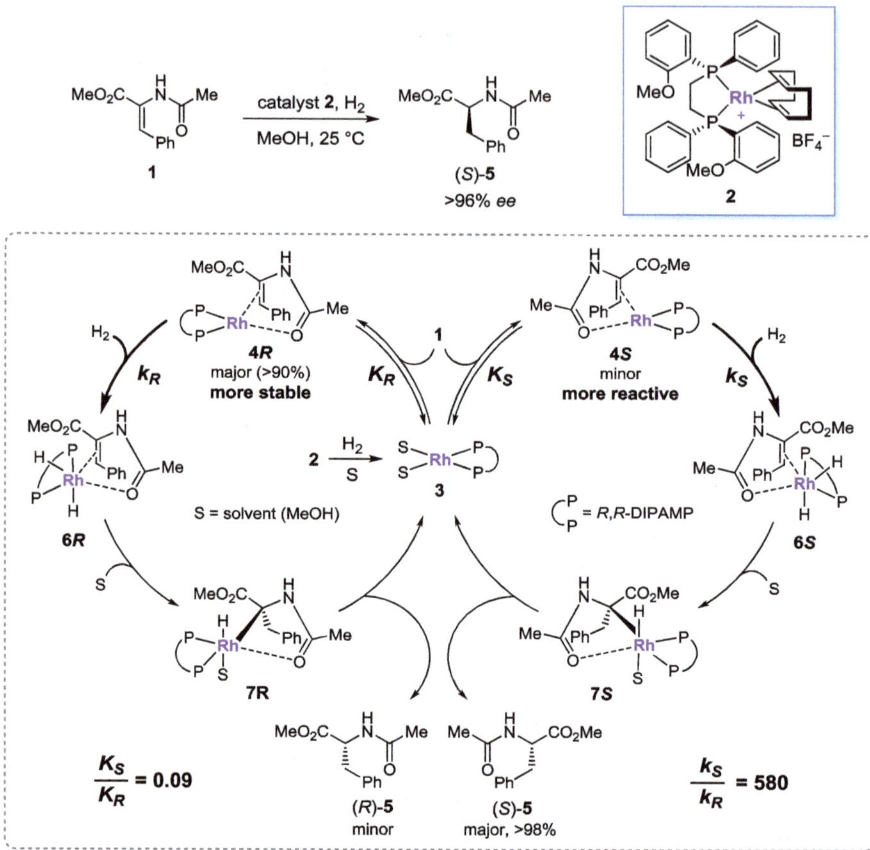

Figure 22. Mechanism of asymmetric hydrogenation of methyl (*Z*)-α-acetamidocinnamate **1** mediated by a chiral rhodium complex **2** with (*R,R*)-DIPAMP ligand [85].

measurements, spectroscopic methods, and several computational studies (e.g., see Refs. [87–89]). Structures of intermediate species have been elucidated and firmly established based on spectroscopy (mostly by NMR) and on single-crystal X-ray analysis of the isolated stable intermediates. Chiral diphosphine ligands at rhodium serve as a source of asymmetric induction. The reaction performance is outstanding. Hydrogenation occurs in a highly enantioselective manner (up to 99% *ee*, depending on the substrate and the chiral ligand combination) at low catalyst loading (0.005 mol%). The catalyst **2** delivers the truly catalytically active species **3** after displacement of 1,5-cyclooctadiene ligand with solvent (MeOH). One of the proposed and widely accepted mechanisms [85] implies the rapid and reversible formation of diastereoisomeric enamide-catalysts adducts ***4R*** and **4*S*** at the first step, with rhodium bound to the opposite faces of the planar olefinic bond. These intermediates give rise to enantiomeric products (*R*)- and (*S*)-**5**, respectively, after completion of the catalytic cycle. Rapid interconversion between ***4R*** and **4*S*** takes place, and more stable adduct ***4R*** dominates (ratio of equilibrium constants $K_S/K_R = 0.09$). The second step is irreversible, oxidative addition of hydrogen leading to dihydride complexes ***6R*** and **6*S***. At room temperature, this step was found to be rate-limiting for the whole catalytic cycle and determines its enantioselectivity. Finally, fast migratory insertion of olefinic carbon into a Rh–H bond occurs followed by the reductive elimination. The latter results in the recovery of catalyst **3** and release of product **5**.

Notably, the predominantly formed (*S*)-enantiomeric product **5** arises from the minor and less-stable enamide-catalyst adduct **4*S***. Such "anti-lock-and-key" behavior is commonly observed in such Rh-catalyzed asymmetric hydrogenations. This is a consequence of the higher reactivity of **4*S*** at the rate-limiting step, in comparison with that of ***4R***. The difference in the respective activation energies is 3.7 kcal·mol^{-1}, which corresponds to the approximately 580-fold faster reactivity of the minor adduct **4*S*** and can compensate for its approximately 10-fold lower concentration. The overall balance of the two factors (concentration difference and consumption rate) favors the more than 50 times faster conversion of **4*S*** than that ***4R*** and results in >96% *ee* in favor of *S*-enantiomer.

The mechanism agrees with the experimentally observed decrease of enantioselectivity with an increase of hydrogen pressure. High concentration of hydrogen accelerates the rate of H_2 oxidative addition step, and so the overall rate and stereochemistry of the reaction becomes more determined by the initial binding of substrate to rhodium catalyst favoring the formation of ***4R*** and eventually *R*-enantiomer. Another interesting outcome is that the enantioselectivity decreases with decreasing temperature since a rapid interconversion between intermediates ***4R*** and **4*S*** becomes "frozen out."

Side reactions, competitive mechanisms, and the background racemic processes. Finally, no side processes leading to loss of catalyst activity

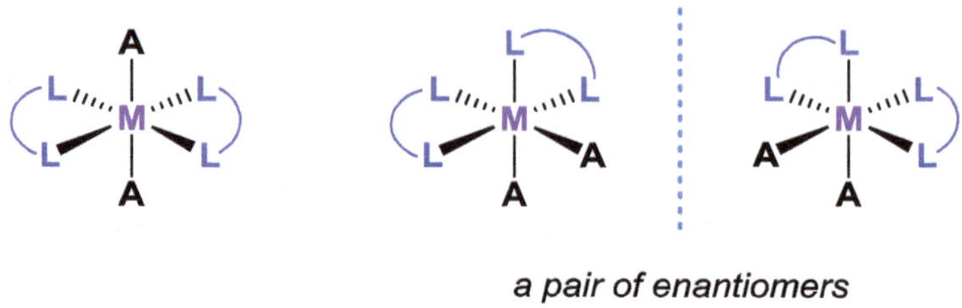

Figure 23. Stereoisomers of an octahedral complex ML_2A_2, where L = a bidentate ligand.

and erosion of enantioselectivity must be present, or at least these must not dominate. Identification and elimination of the destructive pathways such as racemization of products, degradation, or racemization of catalyst, etc. can be very challenging. It is worth mentioning that multiple catalytic species could be generated, e.g., by mixing a metal salt and a chiral ligand, which could catalyze the respective competitive reaction pathways leading to the opposite enantiomeric products. This implies not only different stoichiometry (e.g., ML vs. ML_2) but also spatial arrangement of ligands. For example, three stereoisomeric complexes exist for an octahedral complex ML_2A_2 (with L a bidentate ligand, Figure 23), including a pair of enantiomers (if L is achiral). Moreover, the generation of multiple catalytic species makes the computational modeling of mechanism fairly complex and reaction selectivity is difficult to control.

On the other hand, stereochemically distinct catalytic species do not necessarily result in a frustratingly poor *ee*. This could be beneficial if a positive NLE (asymmetric amplification) is observed since it allows high *ee* values to be reached even with optically impure catalysts. For example, chiral amino alcohol-catalyzed dialkylzinc addition to aldehydes exhibits a large positive NLE (Figure 24) [90, 91]. Thus, (−)-3-*exo*-(dimethylamino) isoborneol (DAIB) with 15% *ee* resulted in the addition product (*S*)-**8** with 95% *ee*. It was found that the intermediate zinc complex was able to form homo and heterodimers with lower reactivity for the later. Thus, the less stable homodimer dissociates more readily and catalyzes the reaction while the more stable heterodimer consumes the minor enantiomer of the ligand, leading to the observed large positive NLE. The opposite scenario of negative NLE (asymmetric depletion) when the *ee* of the product is lower than that of catalyst, can also take place [48].

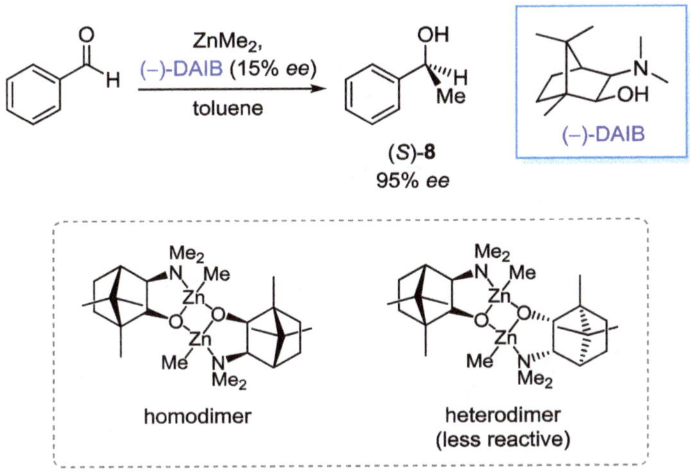

Figure 24. Positive nonlinear effect in the enantioselective addition of $ZnMe_2$ to benzaldehyde [90, 91].

It has already been mentioned above that the activation barrier of a non-catalytic reaction must be higher than that of a catalytic one, to eliminate the possibility of a background racemic process. However, the latter can also be catalyzed by an achiral species with a catalytic activity. For example, in asymmetric additions of organometallic reagents to carbonyl compounds such species are usually uncomplexed metal salts. Thus, asymmetric addition of diphenylzinc to 2-naphthaldehyde occurred with high enantioselectivity only with pure organometallic reagent, while the diphenylzinc generated *in situ* from phenyllithium and $ZnCl_2$ delivered almost racemic carbinol **9** (Figure 25) [92]. The disappointingly low enantioselectivity is an outcome of much faster racemic reaction catalyzed by lithium chloride. The undesired process can be inhibited and high enantioselectivity restored by the addition of *N*,*N*,*N*′,*N*′-tetraethylethylenediamine (TEEDA) which traps the lithium salt into a catalytically inactive complex.

To conclude the discussion: the design of highly enantioselective catalytic processes is fairly challenging and numerous issues must be considered. Judicious mechanistic experiments supported by computational studies can provide a useful guide and assist in the rational process development. The underlying mechanisms of asymmetric induction are

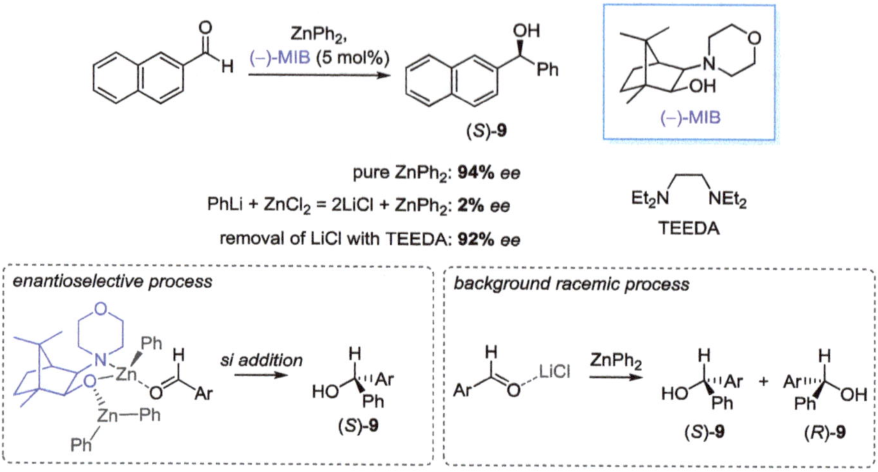

Figure 25. Enantioselective addition of $ZnPh_2$ to 2-naphthaldehyde; background racemic process and its suppression [92].

presented for each branch of asymmetric catalysis in the following Subsections 1.5.2–1.5.4. The chirogenesis in these transformations is commonly attributed to a combination of stabilizing and/or destabilizing steric and electronic interactions taking place in the transition states leading to opposite enantiomers.

1.5.2 Chirogenesis in Metal Catalysis

The mainstream approach to enable asymmetric induction in the metal-catalyzed reactions is the use of chiral ligands. Representative structures of the most common ligands are presented in Figure 26. Many of these are derived from the chiral pool, e.g., amino and hydroxy acids, or are the natural compounds themselves (e.g., the alkaloid sparteine). In terms of symmetry elements, the predominant fraction of the chiral ligands have C_2 and C_1-symmetry, while other types like C_3 and D_2-symmetric ligands are far less common [93–95].

The abundance of C_2-symmetric ligands originates from the fact that C_2-symmetry reduces the number of competitive diastereomeric transition states and therefore facilitates asymmetric induction and ligand design (Figure 27) [96, 97]. For the latter, steric interactions in C_2-complexes can

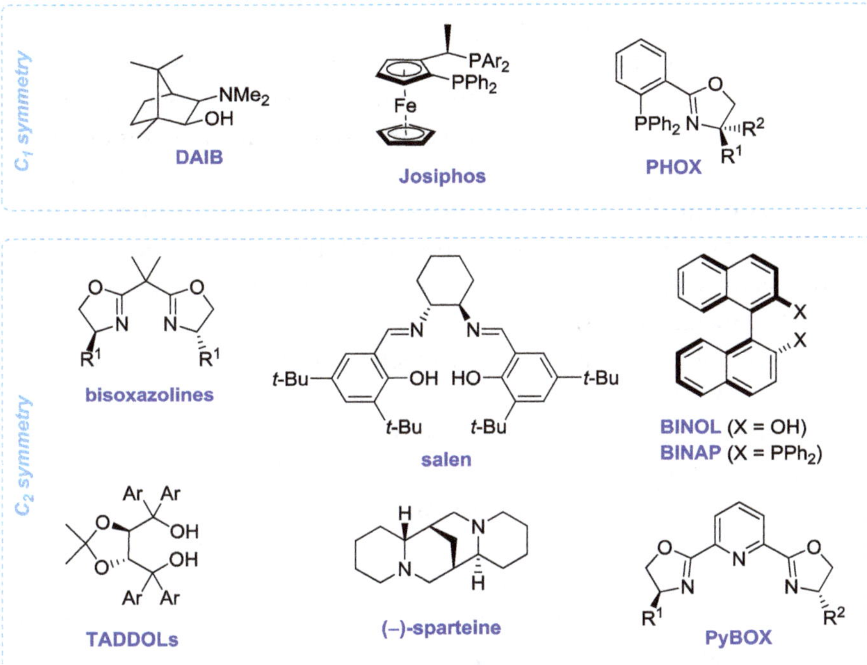

Figure 26. Representative examples of chiral ligands.

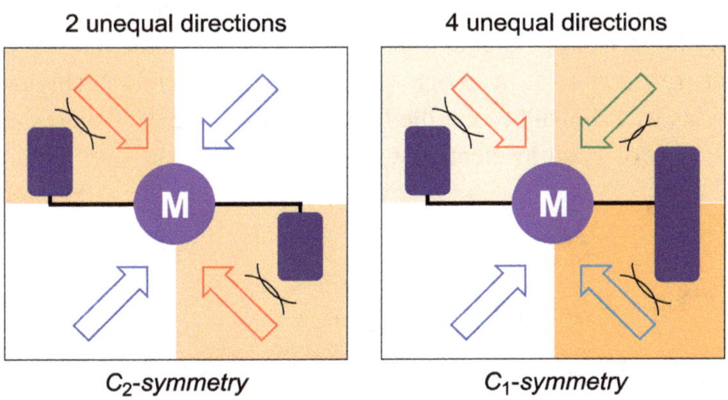

Figure 27. Dissimilarity between C_1 and C_2-symmetrical metal complexes. Orange background represents hindered quadrants.

be easily analyzed with the aid of a quadrant diagram which contains two non-adjacent hindered quadrants. Nevertheless, C_2-symmetry is not an obligatory prerequisite for successful asymmetric induction and C_1-ligands might also perform well. Although, the design of C_1-ligands is more intricate, it adds additional flexibility in the fine-tuning of steric and electronic interactions within diastereomeric transition states [78]. As computational studies facilitate ligand design and become a more accessible tool [84], the previously supplanted C_1-ligands steadily expand their significance in asymmetric metal catalysis [98].

The venerable Rh-catalyzed asymmetric hydrogenation affords high enantioselectivities (>95% *ee*) only for specific combinations of chiral ligands and substrates. Historically, the privileged chiral ligands were the chelating C_2-symmetric diphosphines (Figure 28), although di- and monophosphine ligands with C_1-symmetry can also provide excellent results [98]. One common feature of the C_2-symmetric diphosphines is the presence of two aryl (or cycloalkyl) substituents on each of the two phosphorus atoms. The ligand chirality arises from central chirality on either phosphorus (DIPAMP) or carbon atoms (ChiraPhos, DuPhos).

In the case of the DIPAMP ligand, the edge-turned phenyl groups appear in more hindered quadrants while face-lying *o*-anisyl groups lie in unhindered quadrants [43]. It could be expected that the approaching olefinic substrate occupies predominantly the less hindered quadrant to minimize the steric repulsions. Indeed, this arrangement explains the predominant formation of a more stable diastereomer **4*R*** (Figure 22). However, as was shown before, the faster reaction rate of **4*S*** is responsible for the origin of enantioselectivity. This "anti-lock-and-key" behavior is

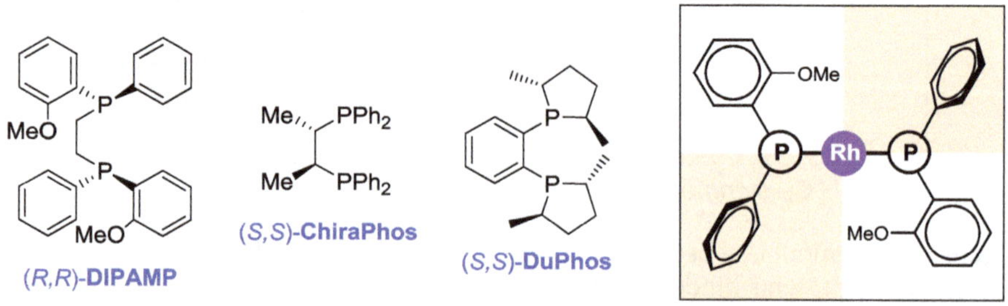

Figure 28. Privileged ligands for Rh-catalyzed asymmetric hydrogenation [43] and a quadrant diagram of the Rh complex **2** with C_2-symmetric phosphine ligand (*R,R*-DIPAMP).

well reproduced by DFT calculations, which can also provide a rationale for the observed dissimilar reactivity of the two diastereomers. Thus, for a rhodium complex with DuPHOS ligand, the differences in stability and reactivity of the corresponding intermediates can be accounted for with simple steric and electronic arguments [88]. It was found that the same steric and electronic factors that destabilize the minor diastereomeric complex actually assist the oxidative addition of hydrogen, while the major diastereomer must considerably distort for the same reaction to occur.

Bimetallic complexes commonly mediate various enantioselective reactions. For example, the Sharpless asymmetric epoxidation is mediated by such species (Figure 12) [47]. Asymmetric addition of organometallic reagents to carbonyl compounds also frequently occurs via bimetallic intermediates, in which one metal center acts as a Lewis acid to perform coordination and activation of a carbonyl group, while the second metal delivers a nucleophile. Thus, the previously mentioned asymmetric addition of dialkylzinc reagents follows such mechanism (Figure 29). The reaction was first reported by Omi and Oguni [99] and then was considerably

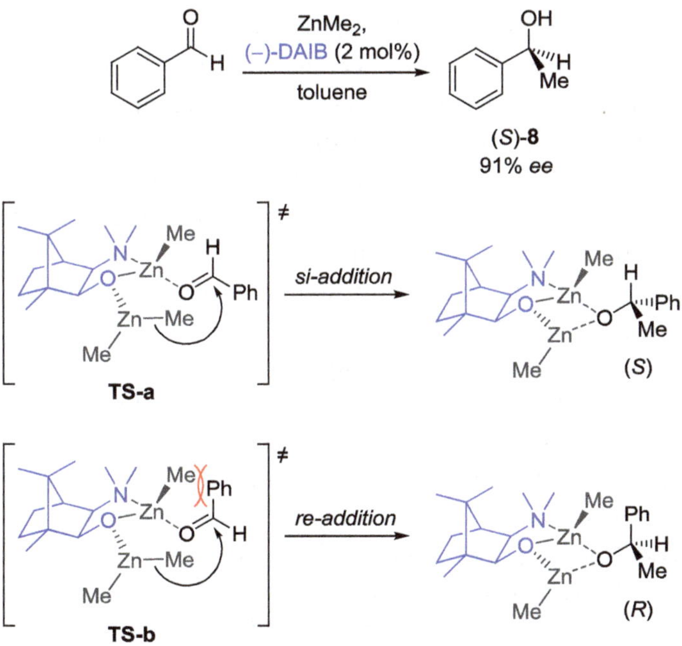

Figure 29. Chirogenesis in the enantioselective addition of diethylzinc to aldehydes.

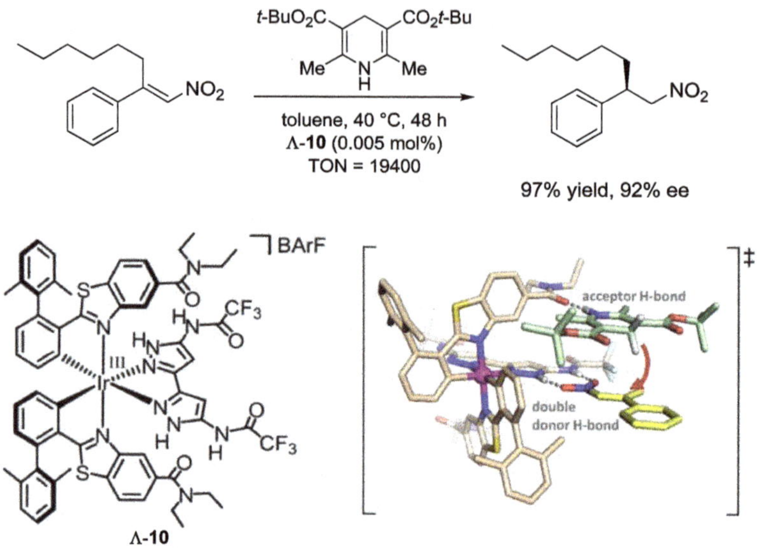

Figure 30. Enantioselective transfer hydrogenation catalyzed by a chiral-at-metal iridium complex. Reproduced from Ref. [104] with permission from Wiley.

improved by Noyori and co-workers [100, 101]. The reaction mechanism has been comprehensively investigated, including by computational studies [102, 103], which elucidated the origin of asymmetric induction. It was found that the transition state **TS-a** leading to the *S*-enantiomeric alcohol **8** is the most energetically preferable among four possible options. The closest *R*-forming transition state **TS-b** is 3.5 kcal·mol^{-1} higher according to DFT calculations. The destabilization mainly arises from nonbonding repulsion between the nonreacting methyl group at zinc and the bulky phenyl group of the aldehyde [102].

It is worth noting that the chirality of metal complexes can arise not only from ligands, but also from a metal atom being a chirality center (see, e.g., **Figure 23** above). Although such metal complexes were discovered a long time ago by Alfred Werner as evidence of his theory of coordination compounds, chiral-at-metal complexes have been rarely used in asymmetric catalysis [104]. Recently, Meggers and co-workers designed a number of efficient enantioselective transformations mediated by chiral-at-metal complexes, thus demonstrating the great promise of these metal species in asymmetric catalysis [105–107]. For example, chiral-at-iridium complex **Λ-10** catalyzes highly enantioselective transfer hydrogenation of nitroalkenes with a Hantzsch ester as the reducing agent [106]. According to the proposed chirogenic model, the catalyst, the substrate, and the reducing reagent form a network of hydrogen bonds that bring both the nitroalkane and the Hantzsch ester into the proper orientation. The transition state, however, does not involve any coordination to the chiral metal center and thus resembles that of organocatalytic reactions, which are presented in the next subsection.

1.5.3 Chirogenesis in Organocatalysis

Asymmetric organocatalytic reactions are mediated by chiral organic molecules. The archetypical organocatalysts **11–17** are presented in Figure 31. Several aspects of metal catalysts and general aspects of catalytic

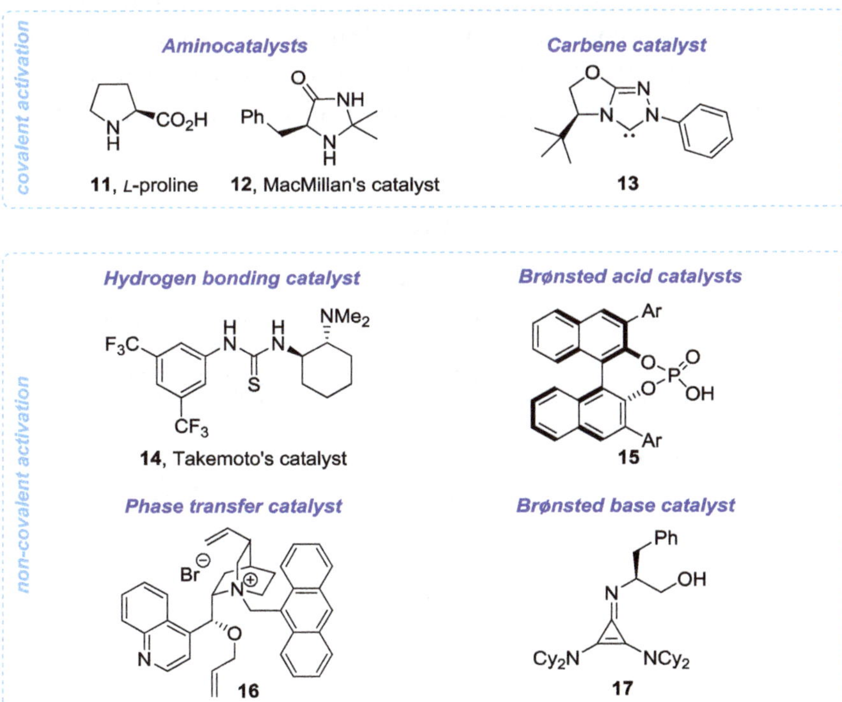

Figure 31. Representative examples of chiral organocatalysts.

reactions are also valid for organocatalysts and therefore, will not be discussed again. Thus, many of the organocatalysts are derived from the natural chiral pool, e.g., amino acids and alkaloids. The same rational principles of catalyst design and the entire catalytic cycle based on mechanistic studies, computational modeling, and QSAR approaches, can be applied [83, 108]. What is notably distinct, is the way that the substrate activation occurs and what interactions are involved in the transition state complexes. In general, the organocatalytic transformations involve the formation of supramolecular catalyst–substrate complexes which contain a network of covalent and/or non-covalent interactions of various types. Whether the latter involve a covalent bonding with the catalyst or rely exclusively on the non-covalent interactions, the organocatalysts can be divided into two big groups.

The first group forms covalently bonded intermediates. Aminocatalysts, such as L-proline (**11**), MacMillan imidazolidinone **12** and their siblings belong to this group. Typically, aminocatalysts activate prochiral carbonyl compounds via the generation of reactive iminium or enamine-type intermediates (Figure 32) [109, 110]. Carbene catalysts (e.g., **13**) represent another group of covalent organocatalysts. Typically, the reactive carbene

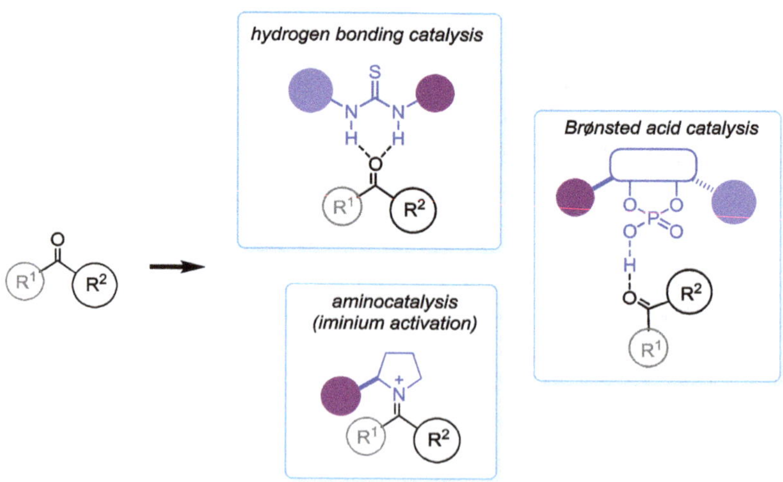

Figure 32. Representative organocatalytic activation modes for a prochiral carbonyl unit.

species are generated *in situ* by deprotonation of various azolium precursors. The activation of carbonyl substrates occurs via the formation of so-called Breslow intermediate [111].

The second large group of catalysts are compounds which enable substrate activation and stereodifferentiation via a network of non-covalent interactions. The most common interaction is hydrogen bonding, which takes place with strong hydrogen bond donors such as thioureas, and some other compounds. Hydrogen bonds also play a central role in the reactions promoted by chiral Brønsted acids [112], as well as supporting roles in other organocatalytic processes. The formation of hydrogen bonds with carbonyl compounds (Figure 32) mimics Lewis acid activation with metals and is heavily exploited.

Chiral phase transfer catalysis is another esteemed organocatalytic method. This approach mostly relies on ion paring and weak π–π interactions with a prochiral substrate [57, 113]. Counterion paring is also significant for Brønsted acid and base catalysts. It is important to note that these catalysts are categorized according to their principal interactions with a substrate, however, all organocatalysts are typically polyfunctionalized compounds and typically possess functional groups such as secondary binding sites with an auxiliary role, e.g., the dimethylamino group in the Takemoto's catalyst **14** could enable additional substrate coordination and serves as a base.

This field is now expanding far beyond the directions outlined above. For example, the use of more exotic non-covalent interactions such as halogen, chalcogen (Group 16), and pnictogen (Group 15) bonding has been commenced recently [62, 114, 115]. Moreover, organocatalysis can be beneficially merged with a number of alternative catalytic approaches [59, 62].

Let us consider the mechanism of chirogenesis in a couple of organocatalytic reactions. The first example covers the already mentioned Hajos–Parrish–Eder–Sauer–Wiechert reaction (Figure 13), which is catalyzed by natural amino acid L-proline. Extensive research has been conducted to elucidate its mechanism and a number of mechanistic models have been proposed [109]. However, the obtained experimental data along with theoretical calculations are in the best agreement with the enamine mechanism occurring via the Houk–List transition state (Figure 33).

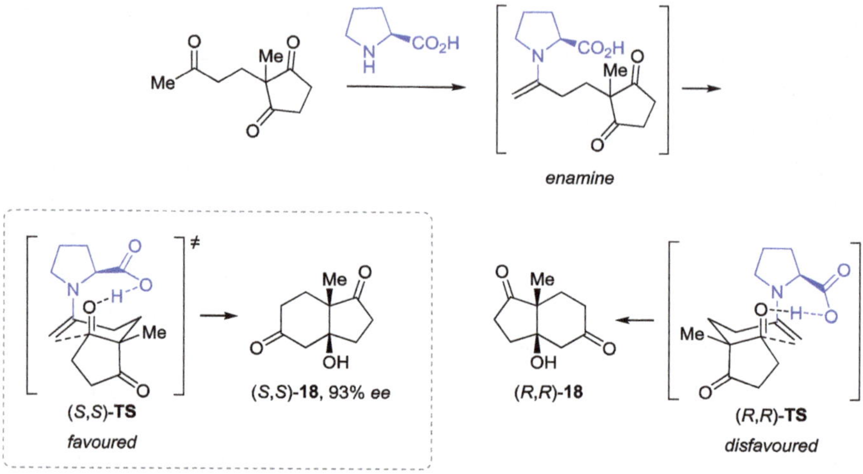

Figure 33. Chirogenesis in Hajos–Parrish–Eder–Sauer–Wiechert reaction [109].

According to DFT calculations [116], the transition state leading to (*S*,*S*)-**18** is favored by more than 3 kcal·mol^{-1} than that of (*R*,*R*)-enantiomer. This is consistent with the high *ee* observed for this reaction. In the less favorable (*R*,*R*)-**TS**, the carboxylic group of L-proline and the enamine double bond are in a *syn*-relationship with respect to the C–N axis, in contrast to *anti*-disposition of the same moieties in (*S*,*S*)-**TS**. The *syn*-disposition results in too close position of the two oxygen atoms linked by a hydrogen bond and to achieve an optimal O–H···O arrangement, the enamine system is forced out of planarity. Furthermore, a $^{\delta+}$NCH···O$^{\delta-}$ stabilizing electrostatic interaction adds additional stabilization to (*S*,*S*)-**TS**.

As the second example, let us consider an organocatalytic reaction which relies exclusively on non-covalent interactions as the source of asymmetric induction. In this respect, the enantioselective α-methylation of indanone **19** mediated by a chiral phase transfer catalyst (**PTC**) is a good illustration (Figure 34) [117]. The reaction delivers the methylated product (*R*)-**20** in excellent yield and high enantiomeric purity.

The suggested structure of the ion-paired intermediate contains a hydrogen bond and a pair of stabilizing π–π interactions between the aryl moieties. This set of interactions enables reliable stereodifferentiation of

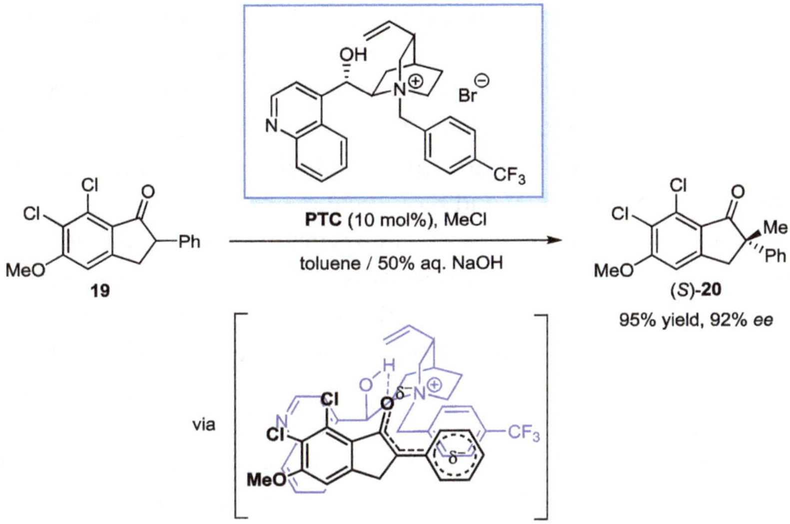

Figure 34. Chirogenesis in asymmetric phase-transfer-catalyzed alkylation [117].

the prochiral faces of the indanone anion, and so the alkylation occurs almost exclusively from the front side which is not shielded by the catalyst.

1.5.4 Chirogenesis in Enzymatic Catalysis

Enzymes are large peptide molecules which are folded to adopt a specific shape. Enzyme-catalyzed transformations are ubiquitous in nature since they occur in living cells. In these reactions, enzymes act as natural supramolecular machines which unify the best features of various catalytic fields. The reaction occurs in an active site, which enables stereospecific binding of a prochiral substrate, in addition to activation of the reactants. Many enzymes contain small non-peptide molecules or metal ions in their active sites (known as cofactors), which assist in catalytic reactions.

Enzymatic catalysis (or biocatalysis) is perhaps the first type of catalysis known to mankind which has been exploited for multiple purposes since the early days of human civilization. The first patent for enzymatically catalyzed asymmetric industrial synthesis was issued as early as the 1930s (production of L-ephedrine) [118]. Biocatalytic methods are widely used in modern organic synthesis [119]. In terms of catalytic activity and specificity, enzymes are outstanding and greatly outperform the current human achievements. Thus, TON approaches 40×10^6 and TOF 4×10^7 s^{-1} for the catalase enzyme. Such high efficiency is the outcome of billions-years long chemical and biological evolution.

As an example of chirogenesis in enzymatic catalysis, the enantioselective reduction of ketones into secondary alcohols mediated by alcohol dehydrogenases (ADHs) follows [120]. The reduction requires a coenzyme such as reduced nicotinamide-adenine dinucleotide (NADH) or its phosphate (NADPH) as a source of hydride anion. The majority of commercially available ADH enzymes deliver the hydride from the *re* face of a prochiral ketone, which result in the (*S*)-alcohol (if a large group L has a higher Cahn–Ingold–Prelog priority, Figure 35). The produced secondary alcohols commonly have *ee*'s above 99%. This high selectivity can be rationalized by looking at the structure of the enzyme active site, e.g., ADH isolated from *Thermoanerobium brockii* (TbSADH), for which the crystal structure was obtained (Figure 36). The active site contains a zinc cation, to which the oxygen of the ketone or alcohol coordinates. It is notable that in the native conformation of the enzyme the zinc ion occupies a position which is too remote for direct interaction with the alcohol group and a displacement of the zinc ion to its catalytic site is presumably

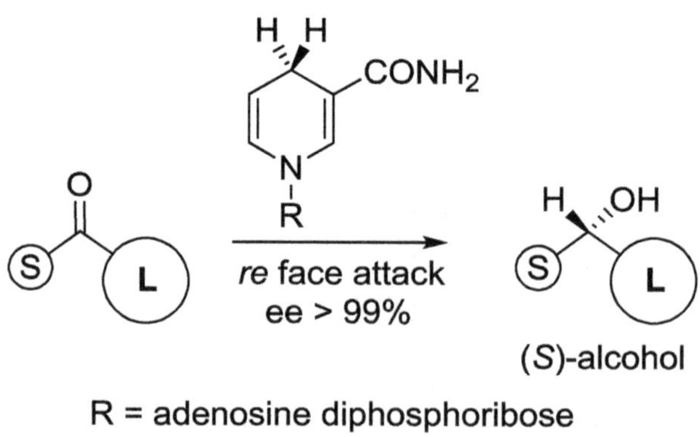

Figure 35. Alcohol-dehydrogenase-mediated reduction of prochiral ketones with NADPH.

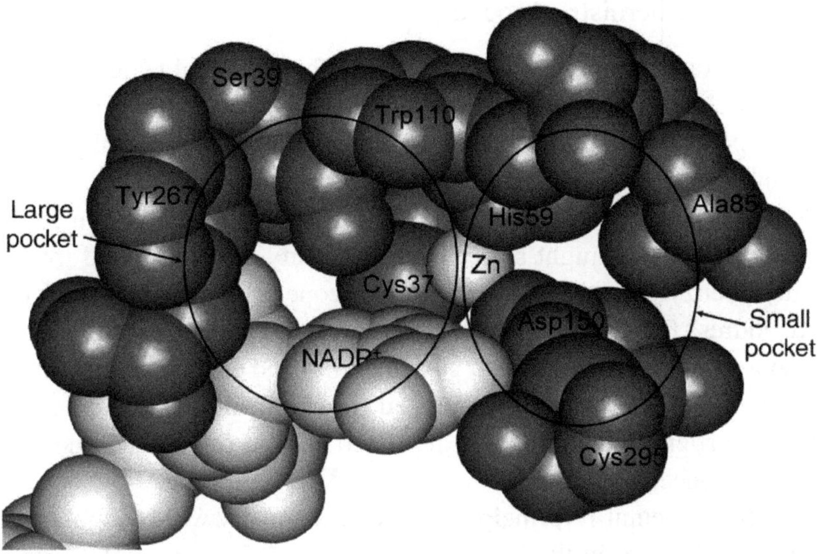

Figure 36. Active site of the secondary alcohol dehydrogenase from *Thermoanerobium brockii* (TbSADH) [120, 122]. (Reproduced from reference [120] with permission from Royal Society of Chemistry)

triggered by NADP [121]. A crevice from the surface to the active site with large and small hydrophobic binding pockets provides access for the substrates and the products. The binding pockets have different affinities toward the substituents of ketone substrates, and therefore can differentiate their prochiral faces.

Engineering of artificial metalloenzymes via incorporation of an abiotic metal cofactor within an existing protein scaffold [61, 123, 124], and controlling the enantioselectivity of enzymatic catalysis by directed evolution [125–127], represent two highly appealing current research directions. For example, Hartwig and co-workers reported replacement of iron with iridium in iron-containing natural proteins like myoglobin [128]. The created artificial iridium-containing enzymes catalyzed reactions which are not catalyzed by the native enzymes, such as C–H insertion and cyclopropanation with carbenes generated from diazo compounds. Moreover, directed evolution of the native protein scaffold allowed the generation of a set of mutant enzymes with altered amino acid residues in the binding pocket. The mutant enzymes were able to form either enantiomer of the products.

1.5.5 Green Chemistry Aspects

In view of practical applications, especially for industrial synthesis, all of the catalytic approaches discussed above have their own chemistry-related pros and cons. Thus, transition-metal-mediated processes usually benefit from low catalyst loadings but could be sensitive to air and moisture, while the opposite behavior is typical for organocatalytic methods. Enzymatic catalysis might suffer from solubility issues, limited thermostability of enzymes, and narrow substrate scope. However, here we would like to address few green chemistry aspects and to compare different catalytic approaches in terms of their sustainability and environmental impact. Commonly, the green-chemistry-related issues of enantioselective catalysis are superficially addressed, if discussed at all. Moreover, some widespread misconceptions exist.

The fundamental 12 principles of green chemistry were formulated by Anastas and Warner in the late 1990s [129, 130]. The principles provide commandments for the creation of a benign-by-design chemical process, paying attention to energy efficiency, safety, and reducing amount of generated waste. The use of catalytic methods rather than stoichiometric approaches is encouraged. The accordance to the principles of green chemistry can be ranked and quantitatively assessed for any chemical process with the aid of green chemistry metrics, for which a number of toolkits have been introduced [131, 132]. For example, for the amount of waste generated, *atom efficiency* (*AE*) provides a quantitative assessment for a chemical reaction, while *process mass intensity* (*PMI*) characterizes the whole process, including isolation and purification stages:

$$AE\,(\%) = \frac{\text{molecular weight of product}}{\text{total molecular weight of reactants}} \times 100 \qquad (7)$$

$$PMI = \frac{\text{total mass in a process}}{\text{mass of product}} \qquad (8)$$

For enantioselective synthesis, the opposite enantiomer is often an undesired product and therefore can be considered as "waste." In this respect, it is obvious that highly enantioselective approaches are beneficial not only in view of their chemical efficacy but also environmental impact. For example, the venerable Rh-catalyzed asymmetric hydrogenation (Figure 22) proceeds with 100% atom economy, it is perfect in terms of enantioselectivity and has extremely low catalyst loadings. However, if we consider a production process, PMI gives a better representation. Considerable input into PMI comes from solvents, responsible for 80–90% of mass consumption and greatly outweighs the contribution of reagents [133]. Therefore, an environmentally friendly approach should deal with solvent sustainability issues. According to the first principle of green chemistry, preventing waste is the best option. Hence, performing the reaction without any solvents is particularly advantageous [134]. Catalytic solvent-free asymmetric processes exist though they are not so numerous at present [135]. In view of the ongoing green revolution, the field of solvent-free chemistry currently experiences a renaissance. The use of mechanochemistry as the way to trigger the reactions between solid reactants is particularly advantageous [136]. However, the area of enantioselective mechanochemical reactions remains in its infancy [137]. In particular, the driving forces of mechanochemistry, especially those related to stereocontrol, are insufficiently understood at present and have not been systematically studied.

A set of different environmental, health and safety issues should also be considered to evaluate the "greenness" of a given process. In respect to metal catalysis, element sustainability plays an important role [138]. Rare metals such as palladium, rhodium, and iridium clearly dominate among the known catalytic enantioselective processes (Figure 37). Although catalyst loadings are typically minute for the noble metal complexes, the steady ongoing dissipation of their traditional supplies highlights the importance of rare metal recycling and search for sustainable replacements. In recent years, the issue of element sustainability has incited intensive research in the catalytic chemistry of earth-abundant metals, which remained for a long time in the shadow of their more successful noble siblings. Such research activity not only helps to mitigate the noble metal depletion risks but is also advantageous in view of the prospects for the discovery of new reactions.

Progress in the catalytic chemistry of earth-abundant metals over the last decades can be nicely illustrated with enantioselective reductions of carbonyl compounds (Figure 38). The original procedure developed by

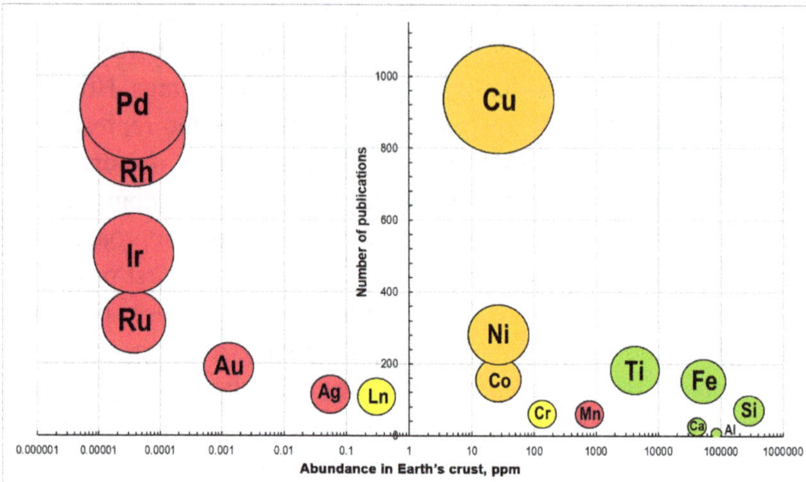

Figure 37. Abundance of metals in the earth's crust [142] versus number of publications on their use in asymmetric synthesis and catalysis, according to a Web of Science database search (performed June 23, 2021, covers publication years 2001–2021). Remaining years until depletion of known reserves are represented by colors according to Ref. [138] (red, 5–50 years; orange, 50–100 years; yellow, 100–500 years). Earth-abundant elements are shown in green. Ln = lanthanides.

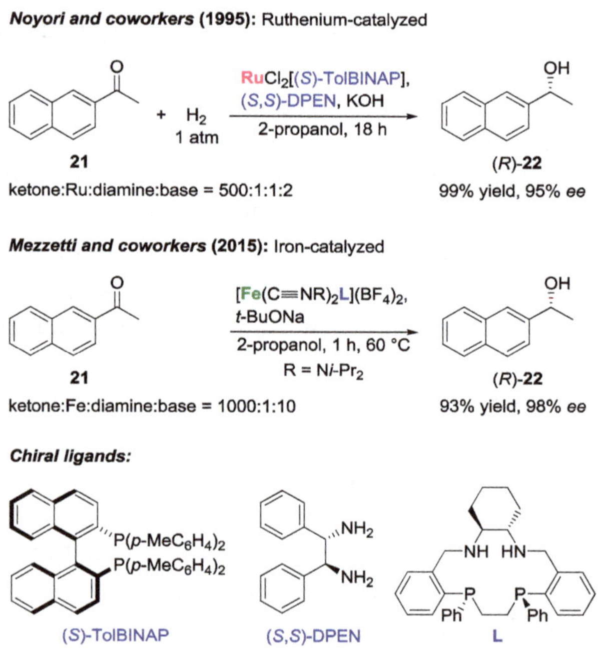

Figure 38. Enantioselective reduction of 2-acetonaphthanone (**21**) catalyzed by chiral ruthenium [139] and iron [140] complexes.

Noyori and co-workers relies on the use of ruthenium complexes [139]. A work of Mezzetti and co-workers illustrates the suitability of iron complexes for the same purpose with even better efficacy [140]. Moreover, the Mazzetti's protocol eliminates the need for flammable hydrogen as reductant, since solvent itself (2-propanol) can serve as reducing agent. Asymmetric catalytic Meerwein–Ponndorf–Verley reduction developed by Wulff and co-workers [141] is another notable discovery and a rare example of asymmetric catalysis with aluminum complexes.

The toxicity of transition metal catalysts is an additional point which has raised ecological and human health concerns. Thus, the accumulation of Pt, Pd, and Rh in biota has been demonstrated in both laboratory and field studies [143]. On the other hand, it should be noted that the toxic properties of transition metals are sometimes exaggerated or unjustifiably generalized [144, 145]. One must remember that toxic properties are attributed to a whole compound rather than its metal constituent. The harmful properties should be evidenced with a reliable toxicological data and environmental fate studies, currently lacking for many of transition metal compounds.

Enantioselective organocatalysis is believed to be a sustainable and world-changing chemical innovation [134]. Organocatalysts are typically derived from abundant and renewable natural feedstocks which can overcome the impact of higher catalyst loadings. Organocatalytic molecules are also often referred to as safer and nontoxic alternatives to transition metal catalysts. However, it should be noted that the evidence for the environmental and health safety of the majority of organocatalysts is almost absent at present. Although such catalysts like L-proline are evidently nontoxic, the cytotoxic and eco-toxic effects have been revealed for bis(trifluoromethyl)phenyl thiourea derivatives [146, 147]. Therefore, as with metal complexes, generalizations about the toxicity and environmental safety of organocatalysts should currently be avoided.

Enzymatic processes are frequently considered as inherently green, since enzymes are derived from sustainable natural sources and enzymatic reactions typically operate in aqueous media under mild conditions [119]. However, such idealistic beliefs could be substantially altered if green metrics analysis is performed through the whole production cycle [148, 149]. Thus, biocatalytic processes, including preparation of biocatalysts, typically consume large amounts of water which contributes significantly to PMI and eventually requires water remediation. Moreover, a feed of supporting reagents and an excess amount of reaction substrate might be required.

Concluding the overview of catalytic methods and their environmental impact, we can state that every approach has its own weak points and nothing is perfect. The consumption of energy and generation of waste is applicable to any chemical process. We can only make these routes more efficient, sustainable, and circular.

It can be seen sometimes from the literature that different branches of enantioselective catalysis are opposed each other in terms of efficiency and their particular drawbacks. In our opinion, it is wiser to consider the complementarity of these approaches, rather than their drawbacks. Multiple existing catalytic tools provide an advantage of flexibility in solving complex problems, typically faced in various applied fields.

1.6 Asymmetric Titanium-Catalyzed Transformations: A Personal Journey

Titanium is the ninth most abundant element in Earth's crust with a plenty of renowned enantioselective applications, such as epoxidations, nucleophilic additions, and cycloadditions [150]. However, the potential of this metal in asymmetric synthesis remains untapped for many synthetically useful transformations. In this subsection, the authors would like to describe their own research experiences devoted to titanium-mediated asymmetric synthesis, which illustrate inspirations and challenges commonly faced during the development of asymmetric reactions.

1.6.1 Chirogenesis in the Asymmetric Oxidation of Ketones

Already in the middle of the 1990s, the search for other uses of the Sharpless catalytic system (a mixture of titanium tetraisopropoxide, diethyl tartrate (DET), and *tert*-butyl hydroperoxide, designed for asymmetric oxidation of an allylic double bond), had started. The direct asymmetric oxidation of ketones was of interest. In 1996, Lopp and co-workers published the first results on this topic, where racemic and prochiral cyclobutanones were found to undergo Baeyer–Villiger oxidation under Sharpless complex catalysis (Figure 39) [151].

These results were well accepted by the chemical audience and that inspired continued research in this direction. In 1997, Lopp and co-workers published their findings that β-hydroxyketones could be oxidized with the same complex affording α, β-dihydroxyketones in up to 97% *ee* (Figure 40) [152, 153]. The isolated yields were higher than 50% and the low enantiomeric purity of the recovered racemic substrate suggested that oxidation may involve epoxidation of a prochiral enol intermediate, in a similar way to oxidation of allylic alcohols according to Sharpless [153].

When continuing the search for other compounds susceptible for oxidation, 1,2-diketones where chosen as the substrates. In 2000, Lopp and co-workers presented the first example of direct asymmetric oxidation of

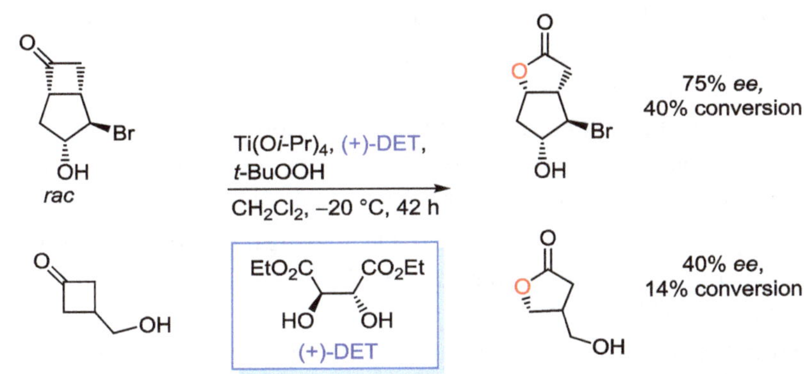

Figure 39. Asymmetric oxidation of cyclobutanones with the Sharpless catalytic system [151].

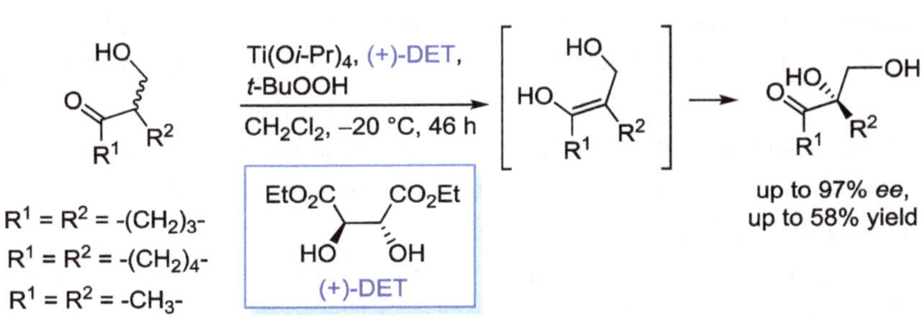

Figure 40. Asymmetric oxidation of β-hydroxyketones [152, 153].

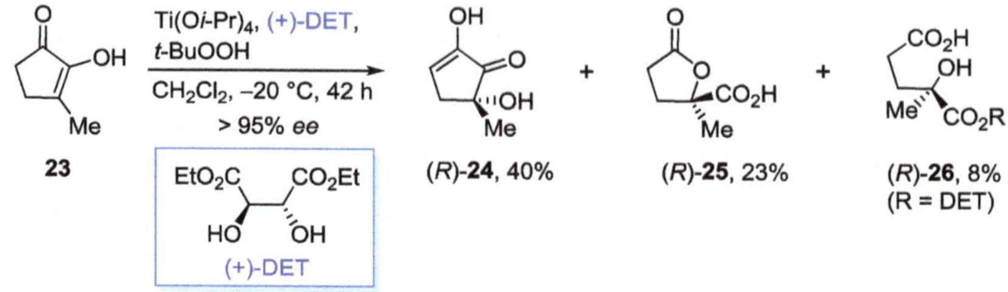

Figure 41. The first example of asymmetric oxidation of 3-alkyl-1,2-cyclopentanediones with the Sharpless catalytic system [154].

3-alkyl-1,2-cyclopentanediones [154]. Thus, oxidation of 3-methyl-1,2-cyclopentanedione (which exists predominantly in enol form **23**) in the presence of (+)-DET as a chiral inducer produced a mixture of α-hydroxylation product (*R*)-**24**, lactone acid (*R*)-**25**, and ring cleaved dicarboxylic acid derivative (*R*)-**26** (Figure 41) [155]. Enantiomeric excesses of more than 95% were determined for the newly formed stereocenters in all the products.

It was possible to alter the yields of oxidation products **24** and **25** by changing the ratio of reagents. Thus, the maximal amount of ring cleavage oxidation products (isolated yield 50%) was obtained with the ratio $Ti(O\textit{i}\text{-}Pr)_4$/DET/*t*-BuOOH = 1/1.6/2.5. Decreasing the amount of *t*-BuOOH to 1.6 eq. resulted in the enhanced yield (40%) of the 3-hydroxylated product (*R*)-**24** [155].

After the subsequent optimization of the reaction conditions, the corresponding γ-lactone acids were obtained in up to 83% yield and in up to 96% *ee* [156]. 1,2-Cyclopentanedione substrates substituted in the 3-position with aryl and functionalized alkyl groups underwent smooth oxidations as well. Oxidation with catalytic amounts of $Ti(O\textit{i}\text{-}Pr)_4$ (30 mol%) and DET (50 mol%) was also achieved, and delivered a range of γ-lactone products in 58%–72% yields and 92%–95% *ee* [157]. The produced γ-lactones were used in the asymmetric synthesis of natural compounds, such as homocitric acid [158], as well as in the preparation of nucleoside analogues [159, 160].

Generation of several oxidation products during the reaction implied that the transformation of 1,2-cyclopentanediones into γ-lactones plausibly occurred as a three-stage cascade reaction. The first oxidation step is chirogenic and determines the stereochemical outcome of the whole sequence, while the subsequent stages involve oxidation of the initially formed chiral precursor into the dicarboxylic acid, followed by its γ-lactonization. A series of judicious mechanistic experiments, including the use of ^{18}O-labeled reagents, allowed to establish the oxidation mechanism as follows (Figure 42) [161].

The reaction begins with asymmetric epoxidation of 1,2-cyclopentanedione with *tert*-butyl hydroperoxide coordinated to the Sharpless titanium–tartrate complex (see below). The first step is rate limiting and determines the stereochemical outcome for the whole process. The highly reactive epoxide **27** produced can be cleaved into the α-hydroxy ketone product **28**, which cannot be oxidized further under the reaction conditions [162]. Alternatively, **27** can be oxidized further with the next equivalent of *t*-BuOOH via a Bayer–Villiger-type rearrangement [161], and eventually produce diacid **31**. Lactonization of the latter results in γ-lactone product **32**. The distribution of ^{18}O atoms in the γ-lactone product (R = Bn) was in full accord with the Bayer–Villiger-type cleavage mechanism and allowed an alternative mechanism involving the intermediate generation of an epoxide to be ruled out.

At the key chirogenic step, epoxidation in the presence of (+)-DET occurred from the *si*-face of the 1,2-cyclopentanedione substrate and

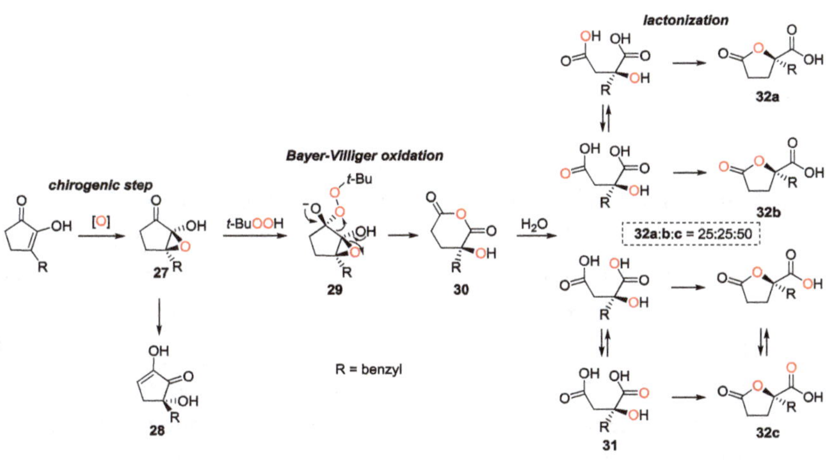

Figure 42. Mechanism of the asymmetric oxidation of 3-substituted 1,2-cyclopentanediones and its chirogenic step. ^{18}O-Labeling experiments confirm Bayer–Villiger-type oxidative cleavage of intermediate **27**. The red color stands for ^{18}O-labeled oxygen atoms [161].

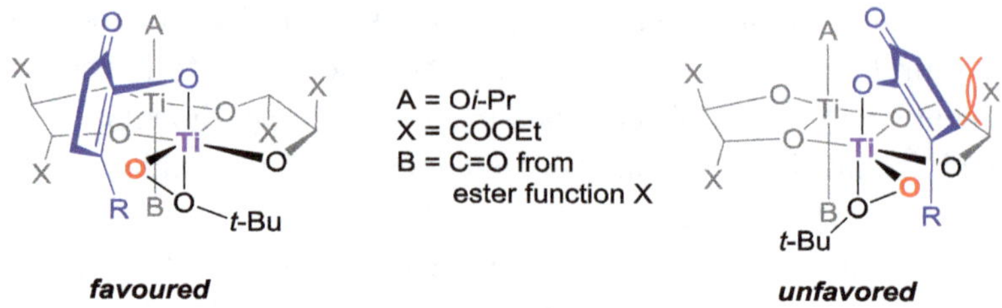

Figure 43. Simplified chirogenic model explaining the enantioface selectivity in the asymmetric oxidation of 1,2-cyclopentanediones [155].

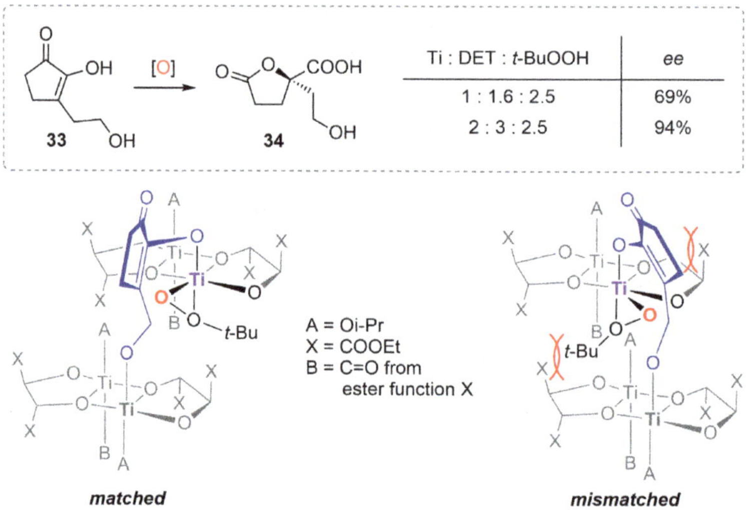

Ti : DET : *t*-BuOOH	*ee*
1 : 1.6 : 2.5	69%
2 : 3 : 2.5	94%

Figure 44. Simplified model of the doubly coordinated complex [162].

eventually resulted in the *R*-configuration for the oxidation products. The observed stereoselectivity can be rationalized in the framework of a simplified model for the Sharpless intermediate complex (Figure 43) [47, 155]. It was assumed that the titanium catalyst forms an enolate-type complex with the substrate and epoxidation predominantly occurred by electrophilic attack of oxygen at the *si*-side of the enolate in the favorable conformation of the substrate–catalyst complex. On the other hand, epoxidation at the opposite side of the substrate is less likely due to unfavorable steric interactions in the corresponding substrate–catalyst complex.

Oxidation of the 2-hydroxyethyl substituted substrate **33** with a free hydroxyl group represented a specific case (Figure 44) [162].

Enantioselectivity in this case was significantly dependent on the substrate/Sharpless complex ratio, and was increased up to 94% *ee* when 2 equivalents of the complex were used. This result indicated that, in the latter case, a complex with chiral catalyst molecules attached to the substrate was formed.

The proposed chirogenic models were merely speculative, and it is likely that more intricately organized species operate. However, experimental structural elucidation is difficult to perform. The computational modeling is also challenging because of the very high number of plausible intermediates involved, considering also multiple stereochemical configurations for each intermediate and their conformational flexibility. For the initial step, an attempt was made to elucidate the structure of the species formed by mixing 3-methyl-1,2-cyclopentanedione (**23**) with $Ti(O\textit{i}\text{-}Pr)_4$ [163]. An NMR study revealed the formation of several titanium enolate complexes, among them a 2:1 complex with two molecules of **23** bound to titanium was dominant. Computational modeling of the complex revealed octahedral coordination of the titanium atom with two isopropoxide ligands and two molecules of **23** acting as bidentate ligands. The two most preferred stereochemical configurations were also identified for this complex.

1.6.2 Chirogenesis in the Asymmetric Kulinkovich Reaction

In 1989, Kulinkovich and co-workers discovered an unprecedented reaction of Grignard reagents with carboxylic esters in the presence of titanium(IV) isopropoxide, affording tertiary cyclopropanols **35** (Figure 45) [164–166]. This discovery commenced the era of titanacyclopropanes [titanium(II)-alkene complexes] **36** in organic synthesis, for which a plethora of versatile synthetic uses have been found after the Kulinkovich's seminal report [167–172]. Despite the high level of diastereocontrol commonly observed in the transformations mediated by titanacyclopropane reagents and their prominent synthetic value, the corresponding enantioselective reactions, especially mediated by catalytic amounts of chiral titanium complexes, remain largely unexplored and represent a notable case of hard-to-achieve enantioselective transformations [173–175].

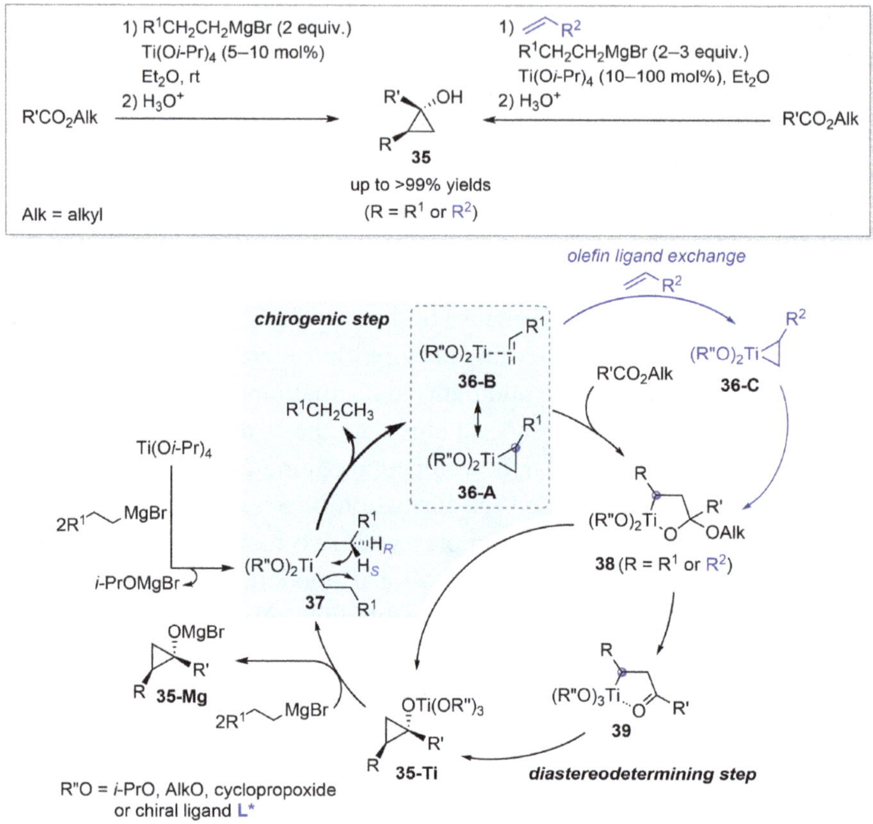

Figure 45. Kulinkovich reaction and its simplified mechanism with chirogenic and diastereodetermining steps.

Mechanistically, the Kulinkovich cyclopropanol synthesis occurs via twofold alkylation of the ester's carbonyl group with the organometallic reagent **36**, which is a resonance hybrid of titanacyclopropane **36-A** and titanium–olefin π-complex **36-B**. This property offers the advantage of olefin ligand exchange (generation of **36-C**), thus greatly the expanding scope of the reaction [176–178]. The reactive titanocyclopropane species are formed in the event of β-hydrogen abstraction in the dialkyltitanium precursor **37**. This step is also a chirogenic step, since it leads to the formation of a stereocenter in the titanacyclopropane complex **36-A**. Sterodiscrimination of the prochiral hydrogens H_R or H_S in the dialkyltitanium precursor **37** can be achieved by the influence of a chiral ligand L*. Therefore, the use of chiral titanium alkoxides as catalysts represents the most evident approach to develop the enantioselective variant.

In 1994, Corey and co-workers reported the first example of an enantioselective Kulinkovich reaction [179]. Cyclopropanation of ethyl acetate with 2-phenylethylmagnesium bromide in the presence of 0.3–1 eq. of titanium(IV) bis-(*R*,*R*)-TADDOLate **40** afforded cyclopropanol (1*S*,2*R*)-**35a** in 65%–72% yield and with 70%–78% *ee* (Figure 46). The proposed chirogenic model assumed preferable formation of (*S*)-2-phenyltitanacyclopropane (*S*)-**36-Ph** with the phenyl substituent in the metallacycle maximally distant from the nearest pseudo-axial aryl group of the TADDOL ligand (i.e., resting in the least hindered quadrant).

Numerous efforts to improve the reaction performance and expand its synthetic scope have met only a limited success. Screening of various chiral ligands in the cyclopropanol synthesis resulted in poor enantiomeric purity of the cyclopropane products [180]. Several chiral titanium alkoxides resulted in 0% *ee*, indicating a dominant competitive racemic reaction [173]. For the synthesis of cyclopropanols, the performance of a catalytic version was found to be highly sensitive to the reaction conditions, such as catalyst loading, nature of solvent, temperature, and even order of addition of reactants, e.g., excess of carboxylic ester substrate had a clearly

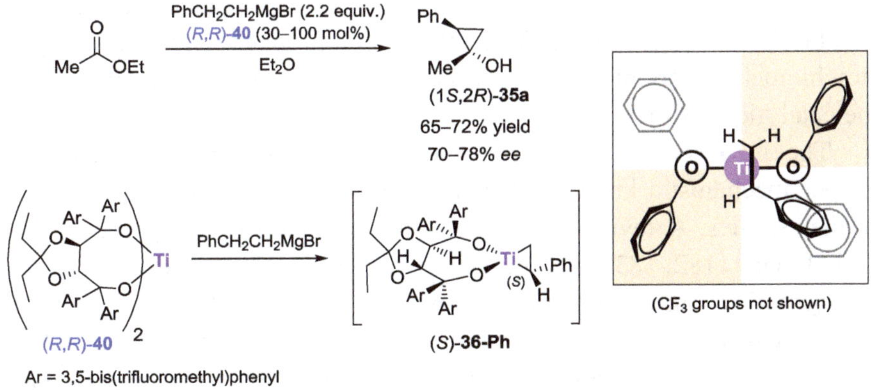

Figure 46. Enantioselective Kulinkovich cyclopropanation of ethyl acetate mediated by titanium alkoxide (*R,R*)-**40** and Corey's chirogenic model [179].

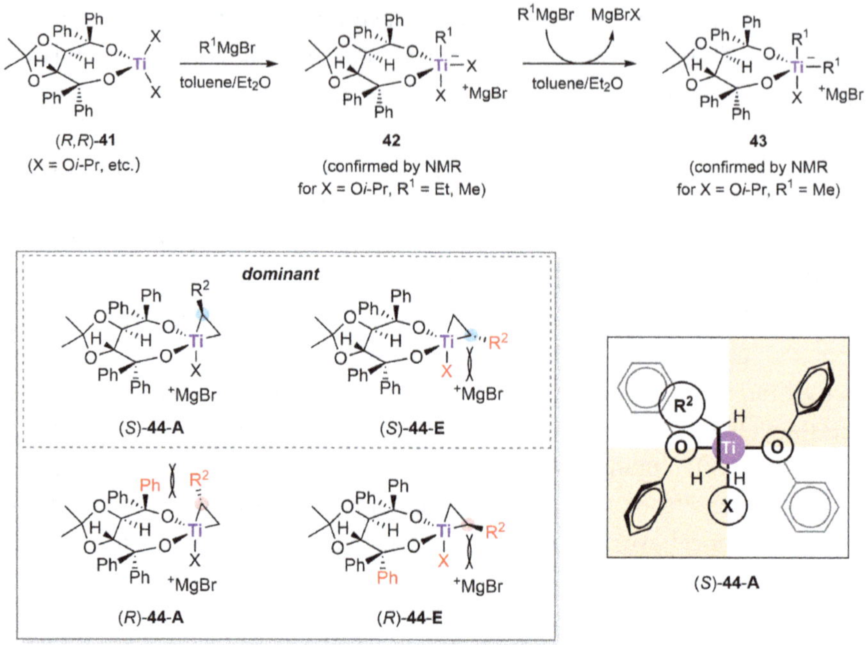

Figure 47. Kulinkovich's chirogenic model assuming the formation of ate complexes [184].

negative impact on the enantioselectivity [181]. It was also found that titanium TADDOLate **41** (X = O*i*-Pr, Figure 47) is a more superior catalyst than the originally employed bis-TADDOLate complex (like alkoxide 40), but far less reactive than Ti(O*i*-Pr)$_4$. Cyclopropanation mediated by a stoichiometric amount of **41** produced cyclopropanols with better *ee* than the catalytic version.

The inconsistent results collected in the early trials implied that deeper mechanistic knowledge had to be obtained as a prerequisite for further studies. A revision of the original mechanism was proposed by Kulinkovich [182, 183], who suggested bimetallic ate complexes (adducts of organotitanium compounds with magnesium alkoxides) as the actual intermediates. Since the coordination sphere of magnesium (presumably linked via μ-oxo bridges) was vague, a conventional "ionic" representation was used to depict these compounds (Figure 47). Kulinkovich's hypothesis has been supported by later spectroscopic [184, 185] and computational studies [186, 187]. Thus, NMR spectroscopy revealed the formation of bimetallic species with a pentacoordinated titanium core, e.g., **42** (X = O*i*-Pr, R^1 = Me, Et) or **43** (X = O*i*-Pr, R^1 = Me), upon mixing of titanium TADDOLate **41** (X = O*i*-Pr) with the corresponding Grignard reagents. Moreover, the dependence of enantioselectivity on the nature of heteroatom ligands X in titanium alkoxides **41** and structure of Grignard reagent (primary vs. secondary) were revealed as additional supporting evidence [184].

The revised chirogenic model was proposed by Kulinkovich (Figure 47) [184]. The model assumed the formation of pentacoordinated titanacyclopropane species **44** with an alkyl substituent R^2 attached to either equatorial (**44-E**) or apical (**44-A**) carbons of the metallacycle, formed in an event of *β*-hydrogen abstraction from the apical alkyl group

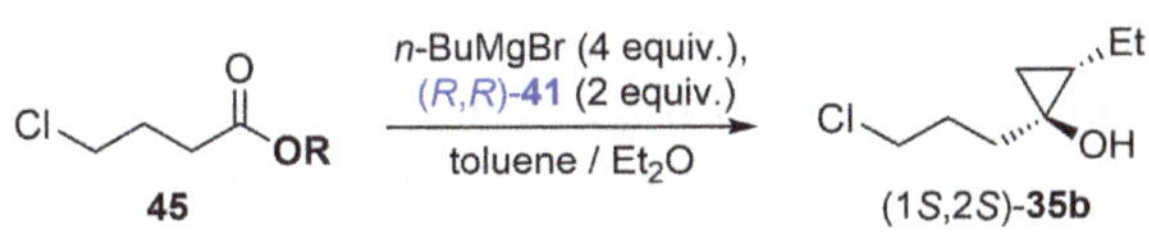

OR	ee
OEt	56
O*i*-Pr	64
OPh	44
OCH(CF$_3$)$_2$	**80**

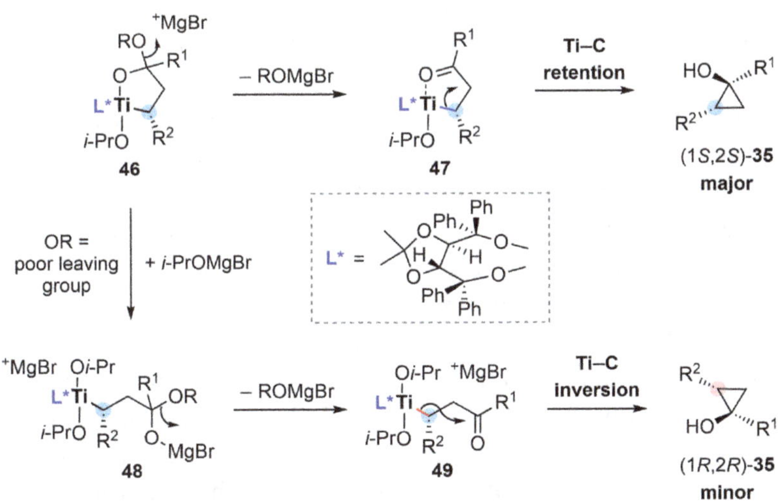

Figure 48. Effect of alkoxide groups in 4-chlorobutyrate esters **45** on enantioselectivity and a suggested mechanistic rationale [184].

R^1 in dialkyltitanium intermediate **43**. The equatorial or apical position of R^2 in **44** depends on whether **43** is produced from a primary (*n*-BuMgBr) or secondary (*i*-PrMgBr) Grignard reagent, respectively. Both (*S*)-isomers **44-E** and **44-A** are less sterically hindered than their (*R*)-counterparts due to the more distant position of the R^2 group from the nearby pseudo-axial phenyls of the ligand in the former. The presence of an additional repulsive $X \cdots R^2$ interaction in **44-E** could account for the diminished *ee* values observed in the case of *n*-BuMgBr.

However, the proposed model did not account for the striking dependence of enantiomeric purity of the cyclopropanol product on the nature of the leaving alkoxide group OR in the 4-chlorobutyrates **45**, which was also revealed in the same work (Figure 45) [184]. The highest enantioselectivity (80% *ee*) was attained with the hexafluoroisopropyl ester, while alkyl esters were less efficient. This discovery suggested an additional process leading to degradation of enantioselectivity during the reaction. It was hypothesized that the dominant (1*S*,2*S*)-enantiomer formed via the cyclopropane ring closure in the corresponding oxatitanacyclopentane **46** or β-titanaketone intermediates **47** with retention of configuration at the stereogenic carbanion center. On the other hand, if the ring closure step is sluggish (e.g., when OR is a poor leaving group), **46** could undergo ring-opening to afford acyclic intermediates **48**, **49** which eventually produce the minor (1*R*,2*R*)-enantiomer of cyclopropanol **35** via inversion of the configuration at the Ti–C bond.

Kananovich and coworkers successfully expanded the scope of the asymmetric Kulinkovich reaction to the cyclopropanation of esters with alkenes, which are able to perform an olefin ligand exchange in the titanacyclopropane [titanium(II)–alkene complexes] complex [185]. A number of functionalized olefins, such as unsaturated alcohols and their silyl-protected derivatives, were successfully involved in the ligand exchange and delivered the corresponding cyclopropane products in up to 87% *ee*. The predominant *S*-configuration of the stereocenter in metallacycle **50** can be rationalized by the framework of Kulinkovich's chirogenic model, adapted for the case of intramolecular ligand exchange with titanium-tethered alkenoxides (Figure 49). Most importantly, the deuterium-labeled olefin **51** offered an appropriate mechanistic probe to test if partial inversion of configuration at the Ti–C bond indeed takes place and if it is

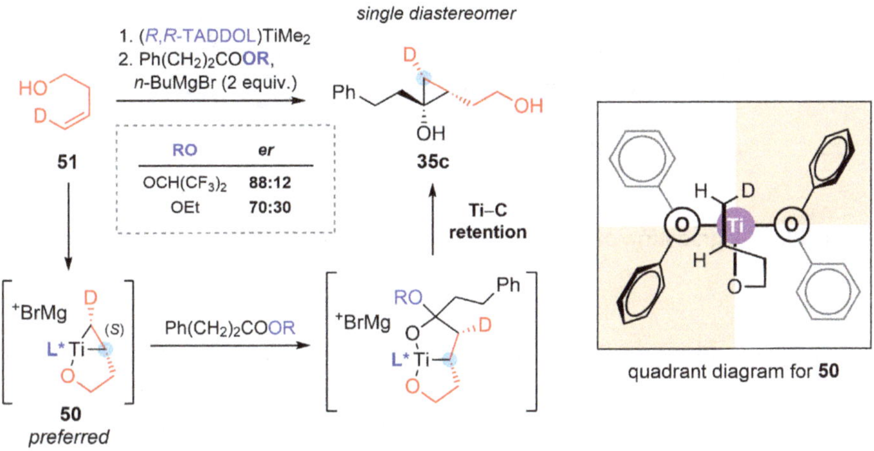

Figure 49. Stereochemistry of the asymmetric cyclopropanation of ethyl and hexafluoropropyl 3-phenylpropionates with homoallylic alcohol **51** [185].

responsible for the erosion of enantioselectivity. The total retention of olefin configuration was observed, while cyclopropanol **35c** was produced with a different enantiomeric purity from the hexafluoropropyl and ethyl 3-phenylpropanoates. This result provided evidence for the total retention of configuration at the Ti–C bond in the cyclopropane-forming step and indicated that another process accounts for the observed degradation of enantioselectivity.

NMR studies of alkyltitanium ate complexes provided several useful insights into the mechanism and indicated the degradation of the bimetallic titanium–magnesium ate species into less sterically hindered and more reactive tetracoordinated titanium compounds as one of the plausible pathways responsible for the erosion of enantioselectivity. For example, the slow reaction rate of titanium *tert*-butoxide complex **41** (X = O*t*-Bu) with MeMgBr and the dissociation of the produced ate complex **52** into dimethyltitanium compound **53** were straightforwardly observed by ^{1}H NMR (Figure 50) [185]. According to NMR, dimethyltitanium compound **53** is the only product of the reaction of dichloride complex **41** (X = Cl) with MeMgBr. The same titanium chloride and *tert*-butoxide complexes afforded a diminished 49% *ee* of the corresponding cyclopropanol product in the reaction of hexafluoropropyl 3-phenylpropanoate with *n*-BuMgBr

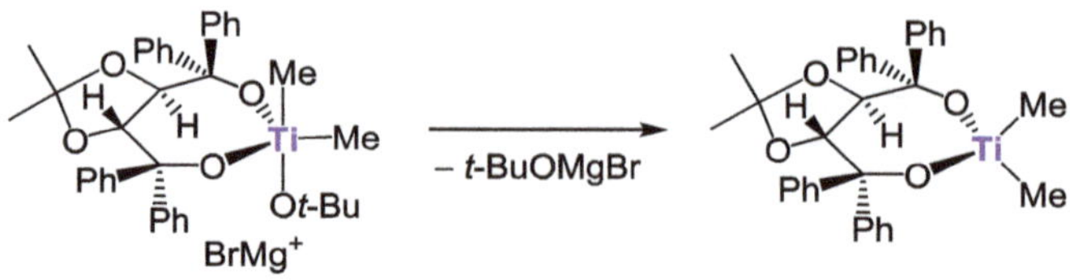

52

53

Figure 50. Degradation of dimethyltitanium ate complex **52** observed by ^{1}H NMR [185].

[188], thus evidencing that non-ate titanium complexes provide less efficient stereoinduction. Similarly, analogous degradation of sterically hindered titanium cyclopropoxides accumulated in the course of the catalytic reaction could be responsible for the diminished efficacy of the catalytic version. It seems likely that complex-forming agents toward magnesium compounds (e.g., carboxylic esters, Lewis bases) also incite the degradation of ate-complexes. Thus, the higher enantioselectivities observed with hexafluoropropyl esters in comparison with ethyl esters could be accounted for by the lower Lewis basicity of the former [185].

The obtained results clearly demonstrate that the TADDOL ligand provides efficient stereoinduction during the formation of titanacyclopropane ate complexes; however, erosion of enantioselectivity occurs due to parasitic processes of degradation of ate complex intermediates. These findings imply that improved design of chiral ligands preventing degradation of ate complexes or mimicking their spatial structure could be the key to designing highly enantioselective catalytic Kulinkovich reactions and related processes mediated by titanacyclopropanes.

1.7 Conclusions

Modern asymmetric synthesis offers a plenty of opportunities for preparation of versatile chiral molecules in a highly enantioselective manner. For that purpose, a large number of chiral reagents, auxiliaries, and catalysts have been developed, many of them derived from natural feedstocks. Although every approach has its own pros and cons, they complement each other and provide the diversity which is essential for solving enormous numbers of synthetic problems, including their industrial adaptations. In spite of these remarkable achievements, there are a number of prospects for further developments and a great number of challenges exist, which require the involvement of knowledge from various branches of chemical and natural sciences for further advances. The routine development of new asymmetric catalytic reactions is highly appealing but arduous. The introduction and experimental validation of new concepts in the field is even more important, since it could lead to the emergence of entirely new catalytic directions and synthetic techniques. Besides the pragmatic problems related to the development of new synthetic methods, such fundamental questions as the origin of prebiotic chirality still remain unanswered. There is no doubt that chirogenic phenomena occurring in organic reactions will remain an inexhaustible source of inspiration for future generations of scientists.

Asymmetric induction is the key phenomenon responsible for chirogenesis in asymmetric transformations. External chiral stimuli eliminate the energetic degeneracy of transition states leading to opposite chirality by providing a dissimilar set of interactions, either attractive or repulsive. The interactions can be commonly identified and their energy inputs estimated by considering simple steric and electronic arguments, either in a speculative manner or with an aid of quantum chemistry calculations. The latter approach provides especially valuable insights and is the only tool which allows the accurate visualization of the respective transition states, providing the original mechanistic inputs are valid.

Although chiral stimulus is an essential precondition for asymmetric reactions to occur, it is not sufficient alone to attain high level of enantiocontrol. This is especially the case for catalytic reactions, in which numerous other factors affecting the catalyst performance and enantiomeric purity of the product should be taken into account. Mechanistic knowledge is essential for the rational development of enantioselective catalytic reactions and the design of chiral catalysts with improved efficacy. Therefore, judicious mechanistic experiments supported by the respective kinetic studies and quantum chemistry calculations are indispensable for success. A world-renowned chemist once told one of the authors: “no money is paid for mechanisms”. Although this statement reflects a sad truth about the modern world and society, we would like to highlight that fundamental understanding of the laws of nature and their underlying driving forces is key for progress in science.

1.8 Acknowledgements

The authors are grateful to Prof. Victor Borovkov for his kind invitation to contribute this chapter. We hope that our contribution totally fits his expectations and nicely completes this comprehensive book covering various aspects of chirogenesis in chemical science. We are grateful to our respected colleagues and gifted students, former and present, for their unconventional support. The authors' personal journeys to the realm of asymmetric synthesis would not be possible without their intellectual inputs and hard experimental work. Dr. Dzmitry Kananovich would like to acknowledge his first scientific advisor Prof. Oleg Kulinkovich, who entered him into the splendid world of cyclopropane chemistry and asymmetric synthesis and guided him for several memorable years. It gives us special pleasure to thank Prof. Richard Taylor (University of York) for insightful critical assessment of our manuscript. Finally, we would like to dedicate this work to the memory of our good friend and colleague Prof. Victor Snieckus (1937–2020).

References

1. Moss, G.P. (1996) Basic terminology of stereochemistry (IUPAC Recommendations 1996). *Pure Appl. Chem.*, **68**, 2193–2222.
2. Borovkov, V.V, Lintuluoto, J.M., and Inoue, Y. (2001) Supramolecular chirogenesis in zinc porphyrins: mechanism, role of guest structure, and application for the absolute configuration determination. *J. Am. Chem. Soc.*, **123**, 2979–2989.
3. Borovkov, V.V, Lintuluoto, J.M., Sugeta, H., Fujiki, M., Arakawa, R., and Inoue, Y. (2002) Supramolecular chirogenesis in zinc porphyrins: equilibria, binding properties, and thermodynamics. *J. Am. Chem. Soc.*, **124**, 2993–3006.
4. Borovkov, V. (2010) Effective supramolecular chirogenesis in ethane-bridged bis-porphyrinoids. *Symmetry*, **2**, 184–200.
5. Koskinen, A. (2012) Introduction, in "*Asymmetric synthesis of natural products*", John Wiley & Sons, Chichester, pp. 1–21.
6. Bryliakov, K.P. (2020) Chemical mechanisms of prebiotic chirality amplification. *Research*, Article ID 5689246.
7. Blackmond, D.G. (2020) Autocatalytic models for the origin of biological homochirality. *Chem. Rev.*, **120**, 4831–4847.
8. Farina, V., Reeves, J.T., Senanayake, C.H., and Song, J.J. (2006) Asymmetric synthesis of active pharmaceutical ingredients. *Chem. Rev.*, **106**, 2734–2793.
9. Blaser, H.-U. (2013) Chirality and its implications for the pharmaceutical industry. *Rend. Fis. Acc. Lincei*, **24**, 213–216.

10. Matsumoto, Y., Naito, M., and Hayashi, T. (1992) Palladium(0)-catalyzed hydroboration of 1-buten-3-ynes: preparation of allenylboranes. *Organometallics*, **11**, 2732–2734.
11. Ruble, J.C., and Fu, G.C. (1996) Chiral π-complexes of heterocycles with transition metals: a versatile new family of nucleophilic catalysts. *J. Org. Chem.*, **61**, 7230–7231.
12. Ribó, J.M., Crusats, J., Sagués, F., Claret, J., and Rubires, R. (2001) Chiral sign induction by vortices during the formation of mesophases in stirred solutions. *Science*, **292**, 2063–2066.
13. Crusats, J., and Moyano, A. (2021) Absolute asymmetric catalysis, from concept to experiment: a narrative. *Synlett*, **32**, 2013–2035.
14. Matsumoto, Y., Naito, M., Uozumi, Y., and Hayashi, T. (1993) Axially chiral allenylboranes: catalytic asymmetric synthesis by palladium-catalysed hydroboration of but-1-en-3-ynes and their reaction with an aldehyde. *J. Chem. Soc., Chem. Commun.*, 1468–1469.
15. Egami, H., and Katsuki, T. (2009) Iron-catalyzed asymmetric aerobic oxidation: oxidative coupling of 2-naphthols. *J. Am. Chem. Soc.*, **131**, 6082–6083.
16. Grabowski, S.J. (2006) Theoretical studies of strong hydrogen bonds. *Annu. Rep. Prog. Chem., Sect. C Phys. Chem.*, **102**, 131–165.
17. Kagan, H.B., and Fiaud, J.C. (1988) Kinetic resolution. *Top. Stereochem.*, 249–330.
18. Huerta, F.F., Minidis, A.B.E., and Bäckvall, J.-E. (2001) Racemisation in asymmetric synthesis. Dynamic kinetic resolution and related processes in enzyme and metal catalysis. *Chem. Soc. Rev.*, **30**, 321–331.
19. Palmans, A.R.A. (2017) Deracemisations under kinetic and thermodynamic control. *Mol. Syst. Des. Eng.*, **2**, 34–46.
20. Clayden, J., Mitjans, D., and Youssef, L.H. (2002) Lithium–sulfoxide–lithium exchange for the asymmetric synthesis of atropisomers under thermodynamic control. *J. Am. Chem. Soc.*, **124**, 5266–5267.
21. Kawasaki, T., Suzuki, K., Shimizu, M., Ishikawa, K., and Soai, K. (2006) Spontaneous absolute asymmetric synthesis in the presence of achiral silica gel in conjunction with asymmetric autocatalysis. *Chirality*, **18**, 479–482.
22. Lente, G. (2012) Genesis — in the beginning: precursors of life, chemical models and early biological evolution, in Seckbach, J., ed., "*Absolute Asymmetric Synthesis and the Origin of Biological Chirality*," Springer Netherlands, Dordrecht, pp. 509–523.
23. Dalidovich, T., Mishra, K.A., Shalima, T., Kudrjašova, M., Kananovich, D.G., and Aav, R. (2020) Mechanochemical synthesis of amides with uronium-based coupling reagents: a method for hexa-amidation of biotin[6]uril. *ACS Sustain. Chem. Eng.*, **8**, 15703–15715.
24. Horeau, A. (1969) Interactions d'enantiomeres en solution; influence sur le pouvoir rotatoire: Purete optique et purete enantiomerique. *Tetrahedron Lett.*, **10**, 3121–3124.
25. Polavarapu, P.L. (2020) Optical purity, enantiomeric excess and the Horeau effect. *Org. Biomol. Chem.*, **18**, 6801–6806.

26. Yamaguchi, S., and Mosher, H.S. (1973) Asymmetric reductions with chiral reagents from lithium aluminum hydride and (+)-(2*S*,3*R*)-4-dimethylamino-3-methyl-1,2-diphenyl-2-butanol. *J. Org. Chem.*, **38**, 1870–1877.
27. Gawley, R.E. (2006) Do the terms "% ee" and "% de" make sense as expressions of stereoisomer composition or stereoselectivity? *J. Org. Chem.*, **71**, 2411–2416.
28. Kagan, H.B., and Gopalaiah, K. (2011) Early history of asymmetric synthesis: who are the scientists who set up the basic principles and the first experiments? *New J. Chem.*, **35**, 1933–1937.
29. Kuhn, W., and Braun, E. (1929) Photochemische Erzeugung optisch aktiver Stoffe. *Naturwissenschaften*, **17**, 227–228.
30. Balavoine, G., Moradpour, A., and Kagan, H.B. (1974) Preparation of chiral compounds with high optical purity by irradiation with circularly polarized light, a model reaction for the prebiotic generation of optical activity. *J. Am. Chem. Soc.*, **96**, 5152–5158.
31. Fischer, E. (1894) Synthesen in der Zuckergruppe II. *Ber. Dtsch. Chem. Ges.*, **27**, 3189–3232.
32. Fischer, E. (1894) Einfluss der Configuration auf die Wirkung der Enzyme. *Ber. Dtsch. Chem. Ges.*, **27**, 2985–2993.
33. Marckwald, W. (1904) Ueber asymmetrische synthese. *Ber. Dtsch. Chem. Ges.*, **37**, 1368–1370.
34. Cram, D.J., and Elhafez, F.A.A. (1952) Studies in stereochemistry. X. The rule of "steric control of asymmetric induction" in the syntheses of acyclic systems. *J. Am. Chem. Soc.*, **74**, 5828–5835.
35. Prelog, V. (1953) Untersuchungen über asymmetrische Synthesen I. Über den sterischen Verlauf der Reaktion von α-Ketosäure-estern optisch aktiver Alkohole mit Grignard'schen Verbindungen. *Helv. Chim. Acta*, **36**, 308–319.
36. McKenzie, A. (1904) CXXVII. — Studies in asymmetric synthesis. I. Reduction of menthyl benzoylformate. II. Action of magnesium alkyl haloids on menthyl benzoylformate. *J. Chem. Soc., Trans.*, **85**, 1249–1262.
37. Corey, E.J., and Ensley, H.E. (1975) Preparation of an optically active prostaglandin intermediate via asymmetric induction. *J. Am. Chem. Soc.*, **97**, 6908–6909.
38. Evans, D.A., Helmchem, G., and Ruping, M. (2007) Asymmetric synthesis –the essentials, in Christman, M., ed., "*Chiral Auxiliaries in Asymmetric Synthesis*," Wiley-VCH, Weinheim, pp. 3–9.
39. Gnas, Y., and Glorius, F. (2006) Chiral auxiliaries — principles and recent applications. *Synthesis*, 1899–1930.
40. Diaz-Muñoz, G., Miranda, I.L., Sartori, S.K., de Rezende, D.C., and Alves Nogueira Diaz, M. (2019) Use of chiral auxiliaries in the asymmetric synthesis of biologically active compounds: a review. *Chirality*, **31**, 776–812.
41. Knowles, W.S., and Sabacky, M.J. (1968) Catalytic asymmetric hydrogenation employing a soluble, optically active, rhodium complex. *Chem. Commun.*, 1445–1446.

42. Nozaki, H., Takaya, H., Moriuti, S., and Noyori, R. (1968) Homogeneous catalysis in the decomposition of diazo compounds by copper chelates: asymmetric carbenoid reactions. *Tetrahedron*, **24**, 3655–3669.
43. Knowles, W.S. (2002) Asymmetric hydrogenations (Nobel lecture). *Angew. Chem. Int. Ed.*, **41**, 1998–2007.
44. Noyori, R. (2002) Asymmetric catalysis: science and opportunities (Nobel lecture). *Angew. Chem. Int. Ed.*, **41**, 2008–2022.
45. Knowles, W.S. (2003) Asymmetric Hydrogenations — The Monsanto L-Dopa Process, in Blaser, H.-U., and Schmidt, E., eds., "Asymmetric Catalysis on Industrial Scale: Challenges, Approaches and Solutions," Wiley-VCH, Weinheim, pp. 21–38.
46. Katsuki, T., and Sharpless, K.B. (1980) The first practical method for asymmetric epoxidation. *J. Am. Chem. Soc.*, **102**, 5974–5976.
47. Finn, M.G., and Sharpless, K.B. (1991) Mechanism of asymmetric epoxidation. 2. Catalyst structure. *J. Am. Chem. Soc.*, **113**, 113–126.
48. Puchot, C., Samuel, O., Dunach, E., Zhao, S., Agami, C., and Kagan, H.B. (1986) Nonlinear effects in asymmetric synthesis. Examples in asymmetric oxidations and aldolization reactions. *J. Am. Chem. Soc.*, **108**, 2353–2357.
49. Soai, K., Shibata, T., Morioka, H., and Choji, K. (1995) Asymmetric autocatalysis and amplification of enantiomeric excess of a chiral molecule. *Nature*, **378**, 767–768.
50. Kawasaki, T., Sato, M., Ishiguro, S., Saito, T., Morishita, Y., Sato, I., Nishino, H., Inoue, Y., and Soai, K. (2005) Enantioselective synthesis of near enantiopure compound by asymmetric autocatalysis triggered by asymmetric photolysis with circularly polarized light. *J. Am. Chem. Soc.*, **127**, 3274–3275.
51. Gehring, T., Busch, M., Schlageter, M., and Weingand, D. (2010) A concise summary of experimental facts about the Soai reaction. *Chirality*, **22**, E173–E182.
52. Blackmond, D.G. (2004) Asymmetric autocatalysis and its implications for the origin of homochirality. *Proc. Natl. Acad. Sci.*, **101**, 5732–5736.
53. Athavale, S.V, Simon, A., Houk, K.N., and Denmark, S.E. (2020) Demystifying the asymmetry-amplifying, autocatalytic behaviour of the Soai reaction through structural, mechanistic and computational studies. *Nat. Chem.*, **12**, 412–423.
54. List, B. (2007) Introduction: organocatalysis. *Chem. Rev.*, **107**, 5413–5415.
55. Lelais, G., and MacMillan, D.W.C. (2007) *New Frontiers in Asymmetric Catalysis*, in Mikami, K., and Lautens, M., eds., "*History and Perspective of Chiral Organic Catalysts*," John Wiley & Sons, Hoboken, pp. 313–358 (Chapter 11).
56. List, B. (2002) Proline-catalyzed asymmetric reactions, *Tetrahedron*, **58**, 5573–5590.
57. Hashimoto, T., and Maruoka, K. (2007) Recent development and application of chiral phase-transfer catalysts. *Chem. Rev.*, **107**, 5656–5682.
58. Taylor, M.S., and Jacobsen, E.N. (2006) Asymmetric catalysis by chiral hydrogen-bond donors. *Angew. Chem. Int. Ed.*, **45**, 1520–1543.
59. Chen, D.-F., Han, Z.-Y., Zhou, X.-L., and Gong, L.-Z. (2014) Asymmetric organocatalysis combined with metal catalysis: concept, proof of concept, and beyond. *Acc. Chem. Res.*, **47**, 2365–2377.

60. Zhou, Q.-L. (2016) Transition-metal catalysis and organocatalysis: where can progress be expected? *Angew. Chem. Int. Ed.*, **55**, 5352–5353.
61. Perez-Rizquez, C., Rodriguez-Otero, A., and Palomo, J.M. (2019) Combining enzymes and organometallic complexes: novel artificial metalloenzymes and hybrid systems for C–H activation chemistry. *Org. Biomol. Chem.*, **17**, 7114–7123.
62. Xiang, S.-H., and Tan, B. (2020) Advances in asymmetric organocatalysis over the last 10 years. *Nat. Commun.*, **11**, article number 3786.
63. Fischer, E. (1890) Ueber die optischen Isomeren des Traubenzuckers, der Gluconsäure und der Zuckersäure. *Ber. Dtsch. Chem. Ges.*, **23**, 2611–2624.
64. Mengel, A., and Reiser, O. (1999) Around and beyond Cram's rule. *Chem. Rev.*, 99, 1191–1224.
65. Koskinen, A. (2012) Asymmetric synthesis, in "*Asymmetric synthesis of natural products*," John Wiley & Sons, Chichester, pp. 55–113 (Chapter 3).
66. Anh, N.T., and Eisenstein, O. (1976) Induction asymetrique 1–2: comparaison ab initio des modeles de cram, de cornforth, de karabatsos et de felkin. *Tetrahedron Lett.*, **17**, 155–158.
67. Reetz, M.T. (1984) Chelation or non-chelation control in addition reactions of chiral α- and β- alkoxy carbonyl compounds [new synthetic methods (44)]. *Angew. Chem. Int. Ed.*, **23**, 556–569.
68. Bartolo, N.D., Read, J.A., Valentín, E.M., and Woerpel, K.A. (2020) Reactions of allylmagnesium reagents with carbonyl compounds and compounds with C=N double bonds: their diastereoselectivities generally cannot be analyzed using the Felkin–Anh and chelation-control models. *Chem. Rev.*, **120**, 1513–1619.
69. Evans, D.A., Ennis, M.D., and Mathre, D.J. (1982) Asymmetric alkylation reactions of chiral imide enolates. A practical approach to the enantioselective synthesis of alpha.-substituted carboxylic acid derivatives. *J. Am. Chem. Soc.*, **104**, 1737–1739.
70. Midland, M.M. (1989) Asymmetric reductions with organoborane reagents. *Chem. Rev.*, **89**, 1553–1561.
71. Midland, M.M. (2001) *B*-3-pinanyl-9-borabicyclo[3.3.1]nonane. *Encycl. Reagents Org. Synth.* https://doi.org/10.1002/047084289X.rp173.
72. (2003) *Handbook of Reagents for Organic Synthesis: Chiral Reagents for Asymmetric Synthesis*, ed. Paquette, L. A., Wiley, Chichester.
73. Noyori, R., Tomino, I., Tanimoto, Y., and Nishizawa, M. (1984) Asymmetric synthesis via axially dissymmetric molecules. 6. Rational designing of efficient chiral reducing agents. Highly enantioselective reduction of aromatic ketones by binaphthol-modified lithium aluminum hydride reagents. *J. Am. Chem. Soc.*, **106**, 6709–6716.
74. Brown, H.C., Chandrasekharan, J., Ramachandran, P.V. (1988) Chiral synthesis via organoboranes. 14. Selective reductions. 41. Diisopinocampheylchloroborane, an exceptionally efficient chiral reducing agent. *J. Am. Chem. Soc.*, **110**, 1539–1546.
75. Gaumont, A.-C., Génisson, Y., Guillen, F., Zgonnik, V., and Plaquevent, J.-C. (2011) Catalytic methods in asymmetric synthesis: advanced materials, techniques, and

applications, in Gruttadauria, M., Giacalone, F., eds., "*Chiral Ionic Liquids for Asymmetric Reactions*," John Wiley & Sons, Hoboken, pp. 323–344 (Chapter 7).
76. Nagata, Y., Takeda, R., and Suginome, M. (2019) Asymmetric catalysis in chiral solvents: chirality transfer with amplification of homochirality through a helical macromolecular scaffold. *ACS Cent. Sci.*, **5**, 1235–1240.
77. Blaser, H.-U., Pfaltz, A., and Wennemers, H. (2012) Chiral compounds. *Ullmann's Encycl. Ind. Chem.* https://doi.org/10.1002/14356007.a18_177.pub2
78. Pfaltz, A., and Drury, W.J. (2004) Design of chiral ligands for asymmetric catalysis: from C_2-symmetric P,P- and N,N-ligands to sterically and electronically nonsymmetrical P,N-ligands. *Proc. Natl. Acad. Sci.*, **101**, 5723–5726.
79. Brown, J.M., and Deeth, R.J. (2009) Is enantioselectivity predictable in asymmetric catalysis? *Angew. Chem. Int. Ed.*, **48**, 4476–4479.
80. Ahn, S., Hong, M., Sundararajan, M., Ess, D.H., and Baik, M.-H. (2019) Design and optimization of catalysts based on mechanistic insights derived from quantum chemical reaction modeling. *Chem. Rev.*, **119**, 6509–6560.
81. Ardkhean, R., Fletcher, S.P., and Paton, R.S. (2020) New directions in the modeling of organometallic reactions. Topics in organometallic chemistry, in Lledós, A., and Ujaque, G., eds., "*Ligand Design for Asymmetric Catalysis: Combining Mechanistic and Chemoinformatics Approaches*," vol. 67, Springer, Cham, pp. 153–189.
82. Reid, J.P., and Sigman, M.S. (2019) Holistic prediction of enantioselectivity in asymmetric catalysis. *Nature*, **571**, 343–348.
83. Zahrt, A.F., Athavale, S.V., and Denmark, S.E. (2020) Quantitative structure–selectivity relationships in enantioselective catalysis: past, present, and future. *Chem. Rev.*, **120**, 1620–1689.
84. Iwamoto, H., Imamoto, T., and Ito, H. (2018) Computational design of high-performance ligand for enantioselective Markovnikov hydroboration of aliphatic terminal alkenes. *Nat. Commun.*, **9**, 2290.
85. Halpern, J. (1982) Mechanism and stereoselectivity of asymmetric hydrogenation. *Science*, **217**, 401–407.
86. Gridnev, I.D., and Imamoto, T. (2004) On the mechanism of stereoselection in Rh-catalyzed asymmetric hydrogenation: a general approach for predicting the sense of enantioselectivity. *Acc. Chem. Res.*, **37**, 633–644.
87. Li, M., Tang, D., Luo, X., and Shen, W. (2005) Mechanism of asymmetric hydrogenation of enamides with $[Rh(BisP^*)]^+$ catalyst: model DFT study. *Int. J. Quantum Chem.*, **102**, 53–63.
88. Feldgus, S., and Landis, C.R. (2000) Large-scale computational modeling of $[Rh(DuPHOS)]^+$-catalyzed hydrogenation of prochiral enamides: reaction pathways and the origin of enantioselection. *J. Am. Chem. Soc.*, **122**, 12714–12727.
89. Mori, S., Vreven, T., and Morokuma, K. (2006) Transition states of binap–rhodium(I)-catalyzed asymmetric hydrogenation: theoretical studies on the origin of the enantioselectivity. *Chem. An Asian J.*, **1**, 391–403.

90. Kitamura, M., Suga, S., Niwa, M., and Noyori, R. (1995) Self and nonself recognition of asymmetric catalysts. nonlinear effects in the amino alcohol-promoted enantioselective addition of dialkylzincs to aldehydes. *J. Am. Chem. Soc.*, **117**, 4832–4842.
91. Kitamura, M., Suga, S., Oka, H., and Noyori, R. (1998) Quantitative analysis of the chiral amplification in the amino alcohol-promoted asymmetric alkylation of aldehydes with dialkylzincs. *J. Am. Chem. Soc.*, **120**, 9800–9809.
92. Kim, J.G., and Walsh, P.J. (2006) From aryl bromides to enantioenriched benzylic alcohols in a single flask: catalytic asymmetric arylation of aldehydes. *Angew. Chem. Int. Ed.*, **45**, 4175–4178.
93. Gibson, S.E., and Castaldi, M.P. (2006) Applications of chiral C_3-symmetric molecules. *Chem. Commun.*, 3045–3062.
94. Li, W.J., and Qiu, S.X. (2010) A novel D_2-symmetrical chiral tetraoxazoline ligand for highly enantioselective hydrosilylation of aromatic ketones. *Adv. Synth. Catal.*, **352**, 1119–1122.
95. Hu, Y., Lang, K., Tao, J., Marshall, M.K., Cheng, Q., Cui, X., Wojtas, L., and Zhang, X.P. (2019) Next-generation D_2-symmetric chiral porphyrins for cobalt(II)-based metalloradical catalysis: catalyst engineering by distal bridging. *Angew. Chem. Int. Ed.*, **58**, 2670–2674.
96. Whitesell, J.K. (1989) C_2 symmetry and asymmetric induction. *Chem. Rev.*, **89**, 1581–1590.
97. Rasappan, R., Laventine, D., and Reiser, O. (2008) Metal-bis(oxazoline) complexes: from coordination chemistry to asymmetric catalysis. *Coord. Chem. Rev.*, **252**, 702–714.
98. Sen, A., and Chikkali, S.H. (2021) C_1-symmetric diphosphorus ligands in metal-catalyzed asymmetric hydrogenation to prepare chiral compounds. *Org. Biomol. Chem.*, **19**, 9095–9137.
99. Oguni, N., and Omi, T. (1984) Enantioselective addition of diethylzinc to benzaldehyde catalyzed by a small amount of chiral 2-amino-1-alcohols. *Tetrahedron Lett.*, **25**, 2823–2824.
100. Kitamura, M., Suga, S., Kawai, K., and Noyori, R. (1986) Catalytic asymmetric induction. Highly enantioselective addition of dialkylzincs to aldehydes. *J. Am. Chem. Soc.*, **108**, 6071–6072.
101. Pu, L., and Yu, H.-B. (2001) Catalytic asymmetric organozinc additions to carbonyl compounds. *Chem. Rev.*, **101**, 757–824.
102. Yamakawa, M., and Noyori, R. (1999) Asymmetric addition of dimethylzinc to benzaldehyde catalyzed by (2*S*)-3-exo-(dimethylamino)isobornenol. A theoretical study on the origin of enantioselection. *Organometallics*, **18**, 128–133.
103. Yamakawa, M., and Noyori, R. (1995) An ab initio molecular orbital study on the amino alcohol-promoted reaction of dialkylzincs and aldehydes. *J. Am. Chem. Soc.*, **117**, 6327–6335.
104. Zhang, L., and Meggers, E. (2017) Stereogenic-only-at-metal asymmetric catalysts. *Chem. Asian J.*, **12**, 2335–2342.

105. Chen, L.-A., Xu, W., Huang, B., Ma, J., Wang, L., Xi, J., Harms, K., Gong, L., and Meggers, E. (2013) Asymmetric catalysis with an inert chiral-at-metal iridium complex. *J. Am. Chem. Soc.*, **135**, 10598–10601.
106. Xu, W., Arieno, M., Löw, H., Huang, K., Xie, X., Cruchter, T., Ma, Q., Xi, J., Huang, B., Wiest, O., Gong, L., and Meggers, E. (2016) Metal-templated design: enantioselective hydrogen-bond-driven catalysis requiring only parts-per-million catalyst loading. *J. Am. Chem. Soc.*, **138**, 8774–8780.
107. Hong, Y., Jarrige, L., Harms, K., and Meggers, E. (2019) Chiral-at-iron catalyst: expanding the chemical space for asymmetric earth-abundant metal catalysis. *J. Am. Chem. Soc.*, **141**, 4569–4572.
108. Li, X., Deng, H., Zhang, B., Li, J., Zhang, L., Luo, S., and Cheng, J.-P. (2010) Physical organic study of structure–activity–enantioselectivity relationships in asymmetric bifunctional thiourea catalysis: hints for the design of new organocatalysts. *Chem. Eur. J.*, **16**, 450–455.
109. Mukherjee, S., Yang, J.W., Hoffmann, S., and List, B. (2007) Asymmetric enamine catalysis. *Chem. Rev.*, **107**, 5471–5569.
110. Erkkilä, A., Majander, I., and Pihko, P.M. (2007) Iminium catalysis. *Chem. Rev.*, **107**, 5416–5470.
111. Enders, D., Niemeier, O., and Henseler, A. (2007) Organocatalysis by *N*-heterocyclic carbenes. *Chem. Rev.*, 107, pp. 5606–5655.
112. Akiyama, T. (2007) Stronger Brønsted acids. *Chem. Rev.*, **107**, 5744–5758.
113. Wang, X., Lan, Q., and Maruoka, K. (2008) Asymmetric Phase-Transfer Catalysis, in Ding, K., and Uozumi, Y., eds. "*Handbook of asymmetric heterogeneous catalysis*," Wiley-VCH, Weinheim, pp. 383–412 (Chapter 11).
114. Bulfield, D., and Huber, S.M. (2016) Halogen bonding in organic synthesis and organocatalysis, *Chem. Eur. J.*, **22**, 14434–14450.
115. Kaasik, M., and Kanger, T. (2020) Supramolecular halogen bonds in asymmetric catalysis. *Front. Chem.*, **8**, 958.
116. Clemente, F.R., and Houk, K.N. (2005) Theoretical studies of stereoselectivities of intramolecular aldol cyclizations catalyzed by amino acids. *J. Am. Chem. Soc.*, **127**, 11294–11302.
117. Dolling, U.H., Davis, P., and Grabowski, E.J.J. (1984) Efficient catalytic asymmetric alkylations. 1. Enantioselective synthesis of (+)-indacrinone via chiral phase-transfer catalysis. *J. Am. Chem. Soc.*, **106**, 446–447.
118. Heckmann, C.M., and Paradisi, F. (2020) Looking back: a short history of the discovery of enzymes and how they became powerful chemical tools. *ChemCatChem*, **12**, 6082–6102.
119. Sheldon, R.A., Brady, D., and Bode, M.L. (2020) The Hitchhiker's guide to biocatalysis: recent advances in the use of enzymes in organic synthesis. *Chem. Sci.*, **11**, 2587–2605.
120. Musa, M.M., and Phillips, R.S. (2011) Recent advances in alcohol dehydrogenase-catalyzed asymmetric production of hydrophobic alcohols. *Catal. Sci. Technol.*, **1**, 1311–1323.

121. Li, C., Heatwole, J., Soelaiman, S., and Shoham, M. (1999) Crystal structure of a thermophilic alcohol dehydrogenase substrate complex suggests determinants of substrate specificity and thermostability. *Proteins Struct. Funct. Bioinform.*, **37**, 619–627.
122. Ziegelmann-Fjeld, K.I., Musa, M.M., Phillips, R.S., Zeikus, J.G., and Vieille, C. (2007) A thermoanaerobacter ethanolicus secondary alcohol dehydrogenase mutant derivative highly active and stereoselective on phenylacetone and benzylacetone. *Protein Eng. Des. Sel.*, **20**, 47–55.
123. Davis, H.J., and Ward, T.R. (2019) Artificial metalloenzymes: challenges and opportunities. *ACS Cent. Sci.*, **5**, 1120–1136.
124. Harnden, K.A., Wang, Y., Vo, L., Zhao, H., and Lu, Y. (2021) Engineering Artificial Metalloenzymes, in Zhao, H., Lee, S.Y., Nielsen, J., and Stephanopoulos, G., eds., "*Protein Engineering: Tools and Applications*," Wiley-VCH, Weinheim, pp. 177–205 (Chapter 8).
125. Reetz, M.T. (2004) Controlling the enantioselectivity of enzymes by directed evolution: practical and theoretical ramifications. *Proc. Natl. Acad. Sci.*, **101**, 5716–5722.
126. Prier, C.K., Zhang, R.K., Buller, A.R., Brinkmann-Chen, S., and Arnold, F.H. (2017) Enantioselective, intermolecular benzylic C–H amination catalysed by an engineered iron-haem enzyme. *Nat. Chem.*, **9**, 629–634.
127. Fasan, R., Jennifer Kan, S.B., and Zhao, H. (2019) A continuing career in biocatalysis: Frances H. Arnold. *ACS Catal.*, **9**, 9775–9788.
128. Key, H.M., Dydio, P., Clark, D.S., and Hartwig, J.F. (2016) Abiological catalysis by artificial haem proteins containing noble metals in place of iron. *Nature*, **534**, 534–537.
129. Anastas, P.T., and Warner, J.C. (1998) *Green Chemistry: Theory and Practice*. Oxford University Press, New York.
130. Anastas, P., and Eghbali, N. (2010) Green chemistry: principles and practice. *Chem. Soc. Rev.*, **39**, 301–312.
131. McElroy, C.R., Constantinou, A., Jones, L.C., Summerton, L., and Clark, J.H. (2015) Towards a holistic approach to metrics for the 21st century pharmaceutical industry. *Green Chem.*, **17**, 3111–3121.
132. Roschangar, F., Li, J., Zhou, Y., Aelterman, W., Borovika, A., Colberg, J., Dickson, D.P., Gallou, F., Hayler, J.D., Koenig, S.G., Kopach, M.E., Kosjek, B., Leahy, D.K., O'Brien, E., Smith, A.G., Henry, M., Cook, J., and Sheldon, R.A. (2022) Improved iGAL 2.0 metric empowers pharmaceutical scientists to make meaningful contributions to united nations sustainable development goal 12. *ACS Sustainable Chem. Eng., 10, 5148–5162.*
133. Constable, D.J.C., Jimenez-Gonzalez, C., and Henderson, R.K. (2007) Perspective on solvent use in the pharmaceutical industry. *Org. Process Res. Dev.*, **11**, 133–137.
134. Gomollón-Bel, F. (2019) Ten chemical innovations that will change our world: IUPAC identifies emerging technologies in chemistry with potential to make our planet more sustainable. *Chem. Int.*, **41**, 12–17.

135. Walsh, P.J., Li, H., and de Parrodi, C.A. (2007) A green chemistry approach to asymmetric catalysis: solvent-free and highly concentrated reactions. *Chem. Rev.*, **107**, 2503–2545.
136. Do, J.L., and Friščić, T. (2017) Chemistry 2.0: developing a new, solvent-free system of chemical synthesis based on mechanochemistry. *Synlett*, **28**, 2066–2092.
137. Williams, M.J., Morrill, L.C., and Browne, D.L. (2022) Mechanochemical organocatalysis: do high enantioselectivites contradict what we might expect? *ChemSusChem*, **15**, e202102157.
138. Hunt, A.J., Farmer, T.J., and Clark, J.H. (2013) Elemental Sustainability and the Importance of Scarce Element Recovery, in Hunt, A., ed., "*Element Recovery and Sustainability*," The Royal Society of Chemistry, Cambridge, pp. 1–28 (Chapter 1).
139. Ohkuma, T., Ooka, H., Hashiguchi, S., Ikariya, T., and Noyori, R. (1995) Practical enantioselective hydrogenation of aromatic ketones. *J. Am. Chem. Soc.*, **117**, 2675–2676.
140. Bigler, R., Huber, R., and Mezzetti, A. (2015) Highly enantioselective transfer hydrogenation of ketones with chiral $(NH)_2P_2$ macrocyclic iron(II) complexes. *Angew. Chem. Int. Ed.*, **54**, 5171–5174.
141. Zheng, L., Yin, X., Mohammadlou, A., Sullivan, R.P., Guan, Y., Staples, R., and Wulff, W.D. (2020) Asymmetric catalytic Meerwein–Ponndorf–Verley reduction of ketones with aluminum(III)-VANOL catalysts. *ACS Catal.*, **10**, 7188–7194.
142. https://www.rsc.org/periodic-table (accessed Jun 23, 2021).
143. Balaram, V. (2020) "Environmental Impact of Platinum, Palladium, and Rhodium Emissions from Autocatalytic Converters — A Brief Review of the Latest Developments," *Handbook of Environmental Materials Management*, Springer, Cham, pp. 1–37.
144. Egorova, K.S., and Ananikov, V.P. (2016) Which metals are green for catalysis? Comparison of the toxicities of Ni, Cu, Fe, Pd, Pt, Rh, and Au salts, *Angew. Chem. Int. Ed.*, **55**, 12150–12162.
145. Egorova, K.S., and Ananikov, V.P. (2017) Toxicity of metal compounds: knowledge and myths. *Organometallics*, **36**, 4071–4090.
146. Nachtergael, A., Coulembier, O., Dubois, P., Helvenstein, M., Duez, P., Blankert, B., and Mespouille, L. (2015) Organocatalysis paradigm revisited: are metal-free catalysts really harmless? *Biomacromolecules*, **16**, 507–514.
147. Konrad, N., Horetski, M., Sihtmäe, M., Truong, K.-N., Osadchuk, I., Burankova, T., Kielmann, M., Adamson, J., Kahru, A., Rissanen, K., Senge, M.O., Borovkov, V., Aav, R., and Kananovich, D. (2021) Thiourea organocatalysts as emerging chiral pollutants: en route to porphyrin-based (chir)optical sensing. *Chemosensors*, **9**, article number 278.
148. Ni, Y., Holtmann, D., and Hollmann, F. (2014) How green is biocatalysis? To calculate is to know. *ChemCatChem*, **6**, 930–943.
149. de María, P., and Hollmann, F. (2015) On the (un)greenness of biocatalysis: some challenging figures and some promising options. *Front. Microbiol.*, **6**, 1257.

150. Bertus, P., Boeda, F., and Pearson-Long, M.S.M. (2016) Titanium-based Catalysts for Asymmetric Transformations, Part 1, in North, M., ed., "*Sustainable Catalysis: With Non-endangered Metals*," The Royal Society of Chemistry, Cambridge, pp. 140–198 (Chapter 7).
151. Lopp, M., Paju, A., Kanger, T., and Pehk, T. (1996) Asymmetric Baeyer-Villiger oxidation of cyclobutanones. *Tetrahedron Lett.*, **37**, 7583–7586.
152. Lopp, M., Paju, A., Kanger, T., and Pehk, T. (1997) Direct asymmetric α-hydroxylation of β-hydroxyketones. *Tetrahedron Lett.*, **38**, 5051–5054.
153. Paju, A., Kanger, T., Pehk, T., and Lopp, M. (2002) Direct asymmetric α-hydroxylation of 2-hydroxymethyl ketones. *Tetrahedron*, **58**, 7321–7326.
154. Paju, A., Kanger, T., Pehk, T., and Lopp, M. (2000) Asymmetric oxidation of 1,2-cyclopentanediones. *Tetrahedron Lett.*, **41**, 6883–6887.
155. Paju, A., Kanger, T., Pehk, T., Müürisepp, A.-M., and Lopp, M. (2002) Asymmetric oxidation of 3-alkyl-1,2-cyclopentanediones. Part 1: 3-hydroxylation of 3-alkyl-1,2-cyclopentanediones. *Tetrahedron: Asymmetry*, **13**, 2439–2448.
156. Paju, A., Laos, M., Jõgi, A., Päri, M., Jäälaid, R., Pehk, T., Kanger, T., and Lopp, M. (2006) Asymmetric synthesis of 2-alkyl-substituted 2-hydroxyglutaric acid γ-lactones. *Tetrahedron Lett.*, **47**, 4491–4493.
157. Paju, A., and Lopp, M. Unpublished results.
158. Paju, A., Kanger, T., Pehk, T., Eek, M., and Lopp, M. (2004) A short enantioselective synthesis of homocitric acid-γ-lactone and 4-hydroxy-homocitric acid-γ-lactones. *Tetrahedron*, **60**, 9081–9084.
159. Jõgi, A., Ilves, M., Paju, A., Pehk, T., Kailas, T., Müürisepp, A.-M., and Lopp, M. (2008) Asymmetric synthesis of 4′-*C*-benzyl-2′,3′-dideoxynucleoside analogues from 3-benzyl-2-hydroxy-2-cyclopenten-1-one. *Tetrahedron: Asymmetry*, **19**, 628–634.
160. Jõgi, A., Paju, A., Pehk, T., Kailas, T., Müürisepp, A.-M., and Lopp, M. (2009) Synthesis of 4′-aryl-2′,3′-dideoxynucleoside analogues. *Tetrahedron*, **65**, 2959–2965.
161. Reile, I., Paju, A., Müürisepp, A.-M., Pehk, T., and Lopp, M. (2011) Oxidation of cyclopentane-1,2-dione: a study with ^{18}O labeled reagents. *Tetrahedron*, **67**, 5942–5948.
162. Paju, A., Kanger, T., Pehk, T., Lindmaa, R., Müürisepp, A.-M., and Lopp, M. (2003) Asymmetric oxidation of 3-alkyl-1,2-cyclopentanediones. Part 2: oxidative ring cleavage of 3-alkyl-1,2-cyclopentanediones: synthesis of 2-alkyl-γ-lactone acids. *Tetrahedron: Asymmetry*, **14**, 1565–1573.
163. Osadchuk, I., Pehk, T., Paju, A., Lopp, M., Öeren, M., and Tamm, T. (2014) Isomers and conformers of complexes of $Ti(O\textit{i}Pr)_4$ with cyclopentane-1,2-dione: NMR study and DFT calculations. *Int. J. Quantum Chem.*, **114**, 1012–1018.
164. Kulinkovich, O.G., Sviridov, S.V., and Vasilevski, D.A. (1991) Titanium(IV) isopropoxide-catalyzed formation of 1-substituted cyclopropanols in the reaction of ethylmagnesium bromide with methyl alkanecarboxylates. *Synthesis*, **1991**, 234.

165. Cha, J.K., and Kulinkovich, O.G. (2012) The Kulinkovich cyclopropanation of carboxylic acid derivatives. *Org. React.*, pp. 1–160.
166. Kulinkovich, O.G., Sviridov, S.V., Vasilevski, D.A., and Pritytskaya, T.S. (1989) Reaktsiya etilmagniybromida s efirami karbonovykh kislot v prisutstvii tetraizopropoksitana. *Zh. Org. Chem.*, **25**, 2244–2245 (In Russian).
167. Kulinkovich, O., Isakov, V., and Kananovich, D. (2008) Stereocontrolled alkylation of unsaturated compounds with alkoxytitanacyclopropane reagents. *Chem. Rec.*, **8**, 269–278.
168. Wolan, A., and Six, Y. (2010) Synthetic transformations mediated by the combination of titanium(IV) alkoxides and Grignard reagents: selectivity issues and recent applications. Part 1: reactions of carbonyl derivatives and nitriles. *Tetrahedron*, **66**, 15–61.
169. Kulinkovich, O.G., and de Meijere, A. (2000) 1,*n*-Dicarbanionic titanium intermediates from monocarbanionic organometallics and their application in organic synthesis. *Chem. Rev.*, **100**, 2789–2834.
170. Wolan, A., and Six, Y. (2010) Synthetic transformations mediated by the combination of titanium(IV) alkoxides and Grignard reagents: selectivity issues and recent applications. Part 2: reactions of alkenes, allenes and alkynes, Tetrahedron, **66**, 3097–3133.
171. Bertus, P., and Szymoniak, J. (2001) New and easy route to primary cyclopropylamines from nitriles. *Chem. Commun.*, 1792–1793.
172. Chaplinski, V., and de Meijere, A. (1996) A versatile new preparation of cyclopropylamines from acid dialkylamides. *Angew. Chem. Int. Ed.*, **35**, 413–414.
173. Konik, Y.A., and Kananovich, D.G. (2020) Asymmetric synthesis with titanacyclopropane reagents: from early results to the recent achievements. *Tetrahedron Lett.*, **61**, 152036.
174. Caillé, J., Setzer, P., Boeda, F., Pearson-Long, M.S.M., and Bertus, P. (2018) Asymmetric titanium-catalyzed cyclopropanation of nitriles with Grignard reagents. *SynOpen*, **2**, 41–49.
175. De Meijere, A., Chaplinski, V., Winsel, H., Kordes, M., Stecker, B., Gazizova, V., Savchenko, A.I., Boese, R., and Schill, F. (2010) Cyclopropylamines from *N,N*-dialkylcarboxamides and Grignard reagents in the presence of titanium tetraisopropoxide or methyltitanium triisopropoxide. *Chem. Eur. J.*, **16**, 13862–13875.
176. Kulinkovich, O.G., Savchenko, A.I., Sviridov, S.V., and Vasilevski, D.A. (1993) Titanium(IV) isopropoxide-catalyzed reaction of ethylmagnesium bromide with ethyl acetate in the presence of styrene. *Mendeleev Commun.*, 230–231.
177. Kasatkin, A., and Sato, F. (1995) Diastereoselective synthesis of trans-1,2-disubstituted cyclopropanols from homoallyl or bis-homoallyl esters via tandem intramolecular nucleophilic acyl substitution and intramolecular carbonyl addition reactions mediated by $Ti(OPr\text{-}i)_4$ / 2 *i*-PrMgBr reagent. *Tetrahedron Lett.*, **36**, 6079–6082.
178. Lee, J., Kang, C.H., Kim, H., and Cha, J.K. (1996) Intramolecular hydroxycyclopropanation of ω-vinyl carboxylic esters. *J. Am. Chem. Soc.*, **118**, 291–292.

179. Corey, E.J., Rao, S.A., and Noe, M.C. (1994) Catalytic diastereoselective synthesis of cis-1,2-disubstituted cyclopropanols from esters using a vicinal dicarbanion equivalent. *J. Am. Chem. Soc.*, **116**, 9345–9346.
180. Racouchot, S., Sylvestre, I., Ollivier, J., Kozyrkov, Y.Y., Pukin, A., Kulinkovich, O. G., and Salaün, J. (2002) Titanium-mediated diastereoselective formation of (*E*)- or (*Z*)-2-substituted 1-vinylcyclopropanols: scope and limitation, applications. *Eur. J. Org. Chem.*, 2160–2176.
181. Konik, Y.A., Kananovich, D.G., and Kulinkovich, O.G. (2013) Enantioselective cyclopropanation of carboxylic esters with alkyl magnesium bromides in the presence of titanium(IV) (4*R*,5*R*)-TADDOLates, *Tetrahedron*, **69**, 6673–6678.
182. Kulinkovich, O.G. (2000) Titanacyclopropanes as versatile intermediates for carbon-carbon bond formation in reactions with unsaturated compounds. *Pure Appl. Chem.*, **72**, 1715–1719.
183. Kulinkovich, O.G., and Kananovich, D.G. (2007) Advanced procedure for the preparation of cis-1,2-dialkylcyclopropanols — modified ate complex mechanism for titanium-mediated cyclopropanation of carboxylic esters with Grignard reagents. *Eur. J. Org. Chem.*, 2121–2132.
184. Kulinkovich, O.G., Kananovich, D.G., Lopp, M., and Snieckus, V. (2014) Insight into the mechanism and stereochemistry of the transformations of alkyltitanium Ate-complexes. An enhanced enantioselectivity in the cyclopropanation of the carboxylic esters with titanacyclopropane reagents. *Adv. Synth. Catal.*, **356**, 3615–3626.
185. Iskryk, M., Barysevich, M., Ošeka, M., Adamson, J., and Kananovich, D. (2019) Asymmetric Kulinkovich hydroxycyclopropanation of alkenes mediated by titanium(IV) TADDOLate complexes. *Synthesis*, **51**, 1935–1948.
186. Bertus, P. (2019) From dialkyltitanium species to titanacyclopropanes: an ab initio study. *Organometallics*, **38**, 4171–4182.
187. Osadchuk, I. (2017) Structures and Catalytic Properties of Titanium and Iridium Based Complexes, PhD thesis, TTÜ Press, Tallinn.
188. Barysevich, M., and Kananovich, D. Unpublished results.

https://doi.org/10.1142/9789811259227_0005

5 Chirogenesis in Polymers and Macromolecules

Puhup Puneet,* Bhanu Nandan† and Michiya Fujiki‡

*Department of Chemical Engineering, National Tsing-Hua University, Hsinchu, Taiwan
†Department of Textile and Fibre Engineering, Indian Institute of Technology Delhi, New Delhi, India
‡Graduate School of Science and Technology, Nara Institute of Science and Technology, Ikoma, Japan

Homochirality is a ubiquitous phenomenon in biomolecules, and its occurrence has been pondered by researchers over a long time, serving as stimulation for research in the field of chiral polymers and macromolecules. In this chapter, we embark on the journey to achieving a fundamental understanding of chirogenesis in polymers and macromolecules through several state-of-the-art chiroptical spectroscopic techniques, namely electronic circular dichroism, vibrational circular dichroism, Raman optical activity, and circularly polarized luminescence spectroscopy. Carefully selected cases are discussed to provide significant insights into the intricacies of absolutes configurational and conformational analysis using chiroptical spectroscopic techniques, both individually and in tandem.

1.1 Introduction

It is worth emphasizing that, when people see themselves in a mirror, they do not actually see their authentic self but a mirror image: the left of themselves becomes the right and vice versa. As trivial as it may sound, if we find a way of viewing ourselves in actuality, there may be significant and surprising dissimilarities as compared to mirror image. We must note that our world and everything in it is asymmetric, i.e., the left-handed version is not exactly equivalent to the right-handed one or, in some cases, the mirror image counterparts may not exist at all [1].

Ever since the discovery of the handedness of molecules (termed as chirality) by Louis Pasteur in 1848, it has become evident that handedness bias in biomolecules dominates all living system in Nature [2–6]. To understand chirogenesis in Nature, a plausible approach would be to analyze asymmetry from the macroscopic level through to fundamental subatomic particles [7–9]. The most relevant analogy for macroscopic chirality could be our left and right hands, which are nonsuperimposable mirror images of each other (Figure 1). However, if observed in detail, the

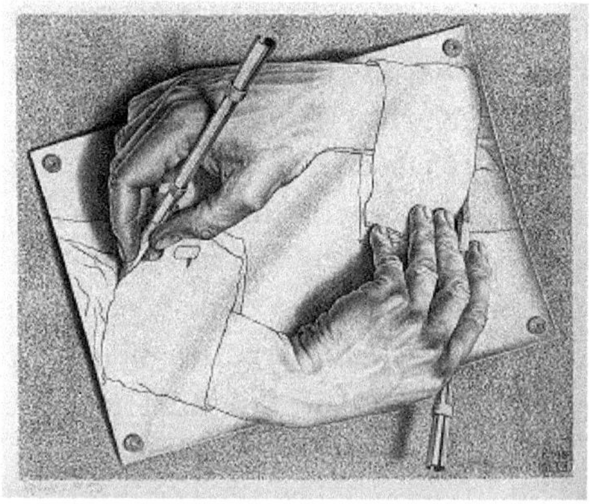

Figure 1. *Drawing Hands* by M.C. Escher (lithograph, 1948) (fair use of a low-resolution image from https://en.wikipedia.org/wiki/Drawing_Hands).

fingerprints of the left hand would not be the same for the right hand, making them not precise mirror images of one another.

Going further into detail, the cells present in left-hand would be indistinguishable from those found in the right hand. However, protein molecules that constitute living cells are predominantly composed of chiral amino acids of *L*- (laevum or left) form exclusively [1–10]. Similarly, nucleic acids preferentially consist of *D* (dextrum or right)-sugars rather than *L*-sugars [1–10]. These facts prompt us to ask vital questions as to the difference between the left- and right-handed chiral molecules and the possible scenario for the chiral bias in Nature. Since the discovery of the fourth fundamental force, the so-called *weak nuclear force*, it was evident that asymmetry is present at a subatomic level as observed for β-decay and *neutrino* emissions [11, 12]. This weak force can also be extrapolated to atoms involved in chiral molecules [13, 14], providing a possible explanation for the first question of the difference between left- and right-handed molecules. However, this is beyond the scope of this chapter, and interested readers may consult other relevant articles.

The chirogenesis in living systems could also be associated with chiral interactions in biomolecules that govern the transfer of chiral information via evolution from small molecules to biomolecules to self-assembled nanoscale chiral architectures to macroscopic chiral objects [9, 10]. The chiral architectures of different length scales have been found to retain or reverse the chiral information. Figure 2 shows the flower of *Brassica*

Figure 2. A flower of *Brassica oleracea* Romanesco (modified from a royalty-free image at Pixta).

oleracea Romanesco, a macroscopic chiral object exhibiting left-handed overall helicity; most intriguingly, it contains left-handed and right-handed subunits.

In attempting to decode this question empirically, synthetic counterparts of biomolecules, i.e., small chiral molecules, polymers, and macromolecules, have been extensively and systematically studied to develop insights into various stages of chirogenesis [15, 16]. The synthetic organic polymers and macromolecules possess an added advantage in providing a molecular backbone with a desirable type and number of functional groups that may have chiral centers or be induced. Chiroptical spectroscopy, such as electronic circular dichroism (ECD), vibrational circular dichroism (VCD), Raman optical activity (ROA), and circularly polarized luminescence (CPL) have been wielded as powerful tools to investigate the details of absolute chiral configurations, molecular conformations, and inter/intramolecular chiral interactions at the ground and excited states of morphological developments [17]. This establishes a broad research area for investigation of chiral polymers and macromolecules and their chiroptics. Hence, rather than providing a complete catalog of all the relevant research works, this chapter focuses on providing a fundamental understanding of different branches of chiroptical spectroscopic techniques and their tandem use for a comprehensive investigation of chiral polymers and macromolecules in selected unique cases.

1.2 A Skirmish of Chiroptical Spectroscopy

The decisive and comprehensive analysis of chiral molecules is of paramount importance in studying biomolecules and pharmaceutical drugs for physiological reasons. Ever since the discovery of optical rotation, i.e., the rotation of plane-polarized light in a chiral medium, chiroptical spectroscopy has taken giant leaps in terms of sophisticated instrumentation, the sensitivity of data accumulation, and theoretical understanding of the ratiocination of the absolute configuration and conformation characteristics of these molecules [18–22]. Individual chiroptical techniques, such as ECD, VCD, ROA, and CPL, provide significant insight into the electronic and vibrational modes and orientation of the molecules, which can be probed in combination to extract precise details regarding absolute configuration and conformational dynamics. Several acceptable theoretical models are often used to support the experimental observations [23].

1.2.1 ECD

ECD spectroscopy, analogously to ultraviolet (UV)–visible spectroscopy, is related to the electronic transition triggered by the absorption of electromagnetic radiation sufficient to overcome the bandgap, i.e., the transition from the ground state (*i*) to the excited state (*j*). A non-zero electric transition dipole moment ($\mu_{ij} \neq 0$) and magnetic transition dipole moment ($m_{ij} \neq 0$), originating from a linear charge dislocation and rotation of electrons, respectively, can contribute to the absorption of electromagnetic radiation quantified by oscillator strength (*f*) [24]. Usually, but by no means essentially, for nearly symmetric achiral materials, the magnitude of μ_{ij} appears much more significant than the m_{ij} component. On the contrary, for chiral systems, the electronic transition could be traced along a twisted path due to the rotation and translation of electric charge providing significant contribution from both μ_{ij} along with m_{ij}, and the CD signal can be directly used to quantify their scalar product, known as rotational strength (R_g).

Differently from UV–visible spectroscopy, the signals from ECD refer to the difference in absorbance between left-handed circularly polarized (CP) light (A^L) and right-handed CP light (A^R) [18–20, 24],

$$\Delta A = A^L - A^R \tag{1}$$

Diastereomeric chiral materials absorb different aliquots of photons from left- and right-handed CP light, resulting in elliptically polarized light. Conventionally, ΔA is measured in ellipticity ($\theta = 32.98\ \Delta A$) in the units of millidegree and can be converted to molar quantity according to the Lambert–Beer law

$$\Delta\varepsilon = \varepsilon^L - \varepsilon^R = \Delta A/cl \tag{2}$$

where $\Delta\varepsilon$ is the difference between molar absorptivity of left-handed CP light (ε^L) and right-handed CP light (ε^R), c is concentration in mol L^{-1}, and l is path length in cm.

Most of the materials absorbed energy could be utilized for the electronic excitation, and materials possessing chromophores that can exhibit transition in the range of 150–1000 nm are measured in practice. Strong electronic bands corresponding to n–π* and π–π* transition have been extensively employed to analyze chiroptical signals, though neighboring functional groups often influence the signals due to delocalization.

1.2.2 VCD

Vibrational spectroscopy [17, 21, 25–26] accounts for the vibrational transitions among vibronic sublevels located within a specified electronic state. The absorption of electromagnetic radiation of a frequency matching the oscillation frequency of the permanent electric dipole moment present in the molecules gives rise to vibrational absorbance (VA). Similar to electronic transitions in ECD, vibronic transitions of VCD signal could be formulated as a difference in absorbance ($\Delta A^{L}_{v,v'}$) between left-handed CP light ($A^{L}_{v,v'}$) and right-handed CP light ($A^{R}_{v,v'}$), responsible for the transition from v to v' states,

$$\Delta A_{v,v'} = A^{L}_{v,v'} - A^{R}_{v,v'} \tag{3}$$

and in molar absorptivity:

$$\Delta\varepsilon_{v,v'} = \varepsilon^{L}_{v,v'} - \varepsilon^{R}_{v,v'} = \Delta A_{v,v'}/cl \tag{4}$$

where $\Delta\varepsilon_{v,v'}$ is the difference between molar absorptivity of left-handed CP light ($\varepsilon^{L}_{v,v'}$) and right-handed CP light ($\varepsilon^{R}_{v,v'}$), c is concentration in mol L^{-1}, and l is path length in cm.

Furthermore, the theoretical assessment for the VCD signals could be mainly correlated to two significant factors that are rotational strength ($R_{v,v}$) and dipole strength ($D^{a}_{v,v}$). For a given transition, a dimensionless quantity called anisotropy ratio, g is given by $\Delta\varepsilon_{v,v}/\varepsilon_{v,v}$ where $\varepsilon_{v,v}$ is the molar absorptivity of VA. The anisotropic factor is a measure of VCD signal intensity related to VA intensity for vibronic transitions. It also gives the ratio of rotational strength to dipole strength:

$$g^{a}_{v,v} = \Delta\varepsilon_{v,v}/\varepsilon_{v,v} = 4R_{v,v}/D^{a}_{v,v} \tag{5}$$

Unlike ECD spectroscopy, the presence of chromophores for electronic transitions is not a prerequisite for VCD spectroscopy and all the functional groups possessing permanent dipole moments could be probed under these techniques for absolute conformational analysis.

1.2.3 ROA

The ROA scattering signals originate from the interference of light waves that are scattered molecular polarizability and optical activity tensors of a molecule [17, 22, 27]. Thus, the interference pattern converted to intensity can be strongly correlated to the degree of circular polarization of the incident light, termed as incident circular polarization along with scattered circular polarization. The authors are referred to detailed descriptions for the theoretical formulations for ROA spectroscopy [17, 27]. A substantially applicable dimensionless term that is experimentally measured in practice is the so-called circular intensity difference (CID), formulated as

$$\Delta = (I^{\mathrm{R}} - I^{\mathrm{L}})/(I^{\mathrm{R}} + I^{\mathrm{L}}) \tag{6}$$

where I^R and I^L are the scattered intensities in the right- and left-CP incident light, respectively.

In principle, ROA measurements can especially be used to extract information about three-dimensional molecular structure with combinations of fine-tuned experimental configurations such as scattering angle to 180°, i.e., backward direction scattering being the most important for studies of biomolecules in aqueous solution.

1.2.4 CPL

The theoretical development of CPL spectroscopy was established based on the existing theoretical grounds of ECD. Initially, Emeis and Oosterhoff validated the isotropic orientation distribution of the emitters and the fact that there is close similarity between the theoretical basis of the CD and CPL signals. Along this line, for a given electronic transition to and from the state vectors $|i\rangle$ and $|j\rangle$ $(i \leftrightarrow j)$, both CD and CPL can be expressed in terms of a different molecular electronic parameter, the rotatory strength (Eq. 7) [28]:

$$R_{ij} = \mathrm{Im}\langle i|\mu|j\rangle\cdot\langle j|m|i\rangle \tag{7}$$

where μ and m are the electric and magnetic dipole moment operators, respectively, and I is the intensity. The sign and magnitude of R_{ij} can be observed to differ for absorption (CD) and emission (CPL) due to the different composition of the state vector |d› and |j› at the ground-state and the excited-state geometries. Therefore, the relative magnitude of R_{ij} (absorption) and R_{ij} (emission) can be extracted when the signature and unique properties of the ground and excited states are compared. In addition, CPL spectroscopy measures the differential left CPL and right CPL of a system. Commonly, the g_{lum} is the quantification of differential emission of right and left CPL.

$$g_{lum} = (I^L - I^R)/[(I^L + I^R)/2] \tag{8}$$

where I^L and I^R are the intensity of the left- and right-handed CP emissions, respectively.

1.3 ECD Analysis of Helical Polymers and Macromolecules

The helix sense of a polymer is a function of the inversion barrier dependent on the steric repulsion interaction between the adjacent units in the polymer segment. If the helix inversion barrier is high enough for the polymer chains to attain a stable helical conformation, they can be referred to as static helical polymers; conversely, helical polymers with feeble inversion barriers are termed as dynamic helical polymers. Based on the pioneering work on helical polymers, clear demarcation of static and dynamic polymers was made, as shown in Figure 3 [15].

1.3.1 Static Helical Polymers and Macromolecules

The first demonstration of a stable helical sense of poly(phenylacetylene) derivatives was prepared by employing a chiral cocatalyst for polymerization

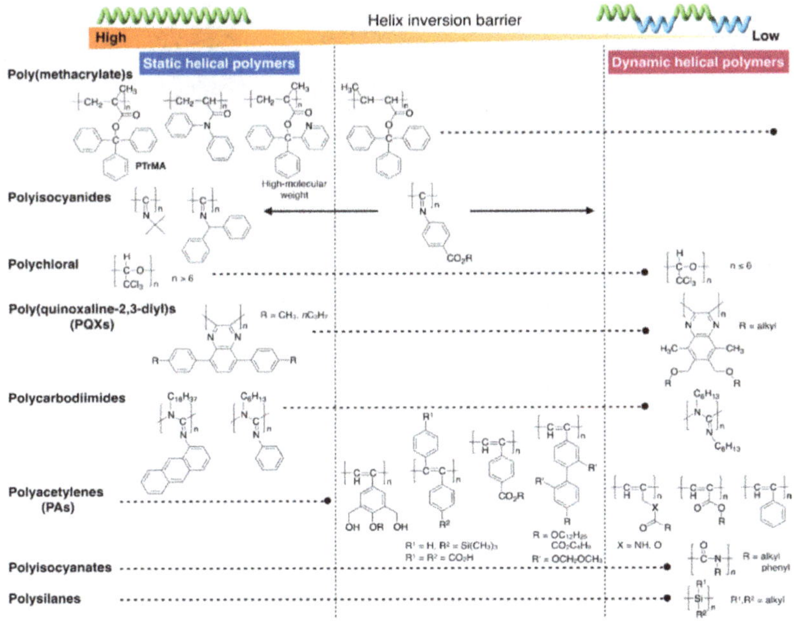

Figure 3. Categorization of helical polymers based on their inversion barriers. Reproduced with copyright permission from Ref. [15].

(Figure 4a) [29]. The resulting polymer exhibited a stable ECD signal in several organic solvents, indicating an exclusive helical sense.

Similarly, asymmetric polymerization on tris(trimethylsilyl)silyl methacrylate with Fl–Li in the presence of chiral ligands such as (−)-Sp, (+)/(−)-DDB (Figure 5a) could produce asymmetric helical conformation [30]. In the Fl–Li/(−)-Sp system, the polymerization-induced CD signal showed linear dependence on the polymerization rate. Moreover, FL–Li/(+)-DDB and FL–Li/(−)-DDB catalyzed polymer systems showed almost mirror image CD signals, indicating a stable helical sense that could be correlated to the chirality of the ligand (Figure 5b).

In another study, supramolecular self-assembly of dendritic dipeptides demonstrated columnar helical porous structures that were thoroughly analyzed with small-angle X-ray scattering and differential scanning calorimetry. Figure 6a shows the libraries of molecular structures of dendritic dipeptides with different lengths of side chains. By contrast, in Figure 6b, porous helical self-assembled architecture has been

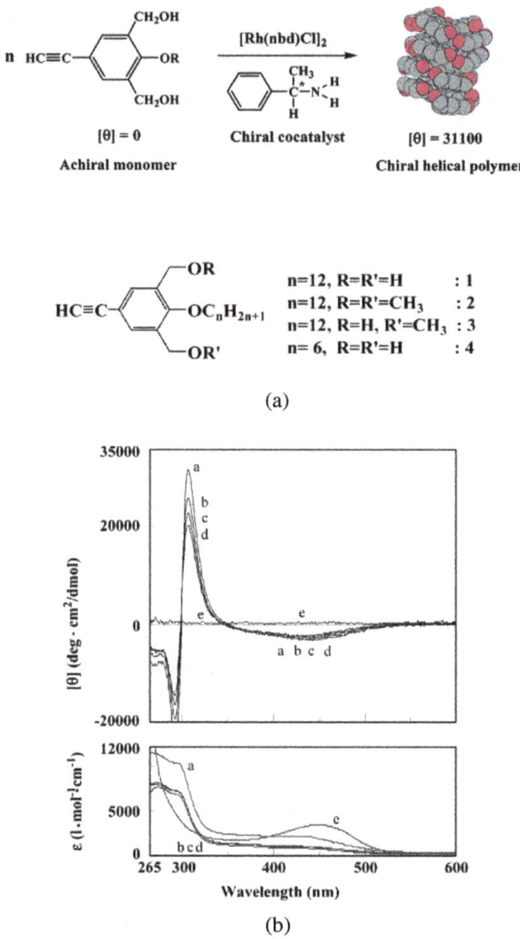

Figure 4. (a) Polymerization scheme of poly(phenyl acetylene) derivatives. (b) circular dichroism and ultraviolet spectra of **poly-1** in several solvents: (a) in CCl_4, (b) in CCl_4/DMSO (50/1), (c) in CCl_4/DMSO (30/1), (d) in CCl_4/DMSO (20/1), and (e) in CCl_4/DMSO (10/1). Reproduced with copyright permission from Ref. [29].

depicted with top and side views [31]. Interestingly, monitoring self-assembly of dendritic dipeptides with ECD and corresponding UV–visible spectroscopy revealed variations in the sign of Cotton bands and strength of ellipticity at various stages of self-assembly with lowering of temperature. Specifically, the negative Cotton band at ~225 nm, attributed to aromatic moieties, kept increasing with the lowering of temperature, suggesting effective modulations in the intramolecular and supramolecular

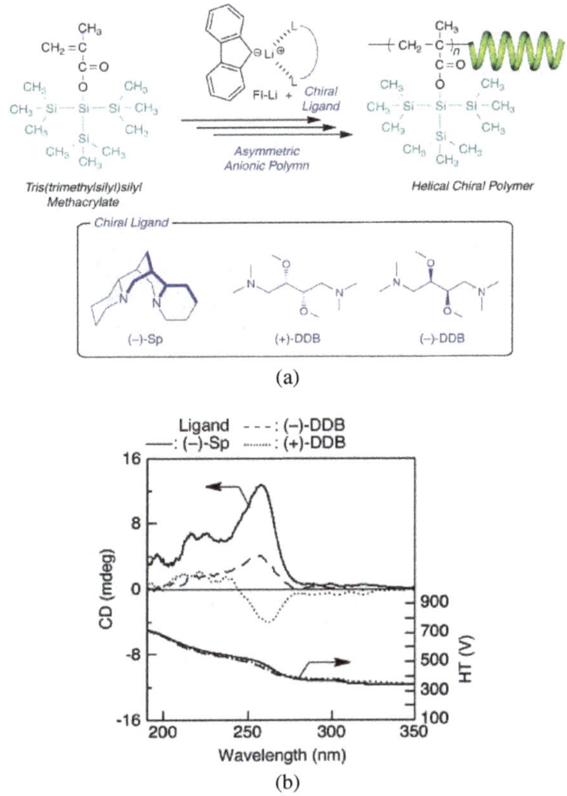

Figure 5. (a) Schematic for the asymmetric synthesis of supersilyl methacrylate via anionic polymerization and (b) its diffuse reflectance circular dichroism spectra acquired with chiral ligands Fl–Li/(–)-Sp, (–)-DDB, or (+)-DDB. Reproduced with copyright permission from Ref. [30].

conformations of aromatic moieties. Alternatively, a unique strategy was examined to induced chirality to the achiral polymer via solvent-to-polymer chirality transfer using a number of poly(diphenylacetylene) (**PDPA**)-conjugated polymers (Figure 7a) with intramolecular stack structure (IaSS) [32]. When achiral PDPA polymers were dissolved in limonene, the solvent chirality was successfully transferred to the side phenyl stack structure illustrated in Figure 7b, leading to intramolecular axial chirality. The phenyl–phenyl IaSS was under thermodynamic control to readily undergo asymmetric changes in chiral limonene, leading to optical activity in the isotropic structure between the main chain and resonant side phenyl rings. The axial chirality was significantly affected by the chain

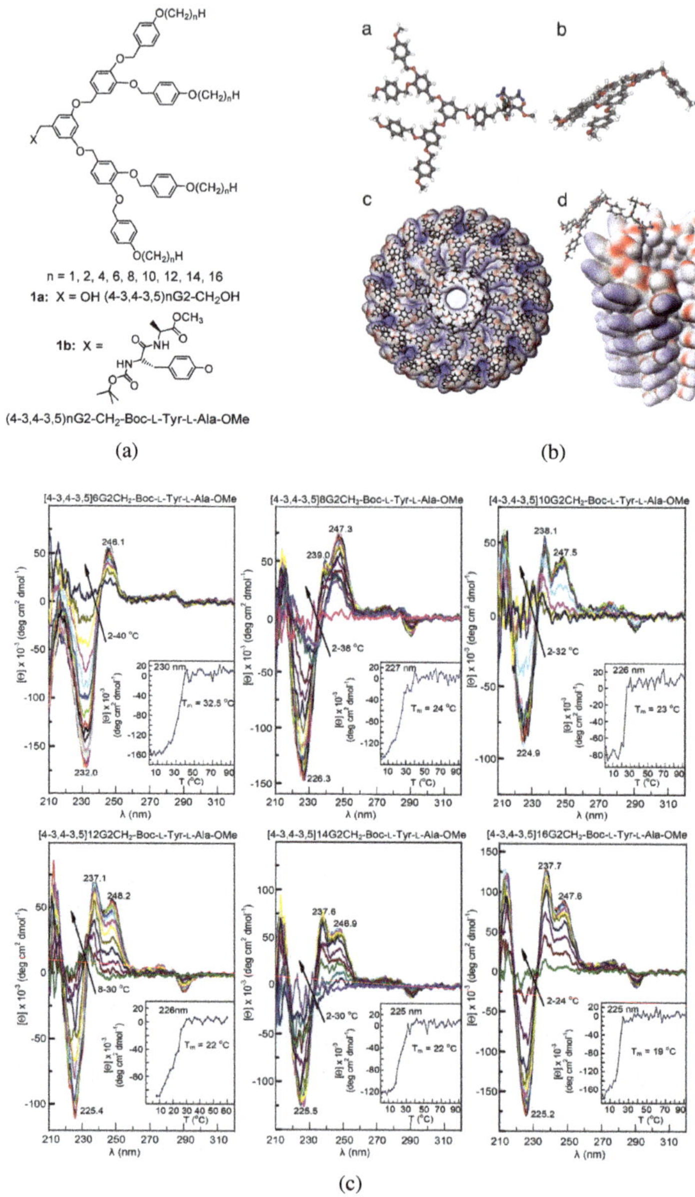

Figure 6. (a) Chemical structures of (4-3, 4-3, 5)$_n$G2-CH$_2$OH (**5a**) and (4-3, 4-3, 5)$_n$G2-CH$_2$–Boc–L-Tyr–L-Ala–OMe (**5b**). (b) Self-assembly of (4-3, 4-3, 5)12G2-CH$_2$–Boc–L-Tyr–L-Ala–OMe into helical pores (a and b). (c) Changes in circular dichroism spectra of (4-3, 4-3, 5)nG2-CH$_2$–Boc–L-Tyr–L-Ala–OMe (n = 6, 8, 10, 12, 14, 16). Arrows indicate trends upon increasing temperature. Reproduced with copyright permission from Ref. [31].

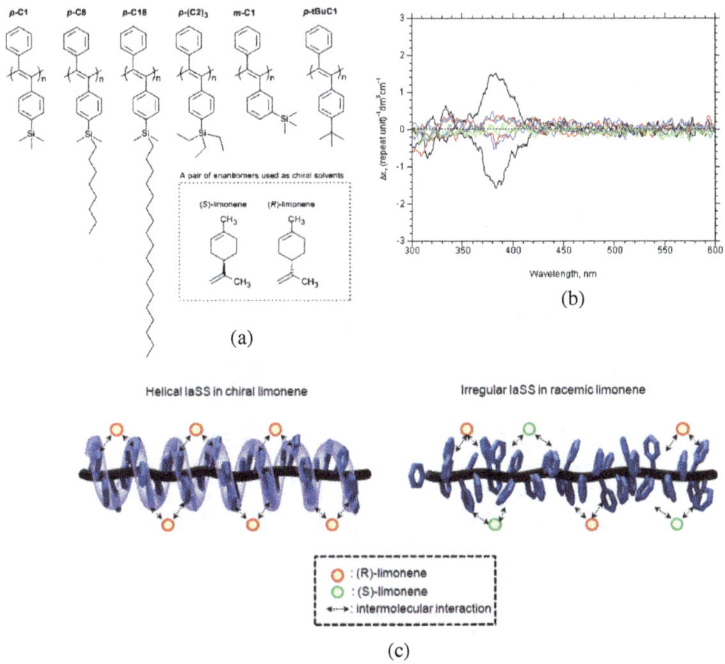

Figure 7. (a) Chemical structures of achiral **PDPA** polymers with chiral solvents. (b) Circular dichroism spectra of **p6-C1**, **p6-C8**, **p6-C18**, and **p6-(C2)**$_3$ in (*R*)-limonene/(*S*)-limonene ($c = 5.0 \times 10^{-4}$ mol L^{-1}). Black: **p6-C1**; red: **p6-C8**; blue: **p6-C18**; green: **p6-(C2)**$_3$; (+)-sign signal at 383 nm: (*S*)-limonene; (–)-sign signal: (*R*)-limonene. (c) Schematic illustration of the proposed intramolecular stack structure (IaSS) of **p6-C1** in chiral and racemic limonenes. Trimethylsilyl groups are omitted for clarity [32].

length and substitution position of the side alkyl groups. The longer alkyl chains and bulkier alkyl group prevented direct intermolecular interactions between the side phenyl rings and the chiral limonene molecules. The **PDPA** with sterically congested, highly stable, and regulated IaSS was unfavorable for efficient solvent-to-polymer chirality transfer, as demonstrated in Figure 7c.

1.3.2 Dynamics in Helical Polymers and Macromolecules

Yashima *et al.* reported a remarkable property of inducing helicity to poly[(4-carboxyphenyl)acetylene] (**poly-7**) by chiral amines that can be retained even after removal of chiral amines, showing a memory effect for absolute configuration [33]. As described in Figure 8a, **poly-1** complex with **(*R*)-2** and **(*S*)-3** yielded a left- and right-handed helix, respectively.

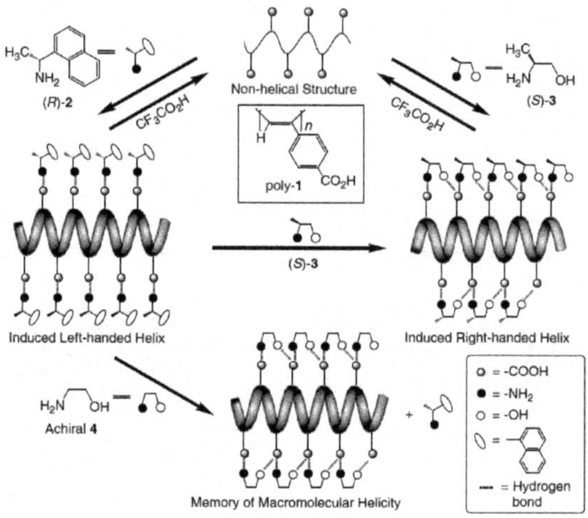

(a)

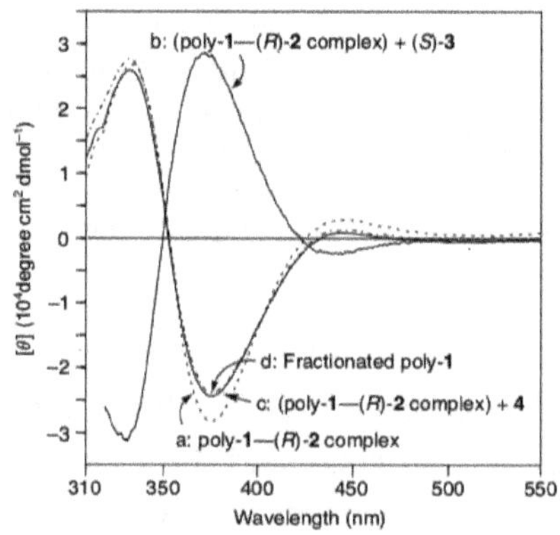

(b)

Figure 8. (a) Sketch of a schematic of induced specific helicity to **poly-7** and helical memory. Helix sense confirmation of **poly-7** is induced with **(*R*)-2** (Left) or **(*S*)-3** (Middle), and the induced macromolecular helicity is memorized by achiral **4** (Bottom). (b) Corresponding variations in circular dichroism (CD) signals of **poly-7** with amines: CD signals of **poly-7-(*R*)-2** complex, a mixture of the **poly-7-(*R*)-2** complex with **(*S*)-3** and **4**, and the isolated **poly-7**. Reproduced with copyright permission from Ref. [33].

The dynamic nature of helicity was rationalized with the addition of **(*S*)-3** molecules to the left-handed helical **poly-1** complex with **(*R*)-2** resulting in an opposite helical sense as observed with ECD spectra in Figure 8b. On the other hand, adding achiral-4 molecules to left-handed helical **poly-1** complex with **(*R*)-2** exhibited retention of configuration revealed by persistent Cotton bands in ECD.

Since the first demonstration of the "Sergeants and Soldiers" effect [34] that involves an amplification of chirality by the incorporation of a small number of chiral dopants and the "majority rule" [35] effect that describes the abundance of chiral units decides the helix sense, numerous studies have been carried out in this context. Recently, Riguera *et al.* employed metal ions of difference valances to complex with copolymers of poly(acetylene) derivatives possessing chiral and achiral moieties (Figure 9) to demonstrate the sergeants and soldiers effect [36].

Essentially, for **poly($8_{0.9}$-*co*-$14_{0.1}$)** to **poly($8_{0.2}$-*co*-$14_{0.8}$)** copolymers, the amplification of ECD signals could be manifested with aliquots of Ba^{2+} ions, i.e., with only 20% of chiral moieties displaying a dominant ECD signal analogous to **poly-8**, indicating right-handed helical sense (Figure 10a and c). By contrast, no effective sergeants and soldiers effect could be seen with the addition of monovalent Li^{+} ions to the solutions of the **poly($8_{0.8}$-*co*-$14_{0.2}$)**; instead, an ECD signal intensity that was 80% of the homopolymer **poly-8** manifested, which corresponds to a left-handed helix (Figure 10a and b). Hence, both left- and right-handed helical senses have been yielded in the case of **poly-8** with the addition of Li^{+} ions and Ba^{2+} ions, respectively.

To further elucidate the factors involved in the chiroptical phenomenon of polymer aggregates composed of helical building blocks, a series of rigid rod helical poly[alkyl-(*S*)-2-methylbutylsilane]s (achiral alkyl side chains = ethyl, *n*-propyl, *n*-butyl, *n*-pentyl, *n*-hexyl) have been investigated (Figure 11a) in mixed solvents [37]. An illustration of polymer aggregates with respect to solvent ratio is depicted in Figure 11b.

As shown in Figure 11c, a positive Cotton band for **17** and a negative Cotton band for **19** in ECD spectra indicate that the chiroptical sign in the ECD spectra of the polysilane aggregates depends on the effect of the achiral side chain length on aggregation. This unique side-chain-length-dependent chiroptical inversion was theoretically predictable using the

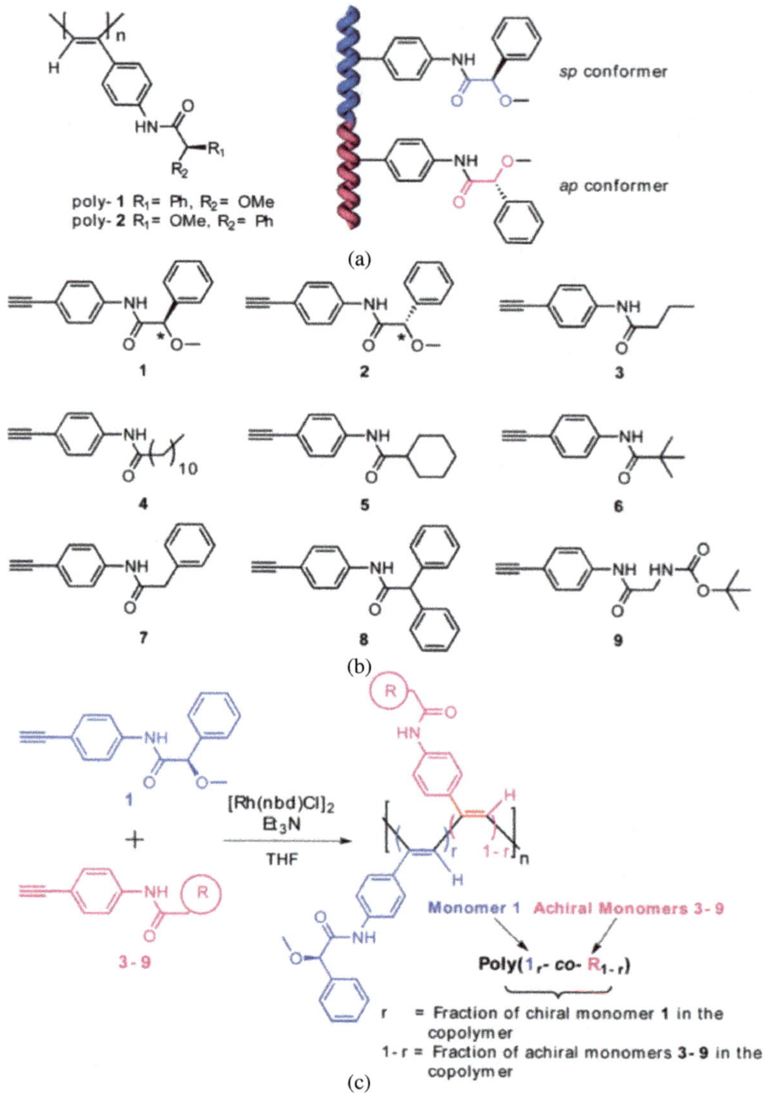

Figure 9. Structures of the synthesized homopolymers, monomers, and copolymers. (a) Structures of **poly-1**, **poly-2**, and the sp/ap conformations. (b) Structures of monomers **1–9**. (c) Synthesis, structure, and notation of the copolymers. Reproduced with copyright permission from Ref. [36].

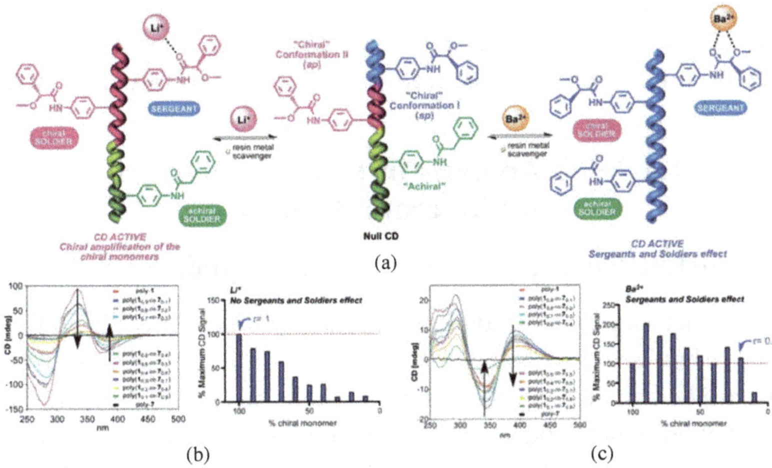

Figure 10. (a) Sergeants and soldiers effect for **poly(8_r-*co*-$14_{(1-r)}$)** copolymers (Right) with Ba^{2+} and chiral amplification of the chiral units (Left) with Li^{+} ions. (b) Variations in circular dichroism (CD) signals and bar graphs as a consequence of chiral amplification for the **poly(8_r-*co*-$14_{(1-r)}$)** series in the presence of Li^{+}. (c) Changes in CD signals and bar graph presenting the chiral amplification via sergeants and soldiers effect of the **poly(8_r-*co*-$14_{(1-r)}$)** series in the presence of Ba^{2+} (*r*-values are highlighted). Reproduced with copyright permission from Ref. [36].

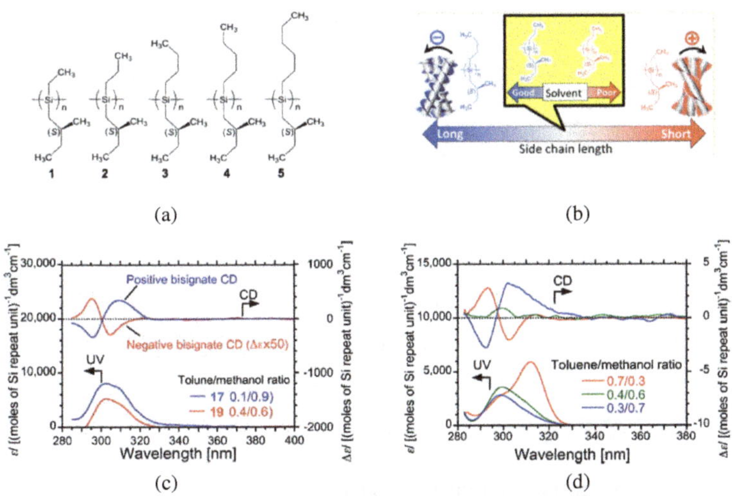

Figure 11. (a) Chemical structures of poly[alkyl-(*S*)-2-methylbutylsilane]s with different achiral alkyl chains: ethyl (**17**), *n*-propyl (**18**), *n*-butyl (**19**), *n*-pentyl (**20**), and *n*-hexyl (**21**). (b) Illustration of polysilane aggregation in good and poor solvents. (c) Comparison of ultraviolet (UV) and inverted circular dichroism (CD) spectra of **17** and **19** aggregates at 20°C. (d) Comparison of CD and UV spectra of **18** aggregate at three volume ratios of toluene/methanol cosolvents. Reproduced with copyright permission from Ref. [37].

novel approach of combining the hard-core cholesteric model and exciton chirality method, whereas aggregates of polymer **18** showed positive and negative Cotton bands in different ratios of good solvent to poor solvent.

1.4 VCD Analysis of Helical Polymers and Macromolecules

1.4.1 Structure Elucidation of Proteins and Peptides

In order to determine the secondary structure of a protein with VCD, two different vibrational bands that correspond to the polypeptide backbone, i.e., the amide band of poly-*L*-lysine, should be taken into account. These two vital vibrational modes could be identified as (a) a combination of C=O stretching vibrations and out-of-phase C–N stretching at 1600–1700 cm^{-1}, and (b) a combined contribution from C–N stretching and N–H bending motions at ~1550 cm^{-1}. In practice, vibration of (a) could be utilized to resolve the secondary structure with VCD, as shown in Figure 12a and b [21, 38, 39]. The α-helix showed a bisignate Cotton band

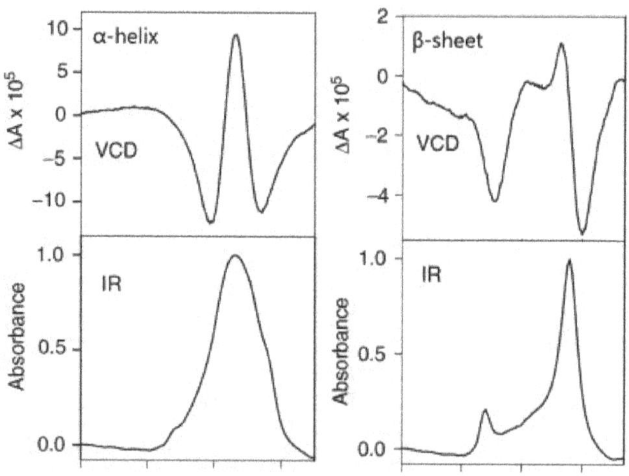

Figure 12. (a) Vibrational circular dichroism (VCD, Top) and infrared (IR) absorption (Bottom) spectra in the amide I region for poly-l-lysine that is mostly α-helical (D_2O at pH ~11). (b) VCD (Top) and IR absorption (Bottom) spectra of β-sheet after heating to 65°C for ~20 min followed by cooling. Reproduced with copyright permission from Ref. [21].

at 1640–1660 cm^{-1}, while β-sheet exhibited two negative bands at 1610 and 1680 cm^{-1}, providing clear distinction of secondary structure based on VCD vibrational bands.

1.4.2 Macromolecular Self-assembly

The macromolecular arrangement of (*S*)-triarylamine trisamide **[(*S*)-TATA]** has been studied owing to the π–π stacking and hydrogen bonding of amides between adjacent molecules [40]. Figure 13 represents the molecular conformation and *P*/*M*-helical stacks of 3:0 and 2:1 **(*S*)-TATA**. An earlier study on the self-assembly of **(*S*)-TATA** concluded that the rate of cooling of a hot solution of **(*S*)-TATA** could determine the stacked structures. This study incorporates three methods of sample preparation, namely by cooling a hot solution on ice, cooling with air, and aging the air-cooled sample for 2 months. Thus, prepared samples were analyzed using ECD and VCD spectroscopic techniques. The ECD profiles

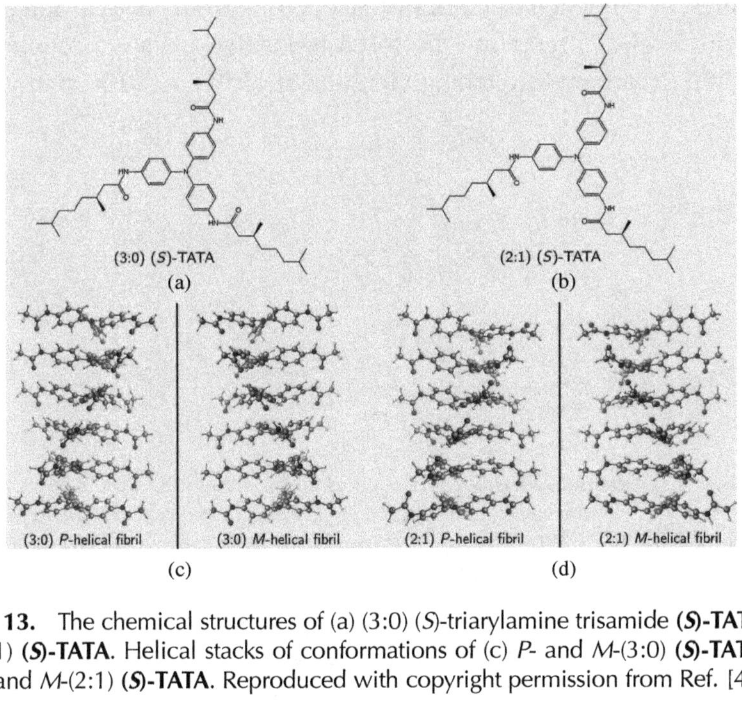

Figure 13. The chemical structures of (a) (3:0) (*S*)-triarylamine trisamide **(*S*)-TATA** and (b) (2:1) **(*S*)-TATA**. Helical stacks of conformations of (c) *P*- and *M*-(3:0) **(*S*)-TATA** and (d) *P*- and *M*-(2:1) **(*S*)-TATA**. Reproduced with copyright permission from Ref. [40].

essentially showed similar bisignate electronic bands with different intensities for all three samples, supporting the speculation of *M*-superhelices development with *P*-fibrils (Figure 14b). Conversely, under VCD analysis, it was observed that the bisignate signal for the air-cooled sample was reversed for ice-cooled samples and aged samples, indicating dynamic Nature and rearrangements of vibronic states corresponding to *M*- and *P*-helices (Figure 14a).

Although a clear assignment of handedness could not be made solely based on experimental evidence, theoretically predicted VCD spectra showed a remarkable resemblance to *P*-helices of hexamer stacks to the experimentally observed air-cooled sample (Figure 14c).

In another instance, Monde *et al.* examined the stereochemical aspects of glycerophospholipids (GPLs) from different sources such as bacterial, eukaryotic, and mitochondrial geneses [41]. The constituent units of GPLs are phosphoester polar headgroups, chiral glycerols, and acyl chains. The exciton-coupled assignment of VCD signals of GPLs exhibited significant differences depending on whether they were obtained from bacteria, eukaryotes, or mitochondria. For instance, GPLs from bacteria and eukaryotes with *sn*-3 configuration disclosed a positive–negative couplet corresponding to carbonyl stretching vibration at 1750 cm^{-1}. By contrast, the

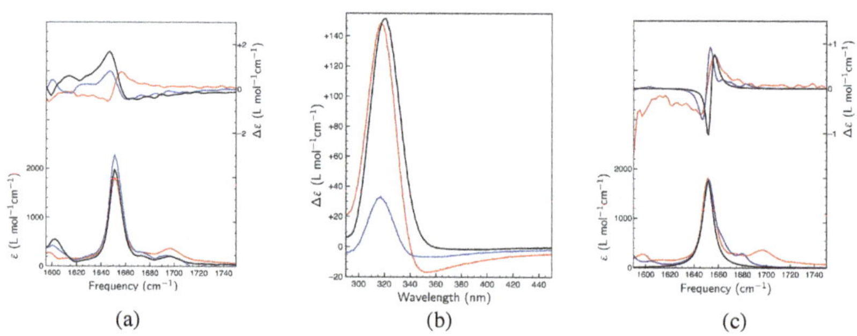

Figure 14. (a) Vibrational absorbance (VA) and vibrational circular dichroism (VCD), and (b) electronic circular dichroism spectra of 1 mM solutions of **(*S*)-TATA** in toluene-d_8 at various conditions (blue spectra correspond to a fresh sample cooled on ice, and the red spectra to a new sample cooled with air, whereas the black spectra correspond to a sample that was cooled with air and aged for 2 months at room temperature). (c) VA and VCD spectra calculated for the hexamer (shown in purple), experimental spectra of an air-cooled 1 mM **(*S*)-TATA** (shown in red), and calculated spectra using a simplified coupled oscillator model (shown in black). Reproduced with copyright permission from Ref. [40].

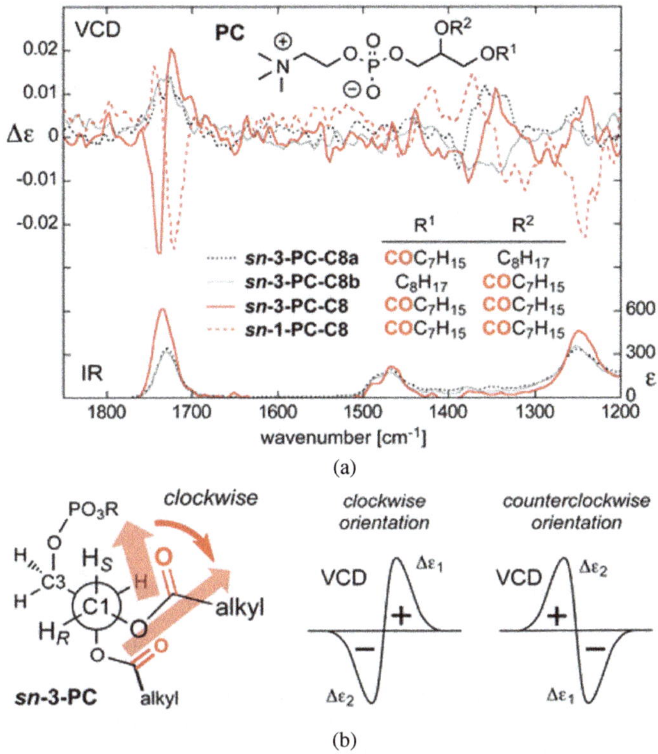

Figure 15. (a) Infrared and corresponding vibrational circular dichroism (VCD) spectra of phosphatidylcholines in $CDCl_3$ [$l = 100$ μm (***sn*-3-PC-C8a** and ***sn*-3-PC-C8b**) or 0.08 M (***sn*-3-PC-C8** and ***sn*-1-PC-C8**)]. (b) Schematic illustration of the orientated carbonyl groups of ***sn*-3-PC** (Left) with the arrangement of the electric transition moments (red arrows parallel to the C=O bonds) and the corresponding sign of the VCD couplet (Right). Reproduced with copyright permission from Ref. [41].

sn-1 configuration of mammalian GPLs conversely displayed a negative–positive couplet for similar vibrational modes (Figure 15).

1.4.3 Self-assembly of Chiral Block Copolymer (BCP*)

Ho *et al*. reported the first demonstration of self-assembly of BCP* of poly(styrene-*b*-*L*-lactide) (PS-PLLA) and poly(styrene-*b*-*D*-lactide) (PS-PDLA) to yield a peculiar helix phase (H*) via microphase separation as shown in TEM micrographs (Figure 16a) [42–45]. Excitingly, the helix phase exhibited helix sense corresponding to the absolute configuration of

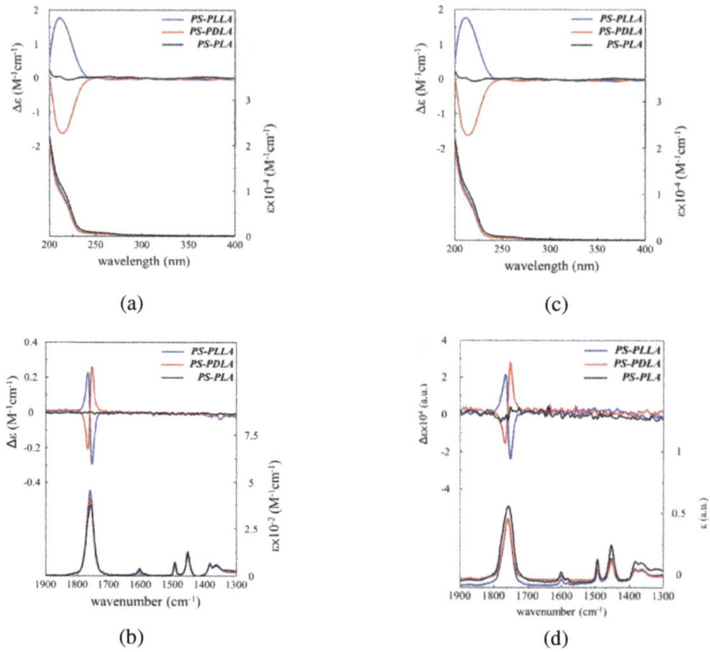

Figure 16. (a) TEM micrograph of **PS-PLLA** H* phase. (b) Circular dichroism and corresponding ultraviolet–visible absorption spectra of polylactide-containing BCPs in dilute acetonitrile solution. (c) Vibrational circular dichroism (VCD) and the corresponding FTIR absorption spectra of polylactide-containing **BCPs** in dilute CH_2Cl_2 solution. (d) VCD and the corresponding FTIR absorption spectra of a polylactide-containing **BCP** thin film. Reproduced with copyright permission from Ref. [42].

constituent PLLA/PDLA chains indicating the effective transfer of chirality through polymer chains to helix phase via homochiral evolution. The ECD spectra showed a mirror-imaged Cotton band at 220 nm attributed to n–π* electronic transitions for PS-PLLA and PS-PDLA in solution signifying intramolecular interactions in one-handed helical conformations of polymer chains (Figure 16c).

VCD measurements were carried out of **PS-PLLA** and **PS-PDLA** in solution. A mirror-imaged split-type Cotton bands were recorded showing a negative signal at 1753 cm^{-1} and a positive signal at 1767 cm^{-1} from **PLLA** that revealed opposite signals for **PDLA**, and no VCD signal appeared for **PS-PLA** (Figure 16b). This signal could account for the characteristic C=O stretching vibrations of lactide units and rationalized to acquire the left-handed helical conformation for **PLLA** and right-handed helical conformation for **PDLA** based on the coupled oscillator model. Moreover, these signals remained intact in the VCD spectrum of film samples (Figure 16a), symptomatic of the retention of acquired conformation in the film state, i.e., H* phase.

1.5 ROA Analysis of Helical Polymers and Macromolecules

1.5.1 Monosaccharides and Europium Complex

The chiral resolution and analysis of monosaccharides has been examined in the presence of europium complexes by measuring the CPL component of ROA signals. The molecular structures of europium complexes and monosaccharides are shown in Figure 17a [46]. The

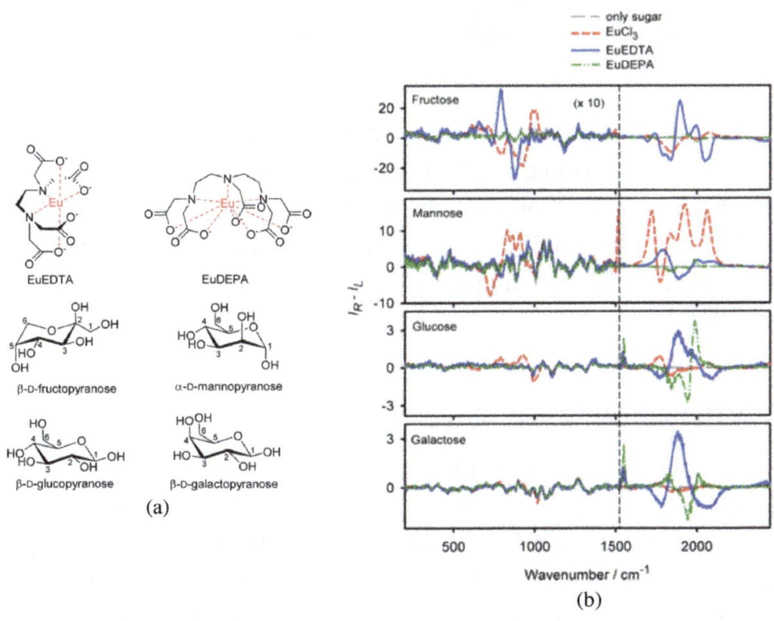

Figure 17. (a) Chemical structures of studied monosaccharides and europium complexes. (b) Raman optical activity spectra of $EuCl_3$, Na^+Eu^{III} EDTA, and Na_2^+ Eu^{III} DEPA solutions in the presence of four monosaccharides showing a strong signal for circularly polarized fluorescence component (right-hand side), specifically correlating to each of the studied sugars. Reproduced with copyright permission from Ref. [46].

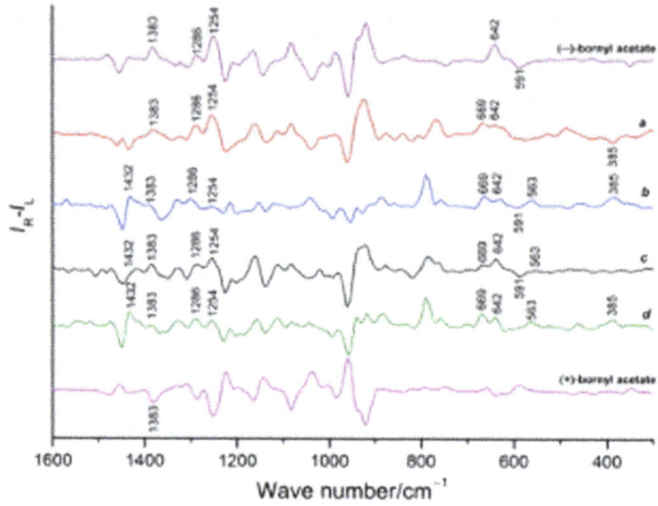

Figure 18. The measured Raman optical activity (ROA) spectra of four samples of pichtae oil (a–d) and corresponding ROA spectra of (–)-bornyl acetate and (+)-bornyl acetate. Reproduced with copyright permission from Ref. [47].

accumulation of europium to saccharides resulted in the generation of new bands at 700–1000 cm^{-1} that are analogous to the CID of neat sugars. Intense bands attributed to europium CPL could be observed in the range of 1500–2450 cm^{-1} that are relatively stronger than the vibrational ROA signal, thereby increasing the accuracy of ROA signal assignments.

1.5.2 Natural Products

The resolution of absolute configuration or the key contributing component in the multicomponent natural product is highly desirable. The pichtae essential oil extracted from Siberian fir needle pine is often employed in home remedies in Eastern Europe and the active component is bornyl acetate. It can be observed from Figure 18 that the ROA profile of four different samples matches favorably with (–)-bornyl acetate rather than (+)-bornyl acetate, insisting that (–)-bornyl acetate is a major component in pichtae essential oil [47].

1.6 CPL Analysis of Helical Polymers and Macromolecules

1.6.1 Solvent-induced Chirality

The (*S*)-limonene (**1*S***) and (*R*)-limonene (**1*R***) are promising candidates as renewable bioresources. The production of optically active poly[(9,9-di-*n*-octylfluoren-2,7-diyl)-*alt*-2,2′-bi-thiophene] (**F8T2**) particles (Scheme 1) with CD and CPL properties (Figure 19) were performed by [48] via solvent chirality transfer using **1*S*** and **1*R***. The CD-silent **F8T2** rapidly produced the particles by solvent chirality transfer at room temperature [48].

In addition, this demonstrated that through weak intermolecular forces, such as CH/π, van der Waals, and π–π interactions, the CD/CPL-active **F8T2** aggregates were effectively produced in a chiral solvent system of chloroform (a good solvent), alkanol (a poor solvent), and limonene (a chiral solvent) (Figure 20). It was also observed that the magnitude of CD/CPL signals was highly dependent on the alkanol and the enantiopurity of the limonene. Furthermore, the order of addition of limonene and methanol to the chloroform solution of **F8T2** significantly influenced the magnitude of the induced CD band.

The renewability of limonenes was examined by reusing the distilled **1*S*** from impure **1*S*** containing chloroform, **F8T2**, and methanol.

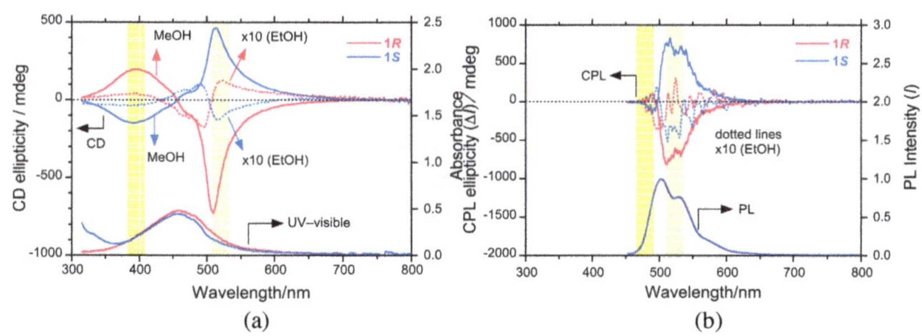

Figure 19. (a) Ultraviolet (UV)–visible and circular dichroism (CD) spectra and (b) PL and circularly polarized luminescence (CPL) spectra of **F8T2** particles (**F8T2**, 1 × 10^{-5} mol L^{-1}) in limonene/chloroform/alcoholic solvent [2.0:0.3:0.7 (v/v/v)]: ethanol (dotted lines) and methanol (solid lines) — **1*R*** (red line) and **1*S*** (blue line). (c) UV–visible and CD spectra of **F8T2** particles (**F8T2**, 1). Modified from the original datasets in Ref. [48].

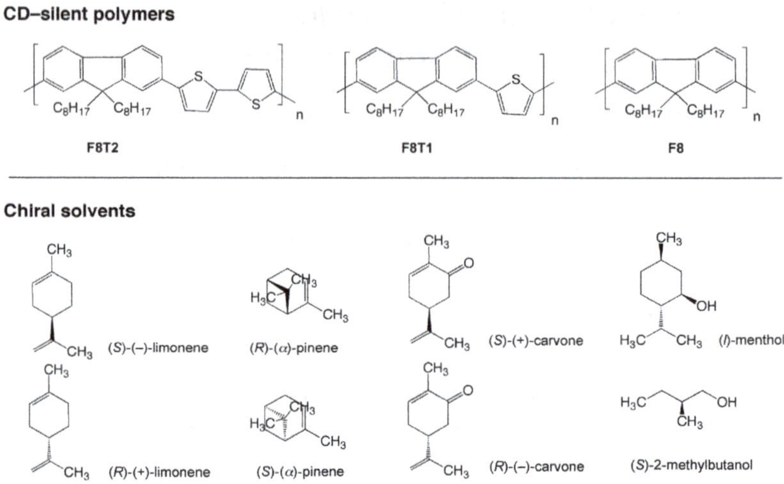

Scheme 1. Chemical structures of achiral polymers, chiral, and achiral solvents [48].

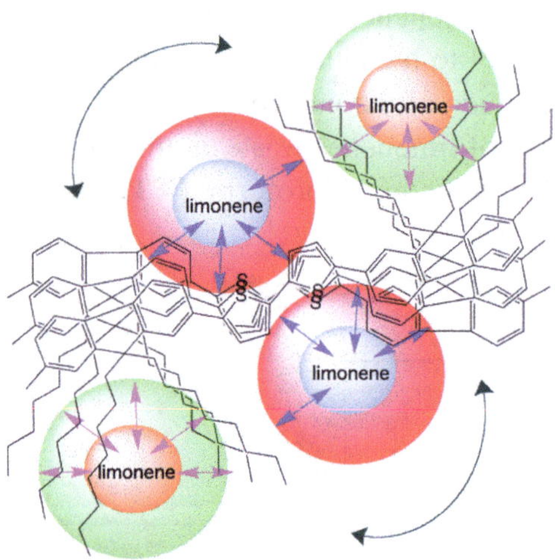

Figure 20. A proposed model structure of helically ordered π–π stacks in **F8T2** assembly with limonene molecules. Modified from Ref. [48].

As a result, the CD/UV–visible signals of the **F8T2** particles obtained from the use of renewed **1*S*** provided CD/UV–visible signals analogous to those used by the fresh **1*S***. In addition, limonene chirality transfer was successfully observed for two CD-silent polymers:

poly(9,9-di-*n*-octylfluorenyl-2,7-diyl) (**F8**) and poly[(9,9-di-*n*-octylfluorenyl-2,7-diyl)-*alt*-thiophene] (**F8T1**) (Scheme 1). The protocol used can provide an environmentally friendly, safe, and mild process to rapidly produce ambidextrous light-emitting polymers at room temperature with minimal loss of starting polymers, which are made from CD-silent polymers without any specific chiral substituents or chiral catalysts.

1.6.2 Circularly Polarized Photon-induced Chirality

It has been speculated that the *r*- or *l*-CP photon source in γ- and X-rays and vacuum UV regions can be responsible for the astrophysical and biological handedness on the planet Earth. The handedness of CP light source may have played an important role in prebiotic evolution biology by interacting with molecules [49–51]. As a practical consequence, CP photon-driven absolute asymmetric synthesis (AAS) can be referred to as photon chirality transfer to photochemically induce chirality in molecules. The literature also shows that by using CP photons as a chiral physical source, achiral nonphotochromic π-conjugated polyfluorene, poly(1-substituted phenylacetylene), diacetylene monomers, and Cu(II) coordinated with succinate and 4,4′-bipyridine can provide the corresponding optically active polymers [52–54]. However, these CP-driven AAS experiments have been conducted primarily on a single photon energy source because it has long been assumed that the chirality of the product is governed solely by the role of the *r*- or *l*-hand of the CP photon and not significantly influenced by CP photon energy.

There is an interesting report on the first CP photon energy-dependent ring-closure reaction of a *cis*-1,2-diarylethyele rotamer in the presence of iodine in toluene to produce nonracemic octahelicene [55]. The preference of the optical activity directed by *l*-CP sources at 290 nm was opposite to that led by *l*-CP sources in the range of 310–410 nm and vice versa, reflecting the bisignate CD signal of octahelicene. Recently, a separate study confirmed the CP photon energy-dependent photodegradation of racemic ^{13}C-alanine films [56]. The CP photon energy sources inverted the preference of the product chirality in the film. This inversion arises from sign characteristics of bisignate CD in the irradiation region [6.19 eV (200 nm) or 6.74 eV (184 nm)] [56].

Similarly, CP irradiation photon-dependent chiroptical inversion was reported in photochromic poly[azobenzene-*alt*-(di-*n*-octylfluorene)] (**F8AZO**) when the same sources of *r*- and *l*-CP photons were applied [57]. An open question is whether photon chirality transfer to achiral non-photochromic colloidal particles dispersed in the optofluidic medium under photophysically and photochemically controlled conditions with reversibility is possible. This system is a model of prebiotic evolution of nonphotochromic biological polymers under heterogeneous conditions and under the far-from-equilibrium open system, thus allowing continuous CP photon energy flow in the daytime. Is an optically active generated substance considered to be a dissipated structure? To answer these questions, selected micrometer-sized particles of **F8T2** were studied [58]. The optically inactive **F8T2** is nonphotochromic but highly photoluminescent with a high Φ_F. The particles (0.6–0.7 μm in diameter) provided ultra-small microreactors to efficiently confine the CP photons by tuning the refractive index (RI) of the chloroform/methanol ($CHCl_3$/MeOH) cosolvent, acting as an optofluidic medium. It should be noted that the wavelength of the incident light in a vacuum decreased markedly and proportionally to the inverse RI in the particle, i.e., the incident light ($\lambda_0 =$ 500 nm in a vacuum) was estimated at approx. 250 nm in the polymer particle, thus allowing the confinement of photons in 0.6–0.7 μm sized particles, when the RI of the particle was assumed to be 2.0 (e.g., RI = 1.9 at 486 nm for **F8AZO**).

By contrast, the speed of light in the polymer was halved (3.0×10^8 m s^{-1}) in a vacuum. Moreover, it was observed that the *l*- and *r*-CP photons induced chiroptical polarization, depolarization, inversion, retention, and switching to **F8T2** particles [58]. The CPL and PL spectra of (a) *h*-**F8T2** particles after *r*- and *l*-CP photon irradiation at 546 nm for 60 min and (b) *l*-**F8T2** particles prepared by the limonene chirality transfer using [(*R*)-limonene (**1*R***) or (*S*)-limonene (**1*S***)]/$CHCl_3$/MeOH [2.0/0.3/0.7 (v/v/v)] tersolvent can be observed in Figure 21.

For *h*-**F8T2** particles induced by CP photon, the magnitudes of CPL, defined as g_{lum}, were -1.1×10^{-3} at 494 nm, $+1.6 \times 10^{-3}$ at 530 nm, and -3.6×10^{-3} at 530 nm for *l*-CP, while they were -1.1×10^{-3} at 494 nm, -2.7×10^{-3} at 530 nm, and $+1.8 \times 10^{-3}$ at 530 nm for *r*-CP. For comparison, in the case of limonene-induced *l*-**F8T2** particles, the magnitudes of

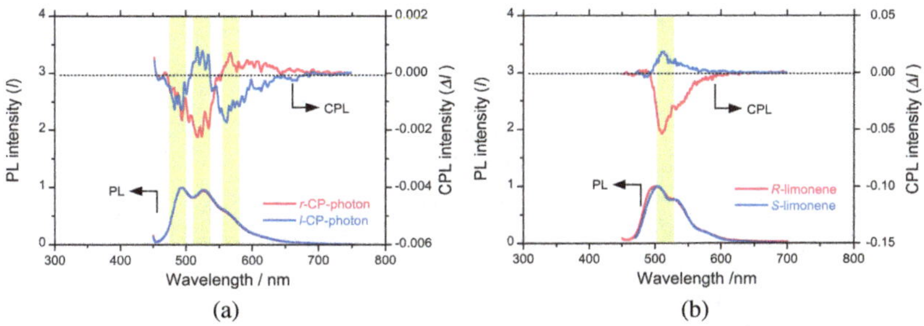

Figure 21. A comparison of circularly polarized luminescence (CPL) spectra excited at 400 nm of (a) *h*-**F8T2** particles induced by *r*- and *l*-CP photons for 60 min in $CHCl_3$/MeOH [2.1/0.9 (v/v)] and (b) *l*-**F8T2** particles produced by limonene chirality transfer in [(*R*)-limonene or (*S*)-limonene]/$CHCl_3$/MeOH [2.0/0.3/0.7 (v/v/v)]. Modified from the original datasets in Ref. [58].

CPL were -5.3×10^{-2} at 509 nm and -4.3×10^{-2} at 533 nm for **1*R***, while they were $+1.9 \times 10^{-3}$ at 510 nm and $+1.0 \times 10^{-2}$ at 535 nm for **1*S***. However, the absolute g_{lum} values of these *h*-**F8T2** particles were considerably weaker by an order of magnitude than those of the limonene-induced *l*-**F8T2** particles (Figure 21). This corresponds to approximately 6%–8% optical purity relative to **1*R***-induced *l*-**F8T2** particles. Nevertheless, it was reported that CP photon irradiation successfully produces CPL-active *h*-**F8T2** particles (Φ_F of 8%) from the corresponding CPL-silent particles, thus showing a Φ_F value of 15% [58]. The decrease in Φ_F value after prolonged CP photon irradiation can be attributed to less efficient emissive fluorenone, thiophene sulfone, and thiophene sulfoxide moieties, which are air auto-oxidation products of the thiophene rings and fluorene of **F8T2**.

1.6.3 Mirror-symmetry Breaking-triggered Self-assembly

An unusual behavior was studied during the mirror symmetry broken self-assembly of hydrogen-bonded helical stacks of threefold symmetrical chiral benzene-1,3,5-tricarboxamide (**BTA**) derivatives [59]. As shown in Figure 22a, the $\mathbf{BTA}^{BA}$ on mirror-symmetry-broken self-assembly induced by sequential heating and cooling predominantly yielded either left-handed (*M*) or right-handed (*P*) helical stacks. These helical stacks could

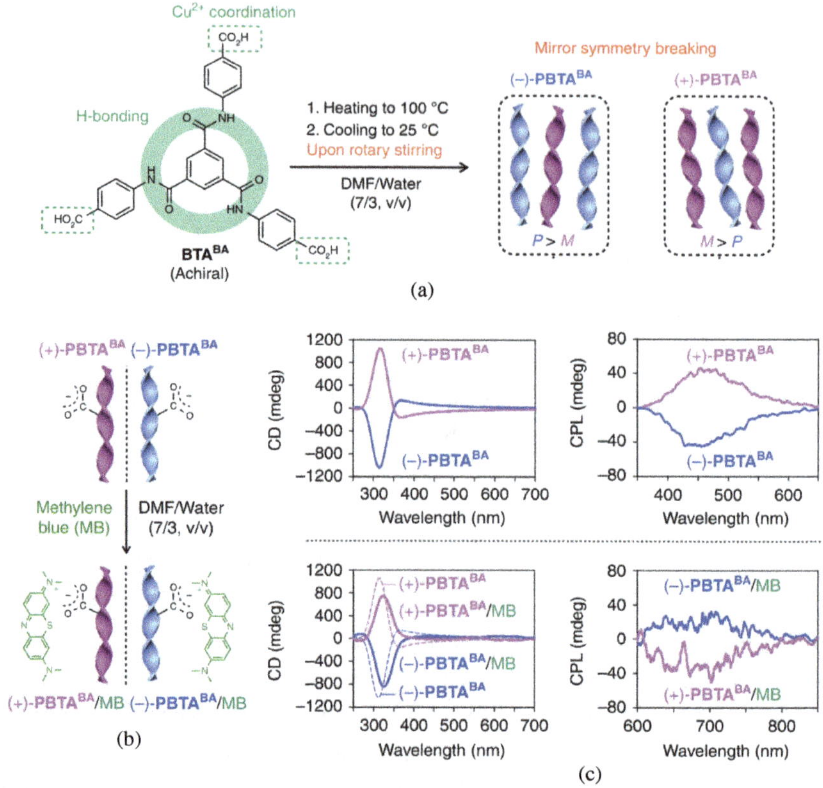

Figure 22. (a) The self-assembly of BTA^{BA} to (–)-**PBTA**^{BA} and (+)-**PBTA**^{BA} one-handed helical constructs via mirror-symmetry breaking; (b) chirality transfer from (–)-**PBTA**^{BA} and (+)-**PBTA**^{BA} to the achiral dye methylene blue (MB) via electrostatic interactions; (c) electronic circular dichroism and circularly polarized luminescence spectra of (–)-**PBTA**^{BA} and (+)-**PBTA**^{BA} with and without the presence of MB [59].

also provide cation-binding sites that are utilized for chirality transfer to achiral dyes via electrostatic interactions (Figure 22b).

The chiroptical properties were studied with ECD and CPL spectroscopy with and without the presence of achiral dye (methylene blue (MB)). Figure 22c shows mirror-image strong ECD signals for (–)-**PBTA**^{BA} and (+)-**PBTA**^{BA} attributed to stable one-handed helical constructs that are also supplemented by corresponding mirror image signals in CPL spectra. The efficient chirality transfer from (–)-**PBTA**^{BA} and (+)-**PBTA**^{BA} to the achiral methylene blue (**MB**) dye were identified via a shift in the ECD signal to higher wavelengths supported by corresponding mirror image CPL signals at a much more significant change of wavelengths at the excited state.

1.7 Conclusions and Future Outlook

The quest to understand ubiquitous chiral bias observed in biomolecules has led to the emergence of synthetic chiral polymers and macromolecules and has been incessant ever since. In this chapter, unique and peculiar aspects of chirality transfer and various molecular stages on the homochiral evolution of polymers and macromolecules have been explored through the results of case studies. Embellishment of the absolute configuration and conformations of polymers was extracted through combinations and individual and efficient chiroptical spectroscopy techniques, including ECD, VCD, ROA, and CPL, providing significant insights into electronic transition and vibrational modes at the ground and excited states.

There are still fissures in the foundation regarding our understanding of hierarchical chirality transfer in biomolecules and the thorough investigation of conformation and configuration of corresponding synthetic polymers and macromolecules. However, this research area has been expanding, with many potential subareas attracting researchers' attention. We have carefully discussed selected topics to throw light on the analysis of chirogenesis in polymers and macromolecules, which may contribute to future innovative research and a more profound understanding, leading to significant insights into chirogenesis.

1.8 Acknowledgments

The authors thank Prof. Victor Borovkov (Tallinn University of Technology, Estonia) for giving us the opportunity to contribute a chapter to this book.

References

1. Wagnière, G.H. (2008) *On Chirality and the Universal Asymmetry: Reflections on Image and Mirror Image*, Wiley, Hoboken, New Jersey, USA, pp. 1–247.

2. Gardner, M. (2005) *The New Ambidextrous Universe: Symmetry and Asymmetry from Mirror Reflections to Superstrings, Third Revised Edition*, Dover Publications, New York, NY, USA.
3. Hanafusa, H., and Akabori, S. (1959) Polymerization of aminoacetonitrile. *Bull. Chem. Soc. J.*, **32**, 626–630.
4. Harada, K., and Fox, S.W. (1964) Thermal synthesis of natural amino-acids from a postulated primitive terrestrial atmosphere. *Nature*, **201**, 335–336.
5. Miller, S.L. (1953) Production of amino acids under possible primitive earth conditions. *Science*, **117**, 528–529.
6. Seckbach, J., Chela-Flores, J., Owen, T., and Raulin, F. (eds) (2004) *Life in the Universe: From the Miller Experiment to the Search for Life on Other Worlds*, Springer, Dordrecht, Netherlands.
7. Guijarro, A., and Yus, M. (2008) *The Origin of Chirality in the Molecules of Life: A Revision from Awareness to the Current Theories and Perspectives of this Unsolved Problem,* RSC Publication, Cambridge, UK.
8. Breslow, R.A. (2011) Likely possible origin of homochirality in amino acids and sugars on prebiotic earth. *Tetrahedron Lett.,* **52**, 2028–2032.
9. Mason, S.F. (1992) *Chemical Evolution. Origin of the Elements, Molecules, and Living Systems*, Clarendon Press, Oxford, UK.
10. Meierhenrich, U. (2008) *Amino Acids and the Asymmetry of Life: Caught in the Act of Formation*, Springer, Berlin, Heidelberg, Germany.
11. Walgate, R. (1983) What will come after the Z0? *Nature*, **303**, 473–473.
12. Rubbia, C. (1985) Experimental observation of the intermediate vector bosons. *Rev. Mod. Phys.,* **57**, 699.
13. Close, F.E. (1976) Parity violation in atoms? *Nature*, **264**, 505–506.
14. Bouchiat, M.A., and Bouchiat, C.C. (1974) Weak neutral currents in atomic physics. *Phys. Lett. B*, **48**, 111–114.
15. Yashima, E., Ousaka, N., Taura, D., Shimomura, K., Ikai, T., and Maeda, K. (2016) Supramolecular helical systems: Helical assemblies of small molecules, foldamers, and polymers with chiral amplification and their functions. *Chem. Rev.*, **116**, 13752–13990.
16. Yashima, E., Maeda, K., Iida, H., Furusho, Y., and Nagai, K. (2009) Helical polymers: Synthesis, structures, and functions. *Chem. Rev.*, **109**, 6102–6211.
17. Berova, N., Polavarapu, P.L., and Woody, R.W. (eds.) (2011) *Comprehensive Chiroptical Spectroscopy: Instrumentation, Methodologies, and Theoretical Simulation*, Wiley, Hoboken, NJ, USA.
18. Lightner, D.A., and Gurst, J.E. (2000) *Organic Conformational Analysis and Stereochemistry from Circular Dichroism Spectroscopy*, Wiley, New York, NY, USA.
19. Fasman, G.D. (ed). (1996) *Circular Dichroism and the Conformational Analysis of Biomolecules*, Springer, Boston, MA, USA.

20. Johnson, W.C. (1996) Circular dichroism instrumentation, in *Circular Dichroism and the Conformational Analysis of Biomolecules* (Fasman G.D. ed.), Springer, Boston, MA, USA, pp. 635–652.
21. Kurouski, D. (2017) Advances of vibrational circular dichroism (VCD) in bioanalytical chemistry. *Anal. Chim. Acta*, **990**, 54–66.
22. Barron, L.D., Hecht, L., and Bell, A.F. (1996) Chapter 19. Vibrational Raman optical activity of biomolecules, in *Circular Dichroism and the Conformational Analysis of Biomolecules* (Fasman G.D. ed.), Springer, Boston, MA, USA, pp. 653–695.
23. Mason, S.F. (1979) *Optical Activity and Chiral Discrimination*, Springer, Dordrecht, Netherlands, pp. 1–24.
24. Berova, N., Bari, L.D., and Pescitelli, G. (2007) Application of electronic circular dichroism in configurational and conformational analysis of organic compounds. *Chem. Soc. Rev.*, **36**, 914–931.
25. Keiderling, T.A. (1996) Vibrational circular dichroism, in *Circular Dichroism and the Conformational Analysis of Biomolecules* (Fasman G.D. ed.), Springer, Boston, MA, USA.
26. Stephens, P. (1985) Theory of vibrational circular dichroism. *J. Phys. Chem.*, **89**, 748–752.
27. Saeideh, O.P., Laurence, D.B., Shaun, T.M., and Ewan, W.B. (2018) Chapter 6 — Raman Optical Activity, in *Chiral Analysis (Second Edition) — Advances in Spectroscopy, Chromatography and Emerging Methods* (Polovarapu, P.E. ed.), Elsevier, Amsterdam, Netherlands, pp. 249–291.
28. Riehl, J.P., and Richardson, F.S. (1986) Circularly polarized luminescence spectroscopy. *Chem. Rev.*, **86**, 1–16.
29. Aoki, T., Kaneko, T., Maruyama, N., Sumi, A., Takahashi, M., Sato, T., and Teraguchi, M. (2003) Helix-sense-selective polymerization of phenylacetylene having two hydroxy groups using a chiral catalytic system. *J. Am. Chem. Soc.*, **125**, 6346–6347.
30. Ishitake, K., Satoh, K., Kamigaito, M., and Okamoto, Y. (2013) Asymmetric anionic polymerization of tris(trimethylsilyl)silyl methacrylate: A highly isotactic helical chiral polymer. *Polym. J.*, **45**, 676–680.
31. Percec, V., Dulcey, A.E., Peterca, M., Ilies, M., Nummelin, S., Sienkowska, M. J., and Heiney, P.A. (2006) Principles of self-assembly of helical pores from dendritic dipeptides. *Proc. Natl. Acad. Sci.*, **103**, 2518–2523.
32. Lee, D., Jin, Y.-J., Kim, H., Suzuki, N., Fujiki, M., Sakaguchi, T., Kim, S.K., Lee, W.-E., and Kwak, G. (2012) Solvent-to-polymer chirality transfer in intramolecular stack structure. *Macromolecules*, **45**, 5349–5386.
33. Yashima, E., Maeda, K., and Okamoto, Y. (1999) Memory of macromolecular helicity assisted by interaction with achiral small molecules. *Nature*, **399**, 449–451.
34. Green, M.M., Reidy, M.P., Johnson, R.D., Darling, G., O'Leary, D.J., and Willson, G. (2002) Macromolecular stereochemistry: The out-of-proportion influence of optically

active comonomers on the conformational characteristics of polyisocyanates. The sergeants and soldiers experiment. *J. Am. Chem. Soc.*, **111**, 6452–6454.
35. Green, M.M., Garetz, B.A., Munoz, B., Chang, H.P., Hoke, S., and Cooks, R.G. (1995) Majority rules in the copolymerization of mirror image isomers. *J. Am. Chem. Soc.*, **117**, 4181–4182.
36. Bergueiro, J., Freire, F., Wendler, E.P., Seco, J., Quiñoá, E., and Riguera, R. (2014) The ON/OFF switching by metal ions of the "sergeants and soldiers" chiral amplification effect on helical poly(phenylacetylene)s. *Chem. Sci.*, **5**, 2170–2176.
37. Suzuki, N., Fujiki, M., Kimpinde-Kalunga, R., and Koe, J.R. (2013) Chiroptical inversion in helical Si–Si bond polymer aggregates. *J. Am. Chem. Soc.*, **135**, 13073–13079.
38. Yasui, S.C., and Keiderling, T.A. (2002) Vibrational circular dichroism of polypeptides. 8. Poly(lysine) conformations as a function of PH in aqueous solution. *J. Am. Chem. Soc.*, **108**, 5576–5581.
39. Yasui, S.C., and Keiderling, T. (1986) Vibrational circular dichroism of polypeptides. 8. Poly(lysine) conformations as a function of ph in aqueous solution. *J. Am. Chem. Soc.*, **108**, 5576–5581.
40. Koenis, M.A.J., Osypenko, A., Fuks, G., Giuseppone, N., Nicu, V.P., Visscher, L., and Buma, W.J. (2019) Self-assembly of supramolecular polymers of N-centered triarylamine trisamides in the light of circular dichroism: Reaching consensus between electrons and nuclei. *J. Am. Chem. Soc.,* **142**, 1020–1028.
41. Taniguchi, T., Manai, D., Shibata, M., Itabashi, Y., and Monde, K. (2015) Stereochemical analysis of glycerophospholipids by vibrational circular dichroism. *J. Am. Chem. Soc.*, **137**, 12191–12194.
42. Ho, R.-M., Li, M.-C., Lin, S.-C., Wang, H.-F., Lee, Y.-D., Hasegawa, H., and Thomas, E.L. (2012) Transfer of chirality from molecule to phase in self-assembled chiral block copolymers. *J. Am. Chem. Soc.*, **134**, 10974–10986.
43. Ho, R.M., Chiang, Y.W., Chen, C.K., Wang, H.W., Hasegawa, H., Akasaka, S., Thomas, E.L., Burger, C., and Hsiao, B.S. (2009) Block copolymers with a twist. *J. Am. Chem. Soc.*, **131**, 18533–18542.
44. Wen, T., Wang, H.-F., Li, M.-C., and Ho, R.-M. (2017) Homochiral evolution in self-assembled chiral polymers and block copolymers. *Acc. Chem. Res.*, **50**, 1011–1021.
45. Wang, H.F., Yang, K.C., Hsu, W.C., Lee, J.Y., Hsu, J.T., Grason, G.M., Thomas, E.L., Tsai, J.C., and Ho, R.M. (2019) Generalizing the effects of chirality on block copolymer assembly. *Proc. Natl. Acad. Sci.*, **116**, 4080–4089.
46. Wu, T., Průša, J., Kessler, J., Dračínský, M., Valenta, J., and Bouř, P. (2016) Detection of sugars via chirality induced in europium(III) compounds. *Anal. Chem.*, **88**, 8878–8885.
47. Chruszcz-Lipska, K., and Blanch, E.W. (2012) *In situ* analysis of chiral components of pichtae essential oil by means of ROA spectroscopy: Experimental and theoretical Raman and ROA spectra of bornyl acetate. *J. Raman Spectrosc.*, **43**, 286–293.

48. Kawagoe, Y., Fujiki, M., and Nakano, Y. (2010) Limonene magic: Noncovalent molecular chirality transfer leading to ambidextrous circularly polarised luminescent π-conjugated polymers. *New J. Chem.*, **34**, 637–647.
49. Feringa, B.L., and van Delden, R.A. (1999) Absolute asymmetric synthesis: The origin, control, and amplification of chirality. *Angew. Chem. Int. Ed. Engl.*, **38**, 3418–3438.
50. Inoue, Y. (1992) Asymmetric photochemical reactions in solution. *Chem. Rev.*, **92**, 741–770.
51. Inoue, Y., and Ramamurthy, V. (eds.) (2004) *Chiral Photochemistry, 1st ed.* CRC Press, Boca Raton, Florida, USA.
52. Wang, Y., Sakamoto, T., and Nakano, T. (2012) Molecular chirality induction to an achiral π-conjugated polymer by circularly polarized light. *Chem. Commun.*, **48**, 1871–1873.
53. Nakano, T. (2014) Tricks of light on helices: Transformation of helical polymers by photoirradiation. *Chem. Rec.*, **14**, 369–385.
54. Wu, S.-T., Cai, Z.-W., Ye, Q.-Y., Weng, C.-H., Huang, X.-H., Hu, X.-L., Huang, C.-C., and Zhuang, N.-F. (2014) Enantioselective synthesis of a chiral coordination polymer with circularly polarized visible laser. *Angew. Chem. Int. Ed.*, **53**, 12860–12864.
55. Bernstein, W.J., Calvin, M., and Buchardt, O. (1972) Absolute asymmetric synthesis. I. Mechanism of the photochemical synthesis of nonracemic helicenes with circularly polarized light. Wavelength dependence of the optical yield of octahelicene. *J. Am. Chem. Soc.*, **94**, 494–498.
56. Meinert, C., Hoffmann, S.V., Cassam-Chenaï, P., Evans, A.C., Giri, C., Nahon, L., and Meierhenrich, U.J. (2014) Photon energy-controlled symmetry breaking with circularly polarized light. *Angew. Chem. Int. Ed.*, **53**, 210–214.
57. Fujiki, M., Yoshida, K., Suzuki, N., Zhang, J., Zhang, W., and Zhu, X. (2013) Mirror symmetry breaking and restoration within nm-sized polymer particles in optofluidic media by pumping circularly polarised light. *RSC Adv.*, **3**, 5213–5219.
58. Fujiki, M., Donguri, Y., Zhao, Y., Nakao, A., Suzuki, N., Yoshida, K., and Zhang, W. (2015) Photon magic: Chiroptical polarisation, depolarisation, inversion, retention and switching of non-photochromic light-emitting polymers in optofluidic medium. *Polym. Chem.*, **6**, 1627–1638.
59. Shen, Z., Sang, Y., Wang, T., Jiang, J., Meng, Y., Jiang, Y., Okuro, K., Aida, T., and Liu, M. (2019) Asymmetric catalysis mediated by a mirror symmetry-broken helical nanoribbon. *Nat. Commun.*, **10**, 1–8.

https://doi.org/10.1142/9789811259227_0006

6 Chirogenesis in Solid State and Spontaneous Resolution

Reiko Oda,[*] Peizhao Liu, Elizabeth Hillard,[†] Patrick Rosa,[†] Sylvain Nlate,[*] Yutaka Okazaki,[‡] Emilie Pouget,[*] Yann Battie[§] and Thierry Buffeteau[¶]

[*]Institute of Chemistry and Biology of Membranes and Nanoobjects, UMR 5248 CNRS, Université de Bordeaux, INP, IECB, Pessac, France
[†]CNRS, Univ. Bordeaux, Bordeaux INP, ICMCB, UMR 5026, F-33600 Pessac, France
[‡]International Advanced Energy Science Research and Education Center (IAESREC), Graduate School of Energy Science, Kyoto University, Yoshida-Honmachi, Kyoto, Japan
[§]Laboratoire de Chimie et de Physique Approche Multi-échelles des Milieux Complexes (LCP-A2MC), Institut de Chimie Physique et Matériaux (ICPM), Metz cedex 3, France
[¶]Institute of Molecular Sciences, UMR 5255, CNRS, Université de Bordeaux, Talence, France

Chirality is a property of asymmetry resulting, for an object, from the non-superposition of its image in a mirror. The notion of symmetry breaking, inherent in the organization of matter, the formation of new structural edifices, and, more fundamentally, weak interactions, is omnipresent. From the physics of elementary particles to molecules of the living world and functional organisms, to climatic phenomena inducing vortices of forces, chirality often plays a crucial role. It is also a conception of geometry exploited in design fields and man-made constructions for its functionality and uniqueness.

In this chapter, we will focus on the chirality observed in solid state matter, that is, chirality based on the solid state organization of atoms or molecules. While there can be an important overlap with inorganic chiral nanostructures or nanoparticles for which there are a number of reviews, the solid state matters treated in this chapter include crystals as well as amorphous solids of both organic and inorganic molecules. As we will discuss below, the study of the chirality of solid materials has mainly been focused on asymmetric ordered and periodic structures. When atoms are considered as a repeating unit, chiral crystals of achiral molecules can be classified as 3D asymmetric periodic structures. Chiral crystal faces of centric crystals and chiral 2D patterns of achiral molecules can be classified as 2D asymmetric periodic structures. Individual helical polymeric chains, chiral carbon nanotubes, and nanoparticles can be classified as 1D asymmetric periodic structures. We should also mention that chiral quasicrystals do not have mirror symmetry or translational symmetry, but have rotational symmetry, showing beautiful chiral ordered structures. We will also describe how chirality can be enhanced by the 2D or 3D organization of building components of solid materials. We will close with a discussion of spectroscopic methods to characterize chiral objects and assemblies.

1.1 Genesis of Chirality in 3D Bulk Materials*

Chiral purification of molecules is a major preoccupation in many fields of natural sciences, ranging from origin of life studies [1], to chemistry, physics, and pharmacy [2]. Today, most newly approved chiral pharmaceuticals are marketed as single enantiomers [3], and thus, the need to produce enantiopure molecules is an important challenge for industry [4–6]. Principal strategies include the transformation of chiral natural products,

---

* Sections 1.1.1 and 1.1.5 were adapted from the thesis of the late Angela Valentin-Perez, University of Bordeaux, 2019.

asymmetric synthesis, and resolution of racemates. The resolution of racemates has proven to be the most economical and straightforward approach, and resolution via crystallization remains a technique of major industrial importance [7–9]. Solid state resolutions can be accomplished by: (i) preferential crystallization, where an initial enrichment may promote a full chiral discrimination in the solid state. In some cases, the crystallization of the preferred enantiomer in a conglomerate system is industrially feasible. (ii) the addition of chiral adducts for the formation and selective crystallization of diastereomers. However, these strategies most often have a maximum yield of 50% of the desired enantiomer. Nevertheless, recent work in solid state deracemization, where a racemic mixture is transformed entirely into a single chirality [10], poses promising perspectives in the production of chiral species with excellent atom economy.

In this section, we will briefly present different methods of resolution and deracemization via crystallization, with a focus on historical milestones.

1.1.1 History

Foundational studies, notably by French physicists, paved the way for the discovery of molecular chirality by L. Pasteur in 1848. In 1808 É.-L.Malus discovered plane-polarized light, light that propagates through a given media with its electric and magnetic field vectors in-phase along the propagation direction [11]. In 1811, F. Arago discovered optical rotation, defined as the rotation of the plane of polarization that occurs when polarized light passes through an optically active substance [11]. In 1817, J.-B. Biot found that some natural compounds, such as sucrose, oil of turpentine, camphor, and tartaric acid, rotated polarized light in solution, in the liquid state or in the vapor phase [12]. Finally, in 1824, A.-J. Fresnel discovered circularly polarized light (CP light) and that optical rotation is due to the different refractive indices for left and right CP light [13]. CP light can be defined as light that propagates through a given media with its electric and magnetic field vectors perfectly out-of-phase along the propagation direction.

In 1844, Biot presented a communication by M. Mitscherlich to the French Academy of Sciences commenting that tartrate rotated polarized light, while paratartrate had no such effect, even if the two molecules otherwise had identical physical properties [14]. The breakthrough regarding the relationship between the two substances came with Pasteur's realization that the crystal forms of optically active sodium ammonium tartrate tetrahydrate and the corresponding salt of paratartrate — now known to be a racemic mixture of D- and L-tartrate — were not identical. After careful examination, Pasteur noted that sodium ammonium paratartrate crystallizes as a mixture of two hemihedral crystals with different hemihedrism directions (Figure 1) [15]. In this way, the two crystals were nonsuperimposable mirror-images (enantiomorphous). Using tweezers, Pasteur manually separated the two types of crystals and measured their optical activity in solution, finding that the two solutions rotated light in the opposite direction, but with the same absolute intensity. In this manner, he demonstrated the relationship between crystal form and physical properties, although the structure of the tartrate molecules at that time was unknown.

Soon thereafter, Pasteur hypothesized that the molecular natures of the tartrates must be different. This led to the proposition that the molecules share the same relationship with each other as the crystals do, that is they are identical, but nonsuperimposable forms [16]. He named one form "right tartaric acid" and the other "left tartaric acid," and expressly related them to differences in their three-dimensional structure. Paratartaric acid was also known as *racemic* acid at the time, and thus the term *racemic* mixture, or racemate, was adopted for equimolar compositions of two enantiomers. Crucially, the discovery of molecular chirality can be

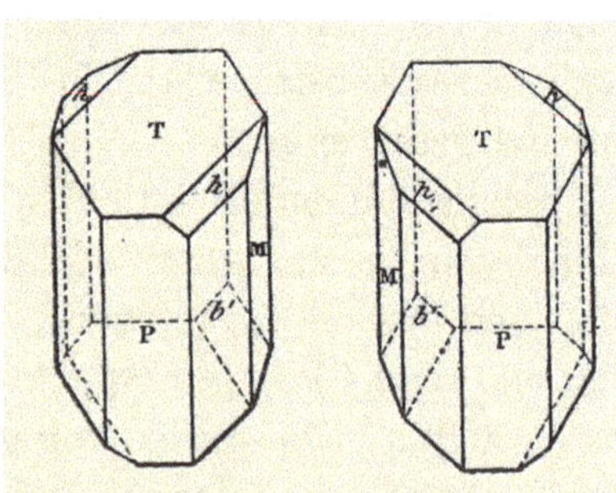

Figure 1. Pasteur's drawing of the crystals of sodium ammonium paratartrate [L. Pasteur; C. R. Acad. Sci. Paris, 1848, 26, 535.]

attributed to the propensity of the paratartrate enantiomers to segregate during crystallization to form a mechanical mixture of enantiomorphic crystals, called a conglomerate.

Conglomerate crystallization also played an important role in the discovery of biological homochirality. In 1851, Pasteur studied the crystals of the naturally occurring amino acid L-asparagine and identified their chiral crystal habit [17]. In 1885, Italian chemist A. Piutti obtained a new form of asparagine by the slow evaporation of the mother liquor from a large-scale synthesis of L-asparagine [18]. He manually separated the crystals of the new species from those of L-asparagine and found that the two crystal habits were enantiomorphous. In addition, the optical rotation of the new compound was found to be equal in absolute magnitude and opposite in direction to that of natural asparagine, while the chemical properties and elemental composition were identical. At that moment, Piutti put forth two hypotheses: that the new substance was either a stereoisomer or a geometric isomer, with the amino group in another position with respect to the natural asparagine. Therefore, Piutti designed a synthetic pathway which allowed him to demonstrate that L-asparagine has the amino group in the α position with respect to the carboxylic acid, and also to establish that his newly isolated form was the mirror image isomer of natural asparagine [19]. Furthermore, he realized that D-asparagine had an intensely sweet taste while the L-form was tasteless [20]. This was the first example of a difference in taste found for enantiomerically related substances and, as anticipated, the first example of enantioselectivity at a biological receptor. This important milestone stimulated additional studies in this area that established the importance of interactions between chiral substances and biological systems. Nowadays it is known that biological systems are also chiral, presenting, for example, amino acids in the L-form or carbohydrates in the D-form.

1.1.2 Conglomerate Crystallization

Pasteur's discovery of molecular chirality can be attributed to considerable good luck. Sodium ammonium tartrate tetrahydrate crystallizes as a conglomerate mixture of hemihedral crystals only below 27°C. At higher temperatures, the two enantiomers crystallize together in centrosymmetric crystals. Had the room in which Pasteur was working been warmer, the discovery of molecular chirality would certainly have been delayed. It was shown by H. W. B. Rooseboom in 1899 that racemic mixtures can crystallize in three principal forms [21]. Most commonly, they crystallize as racemic, centrosymmetric crystals containing both enantiomers in a periodic array (Figure 2). This preponderance of racemic crystals has been attributed by O. Wallach to a denser, and thus more energetically stable, packing in heterochiral crystals [22]. While "Wallach's Rule" has been supported by many studies [23], there also exist numerous exceptions [24].

The formation of conglomerates has been estimated to occur in less than 10% of racemic mixtures [25], although theoretical calculations suggest that the prevalence of conglomerates is likely approximately twice higher [26]. The rarest example is when both enantiomers are present, but randomly distributed throughout the crystal, forming a total or partial solid solution [27]. The crystallization behavior of racemic mixtures is usefully described by phase diagrams, either binary for a pair of optical isomers, or ternary for the pair of isomers and an optically inactive solvent. Many of these diagrams have been verified experimentally, and we refer readers to excellent reviews on the subject [28].

In general, it is not possible to predict the solid state outcome of a racemic mixture, although conglomerates appear to be favored in salts [29], and are highly dependent on the nature of the anion in cationic coordination compounds [30]. As we will see shortly, the presence of

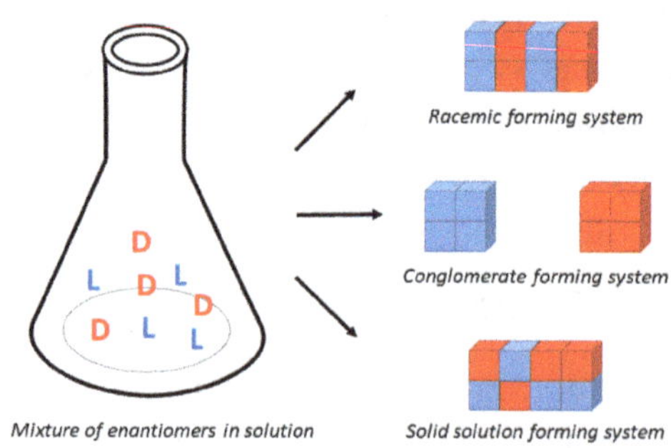

Figure 2. Different outcomes of crystallization of racemic mixtures of enantiomers.

conglomerate crystallization is useful for resolution and deracemization processes, and the rapid identification of conglomerate-producing systems is of great interest. As chiral compounds can only crystallize in one of 65 space groups, known as Sohncke groups [31], conglomerates are often capable of second harmonic generation (SHG). Screening for SHG in a recrystallized racemic mixture is thus a suitable method to determine whether the system forms a conglomerate or a racemic system [32]. When conglomerates are not available for the desired compounds, the synthesis of conglomerate co-crystals can alleviate this problem. An early example was the observation that tri-*o*-thymotide crystallizes as a racemate from methanol, but as a conglomerate as the hexane, benzene, or chloroform solvate [33]. Methods for crystal structure library [34] or phase diagram [35] screening in the prediction of co-crystals have been reported.

The examples of enantiomorphism in the solid state discussed above are due to the intrinsic chirality of the molecule, for example, the presence of stereogenic carbon atoms. However, enantiomorphism can also arise from a dissymmetric arrangement of achiral molecules in the crystal packing. Such compounds are thus completely optically inactive in the liquid or solution state, but their crystallization leads to optically active crystals. The most common examples are sodium chlorate and quartz. The β-phase of quartz crystallizes in space group $P6_421$ or $P6_221$, depending on the handedness of the supramolecular arrangement around the 6-fold axis, while the low-temperature form of α-quartz crystallizes in either space group $P3_221$ or $P3_121$ depending on the arrangement around the 3-fold axis [36]. In sodium chlorate, it is an opposite rotation around a 2_1 screw axis that provides the two crystal enantiomorphs [37]. In a similar fashion, chiral polymers can arise from achiral monomers [38]; this topic will be discussed in detail in a following section of this chapter.

While an enantiopure molecule must crystallize entirely as either right-handed or left-handed crystals, there is no thermodynamic reason why quartz or sodium chlorate should prefer one enantiomorph over the other. Thus the average distribution of several crystallizations should tend toward 50:50 of both enantiomorphic crystals. While it had been shown that a collection of $NaClO_3$ crystals, when ground into a powder, was optically inactive, in 1898 F. Kipping and W. Pope demonstrated that the *number* of crystals of each handedness converged with increasing crystal population [39]. Since then, several studies on conglomerate formation of sodium chlorate have been performed, in part due to its crystallization in a cubic crystal system as large, transparent crystals. These features allow $NaClO_3$ and related compounds, such as $NaBrO_3$, to be studied by polarized optical microscopy, without interference from linear optical effects. While Pasteur was able to visually separate crystalline enantiomers, this kind of manual triage is nontrivial, as the crystals must present clear hemihedral faces. Methods to preferentially crystallize one or the other enantiomer would render this separation step unnecessary.

1.1.3 Preferential Crystallization

The presence of conglomerate crystallization opens several pathways to enrich the enantiomeric excess in the solid state [28b, 28d]. In 1866, D. Gernez, a student of Pasteur, accomplished a selective crystallization of sodium ammonium tartrate tetrahydrate by seeding a supersaturated solution of the racemic mixture with an enantiopure crystal. If a right-handed crystal was used, right-handed tartrate was crystallized, and likewise for the left-handed enantiomer [40]. Seeding of racemic mixtures with enantiopure crystals, also called entrainment or preferential crystallization (PC), is now a common way of obtaining enantiomerically enriched material [28d, 41], and has been notably used in the industrial preparation of chloramphenicol, L-dopa, and (−)-menthol [42]. This process typically requires several cycles, where the crystallization of one enantiomorph leads to an excess of the other enantiomorph in solution, which is then precipitated using the appropriate seed, and so on [43]. As this as an arduous process, refinements have made PC more efficient for industrial applications [28b, 28c, 44]. We bring the interested reader's attention to two reviews on PC [45].

In the case of solution-stable enantiomers, the theoretical maximum yield cannot exceed 50%. To increase the yield of the desired enantiomer, racemization in solution can be provoked, sometimes referred to as second-order asymmetric transformation [46]. During the crystallization of one enantiomer, its antipode is in excess in solution and converts to the desired enantiomer, and thus the yield can approach 100% enantiomeric excess in the solid state. Because the racemization in solution removes any excess of the undesired enantiomer, the probability of the nucleation of the counter-enantiomer decreases and this process can be carried out at lower supersaturation levels than classical PC. This is particularly interesting for industrial applications [47], and the discovery of racemization agents is an active field of study [48]. The transformation of a racemic mixture to one single enantiomer is called deracemization, and is distinct from the resolution techniques described above.

1.1.4 Deracemization

Deracemization refers to the transformation of a chirally undifferentiated medium to a single enantiomer in 100% yield. Different deracemization methods have been established, using chiral catalysts, often enzymes [49], with the application of photochemical [50] or redox [51] perturbations. This is a vast subject and, as it is not necessarily dependent on crystallization, it will not be treated here. In 1954, E. Havinga laid out the three required conditions for deracemization via crystallization to occur: (i) no racemic crystal formation, (ii) the substance must be racemized in solution, and (iii) crystal growth must be faster than the formation of crystal nuclei [52]. In an article entitled "Spontaneous Formation of Optically Active Substances," Havinga demonstrated this phenomenon using the chiral ammonium salt methyl-ethyl-anilinium iodide, which spontaneously crystallized from supersaturated solutions in sealed tubes. This phenomenon is based on the fact that the free energy of a conglomerate suspension will decrease with decreasing surface area of the crystal. Therefore, the Ostwald ripening process, the production of large crystals at the expense of small ones that dissolve, will eventually lead to the production of a single enantiopure crystal [53]. While this process is very slow, it can be hastened by grinding [54], temperature cycling [55], exposure to ultrasound [56], pressure [57], or microwaves [58].

A century after Kipping and Pope showed that unstirred sodium chlorate solutions give rise to an equal distribution of both enantiomorphs, D. Kondepudi and coworkers demonstrated that an enantiomeric excess could be obtained when the crystallization took place while stirring [59]. Each crystallization is stochastic, such that prevailing enantiomer is randomly obtained. The theory was put forth that the chirality of the first appearing crystal (by primary nucleation), the "mother crystal," seeds the chirality of the subsequent "daughter" crystals (by secondary nucleation) [60]. Stirring allows a cleaving of the mother crystal to generate secondary nucleation sites, which then reduces the concentration of the solution such that the formation of additional mother crystals via primary nucleation is suppressed.

C. Viedma later refined this theory by showing that extremely rapid primary nucleation of $NaClO_3$ in a highly oversaturated solution also results in symmetry breaking, with no need of a single mother crystal [61a]. Later he showed that stirring a conglomerate of $NaClO_3$ suspended in water in the presence of glass beads also results in a 100% ee of one or the other enantiomorph [61b]. This was attributed to the equilibrium between solid and solution phases, such that the smallest crystallites dissolve, thus favoring the handedness of the larger crystals, eventually leading to complete symmetry breaking over several dissolution/recrystallization cycles. This process is now called Viedma ripening [54] and is an important method in the deracemizations of chiral organic compounds (in the presence of a racemizing agent) [62], as well as other conglomerate-forming systems [63]. Different mathematical models taking into account the influence of crystal growth, attrition and secondary nucleation have been proposed [64]. A general schematic for Viedma ripening is shown in Figure 3.

Another related deracemization technique that relies on racemization in solution and auto-amplification of small enantiomeric excesses in the solid phase is temperature cycling induced deracemization (TCID), discovered by G. Coquerel in 2013 [64d]. In TCID, starting from a suspension of

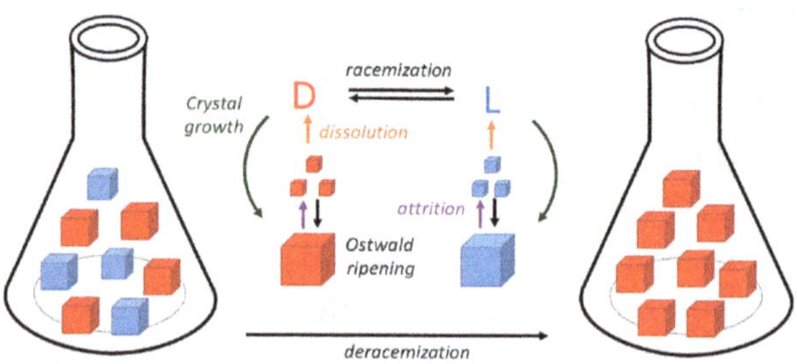

Figure 3. General scheme for deracemization processes by crystallization of conglomerates.

mirror-imaged crystals in equilibrium with their common saturated solution in which both enantiomers interconvert freely, the thermal energy applied to the suspension in a (quasi-)steady state produces dissolution/crystallization cycles, and the racemic mixture is progressively converted into an enantiopure material [8, 65]. This process has been modeled computationally taking into account various parameters such as crystal size distribution, secondary nucleation, agglomeration, crystal growth, dissolution, and racemization processes [66], although the exact mechanism of this process is still unknown. Recently, a method of combining deracemization of the medicine praziquantel with temperature cycling and conglomerate crystallization under flow conditions has been reported [67] and scale-ups of the TCID process have been accomplished [68].

The previously discussed deracemization techniques do not require the addition of any chiral auxiliary, but the final chirality of the crystalline product cannot be controlled. In order to select the desired enantiomer, the addition of small quantities of non-co-crystallizing chiral substances or the use of chiral solvents can bias the reaction [69]. It was shown as early as 1898 that adding a chiral carboxylate to the crystallization of achiral $NaClO_3$ resulted in a preference for one crystalline enantiomorph over the other [39]. Chiral impurities in the environment have been suspected in some cases of favoring *d*-$NaClO_3$ [70] when no exogenous auxiliary is added. This kind of preference for one crystal handedness has been found to be due to differential interactions at the crystal surface which can inhibit crystal growth or promote nucleation [71]. Another approach is the addition of a stoichiometric quantity of a chiral additive, which preferentially interacts with one enantiomer, resulting in the formation of crystalline diastereomers. This resolution technique will be discussed in the following section on Pasteurian resoluton, with an emphasis on chiral anions.

1.1.5 Pasteurian Resolution

Pasteurian resolution is a method to obtain enantiomerically pure molecules from racemic mixtures via the formation and selective crystallization of diastereomeric salts. With different physico-chemical properties between diastereomers, particularly solubility, they can be separated by a selective crystallization step and there is no need for conglomerate formation [72].

This method was first applied by Pasteur in 1853 in the separation of a racemic mixture of tartaric acid [73]. The addition of derivatives of natural chiral bases quinotoxin and cinchotoxin allowed the crystallization of the right and left tartrates, respectively, while the other enantiomer remained in solution.

The resolution of amino acid derivatives by this method was first described by E. Fischer in 1899, using brucine and strychnine [74]. Pasteurian resolution continues to be broadly applied in industry for organic systems using chiral acids or bases from the chiral pool of natural products [75]. Basic resolving agents include the widely used α-methylbenzylamine and quinine, cinchonine, cinchonidine, ephedrine, morphine, strychnine, brucine, etc. [25, 42, 76]. Acidic resolving agents for racemic bases include mainly tartaric acid and derivatives, as well as mandelic acid [77], L-malic acid [78], 2-naphthylglycolic acid, (*S*)-phenylpropionic acid, and camphorsulfonic acid and derivatives [42, 76].

The conjugate bases of such acidic resolving agents have also been used for the resolution of chiral coordination compounds. The first such resolution was performed in 1911 by V. L. King, a student of A. Werner, who resolved cis-$[Co(en)_2(NH_3)Cl]^{2+}$ with 3-bromocamphor-9-sulfonate [79]. As coordination complexes are most often cationic, a series of natural and synthetic anions have been developed for their crystalline resolution. In order to have efficient ion pairing, the chiral anions need to fulfill the following conditions, as enumerated by J. Lacour and coworkers [80]:

1) The stereogenic element(s) should be close to the charged complex to favor enantioselective ion-pairing interactions.
2) The anion should be chemically and configurationally stable.
3) The chiral anions should be easily accessible by synthetic procedures and, if possible, in an asymmetric manner.

The earliest examples of chiral anions employed for enantiomeric resolution of coordination compounds are the conjugate bases of natural chiral carboxylic and sulfonic acids, such as tartaric, mandelic, or 10-camphorsulfonic acids (Figure 4) [81]. Nevertheless, these anions possess a rather large number of potential conformations. To obtain a more rigid structure, they can be coordinated to metal ions, yielding an

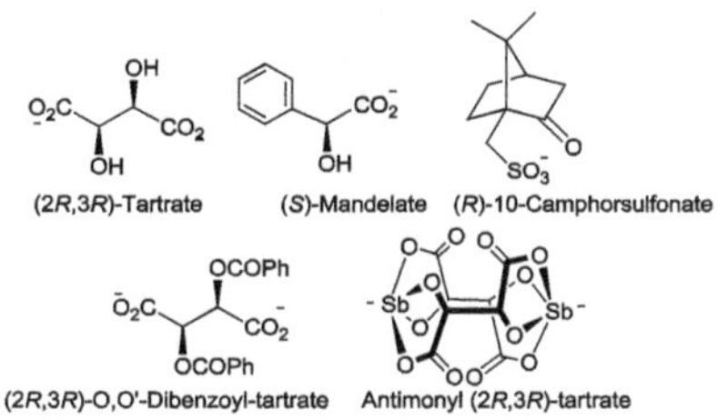

Figure 4. Examples of chiral anions obtained from natural chiral carboxylic and sulfonic acids.

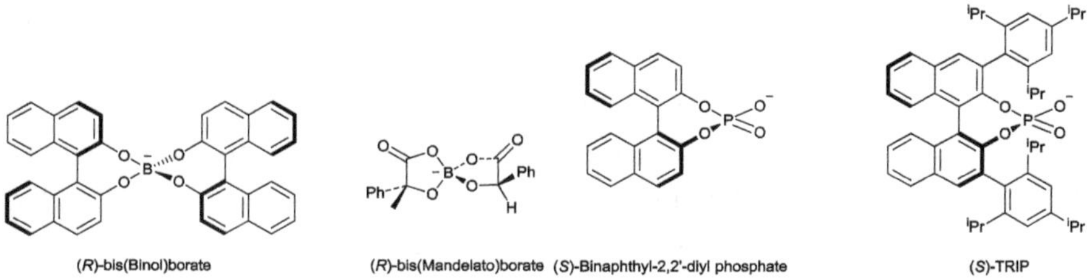

Figure 5. Examples of chiral borate anions and chiral derivatives of phosphoric acid.

anionic complex. The most well-known example is the antimonyl (2*R*,3*R*)-tartrate anion (Figure 4), commercially available as the potassium salt, which has been extensively used for chiral resolution in coordination chemistry [82].

Chiral borate anions can also be synthesized by employing chiral ligands. H. Yamamoto and coworkers first synthesized this type of anion (Figure 5), employing [1,1′]-binaphthalenyl-2,2′-diol (popularly known as BINOL) as the ligand [83]. Chiral resolutions using sodium salts of the (*R*)- or (*S*)-bis(mandelato)borate anion were reported by I. Williams and coworkers for the separation of the enantiomers of the alkaloid tetrahydropalmatine, 1,2-diaminopropane, and the complex $[Co(phen)_3]^{3+}$ (phen = 1,10-phenanthroline) [84].

Conjugate bases of phosphoric acid derivatives frequently provide chiral anions. For many years, the (*S*)-(+)- or (*R*)-(–)-binaphthyl-2,2′-diyl phosphate (Figure 5), which presents planar chirality, was the only such anion used in asymmetric anion-mediated processes [85]. Nowadays, there are many derivatives of this phosphate with substitution at the

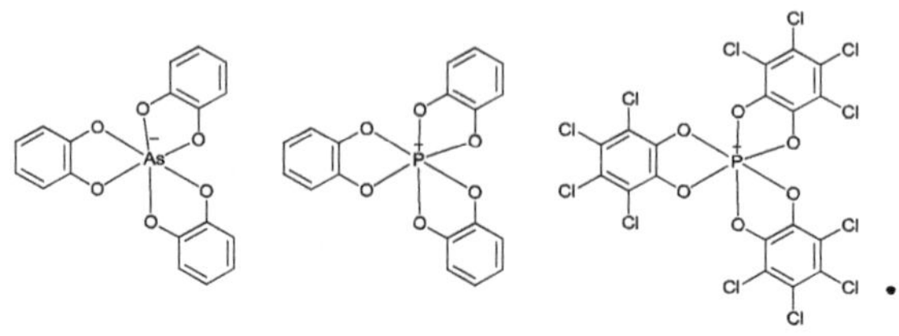

Figure 6. Examples of chiral octahedral anionic complexes of P(V) and As(V).

3,3′-positions. One example is the so-called TRIP anion (Figure 5), synthesized by B. List and coworkers, which is characterized by the presence of very large 2,4,6-triisopropylphenyl substituents at the binaphtyl 3,3′-positions [86]. The resolution of the enantiomers of a dicopper(I) molecular trefoil with binaphthyl-2,2′-diyl phosphate was reported by J.-P. Sauvage and colleagues in 1996 [87].

Octahedral anionic complexes of hexacoordinated phosphorus and arsenic atoms, presenting helicoidal chirality, are also often employed to perform chiral resolutions. These compounds are synthesized as racemates using achiral ligands, and are typically resolved using chiral alkaloids [88]. The TRISCAS (tris(catecholato)arsenate(V)), TRISCAT (tris(catecholato)phosphate(V), and TRISPHAT (tris(tetrachloro-1,2-benzenediolato)phosphate(V) anions are of particular importance (Figure 6). TRISCAS was synthesized for the first time in 1919 by R. Weinland and J. Heinzler [89], and the complex was resolved into its optical enantiomers by A. Rosenheim and W. Plato in 1925, who proposed the tris-chelate structure of the anion [88c], which was confirmed by single-crystal X-ray diffraction in 1972 [90]. It has been employed for the chiral resolution of asymmetric boron cations [91] and helicate dimer and trimer copper (I) complexes [92]. The X-ray crystal structure of the racemic triethylammonium salt of TRISCAT was published in 1973 [93], but its resolution was not performed until 1979. Nonetheless, the resolved anion racemizes rather quickly in solution depending on the nature of the solvent and acidity of the medium [94] For this reason, it is not a very useful chiral anion to perform enantiomeric resolution, but these racemization problems are solved in its perchlorinated TRISPHAT monoanion derivative. Its X-ray crystal structure was reported for the diethylammonium salt in 1992 [95], and its chiral resolution was managed in 1998 by Lacour *et al.* [88b]. This anion has been used, for example, as a resolving agent of M(II) complexes with ligands based on bipyridine and phenanthroline [96].

1.1.6 Helical Conformations of Achiral Polymers in Solids

In the previous section, we discussed the rich variety of crystalline arrangements of chiral molecules. Dissymmetric arrangement of achiral units, such as atoms, can also be the origin of enantiomorphism in solid crystals [39, 97]. As previously mentioned, and shown in Figure 7a and c, helical arrangements of atoms are found in both crystals of α-quartz [98] and sodium chlorate [99]. Similar dissymmetry can also be found in the helical conformation of achiral polymers. In 1954, C.W. Bunn and E.R. Howells first revealed the helical structure of polytetrafluoroethylene (PTFE) in the solid crystal by X-ray diffraction studies (Figure 7b) [100]. After this report, detailed structures of PTFE crystals have been studied

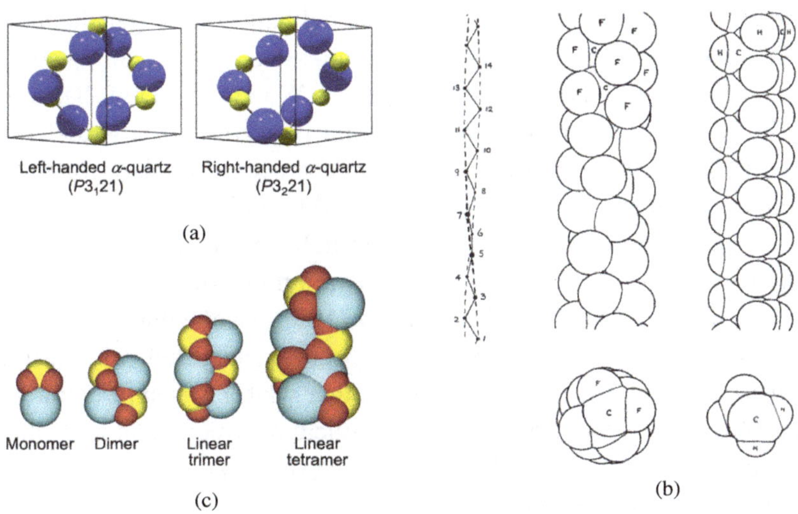

Figure 7. (a) Crystal structures of left- ($P3_121$ space group) and right-handed ($P3_221$ space group) α-quartz. O atoms: large blue, Si atoms: small yellow. (b) Left: twisted zigzag chain found in fluorocarbons. Center: fluorocarbon molecule (side and end views). Right: hydrocarbon molecule. (c) Isotactic oligomerization of sodium chlorate. (a) Adapted with permission from Ref. [98]. (b) Reprinted with permission from Ref. [5, 100]. (c) Adapted with permission from Ref. [99].

by various analytical techniques such as X-ray crystal structural analysis, thermal analysis, and computer simulation, leading to the discovery of the 13/6 helical conformation in the phase II and the 15/7 helical conformation in the phase II and IV [101]. From a historical standpoint, it is interesting to note that the helical conformation of PTFE was reported in the same decade as the right-handed α-helix of polypeptides [102], the right-handed double-stranded helix of DNA [103], and the helical conformation of isotactic polypropylene in the crystal state [104]. The most thermodynamically stable conformation of hydrocarbon chains is the flat zigzag (all *trans*) conformation. However, in the case of fluorocarbons, this zigzag conformation of the polymer chain is twisted. This was attributed to steric hindrance due to the larger van der Waals diameter of the fluorine atom (2.70 Å), compared the hydrogen atom (2.40 Å), whereas the span of the zigzag (distance between the carbon with its next neighbor carbon) is 2.54 Å. Similarly, numerous isotactic vinyl polymers having a helical conformation in the solid crystal have been reported [105].

In such structures, even though steric hindrance forces helical conformations, there is no preference for the helical handedness (left-handed or right-handed) taken by such polymers unless they have chiral sources (chiral center, chiral axis, chiral plane, etc.). When the energy barrier for helix inversion is relatively low, switching between the left- and right-handed helical conformations is a dynamic process in solution and they remain as a racemic mixture [106]. R.J.M. Nolte *et al.* synthesized poly(*tert*-butyl isocyanide) having a high helix inversion barrier in solution and succeed in the enantiomeric separation of synthetic helical polymers by chiral chromatography in 1974 [107]. Meanwhile, Y. Okamoto and coworkers succeeded to synthesize a one-handed helical conformation of isotactic poly(triphenylmethyl methacrylate) by using chiral catalysts [108]. Indeed, the formation of the helical polymer chain is attributed to the ensemble of intramolecular factors (internal rotation potentials around the single bond, repulsions and van der Waals interactions between nonbonded atoms, electrostatic interactions, hydrogen bonds, etc.), and intermolecular factors (hydrogen bonds, intermolecular interactions in the crystal lattice, etc.). A number of helical polymers having well-designed molecular structures have been reported [109] and the strategies have been expanded to various research fields including chiral supramolecular polymers [110]. Carbon nanotubes might also be thought of as achiral unit-based helical polymers, as carbon atoms or benzene rings are considered as the repeating units. A single-walled carbon nanotube (SWCNT) can be considered as a cylinder rolled up from a graphene monolayer with a chiral vector $\mathbf{C}_\mathrm{h} = n\mathbf{a}_1 + m\mathbf{a}_2 \equiv (n, m)$, where $\mathbf{a}_1$ and $\mathbf{a}_2$ are unit vectors and (n, m) defines the chirality (or helicity) of the SWCNT. Left- and right-handed SWCNTs show negative and positive values of $n - m$, respectively [111]. Chiral SWCNTs have been obtained both by chirality-controlled growth methods and solution-based separation approaches.

Such helical polymers, dispersed in solvents, can be deposited on substrates by methods such as drop casting [112], spin coating [113], and spray coating [114] giving access to homochiral solid materials (Figure 8a).

It is also possible to synthesize helical polymers with a selected handedness by use of an external inducer, and to simultaneously solidify them in order to arrest the helicity inversion dynamics (Figure 8b). K. Akagi *et al.* reported the synthesis of helical polyacetylene by using a chiral nematic liquid crystal as a chiral reaction field [115] in which the handedness of helical polyacetylene can be controlled by the chirality of the liquid crystal. The dihedral angle between the repeating units (-CH=CH-) of helical polyacetylene ranged from 0.02° to 0.23°. This very small dihedral angle maintains the planarity of π-conjugated polymer chain with a

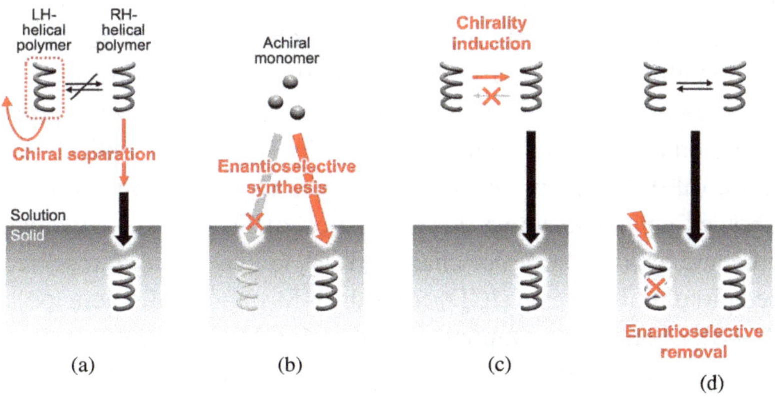

Figure 8. Various preparation approaches of chiral solid materials from helical polymers. (a) Chiral separation, (b) enantioselective synthesis, (c) chirality induction, and (d) enantioselective removal.

high *trans* content (90%) resulting in high electron conductivity, and to form one-handed helical structures with a nonzero dihedral angle [116].

Instead of using chiral liquid crystals, CP light can be used to synthesize enantiopure polymers. M. Iwamoto *et al.* reported that chiral polydiacetylene (PDA) films were produced by using ultra-violet CP light for photopolymerization [117]. In this case, CP light was used as the only chiral source without any chiral molecules in the system, and the sign of the circular dichroism (CD) signals of the obtained PDA films was dependent on the selection of left- and right-handed CP light irradiation. S.-T. Wu and coworkers applied this technique to enantioselective synthesis of chiral coordination polymers [118]. They synthesized [{Cu(succinate)(4,4′-bipyridine)}$_n$]·4H$_2$O$_n$ having left- and right-handed helical structures by irradiation of left- and right-handed CP light in the visible range, respectively.

Some researchers took another approach in order to create chiral polymer solids from an achiral polymer solution taking advantage of its dynamic helix inversion behavior in solution (Figure 8c). For example, the formation of solid films from achiral polymers with chiral dopants can give access to stable chiral structures having memory effects.

E. Yashima *et al.* reported that only one helical handedness of charged poly(phenylacetylene) derivatives can be induced by small chiral guest molecules, and can be memorized in a film by the *Layer-by-Layer (LbL)* assembly technique [119]. R.B. Kaner *et al.* prepared the optically active polyaniline/N-methylpyrrolidinone (NMP) solution by doping with *S*- or *R*-camphorsulfonic acid (CSA). Subsequently, optically active thin films of *R*- or *S*-camphorsulfonic acid (CSA)-doped polyaniline were prepared by casting on a glass plate. The obtained film retained its optical activity after removal of CSA. Interestingly, this optically active polyaniline thin film showed enantioselective discrimination of D- and L-phenylalanine, as confirmed by a visible color change [120]. G. Guerra *et al.* reported the induced circular dichroism (ICD) from the δ-form of syndiotactic polystyrene (s-PS), obtained by evaporation from chiral volatile solvents such as limonene and carvone, even after removal of these chiral compounds [121]. The helical conformation of δ-s-PS memorized by co-crystallization with chiral molecules shows high thermal stability up to 240°C. The imprinted chiral nano-space, which the chiral molecules occupied during the co-crystallization process, can be used for chirality induction in achiral chromophore guests such as azulene [122]. A number of examples of chirality induction by co-crystallization with chiral compounds are summarized in review articles [123].

CP light can be also used as an external force for chirality induction [124]. G. Iftime and coworkers observed ICD signals from an initially achiral azobenzene liquid crystalline polymer in the solid state after irradiation with CP light [125]. The sign of the ICD can be switched by selecting the handedness of irradiated CP light. Induction of a preferred-handed helical conformation of π-conjugated polymers, such as poly(9,9-dioctylfluorene-2,7-diyl) (PDOF) and its derivatives, has been well studied by the groups of Nakano [126] and Fujiki [127].

Finally, enantioselective removal of helical polymers from racemic solids might also be included in the key methods for preparing enantiopure helical polymer-based-solid materials (Figure 8d). M. Teraguchi *et al.* reported the helix-sense-selective photodegradation of a racemic poly[4-dodecyloxy-3,5-bis(hydroxymethyl)phenylacetylene] (poly(DoDHPA)) membrane by irradiation of single-handed with CP light (Figure 8d) [128].

1.1.7 Generation of Chirality in Amorphous Solids

As discussed above, the study of the chirality of solid materials has mainly been focused on asymmetric ordered and periodic structures. When atoms are considered as a repeating unit, chiral crystals of achiral molecules can be classified as 3D asymmetric periodic structures (Figure 9b). Chiral crystal faces of centric crystals and chiral 2D patterns of achiral molecules can be classified as 2D asymmetric periodic structures (Figure 9c). Individual helical polymers, chiral carbon nanotubes, and nanoparticles can be classified as 1D asymmetric periodic structures (Figure 9d). We should also mention that chiral quasicrystals do not have mirror symmetry or translational symmetry, but have rotational symmetry, showing beautiful chiral ordered structures (Figure 9e and f).

On the other hand, noncrystalline (non-ordered and nonperiodic structures) materials also have the potential to show chiral properties. Considering atoms as the smallest unit, the simplest element of molecular

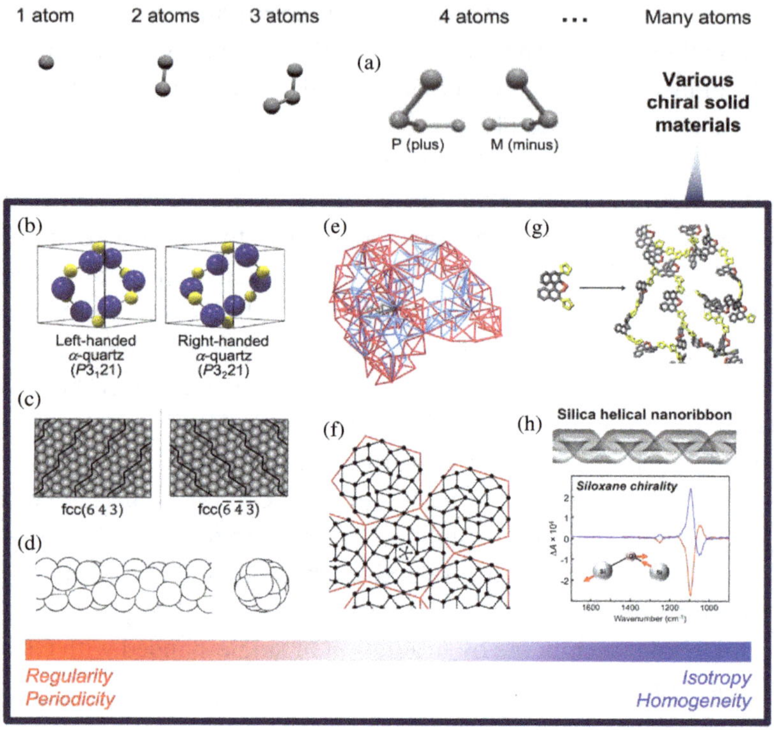

Figure 9. Generation of chirality from the arrangement of atoms in solid materials. (a) Arrangements of four atoms having plus (left) and minus (right) torsion angles. (b) 3D asymmetric periodic structure found in the crystal structure of quartz. (c) 2D asymmetric periodic structure found in the ball models of fcc(643) and fcc($\bar{6}\bar{4}\bar{3}$) surfaces. (d) 1D asymmetric periodic structure found in the fluorocarbons. (e) Chiral 3D icosahedral quasicrystalline structure. (f) The dodecahedron 2D network decorated with the chiral pentagonal quasilattice. (g) Locally chiral and globally isotropic 3D structure found in the microporous BINOL-based polymer network. (h) Schematic illustration of amorphous silica helical nanoribbon (top) and its VCD signals at Si-O-Si asymmetric stretching vibrational band (bottom). (a) Adapted with permission from Ref. [129]. (b) Adapted with permission from Ref. [98]. (c) Adapted with permission from Ref. [138]. (d) Adapted with permission from Ref. [5, 100]. (e) Adapted with permission from Ref. [139]. (f) Adapted with permission from Ref. [140]. (g) Adapted with permission from Ref. [133]. (h) Adapted with permission from Ref. [136].

chirality is the torsion angle θ ($-180° < \theta < 180°$) made from four atoms as shown in Figure 9a) [129]. Note that even one atom [130], monoatomic cations [131], and monoatomic anions [132] can show induced chirality under the support of chiral inducer, such as chiral molecules and chiral molecular assemblies, constructed from more than four atoms. C. Bleschke *et al.* synthesized enantiopure microporous 1,1′-bi-2-naphthol (BINOL)-based polymer networks (Figure 9g) [133], and the obtained microporous polymer material showed high enantioselectivity as a heterogeneous organocatalyst. The interesting point is that this material has local chirality and global isotropy (homogeneity). Similar examples can be easily found in the solution phase, but only a few examples have been reported in solid systems. A good example is that of amorphous (noncrystalline) chiral metal oxides synthesized by the sol-gel condensation of metal alkoxide in the presence of chiral organic molecules as templates. This approach was first reported by Shinkai *et al.* [134]. After this report, numerous helical (morphologically chiral) silica nano-materials have been synthesized and been studied for their chiroptical properties [135]. Interestingly, in some cases, the obtained helical (or twisted) nano silicates show strong vibrational circular dichroism (VCD) signals at Si-O-Si asymmetric stretching vibrational band even after removal of organics by calcination (Figure 9h) [136]. XRD and nano beam electron diffraction results confirmed that the obtained helical (and twisted) nano silica materials are amorphous (noncrystalline). These results indicate that the silica helices have a local chiral arrangement of Si and O atoms, similar as shown in Figure 9a, but also a globally isotropic siloxane network. The lack of regularity, including periodicity and translational symmetry, provides global isotropy (homogeneity) [137]. Such a viewpoint might lead to further developments in the field of chiral solid materials.

1.2 Generation of Chiral Surfaces

1.2.1 Chiral Metal Surfaces for Enantioselective Recognition

As it is well known, a wide variety of substances have their own stereoisomers, such as pharmaceuticals, nutraceuticals, and agricultural chemicals. In an achiral environment, enantiomers of chiral compounds generally exhibit similar physical and chemical properties [141]. However, when they interact with other chiral compounds or chiral surfaces, the enantiomers often exhibit different biological and pharmacological responses [142], and many pharmaceutical products and agrochemicals must be enantiopure to be effective [143]. It is important to realize that more than 50% of the drugs currently in pharmaceutical use have at least one chiral center in their structures and most of the top-selling drugs in the market are used in their enantiopure form [144]. Perhaps the most dramatic example indicating the importance of enantiopure compounds in the pharmaceutical industry is the case of thalidomide, widely used in the 1960s by pregnant women as a sedative and to relieve morning sickness [145]. Whereas the D-enantiomer is harmless, with desirable tranquilizing properties, its L-enantiomer is teratogenic and leads to malformations in embryos.

Therefore, the separation of racemic molecules is of substantial significance not only in basic sciences but also for technical applications, such as fine chemicals and drug development. However, precisely because of the similarity of their physical and chemical properties, their resolution in an achiral environment remains a challenge [141, 146]. Many single-enantiomer drugs are obtained either by separation methods, often involving chromatographic columns [147], or via asymmetric catalysis, which ideally should be performed heterogeneously to minimize problems associated with solubility and with the purification of the products [148]. Both of those methods involve chemistry on solid surfaces. Resolution methods based on the recognition or separation of enantiomers remain an attractive choice to obtain enantiopure compounds.

The most commonly used resolution method is chiral chromatography [147] capable of enantioselective recognition. Other efficient techniques include crystal resolution (as discussed previously), liquid–liquid extraction, and membrane separation which are also used for separation or preparation purposes [149]. Each of these methods has its unique capabilities for enantiomeric recognition, separation, or quantification.

1.2.2 Chirality on Solid Crystal Surfaces

Chiral solid surfaces represent highly useful environments for a variety of heterogeneous enantioselective processes to produce enantiopure bioactive compounds [146c, 150]. Indeed, there are many inherently chiral inorganic materials, such as chiral metal surfaces, which have demonstrated a wide variety of enantiospecific properties and enantiospecific interactions with chiral adsorbates [151]. The design of various types of chiral surfaces using single-crystal Cu surfaces has been extensively studied in recent years with the aim of identifying materials having surfaces with optimal enantiospecific interactions with the chiral compounds of interest [152]. Such surfaces can have potential applications in analytical chemistry, chiroptical measurements, chemical, and electrochemical synthesis as well as separation.

1.2.2.1 *Chirality of crystal surfaces due to crystal structure*

The surface of acentric (without a center of symmetry) crystals is chiral. Among numerous chiral crystals including over 210 acentric crystals of metal oxides [153], natural quartz, which is known to be the most abundant chiral oxides, has been often used for the investigation of chiral surfaces (Figure 10a and b). For example, powders of enantiomorphic quartz show enantioselective adsorption of various chiral compounds such as metal complexes [154] and amino acids [155]. However, it was pointed out that: (1) different crystallographic planes of quartz show different absorption behaviors including non-enantioselective absorption [150b, 156]. Therefore, potentially strong enantioselective absorption effects on some of the crystallographic planes might be weakened or even disappear by the mixing of different crystallographic planes resulting from the powdering process, and (2) most natural quartz crystals are internally twinned by inversion [150b]. Therefore, powdering natural crystals will necessarily blend both left- and right-handed domains [150b]. Therefore, it is crucial that a single crystallographic plane of quartz is used for studying enantioselective adsorption.

Achiral crystals having a symmetrical structure, such as face-centered cubic (fcc), body-centered cubic (bcc), or hexagonal close-packed (hcp) structures (figure 10d) can provide chiral surfaces when their surfaces lack mirror symmetry (known as a chiral kink, chiral step (figure 10e), or chiral terrasse (figure 10f)) [157]. For example, A.M. Cody and R. D. Cody reported that gypsum ($CaSO_4{\cdot}2H_2O$) shows strong asymmetric crystal growth in the presence of various chiral molecules [158]. C.A. Orme *et al.* reported a similar asymmetric crystal growth phenomenon of the centric crystal of calcite ($CaCO_3$) (figure 10c) in close interaction with amino acids [159]. Such mirror symmetry-broken (chiral) surfaces of

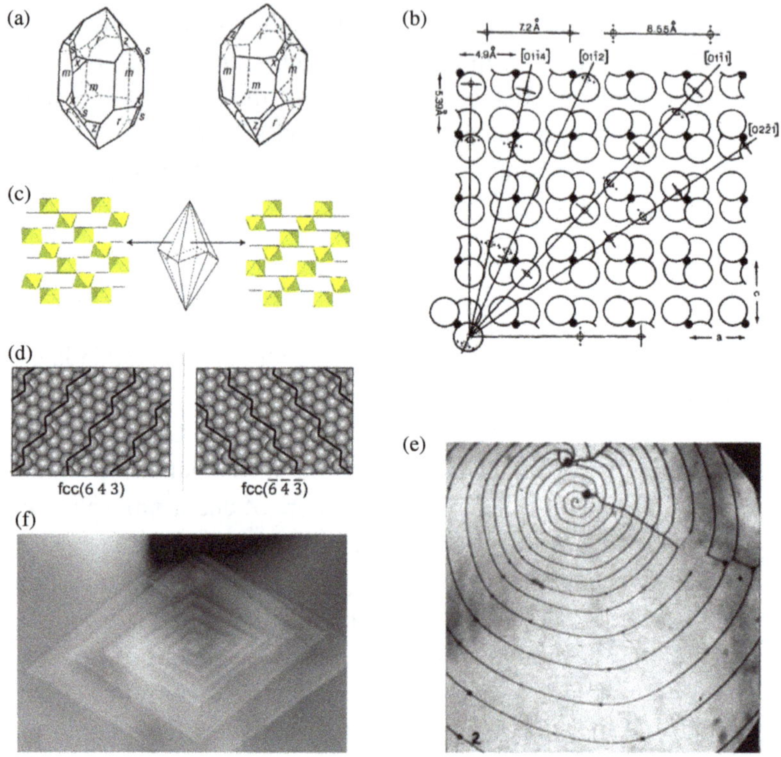

Figure 10. Chirality on crystalline solid surfaces. (a) Left- and right-handed variants of quartz crystals. (b) The (10$\bar{1}$1) crystal face of L-quartz. (c) The common {21$\bar{3}$1} trigonal scalenohedral (dogtooth) form of calcite (center). features adjacent crystal faces with enantiomorphic surface structures. The markedly acentric surface structures of both the (3121) face (left) and the (2131) face (right) consist of corner-linked chains of CaO_6 octahedra, cross-linked by planar CO_3 groups, which are seen almost on edge. L-aspartic acid is observed to adsorb preferentially on the (3121) face, whereas D-aspartic acid adsorbs preferentially on the (2131) face. (d) Ball models of fcc(643) and fcc($\bar{6}\bar{4}\bar{3}$) surfaces. (e) Spiral-patterned step on the crystal surface of carborundum (SiC). (f) AFM image of a spiral terrace of a polyethylene single-crystal. (a and c) Reprinted with permission from Ref. [150b]. (b) Reprinted with permission from Ref. [155b]. (d) Adapted with permission from Ref. [138]. (e) Reprinted with permission from Ref. [163]. (f) Adapted with permission from Ref. [162b]

centric calcite show enantioselective adsorption depending on the exposed facet [160]. Similar chiral surfaces of various materials, including metals [138, 161], polymers [162], and other inorganics [163], have also been studied. The study of chiral surfaces and their enantioselective behavior has been carried out by various analytical techniques such as scanning tunneling microscopy (STM), atomic force microscopy (AFM), low-energy electron diffraction (LEED) [161a, 164], SHG [165], electrochemical methods [166], the quartz crystal microbalance (QCM) method [167], density functional theory (DFT) calculations, and others.

The introduction of chirality on achiral metal surfaces can be achieved by adsorption both of chiral and achiral molecules [168]. The transformation of the surface nature of metals or metal oxides in order to bring in chiral features can be achieved when they are in interaction with chiral or achiral molecules (organized chirally) which then create the chiral environment.

1.2.2.2 *Adsorption of chiral molecules on solid surfaces*

It has been shown that enantiopure chiral adsorbates can develop several ordered structures on various chiral and achiral surfaces, the structures of which depend on the coverage of the adsorbates and the temperature [169]. This reflects a delicate balance among different driving forces, such as intermolecular interactions, which often include hydrogen bonds, and bonds with surface functions [170]. For example, at low surface coverage, many homochiral adsorbates tend to agglomerate to form small, well-defined supramolecular aggregates consisting of a few molecules (monomers, dimers, trimers, and tetramers) in short linear structures. The majority of the molecules are found in dimers [171], while at high coverage, these homochiral adsorbates are found with tetrameric and pentameric units, sometimes with reduced symmetry due to the formation of linear chains, herringbone patterns, or other supramolecular arrangements (Figure 11) [168c, 172].

When using a racemic mixture of chiral absorbates in solution, either they can form racemates from the pairing of enantiomers [173], especially if the two enantiomers strongly interact with each other, or they can show phase separation into homochiral domains (conglomerates) [174]. In general, the formation of homochiral conglomerates on surfaces results from the self-assembly of small homochiral cluster units. These homochiral clusters have higher stability than the corresponding heterochiral clusters, leading to conglomerate formation. Using DFT calculations, it was

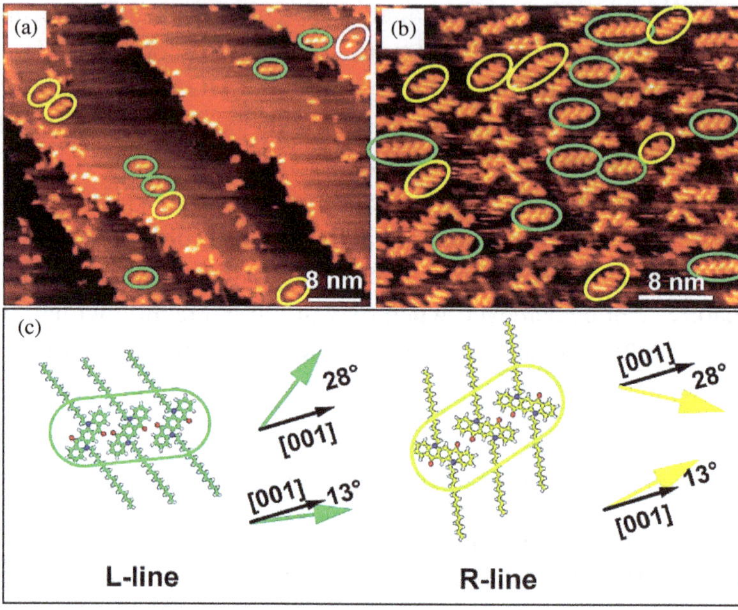

Figure 11. Adsorption of a prochiral quinacridone derivative (QA16C) with two alkyl chains of 16 carbon atoms on a Cu(110) surface: (a and b) STM images obtained at 150 K on Cu(110) at low (QA16C) coverages. One-dimensional structures of QA16C with different orientations are observed. Monomers, dimers, trimers, tetramers, and pentamers are identified. (c) Schematic representation of backbone and line orientations. (a) $U_{bias} = -1.07$ V, and $I_t = 0.50$ nA. (b) $U_{bias} = -1.29$ V, and $I_t = 0.08$ nA. Adapted from Ref. [171] with permission, Copyright 2010 American Society.

revealed that the small clusters can be obtained by strong hydrogen bonding, while the larger, racemic supramolecular arrangements appear to be controlled by weaker van der Waals interactions. The initial small homochiral clusters disappear with increasing surface coverage. The formation of well-defined aggregates and self-assembled patterns on the material surface determines the degree of enantiosegregation of the final monolayers made with racemic mixtures of molecules. A few cases where homochiral and heterochiral structures of the same molecule are formed on the surface have also been reported [173b], possibly resulting from a balance between intermolecular interactions mediated by hydrogen bonds and the strength of adsorbate-surface bonds. Thus, the final structures that form on the surface are the result of a subtle balance between several competing forces of similar magnitude.

Chirality can also be generated by the absorption of achiral or prochiral molecules on achiral surfaces. R.A. Wolkow *et al.* observed chirally adsorbed *trans*-2-butene, which is prochiral molecule, on the achiral silicon(100) surface (Figure 12a) [175]. Similar phenomena were found in the absorption of glycine molecules on the Cu(100) surface [176] and 4-trans-2-(pyrid-4-yl-vinyl) benzoic acid (PVBA) on Pd(110) [177]. Meanwhile, Q. Chen and N.V. Richardson reported an example of surface chirality generation by the adsorption of 1D molecular aggregates. When adenine is adsorbed on a Cu(110) surface with a low coverage at room temperature and annealed at 370 K, the molecules form dimer-based short chain-like aggregates aligned ±19.5° from the [100] symmetry direction of the Cu(110) surface (Figure 12b and c) [178]. T.G. Gopakumar reported that Ni-tetramethyl-tetraazaannulene (Ni-TMTAA), which is an achiral and non-prochiral molecule, forms right- (*R*-form) or left-handed (*S*-form) propeller-shaped chiral trimers to produce well-patterned homochiral 2D surface motifs on the Au(111) surface [179]. In the case of prochiral 1-nitronaphthalene (NN), the deposition of NN in the range of 0.05–0.2 monolayers (ML, 1 ML corresponds to a close-packed molecular layer) on the Au(111) surface leads to chiral (C_2 symmetric) decamer formation (Figure 12d) [180]. The handedness of such chiral decamers is determined by a combination of high-resolution STM observation and local-density calculations. They concluded that chiral clusters are stabilized by intermolecular hydrogen bonding between the O atom of the NO_2 group and the H atom of the naphthalene moiety. The idea of adsorption of prochiral molecular assemblies can be expanded to 2D chiral packing. C.B. France and B.A. Parkinson reported the 2D chiral packing domain of naphtho[2,3-a]pyrene (NP) on the Au(111) surface (Figure 12e) [181]. Even when adsorbent molecules are highly symmetric and achiral (non-prochiral), surface chirality can also be generated by forming chiral patterns from them. For example, M. Neuber and F. Schneider *et al.* reported that achiral (D_{6h} symmetric) benzene molecules form a chiral 2D pattern when adsorbed on an Rh(111) surface (Figure 12f) [164a]. Many similar examples, such as adsorption of C_{2h} symmetric anthracene derivatives on a graphite surface [182] and D_{2h} symmetric dicarbonitrile-polyphenyl

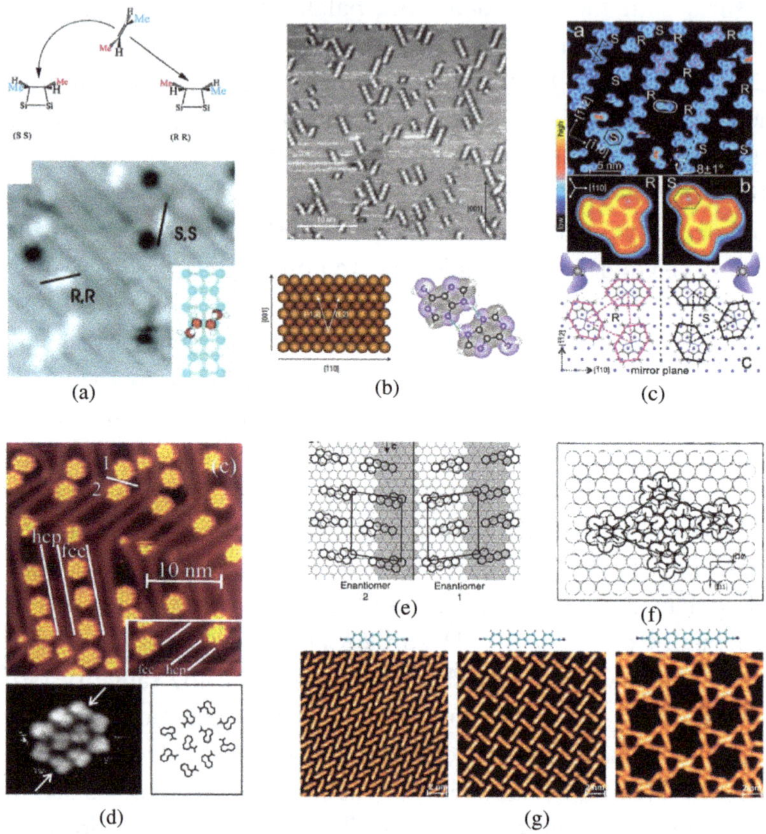

Figure 12. Generation of surface chirality by adsorption of achiral molecules on achiral crystalline solid surfaces. (a) Top: Two distinct faces of planar trans-2-butene. Bottom: STM image of trans-2-butene on a Si(100) surface. (b) STM image of adenine chains on a Cu(110) surface. (c) Top: STM image of Ni-TMTAA on Au(111). Middle: Pseudo-three-dimensional presentation of a constant-current STM image of R and S trimers of Ni-TMTAA. Bottom: Sketch of R (magenta) and S (black) trimers of Ni-TMTAA. (d) STM image of an Au(111) surface with adsorbed 1-nitronaphthalene (NN). Magnified STM image of the NN decamer (bottom left) and its theoretical modeling (bottom right). (e) 2D chiral packing model of naphtho[2,3-a]pyrene (NP) on a Au(111) surface. (f) Proposed chiral structure of densely packed benzene molecules adsorbed on an Rh(111) surface. (g) STM images of various 2D chiral patterns formed by a dicarbonitrile triphenyl molecule (left), a dicarbonitrile tetraphenyl molecule (center), and a dicarbonitrile pentaphenyl molecule (right) on an Ag(111) surface. (a) Reprinted with permission from Ref. [175]. (b and c) Reprinted with permission from Ref. [178a]. (d) Reprinted with permission from Ref. [180b]. (e) Reprinted with permission from Ref. [181]. (f) Reprinted with permission from Ref. [164a]. (g) Reprinted with permission from Ref. [183].

molecules on an Ag(111) surface (Figure 12g) [183] have been reported. In such cases, the formation of these homochiral domains of achiral/prochiral molecules on achiral surfaces occurs with the same probabilities.

S.M. Barlow and R. Raval suggested that surface-induced chiral systems of achiral or prochiral molecules can be classified into two categories: (1) adsorption-induced chiral motifs (point chirality of individual molecules) and (2) adsorption-induced chirally ordered domains (organizational chirality) [184]. There are 10 crystallographic 2D point groups, compare to the 32 crystallographic 3D point groups. Among which, 5 chiral 2D point groups are permitted. In the case of organizationally chiral systems, among 17 possible 2D space groups compared to the 230 crystallographic 3D space groups, 5 of these 2D space groups are chiral [164b]. Therefore, the selection of various adsorbing molecules, solid materials, and crystal faces of solid surface may provide a diversity of chiral surfaces.

1.2.2.3 *Grafting of chiral molecules onto a metal surface*

Chirality can be imparted to achiral surfaces by covalently grafting chiral modifiers [185]. When a chiral molecule interacts strongly with an achiral flat metal surface the interaction could in principle lead to a reconstruction of the metal surface. In general, the binding of adsorbates will induce a restructuring of the surface only if the gain in the adsorption energy of the molecule on the disrupted surface, compared to adsorption on the non-restructured surface, overcompensates the energy required to break the metal bonds. On oxides, chiral molecules can bind with reactive sites on the surface, typically with terminal hydroxo groups. This approach has been widely used for the development of stationary phases of enantioselective chromatography columns [147, 186], as well as for anchoring biomaterials such as RNA strands, enzymes, and proteins, on surfaces in order to develop biosensors, bioassays, and nanoparticle carriers [187]. Furthermore, this approach has also been widely used to create chiral catalysts by immobilizing homogeneous chiral catalysts on solid oxide surfaces [188] in order to design reusable heterogeneous catalysts. In some cases, immobilized catalysts may exhibit higher enantioselectivity than their free counterparts [188b]. If the molecule used as a chiral modifier is complex enough, it is possible that it creates a chiral environment where the reactant can anchor and be chemically converted enantioselectively [138, 166a, 166c].

1.2.2.4 *Molecular imprinting by using chiral molecules as templates*

To encode chiral information in metals, imprinting with chiral molecules as the template is one of the most promising strategies to create and retain chiral structures or chirally imprinted cavities inside bulk metals even after removal of the templates [189]. It is well known that there is a dynamic relationship between the atomistic structure of a metal surface and molecules adsorbed on that surface [190]. Adsorbates can induce a variety of structural changes on the surfaces of metallic substrates. When the adsorbate is chiral, the atoms on the surface of metal substrate can be restructured to give chiral surfaces. This is the essence of chiral imprinting; transmission of molecular chirality to the structure of a metal surface [191].

One can imagine that once a chiral imprinting agent has induced surface reconstruction, it could, in principle, be removed from the surface, leaving an intrinsically chiral and clean metal surface giving access to the new chiral or prochiral adsorbates to directly interact with it. However, in most of the observed cases of chiral imprinting, the organic imprinting agent remains adsorbed on the imprinted metal surface limiting the direct contact of other adsorbing species with the intrinsically chiral metal surface and, in this regard, the imprinted surface is similar to that of a chirally modified achiral metal surface. The chirality of the metal surface under the adsorbed chiral imprinting agent could still amplify the enantiospecific interactions of chiral adsorbates with the imprinted surface [191a]. Therefore, one of the interesting challenges of chiral imprinting as a means of preparing intrinsically chiral materials is the displacement of chiral imprinting agents, while keeping the chiral nature of the surface, which is a difficult task. Indeed, the interaction between the surface and the adsorbate can be quite strong. One possible method to remove the chiral adsorbate could be by thermal decomposition. However, this usually brings contamination onto the surface. In addition, heating of the imprinted metal surface typically results in lifting or loss of the chiral reconstruction to regenerate its originally achiral structure. The ideal process for removal of a chiral imprinting agent would be to leave the surface in its chirally imprinted, high-Miller-index state without contamination. This probably requires a low-temperature process. For example, exposure of the imprinted surface to an adsorbate with a heat (energy) of adsorption that is greater than that of the chiral imprinting reagent could displace the imprinting reagent into the gas phase [151b, 151c]. The ideal displacing adsorbate after the replacement of the imprinting agent should then have a weak barrier to decomposition into fragments, which will readily desorb from the surface.

Another approach to imprint chiral information on a metal surface is based on the chemical reduction of a metal-ligand complex in order to create organic-doped metals. For example, hybrid metal/organic imprinted materials have been successfully prepared at room temperature using chemical reduction in the presence of the organic molecules which serve as templates while entrapped within the metal matrix [192]. Several chiral metal structures have been obtained in this way [193], retaining the chiral information even after removal of the template [189, 194]. Mesoporous core-shell Pd@Pt bimetallic nanoparticles have also been successfully synthesized by the chemical reduction approach. The resulting materials exhibit a very high specific surface area with no aggregation problem, as is often the case with metal nanoparticles, as well as a high potential for the enantioselective recognition of the corresponding chiral compounds [195]. Chirally encoded cavities on metals and metal oxides have been successfully generated in CuO films deposited onto several achiral metal surfaces, for example, Au(001) and Cu(111), by using tartrate enantiomers as chiral templates [196]. The stereochemical information is maintained in the CuO film even after the removal of the chiral molecules. It was found that CuO films imprinted with (+)-tartaric acid and (–)-tartaric acid show opposite configurations, having a $(1\bar{1}\bar{1})$ orientation with an enantiomeric excess of 95% in the first case, and the latter one exhibiting a $(\bar{1}11)$ orientation with an enantiomeric excess of 93% [196a].

It has been shown that the combination of encoding molecular structures and mesoporous features has distinct advantages, such as a significantly improved enantioselectivity due to an extremely high active surface [195]. This new concept has been successfully applied to various types of chiral imprinted cavities using several chiral template molecules.

1.3 Chiral Nanocrystals and Nanoparticles

In the previous sections, we have looked at the generation and expression of chirality in 3D bulk materials and on the solid surfaces. Here, we will discuss how chirality can be induced in semiconductor or metallic nanocrystals. This approach has attracted a lot of attention due to the excellent physical and chemical properties of the nano-objects. There are various ways to obtain chiral signals from nanocrystals (NCs) or nanoparticles (NPs). C. Gautier and T. Burgi [197] have described three categories of chiral metal NPs: (a) chiral ligands are used to stabilize the NPs creating a chiral "footprint" on the surface of the metal, (b) the optical activity arises from an intrinsically chiral inorganic core shape, and (c) the inorganic core can be achiral and the optical activity is induced by a chiral environment such as a chiral organic shell or chiral electrostatic field. This classification is based on the chiral source of the chiral optical activity: particle structure, environment, and surface, respectively.

This classification was further developed by N.A. Kotov *et al.* [198] in an extensive review describing four types of chemical and physical origins of chirality for NPs based on differences in the formation process: (a) NPs with a chiral surface of the inorganic core, which is normally induced through adsorption of chiral molecules, (b) NPs with a chiral pattern of the surface ligands, (c) the inorganic core of metal of the NP has chiral shape itself, and (d) polarization effects in the inorganic core, which can be induced by chiral molecules or assembly of NPs.

In this section, we will specifically focus on the chirality of NPs induced by chiral ligands or lattices and chirality induced by the chiral shape of individual particle. We will discuss the hierarchical organization of particles in the next section

1.3.1 Chirality Induced by Chiral Ligands

Chiral ligands have been used to endow chirality to nanocrystals of noble metal such as Au nanoclusters [199] and Ag NPs [200]. In 2005 [201], a pair of gold nanocluster enantiomers was first reported. The gold nanoclusters capped with S- or R-penicillamine show mirror image CD spectra, and their dissymmetry g-factors decreased with increasing gold nanocluster size. The mechanism of chirality induction by chiral ligands is supposed to involve chiral footprints or a local chiral distortion of the surface atoms. In order to have a better understanding of the chirality induction, the binding model between chiral ligands and Au NPs has been studied using infrared (IR) and VCD spectroscopy [202]. Thiols and carboxylic acid groups were shown to co-interact with the gold particle leading to the chiral footprint, which is the origin of the observed optical activity.

Semiconductor nanocrystals such as quantum dots (QDs) with chiral ligands also attract a lot of attention due to their excellent absorption and emission properties. In 2007 [203], Gun'ko first reported optically active CdS (QDs) combined with S- or R-penicillamine. Many subsequent reports can be found on the investigation of the mechanism of chirality induction from chiral ligands to QDs. In 2009 [204], Nakashima *et al.* reported the observation of CD signals from CdS, CdSe, or CdTe QDs capped with D- or L-cysteinemethylester hydrochloride. Interestingly, neither CdSe nor CdTe QDs showed CD signals in the range of the first exciton peak (CdSe at 470 nm and CdTe at 540 nm). This indicates that the observed CD signals cannot be attributed to the first excitonic transition in QDs cores but rather to the surface Cd atoms coordinated by the ligands S atoms, and thus the chirality of QDs is supposed to originate from the distorted Cd-S-ligands surface. Furthermore, the chirality of the nanocrystals (NC) surface was maintained even after chiral ligands were exchanged with an achiral thiol, providing a chiral memory effect (Figure 13a).

In 2015 [205], Gun'ko proposed that CdSe/ZnS-based QDs have intrinsic chirality caused by the presence of natural chiral defects, even when they are prepared without using chiral ligands. As shown in Figure 13b, the TEM images of as-prepared QDs show the possible presence of left- and right-handed screw dislocations. The QDs are supposed to form racemic mixtures of left- and right-handed QDs and also non-chiral QDs. They have shown that the racemic mixture of such nanocrystals can be separated using water-soluble chiral ligands mixed with oil-soluble QDs through phase transfer methods.

In 2016 [206], Balaz found that the CD signals could be inversed only by an alteration of the chiral ligand structure. The QDs combined with two

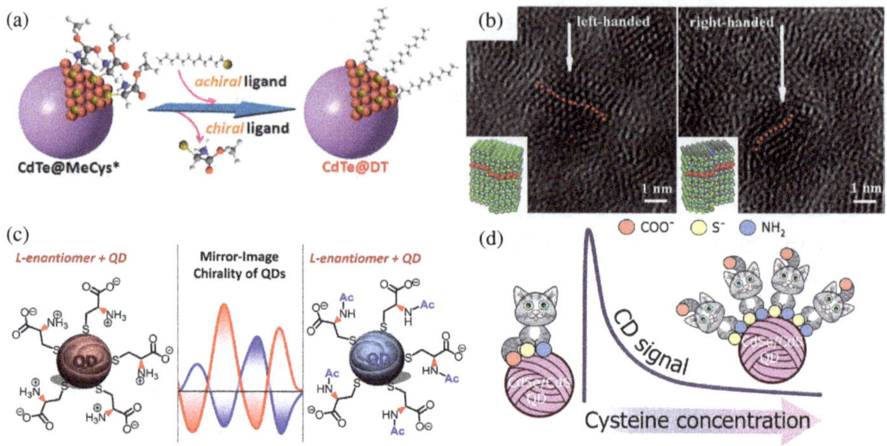

Figure 13. (a) Schematic of the chiral memory effect by ligand-exchange reactions from chiral ligands D- or L-cysteinemethylester hydrochloride (MeCys) to 1-dodecanethiol (DT). (b) TEM images of the CdSe/ZnS QDs. The arrows indicate possible screw dislocations and the red dotted lines indicate the direction of the dislocations. The insert pictures indicate the atomistic models of CdSe/ZnS QDs with right and left screw dislocations, where the dislocations are set in the (010) plane of the CdSe core of nanocrystals with (–) Burgers vector and (+) Burger vector. (c) Same configuration but different binding arrangements of N-acetyl-L-cysteine and L-homocysteine-QDs resulting in mirror CD images. (d) The strong influence of the concentration of a chiral amino acid (cysteine) on its binding modes upon the surface of CdSe/CdS QDs, resulting in varying chiroptical activity and corresponding CD signals.

only slightly different ligands both having the same configuration (L-acetylcysteine vs L-cysteine) configurations show mirror CD images (Figure 13c). It was demonstrated that the CD signals are determined not only by the absolute configuration but also by the binding modes between the chiral ligands and the QD surface.

An inversion of CD signals of HgS QDs upon heating was reported with N-acetyl-L-cysteine [207]. This inversion was explained by Fourier transform infrared (FTIR) spectroscopy and theoretical calculations as being the result of the modification of the coordination between chiral ligands and the surface of the QDs upon heating. Gun'ko *et al.* [208] then reported that the binding mode of cysteine on CdSe/CdS QDs strongly depends on the concentration of the ligands (Figure 13d). With lower cysteine concentrations, the cysteine coordinated to the QD surface in a tridentate mode with enhanced CD signals, while, at higher cysteine concentrations, the tridentate coordination mode changed to a bidentate coordination mode, resulting in a decrease in the CD signals. Indeed, the chiral ligands have a big influence on the chirality induction process.

Beyond the chirality induction based on chiral ligands, the inherent structural properties of nanocrystals such as their size, shape, and crystal structure also show an impact on chirality induction. In 2011, Tang [209] and Markovich [210] reported simultaneously that CdSe QDs combined with chiral ligands show optical activity, which depends on the QD size. As shown in Figure 14d, the dissymmetry g-factor decreased with increasing QD size, which is likely due to the decreasing surface-to-volume ratio. In 2017 [211], Tang reported that CdSe rods with different aspect ratios having L/D-cysteine ligands show the shape dependence of the g-factor, which increases with increasing aspect ratio (sphere to rod), reaching a saturation value with an aspect ratio at about 3.7, based on theoretical calculations. Recently, CdSe/CdS with various morphologies such as quantum rods, nanoflowers, tadpoles, and tetrapods, have also been reported [212]. In 2018 [213], Tang reported that the CdSe nano-platelets having wurtzite (hexagonal) and zincblende (cubic) crystal structures showed different behaviors in terms of chirality induction from their L/D-cysteine ligands. CdSe nano-platelets (NPLs) with a hexagonal crystal structure showed 10 times higher CD spectra compared to those having a cubic crystal structure. This was explained as arising from their different dipole moments and polarization.

In general, bare core QDs suffer lower photoluminescence quantum yield (PLQY) due to their surface traps. QDs with core-shell structure like a CdSe core and a CdS shell could boost the PLQY, which could be widely used in lighting and displays. QDs with a CdSe/CdS core-shell structure showed g-factors and PLQYs which depended on the CdS shell thickness; PLQY increased with CdS shell thickness, while the g-factor decreased due to the increased distance between chiral ligands and CdSe core (Figure 14d) [214].

Recently, perovskite nanocrystals with chiral ligands have been reported to show promising CD and CPL properties [215]. The best

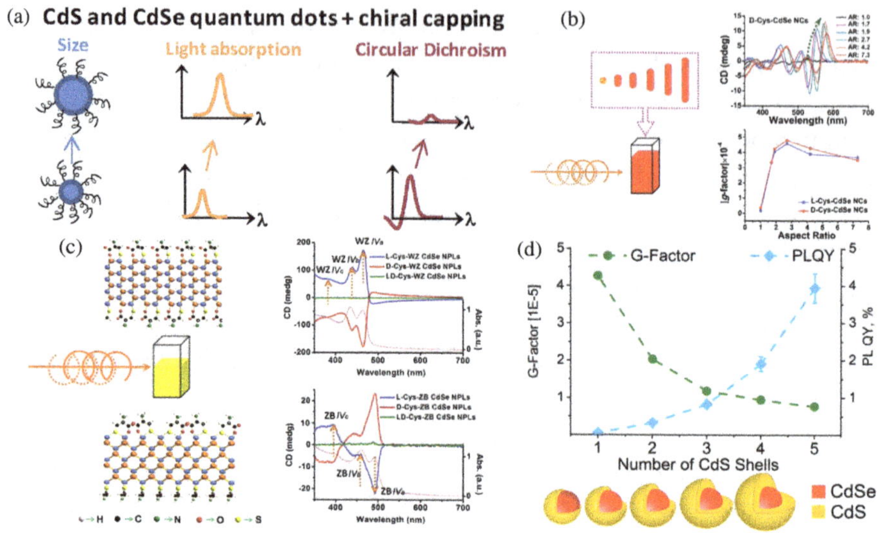

Figure 14. (a) Size effect of the optical activity of CdSe and CdS QDs induced by chiral ligands. The increasing size of QDs results in stronger, red-shifted absorption but lower CD signals. (b) Shape effect of the optical activity of CdSe quantum rods induced by chiral ligands. The dissymmetric g-factor increases with the aspect ratio before reaching a plateau. (c) Crystal structure effect of the optical activity of CdSe nano-platelets (NPLs). The hexagonal crystal structured CdSe NPLs capped with L- or D-cysteine ligands show a distinct CD shape and a much stronger CD signal compared with cubic crystal structured CdSe NPLs. (d) Shell-thickness effect of the optical activity of CdSe/CdS core-shell structured QDs induced by chiral ligands. The increased shell thickness results in enhanced fluorescence but lower induced dissymmetric g-factor.

dissymmetry g-factor approaches 0.02, which is among the highest in the nanoparticle/chiral ligand system.

1.3.2 Chiral Shape of Inorganic Materials

Chiral biomineralized structures in Nature represent an inexhaustible source of inspiration showing hierarchical organization from the nanoscale to the macroscale. Beautiful examples like snail shells, the narwhal tusk, or coccoliths can be observed. Interestingly, the connection between the largely homochiral biomolecules and the larger-scale assemblies and biomineral structures remains largely unknown [216]. A number of studies on bio-inspired minerals or inorganic materials with chiral shapes have been published in the last 20 years, including SiO_2 [217], TiO_2 [218], ZnO [219], CdS [220], Ag [221] films, CuO nanoflowers [222], or glutamic acid [159, 223]. Jiang *et al* have shown that chiral, hierarchically organized architectures for calcium carbonate can be controlled simply by adding aspartic acid. Toroidal suprastructures having a "right-handed" spiral morphology are induced by L-Asp and a "left-handed" morphology is induced by D-Asp. The proposed growth mechanism is based on the tilting of the NPLs due to their binding with chiral amino acids: cooperative tilting of the ensemble of the NPs (or nano-platelets NPLs), amplified over several length scales, creates oriented mineral platelets and chiral vaterite suprastructures (Figure 15) [224].

The structural complexity of such chiral materials has recently been studied with model systems like gold-cysteine platelets. The use of polydisperse platelets has shown that the evolution of multiparticle systems depends on particle symmetry and asymmetry more than on size. Such studies open a pathway to a large family of chiral colloids with complex architectures and chiroptical and chemical properties [224].

In 2014 [225], tellurium and selenium NPs having chiral shapes were synthesized, assisted by chiral thiolated biomolecules, with ECD

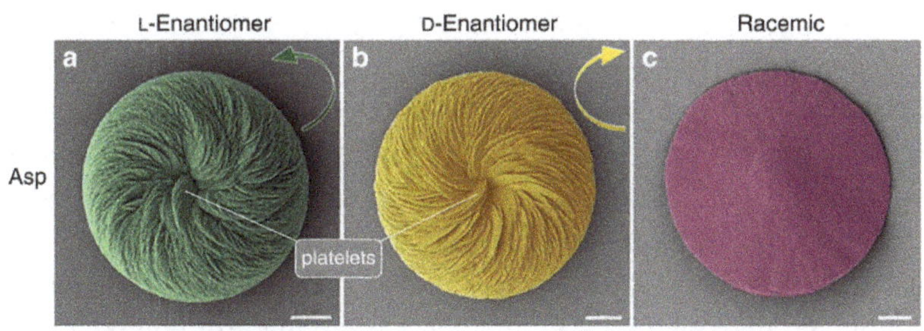

Figure 15. SEM images of calcium carbonate (vaterite) toroids grown in 20 mM L- or D-aspartic acid (a and b, respectively) or in a racemic mixture (c). The L-enantiomer produces right-handed chiral toroids (green), the D-enantiomer left-handed toroids (yellow), and the racemic mixture non-chiral platelets (purple). Scale bars are 6 μm (a and b) and 8 μm (c). Adapted with permission from [224].

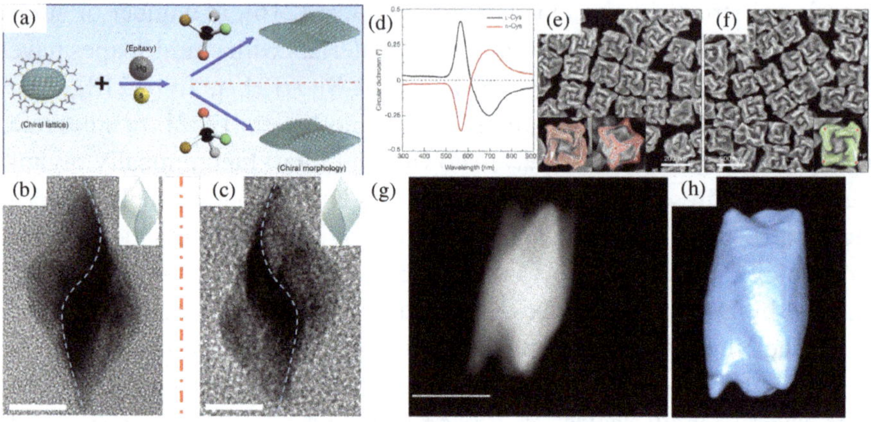

Figure 16. (a) Schematic of the growth process based on the epitaxial principle with involvement of chiral molecules to tailor the morphology chirality. (b) and (c) Typical TEM images of prevailing individual twisted triangular bipyramid nanostructures. Blue-dashed curves are added to guide the eyes for different twisting orientations in the chiral nanostructure. Scale bar, 20 nm. Inset pictures indicate the structural models of two mirrored nanostructures. (d) Circular dichroism spectra of chiral NPLs synthesized using L-cysteine (black) and D-cysteine (red). (e) SEM image of L-cysteine Au NPs. The highlighting in the insets illustrates the fact that the edges (solid lines) are tilted by an angle ϕ with respect to the vertices (red dots) and cubic outline (dashed lines), as viewed along the [110] (left) and [111] (right) directions. (f) SEM image of D-cysteine Au NPs. The inset highlights the tilted edges (solid lines), cubic outline (dashed lines), and tilt angle ($+\phi$). (g) Dark-field STEM image and (h) tomographic reconstruction of the tellurium particle. Scale bar, 100 nm.

dissymmetry g-factors up to 0.03. Figure 16g and h shows the dark-field STEM images and tomographic reconstruction of the tellurium particle, respectively. The chiral tellurium nanostructures were transformed into chiral gold and silver telluride nanostructures with g-factors up to 0.015. In 2017 [226], chiral HgS nanocrystals were synthesized using chiral ligands. Starting from achiral α-HgS nanocrystals as seeds, the successive ion layer adsorption and reaction (SILAR) method was applied for epitaxial growth of HgS, which lead to twisted bipyramid nanocrystals. The prevailing morphology from epitaxial growth is determined by the chiral molecules utilized and is independent of the crystallographic chirality of the seed NPs. Figure 16b and c shows the right- and left-handed twisted bipyramid nanostructures with D- or L-penicillamine molecules as chiral morphology modifiers during the epitaxial growth process, respectively. In 2018 [227], chiral gold NPs, which showed strong optical activity with a large dissymmetry g-factor of 0.2, were synthesized by using chiral amino acids and peptides to control the handedness. As shown in Figure 16d, L- and D-cysteine were used to synthesize right- and left-handed helicoid cubes, which showed mirror CD images. Figure 16e and f shows the SEM images of right-handed and left-handed helicoid cubes, respectively. A number of reports on chiral gold NPs followed [228], and the chiral morphology of palladium NPs was obtained by the same protocol [229].

Inorganic nanocrystals having chiral shape exhibit promising potential applications due to the large g-factors of their CD signals. Designing fluorescent nanocrystals having a chiral shape represents a new challenge with extremely promising applications.

1.4 Chiral Hierarchical Organization of Nanostructures

1.4.1 Organization of Achiral Objects

Organized ensembles of nanostructures can show collective and often enhanced properties, which are different from those displayed by individuals or mixtures of disorganized samples in the bulk. Therefore, the preparation of controlled aligned/organized NPs or molecular assemblies has been the subject of increasing research efforts [230].

In particular, the alignment of anisotropic nanoscale objects is one key prerequisite in materials science when fabricating materials and devices with anisotropic physical properties. For example the absorbance, refractive index, conductivity, and tensile strength are known to differ when measured along different directions in 3D space [231]. Techniques developed for this purpose [232] include the use of evaporation [233], space confinement [234], shear force [235], electric field [236], magnetic field [237], mechanical stretching [238], or a mixture of them [239]. For example, the hierarchical organization via convective evaporation has generated beautifully organized patterned surfaces from non-chiral inorganic anisotropic objects such as nanowires and nanorods [240].

1.4.2 Chiral Organization of Achiral Objects

The above chiral assemblies are mainly related to the chiral source such as chiral ligands or CP light. Other chiral assemblies were also reported without any chiral source. In 2014 [241], Klajn reported the helical self-assembly of magnetite Fe_3O_4 nanocubes at air/liquid interfaces (Figure 17a). These Fe_3O_4 NPs have a cubic shape with average edge length around 13 nm. Without external magnetic fields, the NPs self-assemble into belts due to the shape anisotropy (favoring face-to-face interactions). When an external magnetic field is applied, the magnetite NPs self-assemble into helical superstructures due to the competition between the shape anisotropy and magnetocrystalline anisotropy

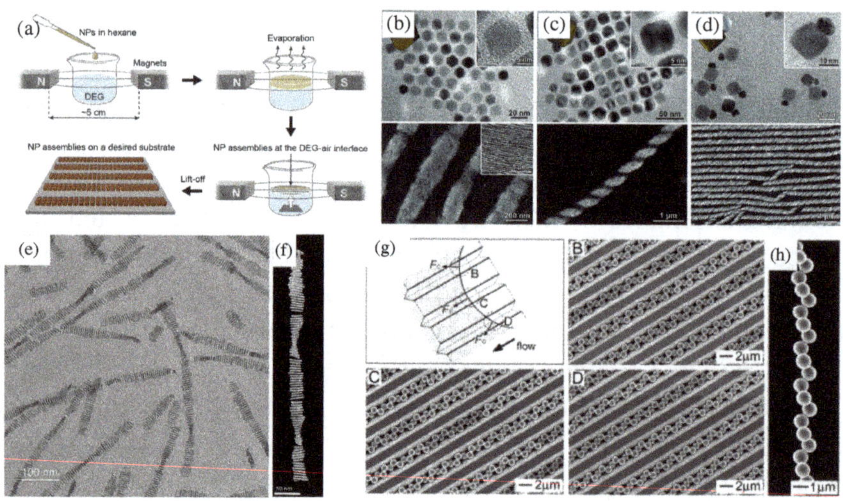

Figure 17. (a) Schematic representation of the experimental setup for self-assembly of magnetic Fe_3O_4 nanocrystals. (b) TEM images of truncated octahedrons (top) and SEM images of 1D belts as they assemble (bottom). (c) TEM images of rounded cubes (top) and an SEM image of the resulting helix (bottom). (d) TEM images of Fe_3O_4-Ag heterodimeric NCs (top) and an SEM image of an ensemble of helices (bottom). (e) TEM image of twisted ribbons of self-assembled CdSe NPLs. (f) 3D model from the tomographic reconstruction of a twisted ribbon. (g) SEM images of helical chains with different chiralities that were obtained with the capillary forces oriented toward different directions relative to the longitudinal axis of the V-grooves. (h) SEM image of a free-standing helical aggregate after it had been welded by thermal annealing and subsequently released from the original V-grooves.

(favoring corner-to-corner interactions). Figure 17b shows how the shape of Fe_3O_4 NPs affects the morphologies of the self-assembly under external magnetic fields, truncated octahedrons self-assemble into belts, while rounded cubes and Fe_3O_4-Ag heterodimeric NCs self-assemble into helical superstructures. These helical self-assemblies also have self-sorting effects and form conglomerate-type mixtures of right-handed and left-handed self-assemblies. Interestingly at high coverage, the entire domain may consist of one-handed self-assemblies, the probability of having one handedness with respect to the other handedness being equal. CdSe nanoplatelets (NPLs) can similarly self-assemble into conglomerate chiral ribbons by adding the achiral ligand oleic acid [242] (Figure 17e).

We also note an interesting report in which Xia *et al.* showed that spherical polystyrene beads with a particle size around several hundred nanometers could organize into helical structures when confined in V-shaped grooves [243] (Figure 17g). The authors reported that a helical organization of these spherical particles was observed when the ratio between the width of the V-grooves and the diameter of the colloids falls between 2.7 and 2.85. The handedness of these helical organizations could be controlled by the relative orientation of the capillary force with respect to the longitudinal axis of the helices, where the capillary force is created by the meniscus of the liquid during the filling of the particles in aqueous suspension into the grooves. The capillary force oriented to the right direction (region B) results in a left-handed helical self-assembly, while the capillary force oriented to the left direction (region D) results in a right-handed helical self-assembly. Meanwhile, when the capillary force is oriented to the middle part (region C), both right- and left-handed helical ribbons could be observed.

Inorganic nanocrystals can also be self-assembled into superstructures by various inter-particle forces, such as van der Waals forces, electrostatic forces, magnetic interactions, molecular surface forces, and entropic effects [244]. Chiral superstructures could also be obtained under certain conditions.

In 2010 [245], Kotov *et al.* showed that CdTe NPs self-assemble into flat ribbons which then transform to twisted ribbons when illuminated by visible light. This chiral self-assembly is driven by the balance between inter-nanoparticle attraction and electrostatic repulsion. However, the self-assembled twisted ribbons are racemic with equal right-handed and left-handed twisted ribbons due to the absence of a chiral source. In order to get nonzero enantiomeric excess (ee) values, right- or left-handed CP light was applied to obtain right- or left-handed twisted ribbons [246]. The CdTe QDs stabilized with the achiral capping agent thioglycolic acid (TGA) could self-assemble into predominantly right- or left-handed twisted ribbons under right or left CP light illumination with an ee value of 0.3 (Figure 18a). Figure 18b and c shows the 3D tomographic reconstruction and SEM images of left- and right-twisted ribbons with around an 800 nm pitch.

Recently, gold NP chiral self-assembly driven by CP light illumination was reported [247]. Subsequently, the nanoparticle chiral self-assembly was optimized to get higher ee values. Interestingly, the CdTe NPs with chiral ligands have a chiral self-sorting effect. The racemic mixture of D- and L-cysteine CdTe assembled structures contained both left- and

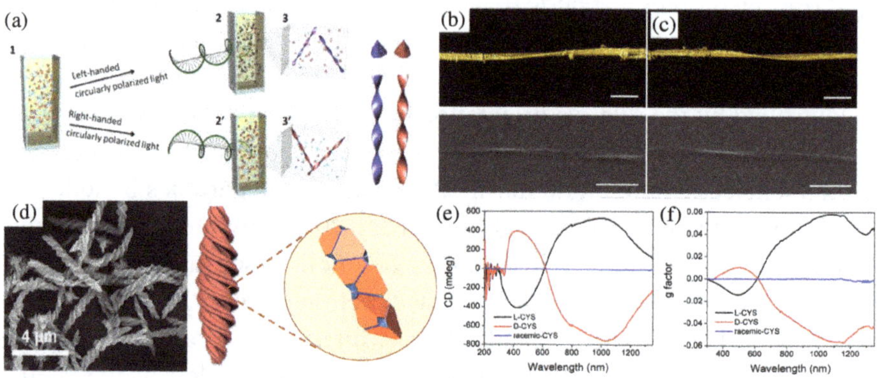

Figure 18. (a) Schematics of the CP light-induced self-assembly process. 1. Racemic mixture of CdTe QDs capped with L- and D-cysteine is prepared. 2. Left-handed CP light selectively activates left-handed NPs. 2′. Right-handed CP light activates right-handed NPs. 3. The excited left-handed NPs are self-assembled into left-handed nanoribbons. 3′. Right-handed NPs are self-assembled into right-handed nanoribbons. (b) and (c) show the surface rendering of 3D TEM tomographic reconstruction and SEM images of single left-handed and right-handed nanoribbons separately. (d) SEM image of CdTe helical superstructure by self-assembly (left) and the origin of enantiopurity in NP assembly into helices (right). (e) and (f) show the CD and g-factor of CdTe helical superstructure self-assembled by L-cys-CdTe NPs, D-cys-CdTe NPs and racemic cys-CdTe NPs.

right-handed helices with no straight ribbons observed (conglomerate-type mixture), that is, the CdTe NPs tend to self-assemble with NPs of the same chirality. The geometry of the helices can be precisely controlled by the solvent [248], pH [249], chiral ligand density, and coordination bridges between NPs [250]. The helical ribbons obtained by CdTe nanocrystal self-assembly can show dissymmetric g-factors as high as 0.06 (Figure 18f).

1.4.3 Organization/Alignment of Chiral Objects

Meanwhile, there are much fewer examples of the alignment of anisotropic chiral particles leading to the design of materials which couple the chirality of the constituent chiral elements and collective properties due to the anisotropic orientation [251].

Self-assemblies of rod-like viruses [252] or cellulose nanocrystals (CNCs) [253] have been used as model systems to design functional materials such as chiral photonic structures. Both are bio-renewable resources that spontaneously organize into chiral liquid crystals with hierarchical structures. Kuncicky *et al.* have shown that a shear-induced alignment can be obtained by pulling a meniscus containing the virus suspension over a substrate for convective nanoparticle assembly.

The periodic, chiral nematic organization of CNC films obtained by self-assembly in water show iridescence and are increasingly used for applications such as cosmetics or photonics. Such chiral organization can also be obtained via controlled evaporation under capillary confinement. Cherpak *et al.* have shown large uniformly aligned chiral photonic films could be obtained from a liquid crystal phase formed by a CNC suspension placed in a thin capillary. The confinement of CNCs during the drying process induced a coexistence of isotropic and chiral phases separated by an interface aligned perpendicular to the long axis of the capillary: the saturation of water vapor in one end of the capillary causes anisotropic drying and promotes unidirectional propagation of the anisotropic phase in large regions that results in chiral CNC solid films with uniformly oriented layered morphology (Figure 19) [254].

Recently, convective evaporation forces have been used in the dip-coating technique, to align synthetic inorganic chiral nanoobjects, namely, silica

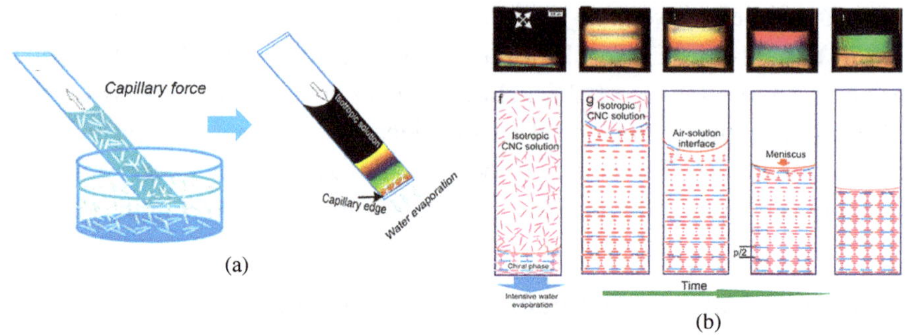

Figure 19. Special confinement process: (a) filling the rectangular capillary with a CNC suspension and immediately transferring to the optical microscope for real-time drying monitoring, and (b) real-time drying monitoring using a polarized optical microscope at 15 min, 120 min, 128 min, 131 min, and 136 min. The polarizer was placed at 45° to the capillary edge. Adapted with permission from Nano Lett. 2018, 18, 11, 6770–6777. Copyright © 2018, American Chemical Society.

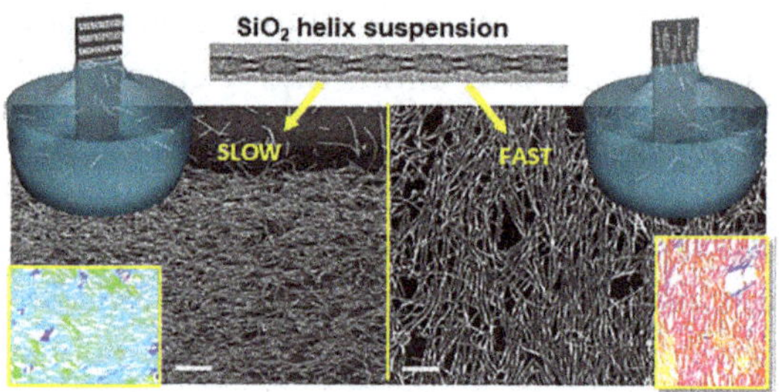

Figure 20. Controlling orientation through withdrawal speed. Left: Horizontal alignment and stick-slip layered organization with low withdrawal speed. Right: Vertical alignment for high withdrawal speed. SEM images. Silica helices orientation measured via Image J giving order parameter S = 0.95 and 0.75, respectively. Scale bars are 1 μm. Adapted with permission from [255].

nanohelices [255]. Nanoscale helices with a large aspect ratio showed an excellent coupling of shear forces in liquid with helical particle motion. The coupling of evaporation forces and physicochemical solution properties induce specific helix alignment, and the stick–slip phenomenon produces a periodical deposition of bands with controllable and regular spacing. (Figure 20).

Multilayered chiral nanostructures can be obtained by top-down lithographic fabrication techniques [256]. While these techniques are powerful in creating organized and highly controlled nanostructures such as arrays and layered plates, they are non-scalable for sizes above 100 nm, and are time-consuming and generally costly. In contrast, bottom-up methods are often faster, less expensive, and more adaptable for arbitrary structural geometries in a larger range of scales [257]. As we have shown previously, chiral films can be made from chirally arranged particles. Among the various deposition methods of particles assemblies, layer-by-layer (LbL) deposition coupled with or without Grazing Incidence Spraying gives an interesting approach for the formation of thin films with hierarchically organized chiral NPs to "design" the spectrum of rotatory optical activity [258].

1.5 Circular Dichroism and Circular Birefringence Measurements in Solid State

The development of spectroscopic techniques (CD, CP luminescence, Raman optical activity, circular photoelectron dichroism, harmonic generation, optical vortices, optofluidic selectivity) and other techniques (X-rays based techniques, mass spectrometry, nuclear magnetic resonance, electron microscopy), is crucial to characterize these new objects, phenomena, and interactions. Among which CD and circular birefringence measurements form powerful analysis methods to quantify the chirality of the systems

1.5.1 Mueller Matrix Formalism

Circular dichroism and birefringence, which are related to the difference between the propagation of left- and right-handed circular polarizations, can be viewed as the fingerprints of chirality. Due to their small magnitudes, their measurements require advanced polarimetric instrumentation. To design an adequate polarimeter, we must evaluate the impact of each optical element on the polarization state of light. In 1948, Hans Mueller developed a matrix formalism to model the evolution of the polarization state of light during its propagation, which is now considered the precursor of modern developments in polarimetry. In addition, the Mueller matrix of a sample contains all the information concerning the sources of dichroism and birefringence. Thus, the measurement of the Mueller matrix is crucial to investigate the optical properties of chiral materials in the solid state. Before introducing the polarimetric set up, we will give a description of the Mueller matrix formalism.

The change of the polarization state of light induced by chiral materials is investigated in the framework of the Mueller matrix formalism [259]. In this formalism, all polarization states of light can be described by the so-called Stokes vector given by:

$$S = \begin{pmatrix} I_{tot} \\ I_X - I_Y \\ I_{45^\circ} - I_{-45^\circ} \\ I_L - I_R \end{pmatrix} \quad (1)$$

Where I_{tot} is the total intensity of light. I_X, I_Y, I_{45°, I_{-45°, I_L and I_R are the intensities measured by an ideal polarizer oriented along two directions, Ox, Oy (perpendicular to the propagation of light direction Oz), the direction at 45° and –45° with respect to the Ox direction, and by left and right circular polarizers, respectively. Thus, the Stokes vector is built with measurable quantities. The Stokes vector, after being normalized by I_{tot}, can be rewritten as:

$$S = \begin{pmatrix} 1 \\ S_1 \\ S_2 \\ S_3 \end{pmatrix} \quad (2)$$

The components of the Stokes vector must respect the following sum rules:

$$1 \geq \sqrt{S_1^2 + S_2^2 + S_3^2} \quad (3)$$

The equality sign in (3) holds only if the light beam is totally polarized. In other words, each polarization state can be represented in 3-dimensional space (S_1, S_2, S_3) limited by a sphere of radius 1, called the Poincaré sphere (Figure 21a). We can define different loci in the Poincaré sphere. Pure polarization states are distributed on the surface of the Poincaré sphere while partially polarized states are located inside. All linear polarization states are located on the equator. The north and the south poles are related to the right and left polarization states, respectively.

Two orthogonal polarizations are centrosymmetrically distributed in this space.

The transmission of light through an anisotropic medium induces a change of polarization state. In other words, we can define two Stokes vectors S_{in} and S_{out}, related to the incident and transmitted polarization, respectively (Figure 21b). These Stokes vectors are linked together by a linear equation:

$$S_{out} = MS_{in} \tag{4}$$

where M is a 4×4 matrix called the Mueller matrix. Its 16 real elements (m_{ij}) connect the input and output Stokes vectors after the interaction of light with an optical medium. In other words, the Mueller matrix can be represented as a transformation operating on the Stokes vectors in the Poincaré sphere. The Mueller matrix contains all information concerning the anisotropy of the medium. Applying the condition (3) to the Mueller matrix, we can introduce a parameter P, which quantifies the degree of polarization of light:

$$P = \frac{\sqrt{Tr\left(M^T M\right) - m_0^2}}{\sqrt{3}m_0} \tag{5}$$

m_{00} is the first element of the Mueller matrix M. The value of P varies from 0 (for non-depolarizing medium) to 1 (for completely depolarizing medium). The Mueller matrix, which describes the propagation of light through a series of optical elements, is equal to the product of the Mueller matrix of each element (M_i) (Figure 21c). If we focus our attention on the incident and the emergent polarization, the overall effect of an entire cascade of N optical elements is described by:

$$S_{out} = M_N M_{N-1} \ldots M_2 M_1 S_{in} = MS_{in} \text{ with } M = \prod_{i=1}^{N} M_i \tag{6}$$

Thus, the Mueller matrix formalism is a suitable tool to describe the evolution of the polarization state of light during the propagation through an optical setup. We can now define the Mueller matrix of usual optical elements such as a linear polarizer (M_P) oriented along the Ox direction,

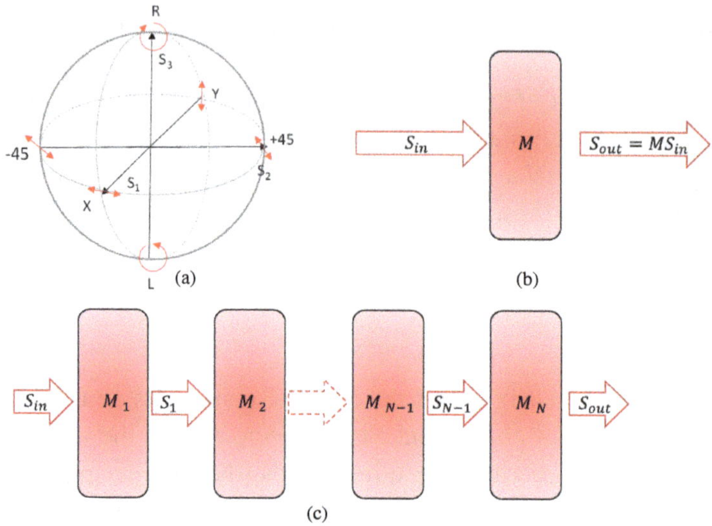

Figure 21. (a) Representation of polarization state of light in the Poincaré sphere. Change of the polarization state after the propagation though (b) a medium and (c) a stack of N media.

and a perfect retarder (M_r) whose fast axis is oriented along the Ox direction:

$$M_p = \frac{1}{2}\begin{pmatrix} 1 & 1 & 0 & 0 \\ 1 & 1 & 0 & 0 \\ 0 & 0 & 0 & 0 \\ 0 & 0 & 0 & 0 \end{pmatrix}, \quad \text{and} \quad M_r = \begin{pmatrix} 1 & 0 & 0 & 0 \\ 0 & 1 & 0 & 0 \\ 0 & 0 & \cos\delta & \sin\delta \\ 0 & 0 & -\sin\delta & \cos\delta \end{pmatrix} \tag{7}$$

Where δ is the phase difference between the fast and slow axis. The rotation of optical elements by an angle θ induces a modification of its Mueller matrix. The Mueller matrix obtained after the rotation M_{rot} is expressed as:

$$M_{rot} = R(-\theta)MR(\theta) \tag{8}$$

with $R(\theta) = \begin{pmatrix} 1 & 0 & 0 & 0 \\ 0 & \cos(2\theta) & \sin(2\theta) & 0 \\ 0 & -\sin(2\theta) & \cos(2\theta) & 0 \\ 0 & 0 & 0 & 1 \end{pmatrix}$ the rotation matrix.

Thus, the Mueller matrix formalism is a powerful tool to describe the change of polarization state of light induced by an optical setup.

1.5.2 Differential Decomposition

Mueller matrices can be phenomenologically interpreted by decomposing them into simple polarimetric parameters. Several decompositions have been proposed in the literature [260]. Differential decomposition is a powerful tool when the light is transmitted by a homogeneous medium. By considering the linear differential propagation equation, the Mueller matrix M of a medium with an optical thickness l is related to a differential matrix:

$$\frac{dM(l)}{dl} = mM(l) \tag{9}$$

By solving equation 9, we find:

$$M(z) = \exp(ml) = \exp(L) \tag{10}$$

L is a 4×4 matrix, which can be decomposed into a sum of a depolarizing matrix (L_u) and a non-depolarizing matrix (L_m) [260a]:

$$L = L_m + L_u, \qquad \text{with} \begin{cases} L_m = \frac{1}{2}\left(L - GL^T G\right) \\ L_u = \frac{1}{2}\left(L + GL^T G\right) \end{cases} \tag{11}$$

where G = diag(1,–1,–1,–1) is the Minkowski metric. The matrix L_m can be expressed as:

$$L_m = \begin{pmatrix} 0 & -LD & -LD' & CD \\ -LD & 0 & CB & LB' \\ -LD' & -CB & 0 & -LB \\ CD & -LB' & LB & 0 \end{pmatrix} \tag{12}$$

Where LD and LD′ are linear dichroisms along Ox-Oy and ±45° axes, LB and LB′ are linear birefringences along Ox-Oy and ±45° axes respectively, CD is circular dichroism and CB is circular birefringence. These polarimetric parameters are related to the complex refractive indices of the medium:

$$LB = 2\pi\bar{\nu}l(n_x - n_y) \tag{13}$$

$$LD = 2\pi\bar{\nu}l(k_x - k_y) = \frac{\ln(10)}{2}(A_x - A_y) \tag{14}$$

$$LB' = 2\pi\bar{\nu}l(n_{45} - n_{-45}) \tag{15}$$

$$LD' = 2\pi\bar{\nu}l(k_{45} - k_{-45}) = \frac{\ln(10)}{2}(A_{45} - A_{-45}) \tag{16}$$

$$CB = 2\pi\bar{\nu}l(n_L - n_R) \tag{17}$$

$$CD = 2\pi\bar{\nu}l(k_L - k_R) = \frac{\ln(10)}{2}(A_L - A_R) \tag{18}$$

Where $\bar{\nu}$ is the wavenumber, n_i is the real part of the refractive index, k_i is the imaginary part of the refractive index and A_i ($i = x, y, 45, -45$, L or R) is the absorbance of the medium for light linearly polarized along the Ox, Oy, +45° and −45° axes, and for left- and right-handed circular polarizations, respectively. For non-depolarizing chiral materials, which have a rotation symmetry, the linear dichroism and birefringence are equal to 0. According to equations 10 and 12, the Mueller matrix of this material through a first order approximation is given by:

$$M = \begin{pmatrix} m_{00} & 0 & 0 & m_{03} \\ 0 & m_{11} & m_{12} & 0 \\ 0 & -m_{12} & m_{11} & 0 \\ m_{03} & 0 & 0 & m_{00} \end{pmatrix} = \begin{pmatrix} 1 & 0 & 0 & CD \\ 0 & 1 & CB & 0 \\ 0 & -CB & 1 & 0 \\ CD & 0 & 0 & 1 \end{pmatrix} \tag{19}$$

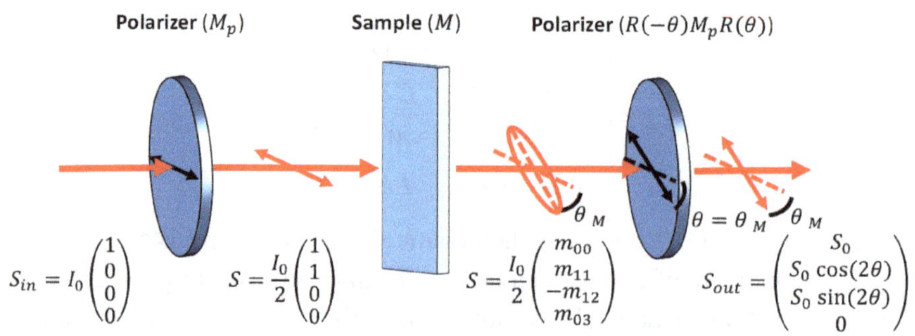

Figure 22. Measurement of optical rotation induced by an active medium.

1.5.3 Determination of CB from the Optical Rotation

According to the definition of CB (equation 17), right and left circular polarizations propagate inside an optically active medium at different speeds. In other words, the phase relationships between the two circularly polarizing waves changes. As linear polarization is a superposition of a right and left circular polarization, the CB induces a rotation of the linearly polarized wave. Before the advent of modern CD polarimeters, the optical activity of chiral molecules was commonly determined by measuring this optical rotation (OR). As illustrated in Figure 22, the simplest OR set up is composed of two linear polarizers, with the sample placed between them. The Stokes vector (S_{out}) of the transmitted light through the optical set up is given by:

$$S_{out} = \begin{pmatrix} S_0 \\ S_1 \\ S_2 \\ S_3 \end{pmatrix} = R(-\theta)M_p R(\theta) M M_p \begin{pmatrix} 1 \\ 0 \\ 0 \\ 0 \end{pmatrix} I_0 \tag{20}$$

By assuming that the chiral material does not exhibit depolarization as well as linear dichroism and birefringence, its Mueller matrix (M) is given by equation 19. The intensity of light measured by the detector at a wavenumber $\bar{\nu}$ is given by the first element of the transmitted Stokes vector (S_0):

$$S_0(\bar{\nu}) = \frac{I_0}{4}(1 + \cos(2\theta) - CB(\bar{\nu})\sin(2\theta)) \tag{21}$$

The second polarizer is rotated to find the angle θ_M, which maximizes the intensity of light measured by the optical sensor. Then, the CB at a given wavenumber $\bar{\nu}$ is directly given by: $CB(\bar{\nu}) = -2\theta_M$

Large amounts of chiral materials are optically active, that is, they exhibit CD and CB. $CB(\bar{\nu})$ *and* $CD(\bar{\nu})$ are linked together by the Kramers–Kronig relations:

$$CD(\bar{\nu}_a) = -\frac{2\bar{\nu}_a}{\pi} P\int_0^{\infty} \frac{CB(\bar{\nu})}{\bar{\nu}^2 - \bar{\nu}_a^2} d\bar{\nu} \tag{22}$$

$$CB(\bar{\nu}_a) = \frac{2}{\pi} P \int_0^{\infty} \frac{\bar{\nu} \cdot CD(\bar{\nu})}{\bar{\nu}^2 - \bar{\nu}_a^2} d\bar{\nu}. \tag{23}$$

We can conclude that $CB(\bar{\nu})$ and $CD(\bar{\nu})$ bring the same information. However, the use of the Kramers–Kronig relation requires knowledge of the CD or the CB at all wavenumbers. As the spectral range of CB measurements is always limited, $CD(\bar{\nu})$ cannot be easily estimated from $CB(\bar{\nu})$. Therefore, it is suitable to use an experimental set up which measures simultaneously the $CD(\bar{\nu})$ and $CB(\bar{\nu})$ spectra.

1.5.4 Determination of Circular Dichroism (CD) and Circular Birefringence (CB) by a UV-visible or Infrared Spectrometer

The CD of an isotropic non-depolarizing medium is related to the difference between the absorption coefficient of left and right circular polarizations. However, the CD of natural materials, which is in the order of 10^{-5}–10^{-4}, is too small to be directly measured from polarized absorption spectroscopy. Most modern instruments are based on a polarization–modulation approach. As shown in Figure 23a, the light emitted by the source passes through a cascade of optical compounds: a polarizer oriented at 45°, a photoelastic modulator (PEM) and the sample. Typical PEMs consist of a piezoelectric transducer mounted to the sides of a fused silica (UV-Visible) or a ZnSe (infrared) crystal. A voltage applied to the transducer creates strains, and so a linear birefringence in the silica (or ZnSe) crystal. The PEM is a perfect retarder whose Mueller matrix M_r is defined by equation 7. The phase difference δ between the fast and slow axis depends on the applied voltage. Thus, the following variable retardation can be created by application of an adequate voltage:

$$\delta(t) = \delta_M(\bar{\nu}) \sin(2\pi f t) \tag{24}$$

where $\delta_M(\bar{\nu})$ is the maximum value of the sine-wave retardation angle at the wavenumber $\bar{\nu}$. This value is adjusted for a given wavenumber $\bar{\nu}$ so as to act as a quarter-wave plate. For other wavenumbers, an efficiency function should be considered ($J_1(\delta_M(\bar{\nu}))$ described below). Generally, the

PEM operates at a frequency of 50 kHz. The Stokes vector (S_{out}) of the transmitted light through the optical set up is given by:

$$\begin{pmatrix} S_0 \\ S_1 \\ S_2 \\ S_3 \end{pmatrix} = MM_r R(-45°) M_p R(45°) \begin{pmatrix} 1 \\ 0 \\ 0 \\ 0 \end{pmatrix} I_0 \qquad (25)$$

The detected intensity is given by S_0:

$$S_0(\bar{\nu},t) = \frac{1}{2} I_0 \left(m_0 + m_{02} \cos\delta(t) - m_{03} \sin\delta(t) \right) \qquad (26)$$

According to equation 19, the detected intensity for a non-depolarizing isotropic chiral material is given by:

$$S_0(\bar{\nu},t) = \frac{1}{2} I_0 (1 - CD(\bar{\nu}) \sin\delta(t)) \qquad (27)$$

The CD appears in equation 27 as the amplitude of the periodic function $sin\delta(t) = \sin(\delta_M(\bar{\nu})\sin(2\pi ft))$. The full sine-wave dependence of the detected intensity is given by:

$$\sin(\delta(t)) = \sin(\delta_M(\bar{\nu})\sin(2\pi ft)) = \sum_{n=odd} 2J_n(\delta_M(\bar{\nu}))\sin(2\pi nft) \qquad (28)$$

where the second equality is a sum over odd-order Bessel functions, J_n, at the odd harmonics of the PEM frequency nf, where n in the summation is equal to only integers starting at 1. The first term in the odd harmonic expansion above is the main CD signal at the detector. This signal, at the fundamental PEM frequency f, is measured by a lock-in amplifier tuned to f as the amplitude of $sin(2\pi ft)$, and is equal to $2J_1(\delta_M(\bar{\nu})\cdot CD(\bar{\nu})$. The measurement of CD with a lock-in amplifier considerably improves the signal-to-noise ratio. This strategy makes CD spectroscopy a routine tool for the determination of the chirality of materials. However, a calibration procedure is necessary to determine the $J\cdot(\delta_M(\bar{\nu}))$ function in Fourier transform infrared spectroscopy [261].

To measure simultaneously the CB and the CD, a second polarizer, oriented along the Ox direction, can be added after the sample [262]. The corresponding experimental set up is depicted in Figure 23b. The transmitted Stokes vector is given by:

$$S_{out} = \begin{pmatrix} S_0 \\ S_1 \\ S_2 \\ S_3 \end{pmatrix} = M_p M M_r R(-45°) M_p R(45°) \begin{pmatrix} 1 \\ 0 \\ 0 \\ 0 \end{pmatrix} I_0, \tag{29}$$

The variation of the detected light intensity is given by S_0:

$$S_0(\bar{\nu},t) = \frac{I_0}{4}\left(m_{00} + m_{10} + (m_{02} + m_{12})\cos\delta\,(t) - (m_{03} + m_{13})\sin\delta\,(t)\right) \tag{30}$$

This equation can be considerably simplified for a non-depolarizing isotropic chiral material:

$$S_0(\bar{\nu},t) = \frac{I_0}{4}\left(1 + CB(\bar{\nu})\cos\delta\,(t) - CD(\bar{\nu})\sin\delta\,(t)\right) \tag{31}$$

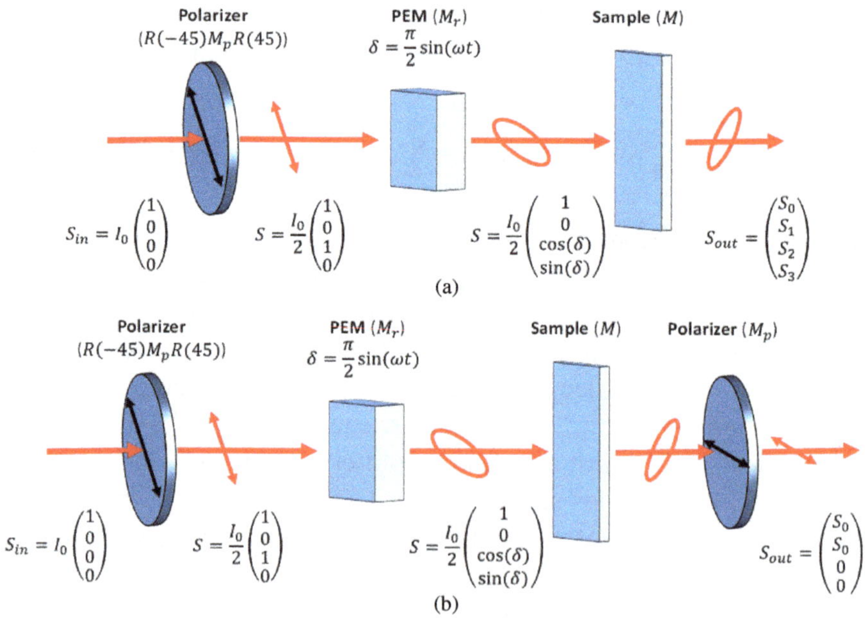

Figure 23. Experimental set up used to measure (a) the CD and (b) the CB.

On the other hand, the CB is related to the amplitude of a periodic function $\cos\delta(t) = \cos(\delta_M(\bar{\nu})\ \sin(2\pi ft))$, which has a frequency *2f*. Thus, $CD(\bar{\nu})$ and $CB(\bar{\nu})$ spectra can be determined by filtering the signal at the frequency *f* and *2f*, respectively.

Since $\cos(\delta(t)) = \cos(\delta_M(\bar{\nu})\ \sin(2\pi ft)) = J_0(\delta_M(\bar{\nu})) + \Sigma_{n=even} 2J_n(\delta_M(\bar{\nu}))$ $\cos(2\pi nft)$, the CB signal measured by a lock-in amplifier tuned to 2f is equal to $2J_2(\delta_M(\bar{\nu}) \cdot \mathrm{CB}(\bar{\nu})$

Despite their high sensitivity, these instruments have several limitations [263]. It can only be used to determine the $CD(\bar{\nu})$ and $CB(\bar{\nu})$ and spectra of a non-depolarizing isotropic sample. However, a large number of natural and artificial chiral materials present anisotropy, particularly in the solid state. These materials thus simultaneously exhibit an optical activity and a linear anisotropy (i.e., linear birefringence (LB) and linear dichroism (LD)). In this case, the Mueller matrix cannot be simply written as equation 19 and the m_{03} and m_{12} elements, determined from the instruments shown in Figure 3, are a complex combination of LB, LB′, LD, LD′, CD, and CB. In other words, the presence of linear dichroism and linear birefringence create artifacts in the measured CD and CB.

1.5.5 Determination of Circular Dichroism (CD) for Samples Exhibiting Linear Birefringence and Linear Dichroism

To determine the measured CD of a chiral material exhibiting LB and LD in the plane of the sample, it is necessary to take into account several imperfections of the optical set up described in Figure 22a, such as the residual static birefringence of the PEM, δ_0, and the different responses of the detector along the *x* and *y* axes. The residual static birefringence of the PEM modifies the Mueller matrix M_r as:

$$M_r = \begin{pmatrix} 1 & 0 & 0 & 0 \\ 0 & 1 & 0 & 0 \\ 0 & 0 & cos(\delta+\delta_0) & sin(\delta+\delta_0) \\ 0 & 0 & -sin(\delta+\delta_0) & cos(\delta+\delta_0) \end{pmatrix}, \tag{32}$$

An additional Mueller matrix is used for the non-perfect detector, $D(\alpha)$, which is given by :

$$D(\alpha)=\left(\left(p_x^2+p_y^2\right) \quad \left(p_x^2-p_y^2\right)cos2\alpha \quad \left(p_x^2-p_y^2\right)sin2\alpha \quad 0\right) \tag{33}$$

where p_x and p_y are the different responses of the detector to radiation polarized along its local x and y axes, respectively, and where these local axes are oriented at an angle α with respect to the Ox axis.

Considering that the maximum linear dichroism is along the Ox direction, referred by the angle $\theta = 0°$ (the θ angle stands for the rotation of the sample around the Oz axis corresponding to the direction of the radiation), the 45° linear birefringence (LB′) and linear dichroism (LD′) are equal to zero in the general Mueller matrix of the sample. Moreover, assuming that (CD, CB) << (LD, LB) << 1 and that sin δ_0 and $(p_x^2 - p_y^2)$ << 1.0, the CD signal measured by a lock-in amplifier tuned to *f* is equal to [264]:

$$2J_1\left(\delta_M\left(\bar{\nu}\right)\right)\Bigg[\left(CD(\bar{\nu})\cos\delta_0 - LD(\bar{\nu})\cos 2\theta\sin\delta_0\right) + \left(\frac{p_x^2-p_y^2}{p_x^2+p_y^2}\right)LB(\bar{\nu})\cos\delta_0\left(\sin 2\theta\cos 2\alpha - \cos 2\theta\sin 2\alpha\right)\Bigg] \tag{34}$$

Equation 34 shows that the measured $CD(\bar{\nu})$ spectrum contains not only the CD of the sample but also linear dichroism and birefringence contributions. The LD contribution is connected to the residual static birefringence introduced by the PEM, whereas the LB contribution is related to polarization imperfections of the detector. This means that it is important to measure LD as well as LB when looking for the true CD spectrum of a sample possessing macroscopic anisotropy. Also, even for achiral samples (i.e., CD=0) a false $CD(\bar{\nu})$ spectrum can be measured when the sample possesses large LD and LB contributions. Generally, the value of $sin\delta_0$ is around 0.001, and consequently the $cos\delta_0$ term is close to 1.

A detailed examination of equation 34 shows that the LD and LB contributions are involved in "$cos2\theta$" and "$sin2\theta$"periodic functions of the sample rotation. Thus, the half-sum of two experimental $CD(\bar{\nu})$ spectra for sample orientations at θ and $\theta \pm 90°$ represents a rapid and simple way to extract the true CD contribution:

$$CD_{true} = \frac{CD(\theta) + CD(\theta \pm 90°)}{2} \quad (35)$$

1.5.6 Mueller matrix Polarimetry

The complete Mueller matrix is required to investigate the optical activity of complex chiral materials. Its determination is achieved by a Mueller matrix polarimeter illustrated in Figure 24. The sample is sandwiched between a polarization state generator (PSG) and a polarizer state analyzer (PSA). Several configurations of PSGs and PSAs are proposed in the literature. The most common are listed in Table 1. The PSG generates a set of input Stokes vectors S_{PSG}, which are transformed by the sample into $S_s = MS_{PSG}$. This Stokes vector is then analyzed by the PSA, which delivers a Stokes vector S_{out} given by:

$$S_{out} = \begin{pmatrix} S_0 \\ S_1 \\ S_2 \\ S_3 \end{pmatrix} = M_{PSA} M M_{PSG} \begin{pmatrix} 1 \\ 0 \\ 0 \\ 0 \end{pmatrix} I_0 \quad (36)$$

where M, M_{PSG}, and M_{PSA} are the Mueller matrices of the sample, the PSG and the PSA. The first element of S_{out} is the intensity of light measured by the detector:

$$S_0 = \begin{pmatrix} m_{00,PSA} & m_{01,PSA} & m_{02,PSA} & m_{03,PSA} \end{pmatrix} \begin{pmatrix} m_{00} & m_{01} & m_{02} & m_{03} \\ m_{10} & m_{11} & m_{12} & m_{13} \\ m_{20} & m_{21} & m_{22} & m_{22} \\ m_{30} & m_{31} & m_{32} & m_{33} \end{pmatrix} \begin{pmatrix} m_{00,PSG} \\ m_{10,PSG} \\ m_{20,PSG} \\ m_{30,PSG} \end{pmatrix}. \quad (37)$$

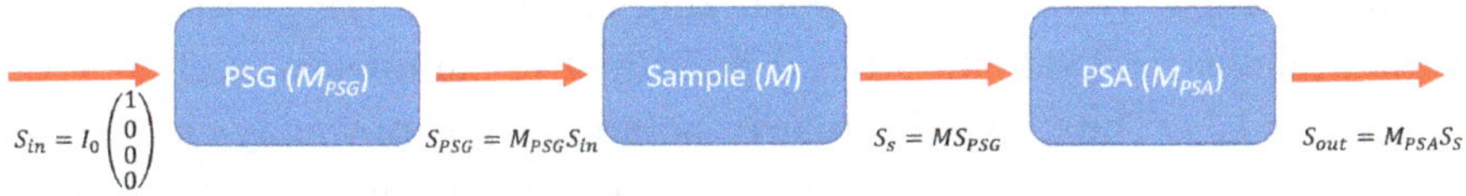

Figure 24. Schematic representation of a Mueller matrix polarimeter.

Table 1. Common structures of PSAs and PSGs.

Configuration	Characteristics	Drawback
A polarizer + a rotating quarter-wave plate (QWP) [265]	• Fixed retardation rotation to change the azimuth of the QWP	• QWP are not completely achromatic • Small modulation frequency due to the mechanical rotation
A polarizer + two liquid crystal cells (LCs) [266]	• Variable retardation for nematic LC • Variable azimuth for ferroelectric LC	• Cannot be used in UV • Temperature dependence • No frequency domain analysis
A polarizer + a rotating PEM [267]	• Fast modulation • Variable retardation • Broad spectral range	• Mechanical rotation • Temperature dependence
A polarizer+ two PEMs [268]	• Fast modulation • Variable retardation • Fixed azimuth • Broad spectral range • No mechanical rotation	• Temperature dependence

m_{ij}, $m_{ij,PSG}$ and $m_{ij,PSA}$, with $i,j = 0,1,2,3$, are the elements of the Mueller matrix of the sample, the PSG and the PSA. Basically, there are two approaches to measure the Mueller matrix of the sample. The first one consists in changing the parameters $m_{ij,PSG}$ and $m_{ij,PSA}$ of the PSG and the PSA at different frequencies. A Fourier transform is performed on the variation of measured intensity. The Mueller matrix elements are then deduced from the amplitudes of all relevant frequency harmonics. In the second approach, several intensity measurements are performed for different configurations of PSG and PSA. According to equation 37, a combination of 4 PSA configurations and 4 PSG configurations are required to obtain all the Mueller matrix elements of the sample.

The differential decomposition of the complete Mueller matrix offers the unique capacity to separate polarizing and depolarizing effects. This method allows the evaluation of all polarimetric parameters such as CD, CB, LB, or LD. Thus, Mueller matrix polarimetry is nowadays an essential tool for investigating the change of polarization of light induced by biological [269] or artificial materials [258b, 270].

References

1. Bonner, W.A. (1991) *Origins Life Evol. Biosphere*, **21**, 59.
2. Ariëns, E.J. (1984) *Eur. J. Clin. Pharmacol.*, **26**, 663
3. Calcaterra, A., *et al.* (2018) *J. Pharm. Biomed. Anal.*, **147**, 323.
4. Francotte, E., *et al.* (2006) *Chirality in Drug Research,* Vol. 33, Wiley-VCH Verlag GmbH & Co.
5. Lin, G.Q., *et al.* (2011) *Chiral Drugs: Chemistry and Biological Action*, John Wiley & Sons, Inc.
6. Lorenz, H., *et al.* (2014) *Angew. Chem. Int. Ed.*, **53**, 1218.
7. Wang, Y., *et al.* (2008) *Org. Process. Res. Dev.*, **12**, 282.
8. Suwannasang, K., *et al.* (2017) *Org. Process. Res. Dev.*, **21**, 623.
9. Bredikhin, A.A., *et al.* (2017) *Chem. Eng. Technol.*, **40**, 1211.
10. J. M. McBride, *et al.* (1991) *Angew. Chem. Int. Ed.*, **30**, 293.
11. Mauskopf, S.H. (1976) *Trans. Am. Philos. Soc.*, **66**, 1.
12. Biot, J.-B. (1817) *Mém. Acad. Sci.*, **2**, 41.
13. Mason, S.F. (1982) *Molecular Optical Activity and the Chiral Discriminations*, Cambridge University Press.
14. Biot, J.-B. (1844) *C. R. Acad. Sci. Paris*, **19**, 719.
15. Pasteur, L. (1848) *Ann. Chim. Phys.*, **24**, 442.
16. Pasteur, L., (1849) *Ann. Chim. Phys. 3e série, tome XXVIII, 56*
17. Pasteur, L. (1852) *Ann. Chim. Phys.*, **34**, 30.
18. Piutti, A. (1886) *Ber. Dtsch. Chem. Ges.*, **19**, 1691.
19. Piutti, A. (1888) *Gazz. Chim. Ital.*, **18**, 457.
20. Gal, J. (2012) *Chirality*, **24**, 959.
21. Rooseboom, H.W.B. (1899) *Z. Physik. Chem.*, **28**, 494.
22. Wallach, O. (1895) *Liebigs Ann. Chem.*, **286**, 90.
23. a) Blazis, V.J., *et al.* (1997) *Acta Crystallogr. B*, **53**, 838; b) Brock, C.P., *et al.* (1991) *J. Am. Chem. Soc.*, **113**, 9811; c) Friščić, T., *et al.* (2008) *Chem. Commun.*, 1644; d) Kitaigorodskii, A.I. (1965) *Acta Crystallogr.*, **18**, 585; e) Wu, X., *et al.*, (2020) *Nat. Commun.*, **11**, 2145.
24. a) Dunitz, J.D., *et al.* (2012) *J. Phys. Chem. B*, **116**, 6740; b) Gavezzotti, A., *et al.* (2014) *J. Org. Chem.*, **79**, 8; c) Jeong, K.S., *et al.* (2009) *Tetrahedron Asymmetry*, **20**, 1736; d) Kennedy, A.R., *et al.* (2011) *Cryst. Growth Des.*, **11**, 1821; e) Patrick, B.O., *et al.* (2006) *Acta Crystallogr. B*, **62**, 488.
25. Jacques, J., *et al.*, (1981) *Enantiomers, Racemates, and Resolutions*, J. Wiley & Sons, Inc.
26. A. Otero-de-la-Roza, *et al.*, (2016). *Cryst. Growth Des.,* 5.
27. a) Diniz, L.F., *et al.* (2019) *Cryst. Growth Des.*, **19**, 4498; b) Huang, J., *et al.* (2006) *J. Am. Chem. Soc.*, **128**, 11985; c) Rekis, T., *et al.* (2018) *Cryst. Growth Des.*, **18**, 264.
28. a) Coquerel, G. (2000) *Enantiomer*, **5**, 481; b) Coquerel, G. (2007) *Novel Optical Resolution Technologies,* K. Sakai, N. Hirayama, and R. Tamura (Eds.), Springer,

pp. 1; c) Coquerel, G., *et al.* (2020) *Symmetry,* **12**, 1796; d) Secor, R.M. (1963) *Chem. Rev*. 63, 297.

29. a) Jacques, J., *et al.* (1981) *Tetrahedron*, **37**, 1727; b) Li, Z.J., *et al.* (1999) *J. Pharm. Sci.*, **88**, 337.
30. Saha, M.K., *et al.* (2002) *Cryst. Growth Des.*, **2**, 205.
31. Flack, H.D. (2003) *Helvetica Chimica Acta*, **86**, 905.
32. Clevers, S., *et al.* (2013) *J. Therm. Anal. Calorimet.*, **112**, 271.
33. Newman, A.C.D., *et al.* (1952) *J. Chem. Soc.*, **3747**.
34. Devogelaer, J.-J., *et al.* (2020) *Angewandte Chemie*, **132**, 21895.
35. Srisanga, S., *et al.* (2010) *Cryst. Growth Des.*, **10**, 1808.
36. Wright, A.F., *et al.* (1981) *J. Solid State Chem.*, **36**, 371.
37. Miyata, M., *et al.* (2015) *Symmetry*, **7**, 1914.
38. Bartus, J., *et al.* (1992) *Polym. Bull.*, **28**, 203.
39. Kipping, F.S., *et al.* (1898) *J. Chem. Soc. Trans.*, **73**, 606.
40. Gernez, D. (1866) *Compt. Rend. Acad. Sci.*, **63**, 843.
41. a) Collet, A., *et al.* (1980) *Chem. Rev.*, **80**, 215; b) Polenske, D., *et al.* (2009) *Chirality*, **21**, 728.
42. Fogassy, E., *et al.* (2006) *Org. Biomol. Chem.*, **4**, 3011.
43. Bolchi, C., *et al.* (2007) *Tetrahedron Asymmetry*, **18**, 1038.
44. a) Binev, D., *et al.* (2016) *Cryst. Growth Des.*, **16**, 1409; b) Dunn, A.S., *et al.* (2020) *Cryst. Growth Des.*, **20**, 7726; c) Eicke, M.J., *et al.* (2013) *Cryst. Growth Des.*, **13**, 1638; d) Galan, K., *et al.* (2015) *Cryst. Growth Des.*, **15**, 1808; e) Levilain, G., *et al.* (2012) *Cryst. Growth Des.*, **12**, 5396.
45. a) Levilain, G., *et al.* (2010) *CrystEngComm*, **12**, 1983; b) Viedma, C., *et al.* (2015) *Handbook of Crystal Growth* (Second Edition) T. Nishinaga (Ed.), Elsevier, pp. 951.
46. a) Jamison, M.M., *et al.* (1942) *J. Chem. Soc.*, 437; b) Oketani, R., *et al.* (2019) *Chem. A Eur. J.*, **25**, 13837.
47. Oketani, R., *et al.* (2019) *Org. Process. Res. Dev.*, **23**, 1197.
48. a) Belletti, G., *et al.* (2020) *Chem. A Eur. J.*, **26**, 839; b) Turner, N.J. (2010) *Curr. Opin. Chem. Biol.*, **14**, 115.
49. Rachwalski, M., *et al.* (2013) *Chem. Soc. Rev.*, **42**, 9268.
50. a) Cheng, F., *et al.* (2020) *Green Chemistry*, **22**, 6815; b) A. Hölzl-Hobmeier, *et al.* (2018) *Nature*, **564**, 240.
51. Yang, B., *et al.* (2020) *Tetrahedron Lett.*, **61**, 152530.
52. Havinga, E. (1954) *Biochim. Biophys. Acta*, **13**, 171.
53. Buhse, T., *et al.* (2021) *Chem. Rev.,* **121**, 2147.
54. Sögütoglu, L.-C., *et al.* (2015) *Chem. Soc. Rev.*, **44**, 6723.
55. Intaraboonrod, K., *et al.* (2020) *Mendeleev Commun.*, **30**, 395.
56. Xiouras, C., *et al.* (2018) *Ultrason. Sonochem.*, **43**, 184.
57. Iggland, M., *et al.* (2014) *Chem. Eng. Sci.*, **111**, 106.

58. Cameli, F., *et al.* (2018) *CrystEngComm*, **20**, 2897.
59. Kondepudi, D.K., *et al.* (1990) *Science*, **250**, 975.
60. a) Kondepudi, D.K., *et al.* (1993) *J. Am. Chem. Soc.*, **115**, 10211; b) Kondepudi, D.K., *et al.* (1992) *Phys. A Statist. Mech. Appl.*, **188**, 113; c) Kondepudi, D.K., *et al.* (1994) *Chem. Phys. Lett.*, **217**, 364.
61. a) Viedma, C. (2004) *J. Cryst. Grow.*, **261**, 118. b) Viedma, C. (2005) Phys. Rev. Lett., 96, 065504.
62. a) Engwerda, A.H.J., *et al.* (2018) *Angew. Chem. Int. Ed.*, **57**, 15441; b) Engwerda, A.H.J., *et al.* (2017) *Cryst. Growth Des.*, **17**, 5583; c) Noorduin, W.L., *et al.* (2009) *Angew. Chem. Int. Ed.*, **48**, 4581; d) Noorduin, W.L., *et al.* (2008) *Angew. Chem. Int. Ed.*, **47**, 6445; e) Shimizu, W., *et al.* (2020) *Cryst. Growth Des.*, **20**, 5676.
63. a) Azeroual, S., *et al.* (2012) *Chem. Commun.*, **48**, 2292; b) Björemark, P.M., *et al.* (2015) *Chem. A Eur. J.*, **21**, 10630; c) McLaughlin, D.T., *et al.* (2014) *Cryst. Growth Des.*, **14**, 1067.
64. a) Cartwright, J.H.E., *et al.* (2007) *Phys. Rev. Lett.*, **98**, 165501; b) Noorduin, W.L., *et al.* (2010) *Angew. Chem. Int. Ed.*, **49**, 8435; c) Saito, Y., *et al.* (2008) *J. Phys. Soc. Japan*, **77**, 113001; d) Suwannasang, K., *et al.* (2013) *Cryst. Growth Des.*, **13**, 3498; e) Uwaha, M. (2004) *J. Phys. Soc. Jpn*, **73**, 2601; f) Uwaha, M. (2008) *J. Phys. Soc. Jpn*, **77**, 083802.
65. a) Breveglieri, F., *et al.* (2020) *Org. Process. Res. Dev.*, **24**, 1515; b) Breveglieri, F., *et al.* (2018) *Cryst. Growth Des.*, **18**, 1873; c) Cameli, F., *et al.* (2020) *Chem. Eur. J.*, **26**, 1344; d) Li, W.W., *et al.* (2016) *Cryst. Growth Des.*, **16**, 5563; e) Maggioni, G.M., *et al.* (2018) *CrystEngComm*, **20**, 3828; f) Oketani, R., *et al.* (2018) *Cryst. Growth Des.*, **18**, 6417; g) Steendam, R.E., *et al.* (2018) *Cryst. Growth Des.*, **18**, 3008; h) Viedma, C., *et al.* (2011) *Chem. Commun.*, **47**, 12786.
66. a) Bodak, B., *et al.* (2018) *Cryst. Growth Des.*, **18**, 7122; b) Katsuno, H., *et al.* (2016) *Phys. Rev. E*, **93**, 013002; c) Suwannasang, K., *et al.* (2014) *Chem. Eng. Technol.*, **37**, 1329; d) Uchin, R., *et al.* (2017) *Chem. Eng. Technol.*, **40**, 1252.
67. Valenti, G., *et al.* (2020) *Angew. Chem. Int. Ed.* 2021, 60, 5279. https://onlinelibrary.wiley.com/doi/abs/10.1002/anie.202013502
68. Suwannasang, K., *et al.* (2016) *Cryst. Growth Des.*, **16**, 6461.
69. a) Kongsamai, P., *et al.* (2018) *Chem. Eng. Technol.*, **41**, 1173; b) Kongsamai, P., *et al.* (2017) *Eur. Phys. J. Spec. Top.*, **226**, 823.
70. Cruz, J.-M., *et al.* (2020) *Chirality*, **32**, 120.
71. a) Addadi, L., *et al.* (1985) *Angew. Chem. Int. Ed.*, **24**, 466; b) Weissbuch, I., *et al.* (1991) *Science*, **253**, 637; c) Weissbuch, I., *et al.* (1987) *J. Am. Chem. Soc.*, **109**, 1869.
72. a) Kozma, D. (2002) *CRC Handbook of Optical Resolutions Via Diastereomeric Salt Formation*. CRC Press; b) Marchand, P., *et al.* (2004) *Tetrahedron Asymmetry*, **15**, 2455.
73. Pasteur, L. (1853) *C. R. Acad. Sci.*, **37**, 162
74. Fischer, E. (1899) *Berichte der deutschen chemischen Gesellschaft*, **32**, 2451.

75. a) Bolchi, C., *et al.* (2003) *Tetrahedron Asymmetry*, **14**, 3779; b) Bolchi, C., *et al.* (2005) *Tetrahedron Asymmetry*, **16**, 1639; c) Borghese, A., *et al.* (2004) *Org. Process. Res. Dev.*, **8**, 532.
76. a) Kinbara, K. (2005) *Synlett*, 732.
77. Faigl, F., *et al.* (2008) *Tetrahedron Asymmetry*, **19**, 519.
78. San, O., *et al.* (2016) *Cryst. Growth Des.*, 16, 307.
79. King, V.L., PhD Thesis thesis, University of Zurich, 1912.
80. Lacour, J., *et al.* (2003) *Chem. Soc. Rev.*, **32**, 373.
81. a) Jeremić, D.A., *et al.* (2010) *Cryst. Growth Des.*, **10**, 559; b) Norden, B. (1972) *Acta Chem. Scand.*, **26**, 126.
82. a) Bark, T., *et al.* (2004) *Chem. A Eur. J.*, **10**, 4839; b) Garbett, K., *et al.* (1966) *J. Chem. Soc. A: Inorgan. Phys., Theor.*, **802**.
83. Ishihara, K., *et al.* (1998) *J. Am. Chem. Soc.*, **120**, 6920.
84. Wong, L.W.Y., *et al.* (2015) *Chem. Commun.*, **51**, 15760.
85. Jacques, J., *et al.* (1989) *Organic Synth.*, **67**, 1.
86. Hoffmann, S., *et al.* (2005) *Angew. Chem. Int. Ed.*, **44**, 7424.
87. Rapenne, G., *et al.* (1996) *J. Am. Chem. Soc.*, **118**, 10932.
88. a) Koenig, M., *et al.* (1979) *J. Chem. Soc. Perkin Trans.*, **2**, 40; b) Lacour, J., *et al.* (1998) *Tetrahedron Lett.*, **39**, 4825; c) Rosenheim, A., *et al.* (1925) *Berichte der deutschen chemischen Gesellschaft (A and B Series)*, **58**, 2000.
89. Weinland, R.F., *et al.* (1919) *Berichte der deutschen chemischen Gesellschaft (A and B Series)*, **52**, 1316.
90. Kobayashi, A., *et al.* (1972) *Acta Crystallogr. B*, **28**, 3446.
91. Ryschkewitsch, G.E., *et al.* (1968) *J. Am. Chem. Soc.*, **90**, 7234.
92. Lemus, L., *et al.* (2013) *Dalton Trans.*, **42**, 11426.
93. Allcock, H.R., *et al.* (1973) *J. Am. Chem. Soc.*, **95**, 3154.
94. Cavezzan, J., *et al.* (1979) *Tetrahedron Lett.*, **20**, 795.
95. Shevchenko, I.V., *et al.* (1992) *Chemische Berichte*, **125**, 1325.
96. a) Gruselle, M., *et al.* (2004) *Mendeleev Commun.*, **14**, 282; b) Lacour, J., *et al.* (2000) *Angew. Chem. Int. Ed.*, **39**, 3695; c) Lacour, J., *et al.* (1998) *Chem. Commun.*, 1733; d) Naim, A., *et al.* (2018) *Inorg. Chem.*, **57**, 14501.
97. Matsuura, T., *et al.* (2005) *J. Photochem. Photobiol. C Photochem. Rev.*, **6**, 7.
98. Winkelmann, A., *et al.* (2015) *Ultramicroscopy*, **149**, 58.
99. Vogl, O. (2011) *J. Polym. Sci. A Poly. Chem.*, **49**, 1299.
100. Bunn, C.W., *et al.* (1954) *Nature*, **174**, 549.
101. a) Clark, E.S. (1999) *Polymer*, **40**, 4659; b) Farmer, B.L., *et al.* (1981) *Polymer*, **22**, 1487; c) Iwasaki, M. (1963) *J. Polym. Sci. A Gen. Pap.*, **1**, 1099; d) Matsushige, K., *et al.* (1977) *Jpn. J. Appl. Phys.*, **16**, 681; e) Weeks, J.J., *et al.* (1981) *Polymer*, **22**, 1480; f) Yamamoto, T. (1985) *J. Polym. Sci. Polym. Phys. Edn*, **23**, 771; g) Yamamoto, T., *et al.* (1986) *Polymer*, **27**, 986.
102. Pauling, L., *et al.* (1951) *Proc. Nat. Acad. Sci. U.S.A.*, **37**, 205.
103. Watson, J.D., *et al.* (1953) *Nature*, **171**, 737.

104. Natta, G., *et al.* (1955) *J. Am. Chem. Soc.*, **77**, 1708.
105. Tadokoro, H. (1979) *Die Makromolekulare Chemie*, **2**, 155.
106. a) Farina, M., *et al.* (1965) *Angew. Chem. Int. Ed.*, **4**, 107; b) Okamoto, Y., *et al.* (1994) *Chem. Rev.*, **94**, 349.
107. Nolte, R.J.M., *et al.* (1974) *J. Am. Chem. Soc.*, **96**, 5932.
108. Okamoto, Y., *et al.* (1979) *J. Am. Chem. Soc.*, **101**, 4763.
109. a) Cornelissen, J.J.L.M., *et al.* (2001) *Chem. Rev.*, **101**, 4039; b) Nakano, T., *et al.* (2001) *Chem. Rev.*, **101**, 4013; c) Yashima, E., *et al.* (2009) *Chem. Rev.*, **109**, 6102.
110. a) Brunsveld, L., *et al.* (2001) *Chem. Rev.*, **101**, 4071; b) Liu, M., *et al.* (2015) *Chem. Rev.*, **115**, 7304; c) Yashima, E., *et al.* (2016) *Chem. Rev.*, **116**, 13752; d) Ariga, K., *et al.* (2020) *Adv. Mater.*, **32**, 1905657.
111. Yang, F., *et al.* (2020) *Chem. Rev.*, **120**, 2693.
112. a) Watanabe, K., *et al.* (2012) *Chem. Mater.*, **24**, 1011, https://doi.org/10.1021/cm2028788; b) San Jose, B. A., *et al.* (2012) *J. Am. Chem. Soc.*, **134**, 19795, https://doi.org/10.1021/ja3086565; c) Nagata, Y., *et al.* (2014) *J. Am. Chem. Soc.*, **136**, 9858, https://doi.org/10.1021/ja504808r
113. a) Oda, M., *et al.* (2002) *Macromolecules*, **35**, 6792, https://doi.org/10.1021/ma020630g; b) Zawadzka, A., *et al.* (2019) *Opt. Mater.*, **96**, 109295, https://www.sciencedirect.com/science/article/pii/S0925346719305051
114. a) Janas, D., *et al.* (2017) *Mater. Des.*, **121**, 119, https://www.sciencedirect.com/science/article/pii/S0264127517301971; b) Krukiewicz, K., *et al.* (2018) *RSC Adv.*, **8**, 30600, http://dx.doi.org/10.1039/C8RA03963A
115. Akagi, K. (1998) *Science*, **282**, 1683.
116. Akagi, K. (2019) *Bull. Chem. Soc. Jpn.*, **92**, 1509.
117. Manaka, T., *et al.* (2006) *Chem. Lett.*, **35**, 1028.
118. Wu, S.-T., *et al.* (2014) *Angew. Chem. Int. Ed.*, **53**, 12860.
119. Maeda, K., *et al.* (2005) *Chem. Commun.*, 4152.
120. a) Guo, H., *et al.* (1999) *Synth. Met.*, **101**, 44; b) Huang, J., *et al.* (2003) *Adv. Mater.*, **15**, 1158.
121. a) Buono, A.M., *et al.* (2007) *J. Am. Chem. Soc.*, **129**, 10992; b) Guadagno, L., *et al.* (2008) *J. Mater. Chem.*, **18**, 567.
122. Rizzo, P., *et al.* (2011) *J. Am. Chem. Soc.*, **133**, 9872.
123. a) Guerra, G., *et al.* (2012) *J. Polym. Sci. B: Polym. Phys.*, **50**, 305; b) Guerra, G., *et al.* (2013) *Rendiconti Lincei*, **24**, 217; c) Rizzo, P., *et al.* (2016) *Chirality*, **28**, 29.
124. Nakano, T. (2014) *Chem. Rec.*, **14**, 369.
125. Iftime, G., *et al.* (2000) *J. Am. Chem. Soc.*, **122**, 12646.
126. Wang, Y., *et al.* (2012) *Chem. Commun.*, **48**, 1871.
127. Fujiki, M., *et al.* (2015) *Polym. Chem.*, **6**, 1627.
128. Miyata, M., *et al.* (2014) *Chem. Lett.*, **43**, 1476.
129. Simonyi, M., *et al.* (2003) *Chirality*, **15**, 680.
130. Bartik, K., *et al.* (2000) *ChemPhysChem*, **4**, 221.
131. Graf, E., *et al.* (1997) *Chem. Commun.*, 1459.

132. Okazaki, Y., *et al.* (2018) *Chem. Commun.*, **54**, 10244, https://hal.archives-ouvertes.fr/hal-02322340files/27/hal-02322340.html
133. Bleschke, C., *et al.* (2011) *Adv. Synth. Catal.*, **353**, 3101.
134. Ono, Y., *et al.* (1998) *Chem. Commun.*, **1477**.
135. a) Kelly, J.A., *et al.* (2014) *Acc. Chem. Res.*, **47**, 1088; b) Marx, S., *et al.* (2007) *Acc. Chem. Res.*, **40**, 768; c) Qiu, H., *et al.* (2011) *Chem. Soc. Rev.*, **40**, 1259; d) van Bommel, K. J. C., *et al.* (2003) *Angew. Chem. Int. Ed.*, **42**, 980.
136. Okazaki, Y., *et al.* (2016) *Nano Lett.*, **16**, 6411, https://doi.org/10.1021/acs.nanolett.6b02858files/51/acs.nanolett.html
137. a) Florescu, M., *et al.* (2009) *Phys. Rev. B*, **80**, 155112; b) Rechtsman, M.C., *et al.* (2008) *Phys. Rev. Lett.*, **101**, 073902.
138. Sholl, D.S., *et al.* (2001) *J. Phys. Chem. B*, **105**, 4771.
139. Chen, Y., *et al.* (2020) *Phys. Rev. Lett.*, **235502**.
140. Konevtsova, O.V., *et al.* (2012) *Phys. Rev. Lett.*, **108**, 038102.
141. Toda, F. (2004) *Enantiomer Separation*, Springer Netherlands.
142. a) Stinson, S.C. (2001) *Chem. Eng. News Arch.*, **79**, 45, https://doi.org/10.1021/cen-v079n020.p045; b) Wagnière, G.H. (2008). *On Chirality and the Universal Asymmetry: Reflections on Image and Mirror Image*, Wiley-Verlag Helvetica Chimica Acta.
143. Smith, S.W. (2009) Toxicological Sciences **110**, 4. https://doi.org/10.1093/toxsci/kfp097
144. Jozwiak, K., Lough, W.J., and Wainer, I.W. (2012) *Drug Stereochemistry* 3ed., CRC Press.
145. De Camp, W.H. (1989) *Chirality*, **1**, 2, https://doi.org/10.1002/chir.530010103
146. a) Chang, C., *et al.* (2012) *TrAC Trends Anal. Chem.*, **39**, 195, https://www.sciencedirect.com/science/article/pii/S0165993612001951; b) Gogoi, A.A.-O., *et al.* (2019) *Molecules*, 24, 1007; doi: 10.3390/molecules24061007; c) Maier, N.M., *et al.* (2001) *J. Chromatogr. A*, **906**, 3, https://www.sciencedirect.com/science/article/pii/S002196730000532X
147. Francotte, E. (2009) *Chimia (Aarau)*, **63**, 867, http://europepmc.org/abstract/MED/28372616; https://doi.org/10.2533/chimia.2009.867
148. Blaser, H.-U., *et al.* (2005) *J. Mol. Catal. A Chem.*, **231**, 1, https://www.sciencedirect.com/science/article/pii/S1381116904009057
149. a) Schuur, B., *et al.* (2011) *Org. Biomol. Chem.*, **9**, 36, http://dx.doi.org/10.1039/C0OB00610F; b) Steensma, M., *et al.* (2006) *Chirality*, **18**, 314, https://doi.org/10.1002/chir.20258
150. a) Ahuja, S. (2001) *Org. Process. Res. Dev.*, **5**, 458, https://doi.org/10.1021/op0100293; b) Hazen, R.M., *et al.* (2003) *Nat. Mater.*, **2**, 367; c) Lang, J.C., *et al.* (2017) *Curr. Opin. Colloid Interface Sci.*, **32**, 94, https://www.sciencedirect.com/science/article/pii/S1359029417301413
151. a) Horvath, J.D., *et al.* (2004) *J. Am. Chem. Soc.*, **126**, 14988, https://doi.org/10.1021/ja045537h; b) Yun, Y., *et al.* (2013) *Angew. Chem. Int. Ed.*, **52**, 3394, https://doi.org/10.1002/anie.201209025; c) Yun, Y., *et al.* (2015) *Langmuir*, **31**, 6055, https://doi.org/10.1021/acs.langmuir.5b00707

152. a) de Alwis, A., *et al.* (2013) *Surf. Sci.*, **608**, 80, https://www.sciencedirect.com/science/article/pii/S0039602812003457; b) Karagoz, B., *et al.* (2019) *Langmuir*, **35**, 16438, https://doi.org/10.1021/acs.langmuir.9b02476
153. Halasyamani, P.S., *et al.* (1998) *Chem. Mater.*, **10**, 2753.
154. a) Karagunis, G., *et al.* (1938) *Nature*, **142**, 162; b) Tsuchida, R., *et al.* (1936) *Bull. Chem. Soc. Jpn.*, **11**, 38.
155. a) Bonner, W.A., *et al.* (1974) *Science*, **186**, 143; b) Hartwig, B.A., *et al.* (1972) *J. Biomed. Mater. Res.*, **6**, 413.
156. Guevremont, J.M., *et al.* (1998) *Am. Mineral.*, **83**, 1246.
157. Jenkins, S.J., *et al.* (2007) *Surf. Sci. Rep.*, **62**, 373.
158. Cody, A.M., *et al.* (1991) *J. Cryst. Grow.*, **113**, 508.
159. Orme, C.A., *et al.* (2001) *Nature*, **411**, 775, https://doi.org/10.1038/35081034
160. Hazen, R.M., *et al.* (2001) *Proc. Nat. Acad. Sci. U.S.A.*, **98**, 5487.
161. a) Kelso, M.V., *et al.* (2018) *J. Am. Chem. Soc.*, **140**, 15812; b) McFadden, C. F., *et al.* (1996) *Langmuir*, **12**, 2483.
162. a) Kajioka, H., *et al.* (2010) *Macromolecules*, **43**, 3837; b) Toda, A., *et al.* (2003) *Polymer*, **44**, 6135; c) Toda, A., *et al.* (2005) *Polymer*, **46**, 8708.
163. Vand, V. (1951) *Nature*, **168**, 783.
164. a) Neuber, M., *et al.* (1995) *J. Phys. Chem.*, **99**, 9160; b) Raval, R. (2009) *Chem. Soc. Rev.*, **38**, 707.
165. a) Kauranen, M., *et al.* (1994) *J. Chem. Phys.*, **101**, 8193; b) T. Petralli-Mallow, *et al.* (1993) *J. Phys. Chem.*, **97**, 1383; c) Verbiest, T., *et al.* (1994) *J. Am. Chem. Soc.*, **116**, 9203.
166. a) Attard, G.A. (2001) *J. Phys. Chem. B*, **105**, 3158, https://doi.org/10.1021/jp0041508; b) Attard, G.A., *et al.* (1999) *J. Phys. Chem. B*, **103**, 1381; c) Hazzazi, O.A., *et al.* (2004) *J. Mol. Catal. A Chem.*, **216**, 247.
167. a) Nakanishi, T., *et al.* (2002) *J. Am. Chem. Soc.*, **124**, 740; b) Nakanishi, T., *et al.* (2004) *Chirality*, **16**, S36.
168. a) Barlow, S.M., *et al.* (2003) *Surf. Sci. Rep.*, **50**, 201, https://www.sciencedirect.com/science/article/pii/S0167572903000153; b) Humblot, V., *et al.* (2004) *J. Am. Chem. Soc.*, **126**, 6460, https://doi.org/10.1021/ja0382056; c) Lorenzo, M.O., *et al.* (1999) *J. Phys. Chem. B*, **103**, 10661, https://doi.org/10.1021/jp992188i; d) Ohmann, R., *et al.* (2011) *ACS Nano*, **5**, 1360, https://doi.org/10.1021/nn103058e
169. a) Darling, G.R., *et al.* (2017) *Phys. Chem. Chem. Phys.*, **19**, 7617; b) Madden, D.C., *et al.* (2014) *Surf. Sci.*, **629**, 81, https://www.sciencedirect.com/science/article/pii/S0039602814000910; c) Madden, D.C., *et al.* (2014) *J. Phys. Chem. C*, **118**, 18589, https://doi.org/10.1021/jp505636v; d) Therrien, A.J., *et al.* (2016) *Chem. Commun.*, **52**, 14282, http://dx.doi.org/10.1039/C6CC05820E
170. a) Forster, M., *et al.* (2016) *Chem. Commun.*, **52**, 14075; b) Sleczkowski, P., *et al.* (2016) *J. Phys. Chem. C*, **120**, 22388, https://doi.org/10.1021/acs.jpcc.6b06800
171. Cun, H., *et al.* (2010) *Langmuir*, **26**, 3402, https://doi.org/10.1021/la903193a

172. a) Mateo Marti, E., *et al.* (2002) *Surf. Sci.*, **501**, 191, https://www.sciencedirect.com/science/article/pii/S003960280102026X; b) Yang, B., *et al.* (2010) *J. Am. Chem. Soc.*, **132**, 10440, https://doi.org/10.1021/ja102989y
173. a) Haq, S., *et al.* (2007) *Langmuir*, **23**, 10694, https://doi.org/10.1021/la700965d; b) Seibel, J., *et al.* (2015) *J. Am. Chem. Soc.*, **137**, 7970, https://doi.org/10.1021/jacs.5b02262
174. a) Chen, Q., *et al.* (2002) *Surf. Sci.*, **497**, 37, https://www.sciencedirect.com/science/article/pii/S003960280101593X; b) Liu, N., *et al.* (2011) *Chem. Commun.*, **47**, 11324, http://dx.doi.org/10.1039/C1CC13317A
175. Lopinski, G.P., *et al.* (1998) *Nature*, **392**, 909.
176. Booth, N.A., *et al.* (1998) *Surf. Sci.*, **397**, 258.
177. Weckesser, J., *et al.* (1999) *J. Chem. Phys.*, **110**, 5351.
178. a) Chen, Q., *et al.* (2003) *Nat. Mater.*, **2**, 324; b) A. McNutt, *et al.* (2003) *Surf. Sci.*, **531**, 131.
179. Gopakumar, T.G., *et al.* (2010) *J. Phys. Chem. C*, **114**, 18247.
180. a) Böhringer, M., *et al.* (1999) *Angew. Chem. Int. Ed.*, **38**, 821; b) Böhringer, M., *et al.* (1999) *Phys. Rev. Lett.*, **83**, 324.
181. France, C.B., *et al.* (2003) *J. Am. Chem. Soc.*, **125**, 12712.
182. Wei, Y., *et al.* (2004) *J. Am. Chem. Soc.*, **126**, 5318.
183. Schlickum, U., *et al.* (2008) *J. Am. Chem. Soc.*, **130**, 11778.
184. Barlow, S.M., *et al.* (2003) *Surf. Sci. Rep.*, **50**, 201.
185. a) Baiker, A. (2005) *Catal. Today*, **100**, 159, https://www.sciencedirect.com/science/article/pii/S0920586104007576; b) Zaera, F. (2008) *J. Phys. Chem. C*, **112**, 16196, https://doi.org/10.1021/jp804588v; c) Gross, E., *et al.* (2013) *J. Am. Chem. Soc.*, **135**, 3881, https://doi.org/10.1021/ja310640b
186. a) Núñez, O., *et al.* (2008) *J. Chromatogr. A*, **1191**, 231; b) Zhao, J., *et al.* (2015) *J. Chromatogr. A*, **1381**, 253, https://www.sciencedirect.com/science/article/pii/S0021967315000540
187. a) Jonkheijm, P., *et al.* (2008) *Angew. Chem. Int. Ed.*, **47**, 9618, https://doi.org/10.1002/anie.200801711; b) Mansur, H.S., *et al.* (2005) *J. Mater. Sci. Mater. Med.*, **16**, 333, https://doi.org/10.1007/s10856-005-0632-y; c) Welch, N.G., *et al.* (2017) *Biointerphases*, **12**, 02D301, https://doi.org/10.1116/1.4978435; d) Yu, Q., *et al.* (2015) *Crit. Rev. Anal. Chem.*, **45**, 62, https://doi.org/10.1080/10408347.2014.881249
188. a) Heitbaum, M., *et al.* (2006) *Angew. Chem. Int. Ed. Engl.*, **45**, 4732; b) P. McMorn, *et al.* (2004) *Chem. Soc. Rev.*, **33**, 108, http://dx.doi.org/10.1039/B200387M
189. Durán Pachón, L., *et al.* (2009) *Nat. Chem.*, **1**, 160, https://doi.org/10.1038/nchem.180
190. Woodruff, D.P. (2002) *Surf. Sci.*, **500**, 147, https://www.sciencedirect.com/science/article/pii/S0039602801015345
191. a) Gellman, A.J., *et al.* (2015) *Catal. Lett.*, **145**, 220, https://doi.org/10.1007/s10562-014-1400-8; b) Lawton, T.J., *et al.* (2013) *J. Phys. Chem. C*, **117**, 22290, https://doi.org/10.1021/jp402015r

192. a) Behar-Levy, H., *et al.* (2002) *Chem. Mater.*, **14**, 1736, https://doi.org/10.1021/cm011558o; b) Nesher, G., *et al.* (2008) *Chem. Mater.*, **20**, 4425, https://doi.org/10.1021/cm800657j; c) Yosef, I., *et al.* (2006) *Chem. Mater.*, **18**, 5890, https://doi.org/10.1021/cm0615368
193. a) H. Behar-Levy, *et al.* (2007) *Adv. Mater.*, **19**, 1207, https://doi.org/10.1002/adma.200601702; b) Goldsmith, M.-R., *et al.* (2006) *Phys. Chem. Chem. Phys.*, **8**, 63, http://dx.doi.org/10.1039/B511563A; c) Yang, H., *et al.* (2014) *Chem. Commun.*, **50**, 8868, http://dx.doi.org/10.1039/C4CC02823F
194. Bohannan, E.W., *et al.* (2004) *J. Am. Chem. Soc.*, **126**, 488, https://doi.org/10.1021/ja039422+
195. a) Fang, Y., *et al.* (2017) *ChemElectroChem*, **4**, 1832, https://doi.org/10.1002/celc.201700257; b) Wattanakit, C., *et al.* (2014) *Nat. Commun.*, **5**, 3325, https://doi.org/10.1038/ncomms4325
196. a) Kothari, H.M., *et al.* (2004) *Chem. Mater.*, **16**, 4232, https://doi.org/10.1021/cm048939x; b) Switzer, J.A., *et al.* (2003) *Nature*, **425**, 490, https://doi.org/10.1038/nature01990
197. Gautier, C., *et al.* (2009) *ChemPhysChem*, **10**, 483, https://doi.org/10.1002/cphc.200800709 https://chemistry-europe.onlinelibrary.wiley.com/doi/pdfdirect/10.1002/cphc.200800709?download=true
198. Ma, W., *et al.* (2017) *Chem. Rev.*, **117**, 8041, https://doi.org/10.1021/acs.chemrev.6b00755
199. a) Gautier, C., *et al.* (2008) *J. Am. Chem. Soc.*, **130**, 7077, https://www.ncbi.nlm.nih.gov/pubmed/18459786; b) Schaaff, T.G., *et al.* (2000) *J. Phys. Chem. B*, **104**, 2630.
200. Li, T., *et al.* (2004) *Nanotechnology*, **15**, S660.
201. Yao, H., *et al.* (2005) *J. Am. Chem. Soc.*, **127**, 15536.
202. a) Gautier, C., *et al.* (2006) *J. Am. Chem. Soc.*, **128**, 11079, https://www.ncbi.nlm.nih.gov/pubmed/16925425; b) Ock, K.S., *et al.* (2013) *Spectrochim. Acta A Mol. Biomol. Spectrosc.*, **102**, 419, https://www.ncbi.nlm.nih.gov/pubmed/23261512
203. Moloney, M.P., *et al.* (2007) *Chem. Commun.*, 3900, https://www.ncbi.nlm.nih.gov/pubmed/17896026
204. Nakashima, T., *et al.* (2009) *J. Am. Chem. Soc.*, **131**, 10342, https://www.ncbi.nlm.nih.gov/pubmed/19588974
205. Mukhina, M.V., *et al.* (2015) *Nano Lett.*, **15**, 2844, http://www.ncbi.nlm.nih.gov/pubmed/25908405
206. Choi, J.K., *et al.* (2016) *ACS Nano*, **10**, 3809, http://www.ncbi.nlm.nih.gov/pubmed/26938741
207. Kuno, J., *et al.* (2018) *Angew. Chem. Int. Ed. Engl.*, **57**, 12022, https://www.ncbi.nlm.nih.gov/pubmed/30054963
208. Kuznetsova, V.A., *et al.* (2019) *ACS Nano*, **13**, 13560, <Go to ISI>://WOS:000500650000129
209. Zhou, Y., *et al.* (2011) *Angew. Chem. Int. Ed. Engl.*, **50**, 11456, http://www.ncbi.nlm.nih.gov/pubmed/22113805

210. Moshe, A.B., *et al.* (2011) *ACS Nano*, **5**, 9034, https://www.ncbi.nlm.nih.gov/pubmed/21967095
211. Gao, X., *et al.* (2017) *J. Am. Chem. Soc.*, **139**, 8734, http://www.ncbi.nlm.nih.gov/pubmed/28582623
212. a) Cheng, J., *et al.* (2018) *ACS Nano*, **12**, 5341, http://www.ncbi.nlm.nih.gov/pubmed/29791135; b) Hao, J., *et al.* (2020) *ACS Nano*, **14**, 10346, https://pubs.acs.org/doi/pdf/10.1021/acsnano.0c03909; c) Shao, X., *et al.* (2020) *Inorg. Chem.*, **59**, 14382, https://www.ncbi.nlm.nih.gov/pubmed/32945662
213. Gao, X., *et al.* (2018) *Nano Lett.*, **18**, 6665, http://www.ncbi.nlm.nih.gov/pubmed/30350652
214. Purcell-Milton, F., *et al.* (2017) *ACS Nano*, **11**, 9207.
215. a) Chen, C., *et al.* (2019) *Nat. Commun.*, **10**, 1927, https://www.ncbi.nlm.nih.gov/pubmed/31028272; b) Chen, W., *et al.* (2019) *J. Phys. Chem. Lett.*, **10**, 3290, http://www.ncbi.nlm.nih.gov/pubmed/31146530; c) He, T., *et al.* (2017) *Appl. Phys. Lett.*, **111**, 151102; d) Long, G., *et al.* (2020) *Nat. Rev. Mater.*, **5**, 423.
216. Jiang, W., *et al.* (2019) *Mater. Horiz.*, **6**, 1974, http://dx.doi.org/10.1039/C9MH00431A
217. a) Che, S., *et al.* (2003) *Nat. Mater.*, **2**, 801, https://www.ncbi.nlm.nih.gov/pubmed/14634644; b) Delclos, T., *et al.* (2008) *Nano Lett.*, **8**, 1929, https://doi.org/10.1021/nl080664nfiles/128/nl080664n.html; c) Paredes, B., *et al.* (2004) *Nature*, **429**, 277, https://www.ncbi.nlm.nih.gov/pubmed/15152247; d) Sugiyasu, K., *et al.* (2002) *Chem. Commun.*, **1212**.
218. Liu, S., *et al.* (2012) *Nat. Commun.*, **3**, 1215, https://www.ncbi.nlm.nih.gov/pubmed/23169056
219. Duan, Y., *et al.* (2015) *Angew. Chem. Int. Ed. Engl.*, **54**, 15170, https://www.ncbi.nlm.nih.gov/pubmed/26489386
220. Sone, E.D., *et al.* (2002) *Angew. Chem. Int. Ed.*, **41**, 1705, https://onlinelibrary.wiley.com/doi/pdfdirect/10.1002/1521-3773%2820020517%2941%3A10%3C1705%3A%3AAID-ANIE1705%3E3.0.CO%3B2-M?download=true
221. Ma, L., *et al.* (2017) *Angew. Chem. Int. Ed. Engl.*, **56**, 8657, https://www.ncbi.nlm.nih.gov/pubmed/28544063
222. Duan, Y., *et al.* (2014) *J. Am. Chem. Soc.*, **136**, 7193, https://www.ncbi.nlm.nih.gov/pubmed/24802633
223. Wu, Y.-J., *et al.* (2012) *Cryst. Growth Des.*, **12**, 547, https://doi.org/10.1021/cg201246m
224. Jiang, W., *et al.* (2017) *Nat. Commun.*, **8**, 1.
225. Ben-Moshe, A., *et al.* (2014) *Nat. Commun.*, **5**, 4302, http://www.ncbi.nlm.nih.gov/pubmed/25001884
226. a) Wang, P.P., *et al.* (2017) *Nat. Commun.*, **8**, 14312, http://www.ncbi.nlm.nih.gov/pubmed/28148957; b) Wang, P.P., *et al.* (2017) *J. Am. Chem. Soc.*, **139**, 6070, http://www.ncbi.nlm.nih.gov/pubmed/28414901

227. Lee, H.E., *et al.* (2018) *Nature*, **556**, 360, https://www.ncbi.nlm.nih.gov/pubmed/29670265
228. a) Cho, N.H., *et al.* (2020) *ACS Nano*, **14**, 3595, https://www.ncbi.nlm.nih.gov/pubmed/32134639; b) Karst, J., *et al.* (2019) *ACS Nano*, **13**, 8659, https://www.ncbi.nlm.nih.gov/pubmed/31294546; c) Kim, H., *et al.* (2020) *Angew. Chem. Int. Ed. Engl.*, **59**, 12976, https://www.ncbi.nlm.nih.gov/pubmed/32337812; d) Yuan, C., *et al.* (2019) *J. Nanosci. Nanotechnol.*, **19**, 2789, http://www.ncbi.nlm.nih.gov/pubmed/30501781
229. Cho, N.H., *et al.* (2019) *Part. Part. Syst. Characterizat.*, **36**.
230. a) Glotzer, S.C. (2004) *Science*, **306**, 419; b) Glotzer, S.C., *et al.* (2007) *Nat. Mater.*, **6**, 557, http://dx.doi.org/10.1038/nmat1949; c) Velev, O.D., *et al.* (2009) *Adv. Mater.*, **21**, 1897; d) Zhang, X., *et al.* (2016) *Adv. Mater.*, **28**, 2475, https://doi.org/10.1002/adma.201504206
231. Blell, R., *et al.* (2017) *ACS Nano*, **11**, 84, https://doi.org/10.1021/acsnano.6b04191
232. Liu, J.-W., *et al.* (2012) *Chem. Rev.*, **112**, 4770.
233. Faustini, M., *et al.* (2010) *J. Phys. Chem. C*, **114**, 7637.
234. a) Han, W., *et al.* (2012) *Angew. Chem. Int. Ed.*, **51**, 1534, https://onlinelibrary.wiley.com/doi/abs/10.1002/anie.201104454; b) Huang, J., *et al.* (2007) *Angew. Chem. Int. Ed.*, **46**, 2414, https://onlinelibrary.wiley.com/doi/abs/10.1002/anie.200604789; c) Petty, M.C. (1992) *Thin Solid Films*, **210–211**, 417, https://www.sciencedirect.com/science/article/pii/004060909290300Z
235. a) Hu, H., *et al.* (2017) *Nanoscal*; b) Huang, Y., *et al.* (2001) *Science*, **291**, 630, https://science.sciencemag.org/content/sci/291/5504/630.full.pdf
236. a) Sarker, B.K., *et al.* (2011) *ACS Nano*, **5**, 6297, https://doi.org/10.1021/nn201314t; b) Wang, B., *et al.* (2010) *Adv. Mater.*, **22**, 3067, https://onlinelibrary.wiley.com/doi/abs/10.1002/adma.201000705; c) Xie, J., *et al.* (2008) *Macromol. Rapid Commun.*, **29**, 1775, https://onlinelibrary.wiley.com/doi/abs/10.1002/marc.200800381
237. a) Hangarter, C.M., *et al.* (2007) *Nanotechnology*, **18**, 205305; b) Shklyarevskiy, I.O., *et al.* (2005) *Langmuir*, **21**, 2108, https://doi.org/10.1021/la047166o
238. a) Jin, L., *et al.* (1998) *Appl. Phys. Lett.*, **73**, 1197, https://aip.scitation.org/doi/abs/10.1063/1.122125; b) Zhang, S., *et al.* (2014) *Biomaterials*, **35**, 3243, https://www.sciencedirect.com/science/article/pii/S0142961213015779
239. Xu, C.Y., *et al.* (2004) *Biomaterials*, **25**, 877, https://www.sciencedirect.com/science/article/pii/S0142961203005933
240. a) Dugyala, V.R., *et al.* (2015) *RSC Adv.*, **5**, 60079; b) Li, P., *et al.* (2016) *Adv. Mater.*, **28**, 2511; c) Srikantharajah, R., *et al.* (2016) *Nanoscale*, **8**, 19882.
241. Singh, G., *et al.* (2014) *Science*, **345**, 1149, https://www.ncbi.nlm.nih.gov/pubmed/25061133
242. Jana, S., *et al.* (2017) *Sci. Adv.*, **3**, e1701483, https://www.ncbi.nlm.nih.gov/pubmed/28924613
243. Yin, Y., *et al.* (2003) *J. Am. Chem. Soc.*, **125**, 2048, https://www.ncbi.nlm.nih.gov/pubmed/12590523

244. a) Bishop, K.J., *et al.* (2009) *Small*, **5**, 1600, http://www.ncbi.nlm.nih.gov/pubmed/19517482; b) Boles, M.A., *et al.* (2016) *Chem. Rev.*, **116**, 11220, http://www.ncbi.nlm.nih.gov/pubmed/27552640
245. Srivastava, S., *et al.* (2010) *Science*, **327**, 1355, https://www.ncbi.nlm.nih.gov/pubmed/20150443
246. Yeom, J., *et al.* (2015) *Nat. Mater.*, **14**, 66, http://www.ncbi.nlm.nih.gov/pubmed/25401922
247. Kim, J.Y., *et al.* (2019) *J. Am. Chem. Soc.*, **141**, 11739, https://www.ncbi.nlm.nih.gov/pubmed/31329438
248. Feng, W., *et al.* (2017) *Sci. Adv.*, 3, e1601159, https://www.ncbi.nlm.nih.gov/pmc/articles/PMC5332156/pdf/1601159.pdf
249. Zhou, Y., *et al.* (2016) *ACS Nano*, **10**, 3248, http://www.ncbi.nlm.nih.gov/pubmed/26900920
250. Yan, J., *et al.* (2019) *Chem. Mater.*, **32**, 476.
251. Jiang, W., *et al.* (2020) *Science*, eaaz7949, https://science.sciencemag.org/content/sci/early/2020/04/08/science.aaz7949.full.pdf
252. a) Kuncicky, D.M., *et al.* (2006) *Small*, **2**, 1462; b) Wargacki, S.P., *et al.* (2008) *Langmuir*, **24**, 5439, https://doi.org/10.1021/la7040778
253. a) Dumanli, A.G., *et al.* (2014) *ACS Appl. Mater. Interfaces*, **6**, 12302, https://doi.org/10.1021/am501995e; b) Tran, A., *et al.* (2020) *Adv. Mater.*, **32**, 1905876, https://onlinelibrary.wiley.com/doi/abs/10.1002/adma.201905876
254. Cherpak, V., *et al.* (2018) *Nano Lett.*, **18**, 6770, https://doi.org/10.1021/acs.nanolett.8b02522
255. Gao, J., *et al.* (2019) *Chem. Mater.*, **32**, 821.
256. a) Dietrich, K., *et al.* (2012) *Adv. Mater.*, 24, OP321, https://onlinelibrary.wiley.com/doi/abs/10.1002/adma.201203424; b) Zhao, Y., *et al.* (2012) *Nat. Commun.*, **3**, 870, https://doi.org/10.1038/ncomms1877
257. Urban, M.J., *et al.* (2019) *Annu. Rev. Phys. Chem.*, **70**, 275.
258. a) Liang, Z., *et al.* (2016) *Faraday Discuss.*, **191**, 141, http://dx.doi.org/10.1039/C6FD00064A; b) Gao, J., *et al.* (2020) *ACS Nano*, **14**, 4111.
259. Gil Perez, J.J., *et al.* (2016) *Polarized Light and the Mueller Matrix Approach,* CRC Press.
260. a) Arteaga, O., *et al.* (2012) *Appl. Opt.*, **51**, 6805, http://ao.osa.org/abstract.cfm?URI=ao-51-28-6805; b) Azzam, R.M.A. (1978) *J. Opt. Soc. Am.*, **68**, 1756, http://www.osapublishing.org/abstract.cfm?URI=josa-68-12-1756; c) Cloude, S. (1990) *Conditions for the Physical Realisability of Matrix Operators in Polarimetry*, Vol. 1166, 177 SPIE; d) Lu, S.-Y., *et al.* (1996) *J. Opt. Soc. Am. A*, **13**, 1106, http://josaa.osa.org/abstract.cfm?URI=josaa-13-5-1106
261. Drake, A.F. (1986) *J. Phys. E Sci. Instrum.*, **19**, 170, http://dx.doi.org/10.1088/0022-3735/19/3/002
262. Nafie, L.A., *et al.* (1982) *Techniques Using Fourier Transform Interferometry*, Vol. 3, Elsevier.

263. a) Modine, F.A., *et al.* (1975) *Appl. Opt.*, **14**, 761, http://ao.osa.org/abstract.cfm?URI=ao-14-3-761; b) Shindo, Y., *et al.* (1989) *Rev. Sci. Instrum.*, **60**, 3633, https://doi.org/10.1063/1.1140467
264. a) Disch, R.L., *et al.* (1969) *Anal. Chem.*, **41**, 82, https://doi.org/10.1021/ac60270a038; b) Shindo, Y., *et al.* (1990) *Biopolymers*, **30**, 405, https://doi.org/10.1002/bip.360300318
265. Buffeteau, T., *et al.* (2005) Appl. Spectroscopy, 59, 732, https://doi.org/10.1366/0003702054280568
266. Collins, R.W., *et al.* (1999) *J. Opt. Soc. Am. A*, **16**, 1997, http://josaa.osa.org/abstract.cfm?URI=josaa-16-8-1997
267. De Martino, A., *et al.* (2003) *Opt. Lett.*, **28**, 616, http://ol.osa.org/abstract.cfm?URI=ol-28-8-616
268. Jellison, J., *et al.*, *Two Modulator Generalized Ellipsometer for Complete Mueller Matrix Measurement*, **5956147**.
269. a) Arwin, H., *et al.* (2019) *J. Vac. Sci. Technol. B*, **38**, 014004, https://doi.org/10.1116/1.5131634; b) A. Mendoza-Galván, *et al.* (2018) *J. Opt.*, **20**, 024001, http://dx.doi.org/10.1088/2040-8986/aa9e7d
270. a) Nikolas, J.P., *et al.* (2008) *J. Nanophotonics*, **2**, 1, https://doi.org/10.1117/1.3062210; b) Oates, T.W.H., *et al.* (2014) *Opt. Mater. Express*, **4**, 2646, http://www.osapublishing.org/ome/abstract.cfm?URI=ome-4-12-2646

https://doi.org/10.1142/9789811259227_0007

7 Chirogenesis in Photochemistry

Ravichandranath Singathi, Sruthy Baburaj, Lakshmy Kannadi Valloli, Jayachandran Parthiban and Jayaraman Sivaguru*

Chemistry Department, Bowling Green State University, Bowling Green, OH, USA

The chapter highlights three distinct approaches to influence the stereochemical outcome of photochemical reactions originating from the excited state. The first approach highlights how one can employ molecular restriction to translate chiral information known as molecular photochirogenesis. The second approach highlights how one can leverage supramolecular environments to influence the outcome of light induced transformations. In the third approach organocatalytic systems are featured to showcase their influence in controlling the outcome of excited state reactions.

1.1 Introduction

Chiral molecules are ubiquitous in many biomolecules (sugar, amino acids, nucleic acids, etc.) and man-made chemicals and active pharmaceutical ingredients. Photochirogenesis or chiral photochemistry has become an

*Corresponding author: sivagj@bgsu.edu

interesting and rapidly growing area to access to enantiomerically pure compounds [1]. In the past decades, light-mediated methods for synthesizing optically active molecules have gained significant growth in pharmaceuticals and agrochemicals industries. Photochirogenesis provide structural and spatio-temporal control on achieving complex structures in organic synthesis [2, 3]. However, when compared to the conventional thermal reactions, achieving stereocontrol involving photochemical transformations are challenging and limits their frequent use in asymmetric synthesis [4]. Photochirogenesis involves interaction of light with molecules, thus producing highly energetic short-lived excited state species with low activation energy barrier(s) and making the process downhill and hence providing little opportunity to differentiate the diastereomeric transition states leading to poor stereo-enhancement in the product(s). Similar to the other methods in achieving asymmetric synthesis, in photochirogenesis a physical or chemical source of chirality will be guided to the reactant(s) in the ground state to drive a desired reaction trajectory so that chirality can be enhanced in the product(s). Such sources of chirality with unique methodologies were developed by various research groups in the fields of chiral photochemistry [5–16]. This chapter will highlight a few of those strategies to achieve chirogenesis during photochemical reaction (Figure 1).

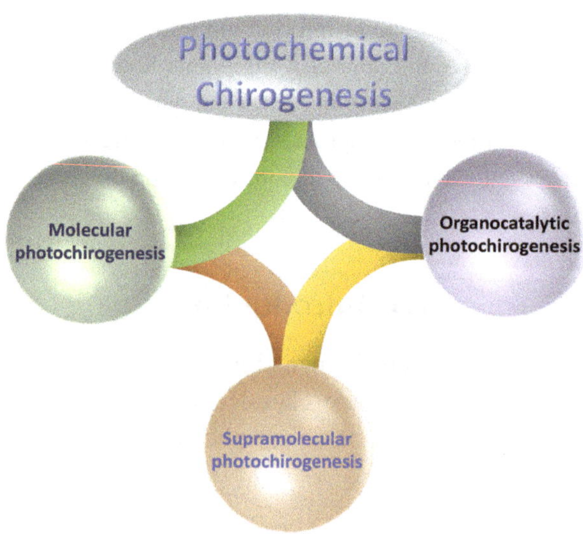

Figure 1. Broad overview of the photochemical chirogenesis approaches.

1.2 Molecular Photochirogenesis

Photochirogenesis involving restricted bond rotations is one of the elegant strategies that involves chiral transfer from a molecule with inherent chirality (non-racemic reactant) to an enantiomerically pure product(s). For simplicity and to detail the conceptual significance, in this chapter we will restrict ourselves to photoreactions of atropisomers to highlight how molecular chirality can influence chirogenesis during phototransformations [17]. Atropisomers are a special class of molecules without any chiral centers but exist as pair of enantiomers because of the restricted rotation around a single bond (chiral axis) [18]. The energy barrier for restricted rotation around the chiral axis can be determined with appropriate substituents that are responsible for steric hindrance. Hence, atropisomeric reactants with pre-organized chirality can transfer chiral information to the desired photoproducts and with enhanced reactivity/selectivity during the photochemical transformation. Sivaguru and coworkers utilized the concept of atropisomerism and conducted investigations on racemization kinetics, photophysical studies, and all the events responsible for asymmetric induction in the products line [19–34]. The axial chirality in non-biaryl atropisomers with pre-organized chirality is responsible for the contribution of rotameric/enantiomeric ratio in the ground state and can be experimentally calculated by studying the higher energy barrier to rotation ($\Delta G^{\ddagger}_{rac}$) or higher half-life ($\tau_{1/2}$) for racemization. By considering concepts of Havinga's non-equilibrating excited-state rotamers (NEER) principle [35, 36], one can understand that same rotameric/enantiomeric ratio could be maintained in excited state as well. The photoreactions when carried starting with optically pure atropisomers can lay the foundation for "axial to point chirality transfer" during the photochemical transformation (Figure 2) [37–39].

1.2.1 Photoreactions of Atropisomeric Chromophores

The design of atropisomeric chromophores for stereospecific reactions can undergo different pathways; some of them include stepwise mechanism through a planarization of intermediates by losing axial chirality, yet providing enantioenriched photoproducts. Pure atropisomeric molecules with bulky substituents not only impart restricted bond rotations, but also induce conformational changes and alters the reactive pathway from the excited state(s).

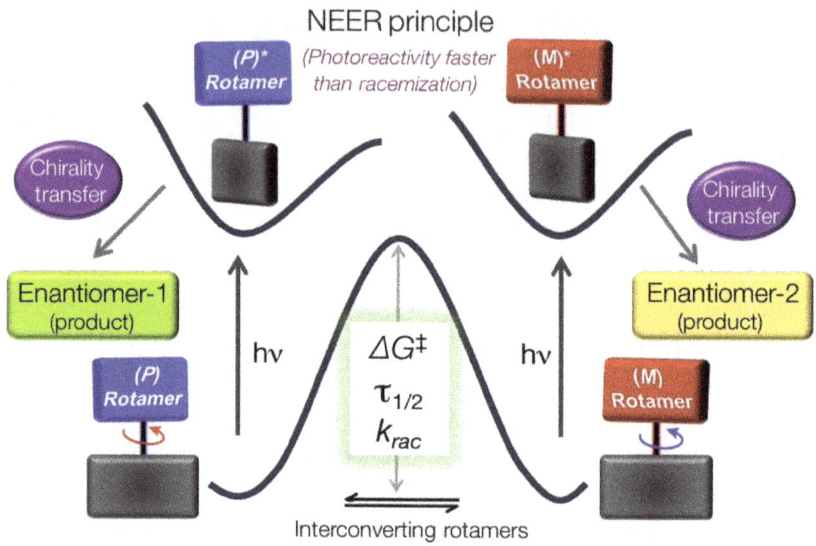

Figure 2. General strategy of "axial to point chirality transfer" in photochemical reactions. Reproduced from Ref. [17] with permission from American Chemical Society.

1.2.1.1 *6π-Photocyclization of acrylanilides*

Sivaguru and coworkers conducted investigations involving axially chiral molecules to impart asymmetric induction leading to desired enantioenriched photoproducts [18, 22, 30]. In 2009, they reported 6π-photocyclization of atropisomeric acrylanilides that undergo *conrotatory* cyclization (ring closure) in solutions resulting in enantioenriched products [19, 20, 40–42]. They investigated achiral *ortho*-substituted N–H acrylanilides **1a** and the corresponding atropisomeric substrates **1b-e**. Both achiral and atropisomeric acrylanilides underwent photoreaction albeit leading to different regioisomeric products (Scheme 1a), where the regioselectivity was determined by their *N*-substitution [20]. The achiral acrylanilide N–H substituted **1a** underwent cyclization at the unsubstituted *ortho*-position on the N-phenyl ring leading to **2a** (with *tert*-butyl group intact), whereas the N-methyl derivative **1b** cyclized at *ortho* position (C_2 carbon bearing the *tert*-butyl group) leading to the cyclized product **2b** along the formation of isobutene by-product (via elimination accompanied by relieving the strain in the molecule; Scheme 1a).

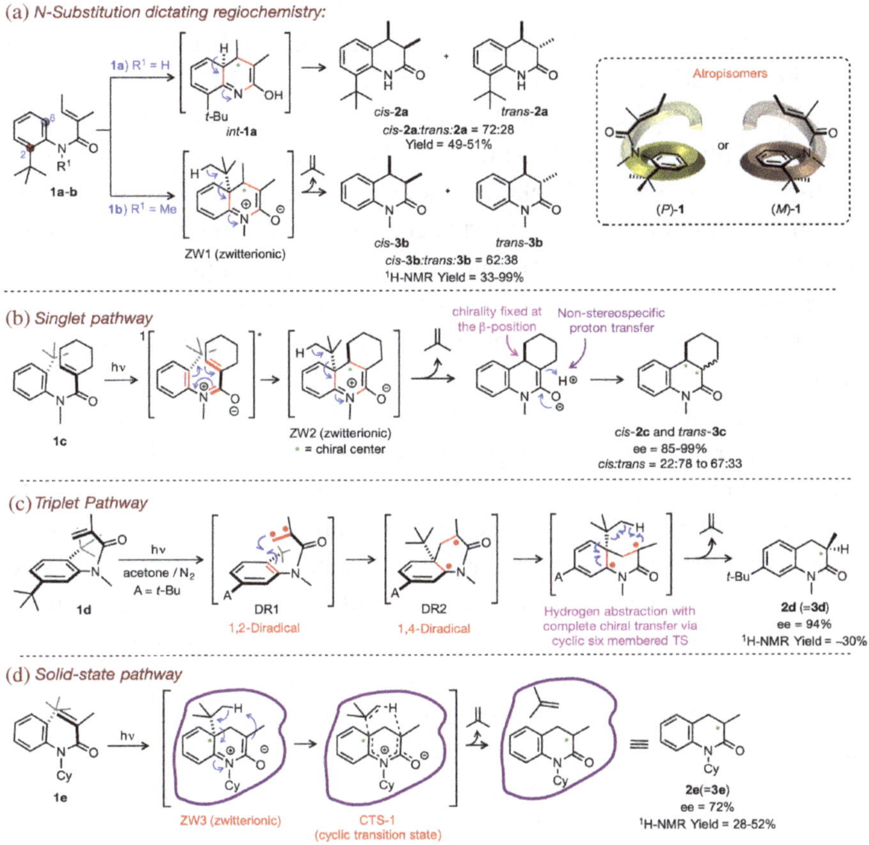

Scheme 1. 6π-Photocyclization of achiral and atropisomeric acrylanilides [19, 20, 23, 24, 29]. Reproduced from Ref. [17] with permission from American Chemical Society.

Comparative studies on the crystal structures of achiral N–H derivative **1a** with N-Me atropisomeric **1b** disclosed that the *N*-methyl substitution on the amide nitrogen in **1b** is responsible for atropisomerism in the molecule. The XRD analysis revealed that the photocyclization in atropisomeric acrylanilides N-Me **1b** occurred on the C_2 carbon to alleviate the developing allylic 1,3-strain (A-strain in intermediate ZW1; Scheme 1a) resulting in the loss of *ortho-tert*-butyl group as isobutene. The keto-enol tautomerization in achiral acrylanilide **1a** (with N–H substitution) leads to the non-existence of allylic 1,3-strain (*int*-**1a**; Scheme 1a) between the *ortho*-*tert*-butyl and *N*-substituents and hence the cyclization occurred at the *ortho* C_6 carbon (without *tert*-butyl) leading to the formation of **2a** [20]. Direct irradiation of enantiopure atropisomeric acrylanilides (e.g., **1c**) afforded photocyclized product with 99% enantiomeric excess (ee) for both the *cis*-**2c** and *trans*-**3c** photoproducts. Based on the above-described mechanistic rationale, atropisomeric acrylanilides **1c** undergo photocyclization through zwitterionic intermediate ZW2 (*via* a singlet excited state) by the elimination of the *ortho*-*tert*-butyl substituent as isobutene by-product. The first cyclization step during the process forms a stereogenic center at the β-position/benzylic position enabling chiral transfer from axial to point chiral, whereas the second step involving H-migration (from solvent or intramolecular H migration) was non-stereospecific which can be observed by the identical ee in photoproducts *cis*-**2c** and *trans*-**2c** [19]. The ee in the photoproducts with direct irradiation (reactivity from singlet excited state) and the first chiral transfer cyclization step of the proposed mechanism detail the importance of β-substituent. This can be concluded for the lack of enantioselectivity in the photoproduct **2d** formed upon direct irradiation for optically pure atropisomeric acrylanilide **1d** that lacked the *tert*-butyl group as β-substituent. Triplet sensitized reactions in α-substituted atropisomeric acrylanilides were reported to sidestep the issue of chiral transfer [23]. Acetone that can act as a solvent and a triplet sensitizer (E_T ~79 kcal×mol^{-1}) was used for atropisomeric acrylanilide **1d**(~77 kcal × mol^{-1}) facilitating the downhill triplet energy transfer and afforded **2d** with 94% enantioselectivity [23]. Photophysical investigations on atropisomeric α-substituted acrylanilides ascertained that the photocyclization occurred from the triplet $\pi\pi^*$ via a radical mechanism involving triplet diradical intermediates DR1 and DR2 (Scheme 1c). This triplet diradical DR2 subsequently underwent a hydrogen atom abstraction from the *ortho*-*tert*-butyl substituent in a stereospecific fashion via a cyclic six membered transition state CTS-1 resulting in the formation of 3,4-dihydroquinolin-2-one photoproduct with 72% ee. Direct irradiation of α-substituted acrylanilide **1d** in trifluoroethanol afforded photocyclized products as racemic mixture (due to reaction originating from the singlet excited state). It was envisioned that photoreactions involving heavy atoms during direct irradiation of α-substituted atropisomeric acrylanilides could enhance the intersystem crossing, resulting in a triplet $\pi\pi^*$

excited state that subsequently guide photocyclization pathway via a radical mechanism [29]. As a result, irradiation of **1d** in trifluoroethanol in the presence of CsF resulted in the desired photoproduct **2d** with 94% ee in solution. Taking advantage of the crystalline nature of α-substituted acrylanilide **1e** the influence of crystalline environment during direct irradiation for 6π-photocyclization resulted in photoproducts **2e** with an ee value of 72% (Scheme 1d) [24]. It was rationalized that the enantioselectivity in photoproduct was due to stereospecific hydrogen migration from the zwitterionic intermediate in the crystalline state (ZW3; Scheme 1d). Single crystal X-ray analysis of **1e** substrate revealed that the distance between the *ortho-tert*-butyl hydrogen and the α-alkenyl carbon to be around 2.8 Å which favored the prevention of allylic 1,3-strain (A-strain) between the N-alkyl and the *ortho-tert*-butyl. This strain prevented the photocyclization at the unsubstituted *ortho*-carbon (C_6 carbon) on the phenyl ring. In addition, it also directed a stereospecific H-atom migration from the zwitterionic intermediate (ZW3) through a cyclic six-membered transition state "CTS-1" leading to the desired photoproduct with 72% ee and isobutene as by-product (Scheme 1d).

1.2.1.2 *Divergent photochemical transformation of enone carboxamides*

Sivaguru and coworkers also evaluated atropisomeric enone carboxamides to study the influence of axial chirality toward the reactivity during photochemical transformations. A comparative study on the reactivity of achiral enone carboxamide **4a** and atropisomeric enone carboxamide **4b–j** with bulky *tert*-butyl group at the *ortho* position under various reaction conditions (Scheme 2) [34]. The achiral carboxamide **4a** (without any substitutions on N-Aryl ring) on direct irradiation at ~350 nm or triplet sensitization in the presence of sensitizer (thioxanthone) at ~420 nm underwent 6π-photocyclization and led to the formation of 3,4-dihydroquinolin-2-one *cis*-**5a** and *trans*-**5a** (Scheme 2). On the other hand, atropisomeric enones with **4b-j** with bulky *tert*-butyl group at the *ortho* position responsible for the steric bias and as a face-shielding group afforded the corresponding spiro-β-lactam photoproducts **6** with ee >95% and *de* >90% (**6/7**). The presence of bulky *ortho-tert*-butyl group directed the reaction to undergo efficient hydrogen abstraction which was similar to atropisomeric

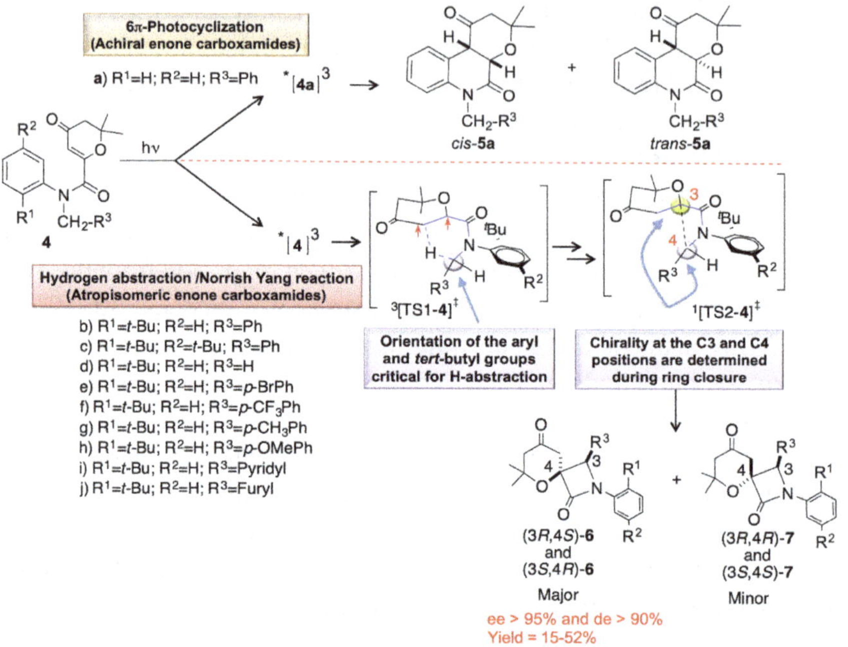

Scheme 2. Divergent phototransformation of enone carboxamides **4** dictated by axial chirality [34]. Reproduced from Refs. [17, 34] with permission from American Chemical Society and John Wiley and Sons, respectively.

acrylanilides (Scheme 2) that showed prevention of developing strain and 6π-photocyclization in atropisomeric enones [34]. The *tert*-butyl group forced the H-atom abstraction (Norrish Yang reaction) of the *N*-methenyl hydrogen due to its optimal bond distance resulting in the formation spiro-β-lactam **6/7**. The relative orientations of benzylic aryl group and the *ortho-tert*-butyl substituent in the enones were investigated by computational studies for rationalizing the formation of spiro-β-lactam from atropisomeric enones (Scheme 2). The observed chemo- and stereoselectivity in the photoproducts was understood from the proposed mechanistic rationale, that detailed the critical role of the above-mentioned two orientations. The crucial step of H-atom abstraction step (via a triplet transition state ^{3}[TS1-**4**]) followed by subsequent ring-closure step (via the singlet transition state ^{1}[TS2-**4**]) was directed by the steric orientation of bulky group at *ortho* position. This investigation highlighted that the atropisomeric enone carboxamides can show divergent photochemical reactivity when compared to their achiral counterparts leading to new reactivity.

1.2.1.3 *Norrish Yang cyclization/Type II reactions of α-oxoamides*

Atropisomeric α-oxoamides with *N*-substitutions and at different irradiation conditions underwent Norrish type II reaction to yield β-lactam or oxazolidin-4-one products (Scheme 3) [21, 43]. For example, photoirradiation of enantiopure α-oxoamides **8a** in chloroform underwent Norrish cyclization and afforded β-lactam **9a** as the major product with >95% conversion and relatively high enantioselectivity. It was reported that these reactions show temperature different enantioselectivity with favorable differential enthalpic ($\Delta\Delta H^{\ddagger}$) contributions at low temperatures and with significant differential entropic ($\Delta\Delta S^{\ddagger}$) contributions at elevated temperatures. The enantioselectivity outcome in the β-lactam product facilitated by temperature dependence was postulated to be due to the competition between bond rotation around C-N axis versus bond formation in the 1,4-diradical (DR3). It was understood that an efficient transfer of axial chirality to point chirality is favored at low temperatures

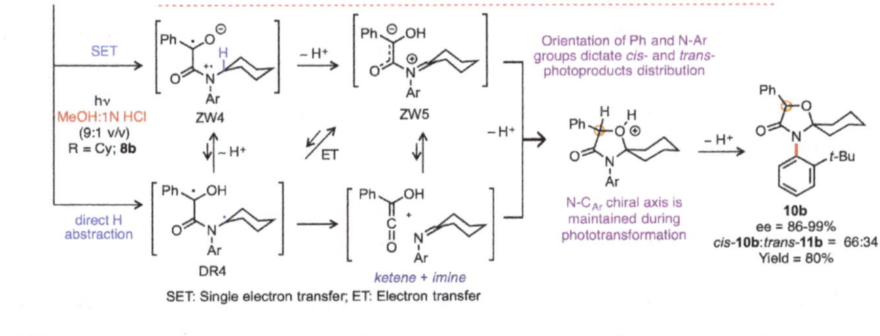

Scheme 3. Atropselective Norrish-Yang photoreaction of atropisomeric α-oxoamides **8** [21, 43]. Reproduced from Refs. [17, 43, 21] with permission from American Chemical Society and Royal Society of Chemistry, respectively.

because of the faster rate of ring closure leading to achieve high ee in β-lactam **9** photoproducts. On the other hand, irradiation of *N*-substituted cyclohexyl (**8b**) in protic medium (MeOH-1N HCl 9:1 v/v) resulted in the formation of oxazolidin-4-ones **10/11** as the major product with high enantioselectivity (Scheme 3) [27]. Depending on the orientation of the *ortho-tert*-butyl group with respect to chiral center at the phenyl ring the diastereomeric *cis*-**10b** and *trans*-**11b** was observed in the photoproduct was resolved by chromatography. The rotational barrier of the N–$C_{(Aryl)}$ chiral axis in oxazolidin-4-one product was found to be small when compared to the reactant **8b** and it was ascertained due to the reduced C–N–C bond angle. This favored the efficient conversion of isolated enantiopure *trans*-**11b** isomer to the corresponding *ent-cis*-**10b** isomer without altering C_5 chiral center and thus enhancing a high enantiomeric resolution of *cis*-**10d** oxazolidin-4-one photoproduct with 98% ee. The proposed mechanism of the phototransformation involved a net hydrogen atom transfer to the excited carbonyl group via a planar zwitterionic intermediate that can be achieved either by direct hydrogen abstraction (Scheme 3, top) or can happen through a sequential two-step process beginning with a single electron transfer followed by a proton transfer (Scheme 3, bottom) [44–46]. Depending on the reaction conditions like substitution on the amide nitrogen or reaction medium, the formed intermediate can be a diradical (DR3/DR4) or a zwitterion (ZW4/ZW5) that undergo cyclization leading to form either β-lactam **9** or oxazolidin-4-one **10** photoproduct. In addition, the formation of **10** can be occurred via a planar zwitterionic (ZW4/ZW5) intermediate or via a ketene-imine geminate pair (when a diradical DR4 is involved) that leads to the formation of oxazolidin-4-one without stereo-isomerization in the intermediate. During the course of net hydrogen atom transfer the N–$C_{(Aryl)}$ chiral axis is maintained through a near planar intermediate(s), and the chiral memory retained during the transformation.

1.2.1.4 *4π-Ring closure of 2-pyridones*

Building on the axial-point chiral transfer strategy, 4π-ring closure reaction of atropisomeric 2-pyridones **12** in solution was evaluated to produce enantioenriched β-lactam **13** photoproduct with good isolated yields

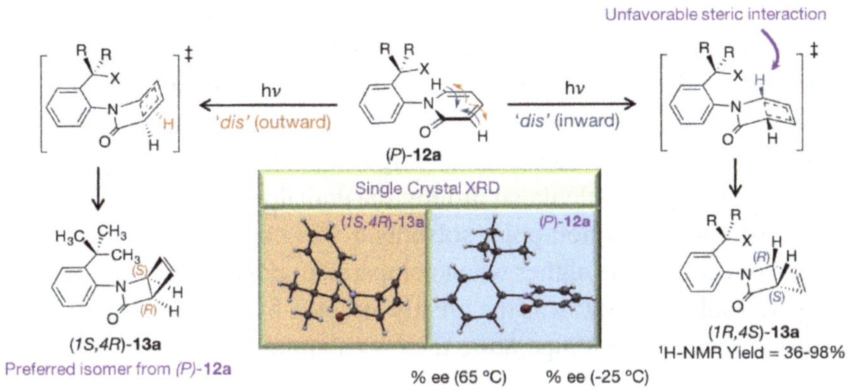

12a = R, X = Me
12b = R = Me; X = OH
12c = R = Ph; X = OH

		% ee (65 °C)	% ee (-25 °C)
12a	Toluene	72	72
	MeOH	86	86
12b	Toluene	21	82
	MeOH	61	93

Scheme 4. Enantiospecific 4π-ring closure of atropisomeric 2-pyridones [25]. Reproduced from Refs. [17, 25] with permission from American Chemical Society.

(Scheme 4) [25]. The carbonyl oxygen in 2-pyridone was designed to engage in H-bonding in addition to providing steric bias that imparted by the axial chirality (as a result of face-shielding group substitution on the *ortho*-substituents at the *N*-Aryl ring) in **12b-c**; (R = Ph or Me; X = OH). Stereospecific 4π-ring closure reaction of **12b-c** was evaluated and the reactivity with non-H-bonding substituents **12a** (R = X = Me) that feature only steric interactions.

The polarity of solvents, H-bonding ability, and the temperature played a major role in providing higher stabilization in terms of the racemization barrier. For the hydroxyl substituted pyridones **12b-c**, the intra and inter molecular H-bonding with polar protic solvent like MeOH enhanced the stability of the axial chirality. This impacted the racemization barriers that had influence on kinetic studies and atroposelective phototransformations. In studying the effect of solvent and temperature on the enantioselectivity in the photoproducts, it was observed from the Eyring plots that in case of **12a**, the 4π-ring closure was primarily entropically governed and has near-zero/minimal enthalpic ($\Delta\Delta H^{\ddagger}$) contribution. In **12a** (that lack of –OH substitution), the N–$C_{(Aryl)}$ bond rotation was predominantly controlled by pure steric interactions, and hence, the ee values were not affected by changing the temperature and solvent conditions. On the other hand, substrates **12b-c** that possess hydroxy functional group for H-bonding, temperature and solvent affects the ee values in the photoproduct. The role of steric/H-bonding groups in the 4π-photocyclization was supported by the calculated differential enthalpic ($\Delta\Delta H^{\ddagger}$) and entropic ($\Delta\Delta S^{\ddagger}$) parameters. In non-polar solvents it was observed that the contribution from $\Delta\Delta H^{\ddagger}$ was higher when compared to polar protic solvents and for a given solvent $\Delta\Delta H^{\ddagger}$ was the greatly affected by the temperature. This suggested that at high temperature the contribution from $\Delta\Delta H^{\ddagger}$ decreased [25].

1.2.1.5 *[2+2]-Photocycloaddition*

Organic reactions involving [2+2]-photocycloaddition is one of the most utilized classical photoreactions employed in the organic synthesis [47]. This reaction methodology provides avenues to access complex carbocyclic and heterocyclic cyclobutane derivatives with good stereo and chemo-selectivity. Apart from using the *ortho-tert*-butyl substituent on *N*-Aryl ring, investigations on [2+2]-photocycloaddition of atropisomeric chromophores were carried out with relatively less bulkier functional groups [26, 30]. By this design change in the atropisomeric substrates, apart from providing axial chirality presented an added advantage was to modify one group to a reactive functional unit that upon chemical transformation results in permanently locked axial chirality in the desired enantioenriched photoproducts with a complex structural units [27]. The photoreaction of 3,4-dihydro-2-pyridones **14** proceeded efficiently and furnished desired enantioenriched tricyclic cyclobutane derivative **15/16** in the presence of triplet-sensitizers like xanthone and acetone (Scheme 5). The nature of R^1 substituent (substituents that govern the axial chirality in the molecule) on the alkyl chain dictates the diastereomeric ratio (dr between **15/16**) in the product (Scheme 5). In atropisomeric enamides **14a** or **14c** (when R^1 = H) an exclusive cyclobutane product **15** was observed. However, in **14b** (when R^1 was methyl substituent) a new diastereomeric *trans*-cyclobutane product **16b** was formed that was confirmed by single crystal XRD studies. It was ascertained that the formation of a new

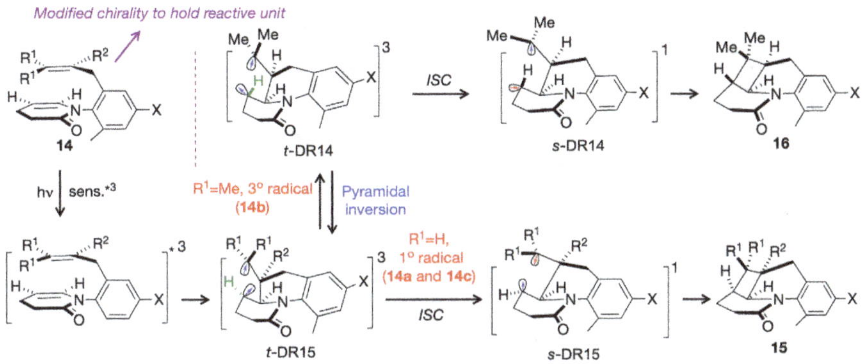

Substrate		Product (selectivity)
R^1 = H; R^2 = CH_3; X = CH_3	(-)**14a**	(*S,S,S*)-**15a** (98% ee; exclusive product)
	(+)**14a**	(*R,R,R*)-**15a** (98% ee; exclusive product)
R^1 = CH_3; R^2 = H; X =H;	(-)**14b**	**15b** (98% ee); **15b**/**16b** = 2.8 :1 to .4.3:1 (solvent dependent)
R^1 = H; R^2 = H; X = H;	(-)**14c**	(*S,S,S*)-**15c** (98% ee; exclusive product)
	(+)**14c**	(*R,R,R*)-**15c** (98% ee; exclusive product)

ISC = Intersystem crossing ^{1}H-NMR Yield = 20-92%

Scheme 5. [2+2]-Photocycloaddition of atropisomeric 3,4-dihydro-2-pyridone **14** [27]. Reproduced from Refs. [17, 27] with permission from American Chemical Society and Royal Society of Chemistry.

diastereomer was a result of formation of a relatively stable 1,4-diradical intermediate which was generated during [2+2]-photocycloaddition. Photophysical studies clearly demonstrated the reaction occurred from the triplet excited state, in which the stability of 1,4-diradical intermediate was dictated by the nature of R^1 substituent (H vs. Me). Depending on the type of substituent a primary 1,4-diradical *t*-DR15 when R^1 = H with a short lifetime and a faster recombination via a singlet diradical (*s*-DR15) and subsequent cyclization resulted **15** as the exclusive product. On the other hand, when with R^1 = Me a tertiary 1,4-diradical *t*-DR14 with a longer lifetime is enough for to lead a pyramidal inversion at the β-carbon of the lactam ring resulting in diastereomer **16b** with a decreased diasteroselectivity. Thus, this opened up a new strategy to produce chiral center(s) with excellent enantiomeric purity in the desired products.

The influence of axial chirality in atroposelective [2+2]-photocycloaddition reactions was extended to evaluate other chromophores such as maleimide **17** (Scheme 6) [32]. The chemoselectivity in maleimides was

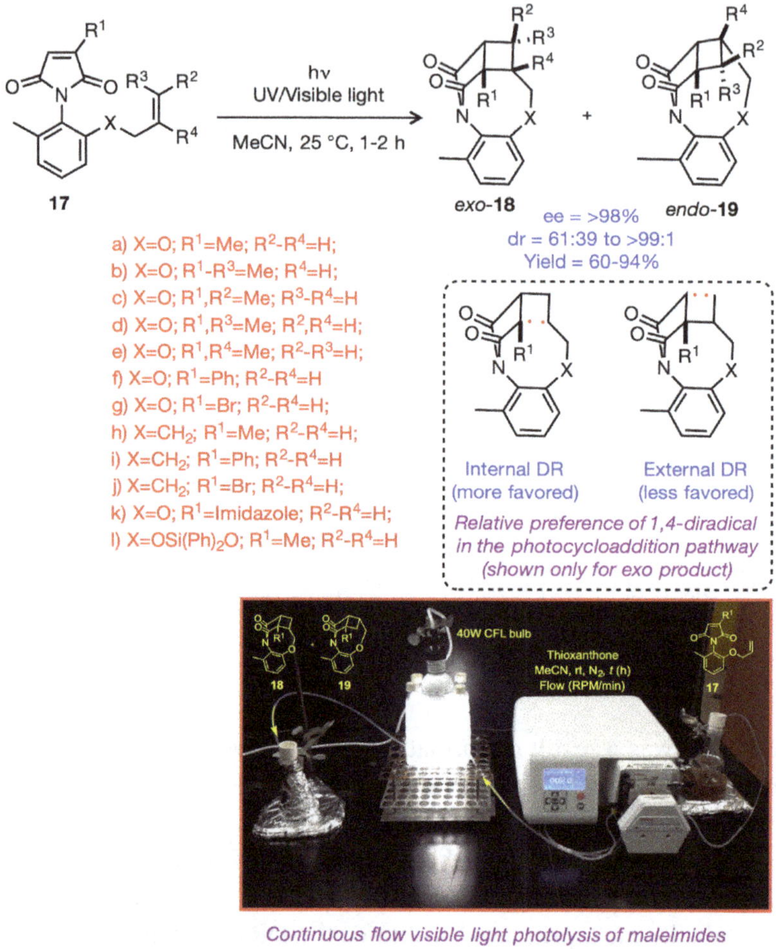

Scheme 6. [2+2]-Photocycloaddition of atropisomeric maleimide [32]. Reproduced from Refs. [17, 32] with permission from American Chemical Society.

completely driven by the chain length of alkenyl tether. For example, subjecting maleimide with a four-carbon (butenyl) tether instead of three-carbon (allyl) tether to photoreaction (direct or sensitized) resulted in [2+2]-photoproduct exclusively (Section 2.6). The role of stable chiral axis and its efficient chirality transfer to a desired enantioenriched photoproduct were studied by varying different substituents like oxygen on to the tether (e.g., oxy-allyl chain) that can be cleaved to form building cyclobutane building blocks after the photoreaction. The photophysical studies revealed that maleimides in the presence of suitable triplet sensitizer (xanthone and thioxanthone) formed the corresponding triplet excited states leading to the formation of *exo*-**18** and *endo*-**19** photoproducts which were characterized by XRD analysis. Photochemical reactions in all the designed substrates resulted in desired photoproducts with >98% ee. It was concluded that the diastereomeric ratio in the photoproduct was dictated by the substitution in the maleimide double bond and the alkenyl tether. For example, compounds with R^1 = Me and Ph on the maleimide double bond resulted in the dr of 79:21 and >99:1, respectively; on the other hand, when allyl-oxy (for R^1 = Me) was used in the place of butenyl tether the diastereomeric ratio was changed from 79:21 to 74:26. Scrambling studies were performed to understand the mechanistic details involved in the formation of cyclobutane photoadduct. It was understood that formation of external bond in the cyclobutane is the first step followed by the sequential recombination of the 1,4-biradical [48]. With the optimized reaction conditions continuous-flow visible-light photocatalysis (Scheme 6) was performed to show the feasibility of large scale reactions. Employing continuous flow conditions, complete and efficient conversion was achieved in 35 min, whereas only similar scale on batch mode resulted in only 23% conversion.

The role of axial chirality in classical photochemical reactions provided excellent atropselectivity in the photoproducts. Using the same strategy, atropisomeric acrylimides **20a-c** were studied to understand straight versus cross photocycloaddition (Scheme 7). On contrast with the reported literature [49, 50], a comparative study on atropisomeric

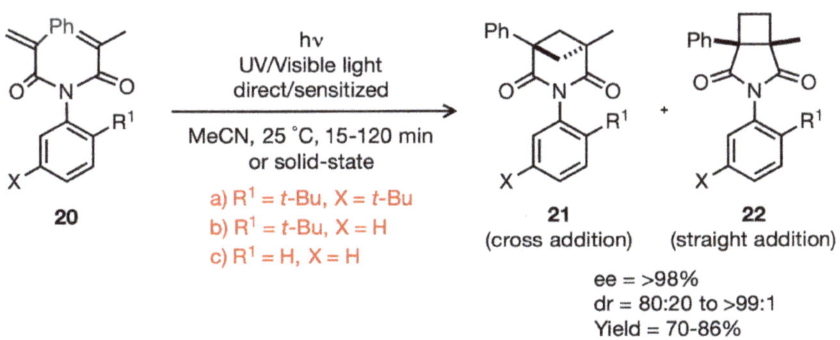

Scheme 7. [2+2]-Photocycloaddition of atropisomeric acrylimides [31]. Reproduced from Refs. [17, 31] with permission from American Chemical Society.

acrylimides (**20a-b**, R[1] = *t*-Bu) with non-atropisomeric acrylimide (**20c**, R[1] = H) resulted in a mixture of cross and straight addition products of **20c** was formed in the ratio of 80:20 [31]. Whilist, **20a-b** in both solid and solution state yielded exclusively the cross photocycloaddition products **21a-b** with >99:1 atropselectivity. On the basis of single-crystal XRD studies of enantiopure **20a** it was rationalized that for cross addition, the formation of photoproduct was a result of minimal atomic movement.

Similarly, evaluation of chemoselectivity in atropisomeric enones **23a-e** was performed to study the ratio of straight and cross [2+2]-photocycloaddition products [51]. Photoreactions on enone amides (**23a-c**) and enone imides (**23d-e**) resulted in exclusive straight addition products (*cis,cis*-**24**/*cis,trans*-**25**; Scheme 8) [52]. Interestingly, the nature of substituent "X", i.e., amide versus imide played a role in determining the ratio of photoproducts **24**:**25**. For example, in enone amide (X = alkyl), a 1:1 mixture of diastereomers was observed, whereas, in enone imide (X = CO) exclusive *cis,cis*-**24** was observed. On the basis of extensive photophysical studies, it was understood that the triplet state reactivity was likely triggered by $\pi^* \rightarrow \pi^*$ interaction between the excited enone and the electron-deficient alkene tether (in case of imide substrates). On a particular note, loss of

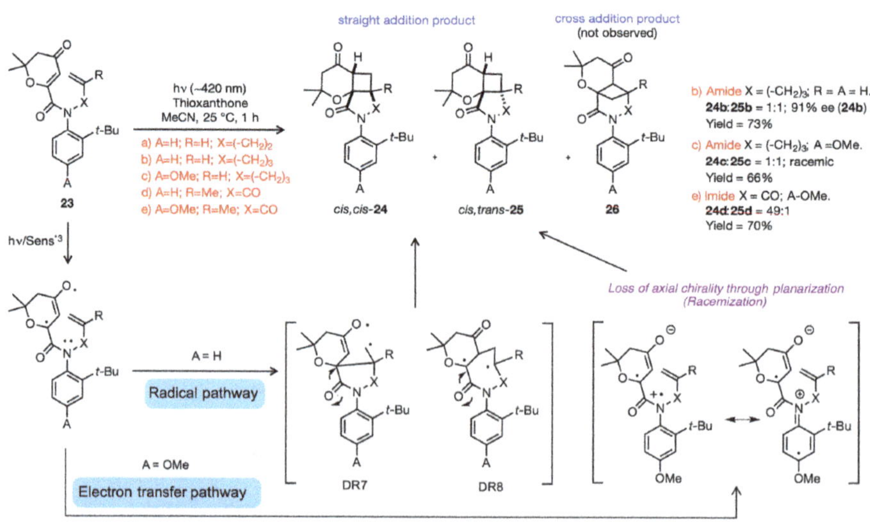

Scheme 8. [2+2]-Photocycloaddition involving atropisomeric enones [52]. Reproduced from Refs. [17, 52] with permission from American Chemical Society.

atropselectivity with enone-amide **23c** was observed in the photoproduct and was ascertained because of the strong electron-donating nature of *para*-methoxy substituent. Thus, it would likely proceed via an electron transfer pathway leading to the loss of axial chirality through planarization of the N–$C_{(Aryl)}$ chiral axis (Scheme 8). Nevertheless, the formation of desired enantio- and diastereo-enriched products was mainly guided by the presence of N–$C_{(Aryl)}$ chiral axis in these atropisomeric enones.

1.2.1.6 *Paternò–Büchi reaction*

Stereospecific Paternò–Büchi reaction [53, 54] of atropisomeric α-oxoamides **27a-c** in solution leads to corresponding diastereomeric bicyclic oxetane products **28** and **29** in moderate to high yields (Scheme 9), respectively [28]. On the other hand, control study on the non-atropisomeric α-oxoamides **27d** (R^2 = H; X = H) in solution resulted in less product formation (30% yield) after 12 h of irradiation. Irradiation of optically pure reactants in the solid state showed reversals of diastereomeric ratios when compared to solution. For example, in the case of atropisomeric oxamide (*M*)-**27b,** photoirradiation in acetonitrile resulted in **28b**:**29b** product ratio of 82:18 with the formation of (*R,R,M*)-**28b** with an ee of 99%, while in solid-state irradiation the product ratio of **28b**:**29b** was 15:85 with the formation of **28b** in 99% ee. Based on the photochemical reactivity of excited ketones with electron-deficient alkenes, the charge transfer interactions between $\pi^*_{C=O}$ of the carbonyl group and the $\pi^*_{C=C}$ of alkene have an important consequence on the stereochemical outcome of the reaction. This orbital interaction led to the formation of a 1,4-diradical intermediate where the first bond is formed between the alkene CH_2 (β-carbon of the alkene) and the carbonyl carbon. The results showed that formation of the diradical intermediate both in solution and in the solid state was completely conformer-dependent and the subsequent radical recombination dictates the diastereomeric ratio in the oxetane products **28** and **29** (Scheme 9, bottom).

1.2.1.7 *[5+2]-Photocycloaddition*

Atropselective [5+2]-photocycloaddition of alkenyl-substituted maleimides **30** led to formation of highly enantio- (ee > 98%) and diastereomer-enriched

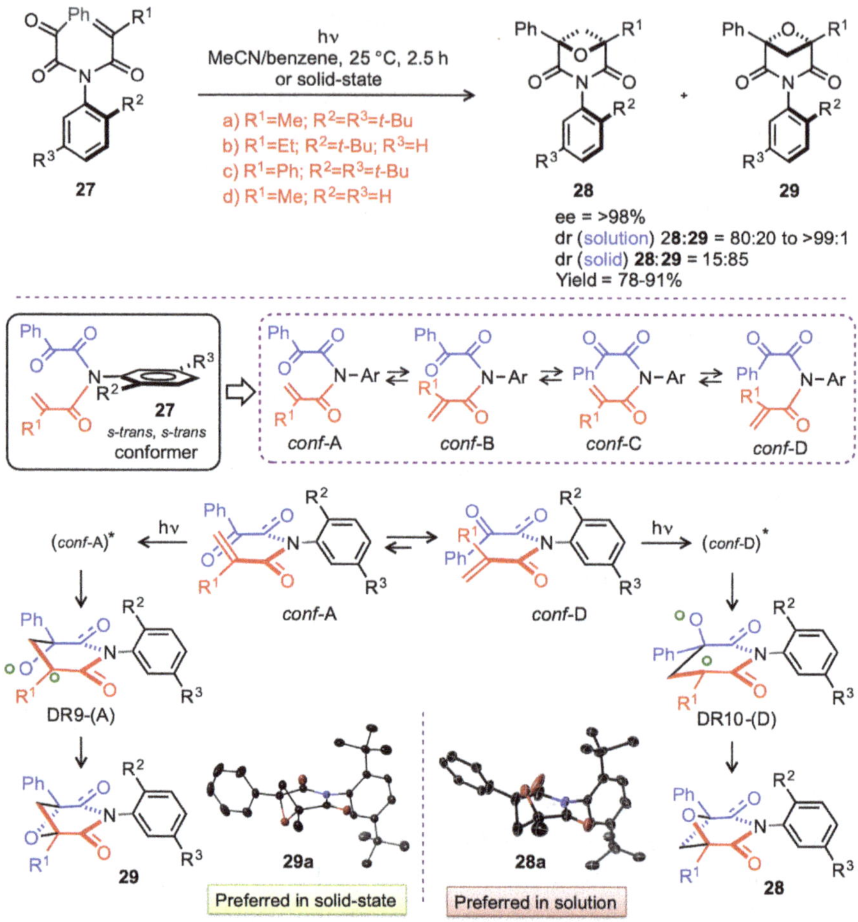

Scheme 9. Paternò–Büchi reaction of atropisomeric α-oxoamides **27** [28]. Reproduced from Refs. [17, 28] with permission from American Chemical Society and Royal Society of Chemistry, respectively.

(dr > 98%) azepinone/indoline products (Scheme 10) [33, 55]. As reported earlier the chemoselectivity of photocycloaddition ([2+2] vs. [5+2]) in alkenyl-substituted maleimides was dictated by chain length of the alkenyl tether. It was concluded that maleimides **30** bearing allyl groups (three-carbon alkene tether), the [5+2]-photocycloaddition was observed exclusively with excellent ee values (>98%) in the photoproduct **31/32**. The formation of diastereomeric photoproducts after the photoreaction was greatly influenced by type of R^1 substituent (electronic effect) on the

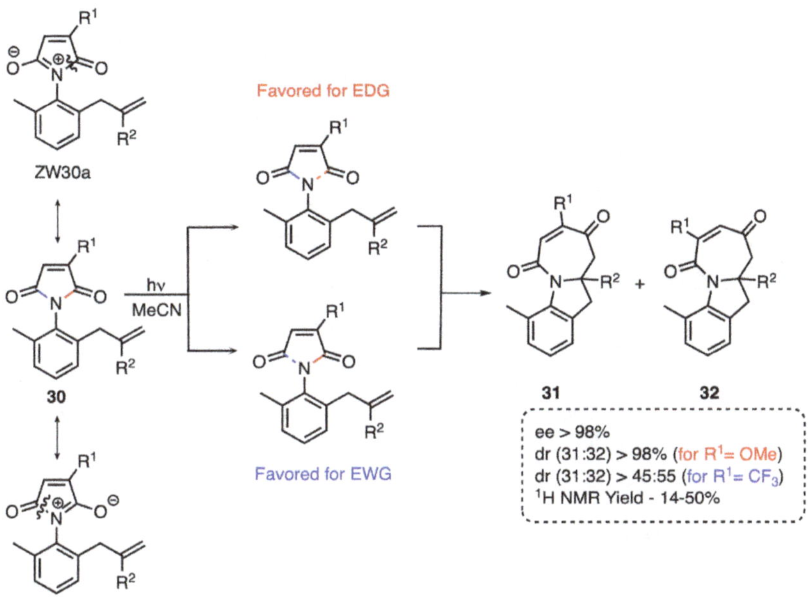

Scheme 10. [5+2]-Photocycloaddition of atropisomeric maleimide derivatives **30** [33]. Reproduced from Refs. [17, 33] with permission from American Chemical Society and John Wiley and Sons, respectively).

atropisomeric maleimide. The reaction occurred through the cleavage of regioisomeric N_{aryl}–CO bonds (blue vs. red bonds in **30** in Scheme 10). The electron-donating (OMe) R^1 substituent in **30** resulted in **31** as the major product (**31**:**32** >98:2) and when the substitution R^1 changes to electron-withdrawing group (EWG) (CF_3) led to the formation of **32** in relatively excess yields. Thus, the stabilization of zwitterionic forms (ZW30a and ZW30b) of maleimide were either strengthened or weakened by one of N_{aryl}–CO bonds (blue vs. red bonds in **30** in Scheme 10) as a result the reversal in the diastereomeric ratio was observed.

1.2.1.8 *Effect of pressure on racemization barrier and stereospecific photoreaction*

Since pressure, volume, and temperature are three fundamental parameters, these physical parameters were exploited to understand and

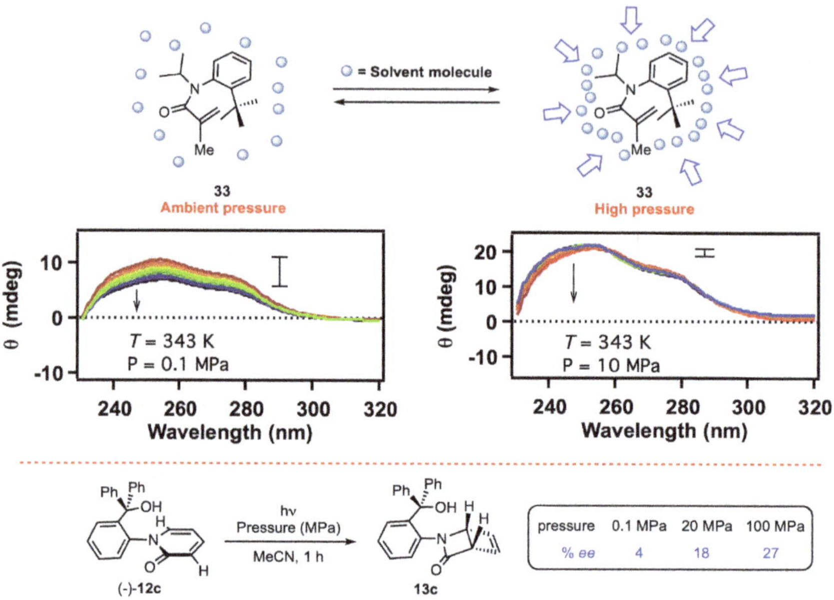

Scheme 11. (a) Racemization kinetics of (*P*)-**33** monitored by CD spectroscopy under ambient (left) and elevated (right) pressures in methylcyclohexane (MCH) at 343 K. (b) Enantiospecific 4π-ring closure of pyridone **12c** under elevated pressure [26]. Reproduced from Refs. [17, 26] with permission from American Chemical Society and John Wiley and Sons, respectively.

investigate the dynamics in atropisomeric systems. In non-biaryl atropisomeric compounds, increasing pressure in the system can slow down the racemization of 33 around axial chirality, was exploited for controlling photochemical transformations performed even at higher temperatures (Scheme 11, top right vs. left) [26]. The reduction in activation volume in the transition state during racemization lowered the rate of racemization. For example, in pyridone **12c**, upon increasing the pressure from 0.1 (ambient pressure) to 100 MPa resulted in the increase of ee 4% to 27% at 70°C in the formation β-lactam photoproduct **13c** (Scheme 11).

1.3 Supramolecular Photochirogenesis

Enzyme-catalyzed reactions demonstrate nature's mastery over non-covalent interactions [56]. Enzymatic reaction is based on sequestering substrates in a reaction-ready configuration by lowering the activation energy for the chemical transformation(s) and expelling the product(s) by unfavorable interactions in the confined environment of the enzymatic pocket. Inspired by this fine-tuning during molecular evolution, chemists developed systems that feature confinement and non-bonding interactions for catalysis [57].

To fine-tune this molecular action beyond the reactive substrates (called guests, G), chemists built molecular architectures "supramolecules" (called hosts, H) that could potentially impact the outcome of the photochemical transformation [58]. In this section we will highlight the role of supramolecules in assisting and controlling light-induced transformation(s) leading to supramolecular photocatalysis. In order to influence the photochemical and photophysical properties of the guest molecules using the non-bonding interactions that develop between the host and guest, three important features [59] have to be manipulated, viz., (a) rigidity in the structure that arises due to interactions between the supramolecular host(s) and the reactive guest(s); (b) the translational, rotational, and vibrational motion(s) of the guest(s) impacted by space and volume restrictions when interacting with the host(s). To translate these non-bonding interactions for efficient supramolecular photocatalytic process precise control of kinetic, thermodynamic, and excited state properties of the guest molecules within the supramolecular environment needs to be manipulated [59]. Various supramolecular assemblies based on organic and inorganic templates were developed to enhance and control photochemical reactivity and selectivity and fine-tune the photophysical properties of the guest molecules [58, 60].

In this section we will focus on the photochemical/supramolecular photocatalytic reactivity of the guest molecules under the influence of host. A variety of well-defined supramolecular hosts like zeolites [60], micelles, dendrimers, etc. offer unique properties and impart excellent control on reactive substrates. In this section we will highlight a few supramolecular hosts (Figure 3), viz. cucurbiturils [59, 61–63], cyclodextrins [64], cavitands [65], co-ordination nanocage (CNC) [66–68], H-bonding templates [69–76], and metallosupramolecular assemblies [77–79]. Readers are encouraged to refer to the use of biomolecules and organized chiral templates for photochirogenesis reviewed by Inoue and coworkers [80] and template effects for controlling photoreactions in solution by Bassani and coworkers [81].

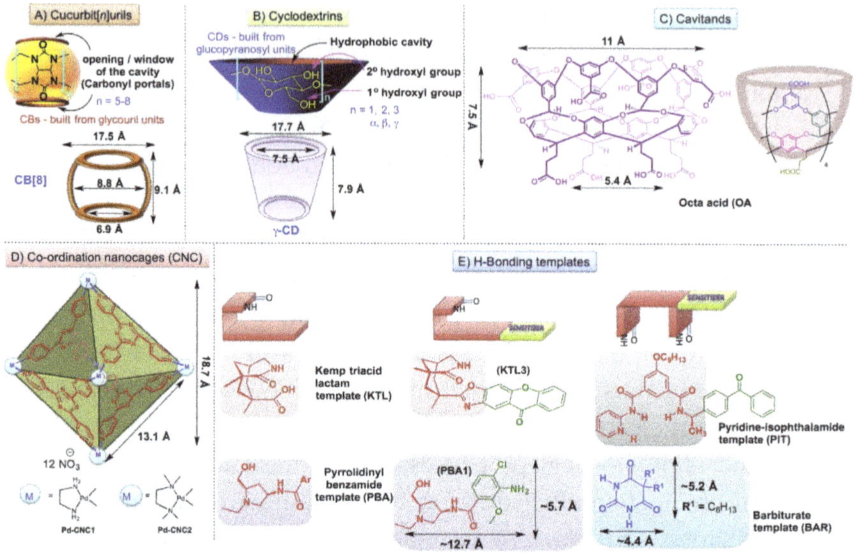

Figure 3. Structural features of various supramolecular containers and supramolecular templates. Reproduced from Ref. [11] with permission from Royal Society of Chemistry.

1.3.1 Influence of Confinement and Non-bonding Interactions in Altering the Excited State Properties

Photo-induced transformations involve reactive excited states that are short-lived and can readily relax to ground state through radiative and/or non-radiative transitions. Hence, a delicate control over molecular dynamics that influence the excited state reactivity and selectivity as well as photophysical events is quintessential. One of the successful strategies developed to channel the excited-state energy toward useful photochemical transformation is by employing organized assemblies that pre-orient reactive molecule(s) in a "reaction-ready" configuration. Governing reactivity using supramolecular scaffolds enables chemists to efficiently carryout light-induced transformations that are generally inefficient in isotropic media (using solvent alone). The confinement of chromophores within supramolecular assemblies opens up avenues to enhance or alter their excited state properties. Physical properties (e.g., solubility) of the host that is employed as supramolecular catalyst should also be considered for the same.

1.3.2 General Considerations about Supramolecular Photocatalysis

Acceleration of the reaction through supramolecular assemblies typically require host–guest interactions which are assisted by non-covalent forces (e.g., Van der Waals, hydrogen-bonding, Coulombic and π–π interactions). The three major criteria which need to be satisfied for efficient supramolecular photocatalysis are (a) pre-orientation of reactant(s) in a "reaction-ready" configuration, (b) the quantum yield of photoreaction of interest within the supramolecular scaffold must be higher than the quantum yield of photoreaction (for both uni- and bi-molecular reactions) in isotropic media, and (c) the dynamic exchange of photoproduct(s) and reactant(s) that enables sequestering of reactant(s) and product release from the supramolecular scaffold leading to turnover and catalysis.

Comprehending the above three fundamental criteria for supramolecular photocatalysis require a clear comprehension of the thermodynamic and kinetic features involving host–guest interactions as well as the efficiency of photochemical transformations of the guest substrate sequestered by the supramolecular host. In this section we will highlight photocatalysis of unimolecular and bimolecular transformations involving host:guest ratios of 1:1 (Figure 4, top), 2:1 (Figure 4, middle), and 1:2 (Figure 4, bottom), respectively. One can envision an extension of this treatment of supramolecular photocatalysis to higher order host–guest complexes (e.g., 2:2 host–guest complex). was observed.2 depicts that irrespective of the type of host–guest complex (1:1, 2:1, or 1:2), the supramolecular host–guest interactions are reflected by the thermodynamic binding constant (K_a) for the reactive guest(s) and thermodynamic dissociation constant (K_d) for the photoproducts is necessary [84].

For a better catalytic turnover, the binding affinity of the reactive guest(s) substrate should be greater or proportional to the binding affinity of the photoproduct(s) with the supramolecular host to prevent the product inhibition during the catalytic processes. For a desirable catalytic turnover, it is essential to ascertain the kinetics of the individual microscopic steps (both forward and reverse rate constants). The efficiency of the photochemical transformation mediated by the supramolecular host can be evaluated using reaction velocity or quantum yield measurements. The thermodynamic parameters of the host–guest complex can be determined

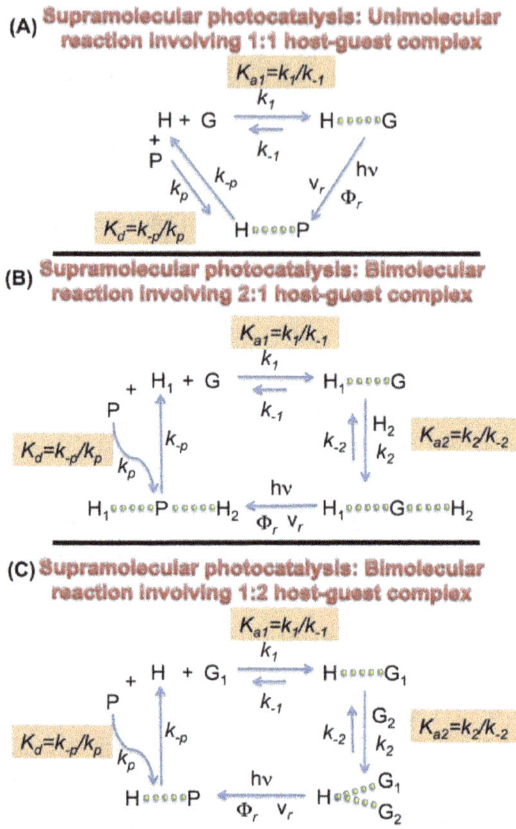

Figure 4. Paradigm for supramolecular photocatalysis and host–guest chemistry. H and G represent the host and guest, respectively. K_a represents the thermodynamic binding affinity of the guest with the host for the individual host–guest complex formation steps. K_d is the dissociation constant for the product from the supramolecular host. k_1, k_{-1}, k_2, and k_{-2} are the forward and reverse rate constants for individual microscopic steps leading the host–guest formation. k_p and $k_{-\mathrm{p}}$ are the rate constants for product association and dissociation with the supramolecular host, respectively. Φ_r and V_r represent the quantum yield and reaction velocity of photoreaction, respectively, mediated by the supramolecular host. Reproduced from Ref. [11] with permission from Royal Society of Chemistry.

experimentally by performing fluorescence titration measurements or by using isothermal calorimetric titrations and the kinetics of the complexation process can also be ascertained by stopped-flow measurements. The forward and reverse rate constants will reflect the thermodynamic binding constants for individual microscopic steps involved in the complexation. Actinometry measurements can be used to determine the quantum yield of the photoreaction and the reaction velocity can be measured by different analytical techniques (e.g., ultraviolet [UV]-visible). Hence, by employing multiple analytical techniques not only helps in understanding host–guest interactions but also provides a base for evaluating the efficiency of the photochemical transformation of interest.

1.3.3 Examples of Different Supramolecular Photocatalysis

1.3.3.1 *Cucurbit[n]urils*

Cucurbit[n]urils (CBs) exhibit a family of macrocyclic compounds which was originally synthesized by Behrend and coworkers [61]. It was synthesized from an acid-catalyzed condensation reaction of formaldehyde with six glycouril units (cucurbit[5]uril; Table 1). Mock and coworkers [85]

Table 1. Structural parameters and physical properties of cyclodextrins (CDs) and cucurbiturils (CBs). Reproduced from Ref. [11] with permission from Royal Society of Chemistry.

Entry	Parameters	Cyclodextrins (CDs)[a]			Cucurbit[n]urils (CBs)[b]		
		α-CD	β-CD	γ-CD	CB[6]	CB[7]	CB[8]
1	No. of units	6	7	8	6	7	8
2	Molecular weight (anhydrous)	972	1135	1297	997	1163	1329
3	Water solubility (g/L)	145	18.5	232	<0.18	<35.0	<0.15
4	Cavity height (Å)	7.9	7.9	7.9	9.1	9.1	9.1
5	Outside diameter (Å)	15.2	16.6	17.7	14.4	16.0	17.5
6	Portal diameter (Å)[a]	4.7	6.0	7.5	3.9	5.4	6.9
7	Inner cavity diameter (Å)[b]	5.3	6.5	8.5	5.8	7.3	8.8
8	Cavity volume (Å^3)	174	262	427	164	279	479
9	Refs.	[64]	[64]	[64]	[59, 62, 63]	[59, 62, 63]	[59, 62, 63]

[a]Inner rim diameter of CDs.
[b]Outer rim diameter of CDs.

Table 2. Structural parameters and physical properties of co-ordination nanocage (CNC), octa acid (OA) cavitands, and H-bonding templates.

Entry	Supramolecular hosts	Structural unit	Dimension (Å)	Refs.
1	CNC[a]	6:4 ratio of metal ions capped with ethylenediamine:1,3,5-tris(4-pyridyl)triazine	~18.7 (length) ~13.1 (facial length)	[66–68]
2	OA[b]	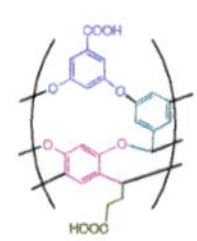	~11 (top opening) ~7.5 (height) ~5.4 (bottom opening)	[65, 82, 83]
3	Kemp triacid lactam (KTL) template[c]	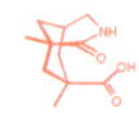	N/A	[69–72]
4	Pyrrolidinyl benzamide (PBA) template[d]	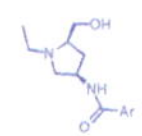	~ 12.7 × ~5.7 (PBA1)	[73]
5	Pyridine isophthalamide (PIT) template[c]	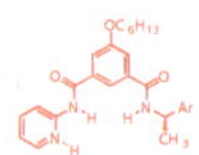	N/A	[74]
6	Barbiturate (BAR) template	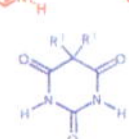	~ 4.4 × ~5.2	[75, 76]

[a]Structural parameters from Refs. [66–68].
[b]Structural parameters from Refs. [65, 82, 83].
[c]NA = structural parameters not available.
[d]Structural parameters from Ref. [73].
Table contents reproduced from Ref. [11] with permission from Royal Society of Chemistry.

coined the name cucurbituril as these supramolecules resembled pumpkins in shape which has the botanical name *Cucurbitaceae*. Kim [86] and Day [87] reported the synthesis of higher homologs of cucurbituril with seven glycouril units known as cucurbit[7]uril (CB[7]) and eight glycouril units (cucurbit[8]uril; CB[8]). In the last decade, functionalized cucurbiturils have also been synthesized [62, 63, 88]. In this section we will concentrate on the ability of CB[8] and CB[7] in photocatalysis as examples of supramolecular catalysis [59]. Kim and coworkers [89] reported enhancing chemical reactivity by employing sub-stoichiometric amounts of CBs. The stereoselective [2+2]-photodimerization of (*E*)-diaminostilbene **34** (Scheme 12) was assessed by employing 50 mol% of CB[8]. A *syn:anti* ratio of 95:5 was observed with a selective formation of corresponding *syn*

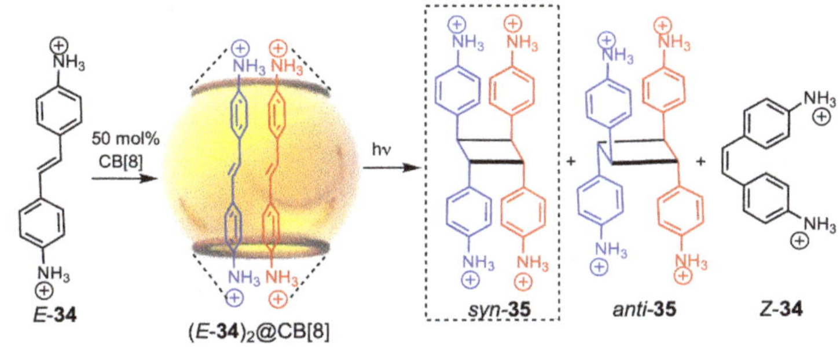

Scheme 12. Stereoselective [2+2]-photodimerization of (*E*)-diaminostilbene mediated by CB[8] in solution. Reproduced from Ref. [11] with permission from Royal Society of Chemistry.

35 photodimer. The confining environment of CB[8] prevented the facile *E/Z* isomerization of stilbene that was observed in isotropic media (solvent). The higher efficiency for the formation of *syn* **35** was attributed to the formation of a stable 1:2 (*E*-**34**)$_2$@CB[8] host–guest complex. The ternary complex has a binding constant of 3.8×10^4 M^{-2} in formic acid-water (2:3 ratio) ascertained by isothermal titration calorimetry measurements. The cationic nature of the reactive guest enabled efficient formation of the host–guest complex, but the release of the photoproduct was not very facile. This necessitated the use of 50 mol% of the supramolecular scaffold. Hence, the utility of CB[8] as a nano-container was successfully demonstrated in which two cationic guest molecules were oriented in close proximity that could promote photoreactions with high control over regiochemistry and stereoselectivity.

Ramamurthy and coworkers employed CB[8] for photodimerization of *trans*-1,2-*bis*(n-pyridyl)ethylene **36a-c** as well as *trans*-stilbazole **36d** in aqueous solution [90, 91]. NMR titrations were used to evaluate the host–guest complexation between *E*-**36a**. The binding constants were found to be 1.03×10^3 (K_{a1}) and 8.6×10^5 M^{-1} (K_{a2}) for the formation of 1:1 and 1:2 host–guest complexes, respectively [91]. Efficient photodimerization was observed in the presence of 50 mol% of CB[8] yielding *syn*-dimer **37**. Control reaction on irradiation in the absence of CB[8] gave the isomerized photoproduct **38**, the intramolecular cyclization product **39**, as well as the hydration product **40** (Scheme 13).

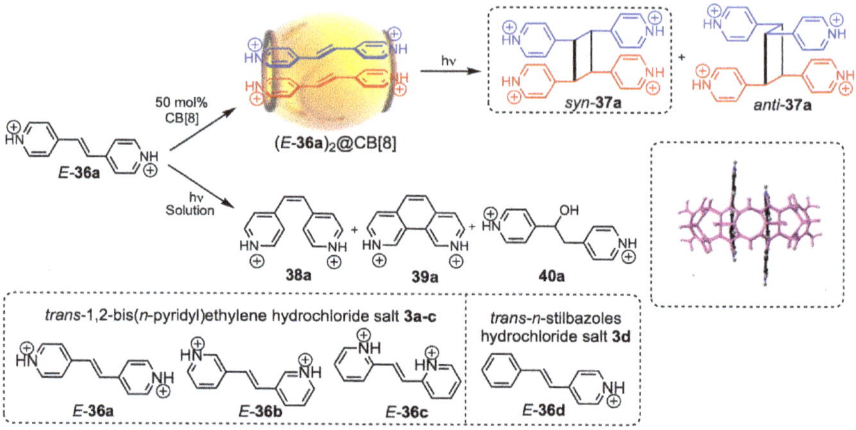

Scheme 13. CB[8]-mediated photodimerization of *trans*-1,2-*bis*(n-pyridyl)ethylene and stilbazole in aqueous media. Single-crystal XRD of the CB[8]-(*E*)-**36a** 1:2 host–guest complex. Reproduced from Ref. [11] with permission from Royal Society of Chemistry.

Two large cationic organic molecules can be efficiently sequestered in CB[8] to influence the photochemical reactivity, CB[7] can be employed for appropriately designed small guest molecules that can be accommodated within its hydrophobic cavity. The contrasting features between CB[8] and CB[7] are their differences in volume (279 $Å^3$ in CB[7] vs. 479 $Å^3$ in CB[8]) and solubility in water (CB[7] is more soluble than CB[8]). Macartney and coworkers [92] reported [4+4]-photodimerization of 2-aminopyridine hydrochloride **41** in the presence of CB[7]. Selective [4+4]-photodimerization of **41** in the presence of CB[7] was observed leading to exclusive formation of the *anti-trans*-dimer **42**. In the absence of CB[7], both *anti-trans* **42** and *syn-trans* **42** photodimers were observed in a ratio of 4:1. The authors explained that the addition of the CB[7] stabilized the photoproduct **42**, thus preventing the rearomatization of **42** to the 2-aminopyridine monomer at room temperature.

Confinement of molecules by cucurbiturils allows for an increased chance of collision between the reactive partners leading to catalysis. The studies above detail one of the common features which is the use of cationic substrates, as the ion–dipole and hydrogen-bonding interactions between cationic groups of guest and portal carbonyl groups of the host, as well as the developing hydrophobic interaction between host and guest favored a thermodynamically stable host–guest complex. But this impedes the facile expulsion of the photoproduct that is formed within the supramolecular cavity, affecting the overall catalysis and turnover. This originates from the strong binding interaction of the guest (now the photoproduct) that is supported by the same ion–dipole and hydrophobic interactions that enabled the host–guest formation between the substrate and the supramolecular host. Despite employing sub-stoichiometric amounts (50 mol%) of CB[8], it does not fully address the issue of low solubility of CB[8] (Table 1) compared to the corresponding cyclodextrin analog (γ-CD). This demanded the requirement of a strategy for sequestering neutral molecules instead of cationic guests within the supramolecular container and controlling their photoreactivity with less than 50 mol% of the host, which can lead to efficient product formation, followed by the expulsion of a neutral photoproduct leading to efficient catalysis [59].

Sivaguru and coworkers reported the photodimerization of neutral molecules (coumarin derivatives) in the presence of CB[8] (catalytic amounts, 10 mol%) in aqueous media [59, 93–95]. 6-Methylcoumarin (**43**, Scheme 15) was used as the model system as its photoreactivity in solution and organized assemblies is well established in literature [93, 95]. The photodimerization of **43** was inefficient in the absence of CB[8] (<10% conversion; 60 min irradiation) in water, leading to the formation of both *syn*- and *anti*- adducts. On the other hand, the same reaction was found to be efficient and clean in the presence of CB[8] in water, leading to the selective formation of *syn*- dimers, with a *syn*-HH **44**/*syn*-HT **46** ratio of about 70:30. The 1:2 host–guest complex (Figure 5a) was successfully

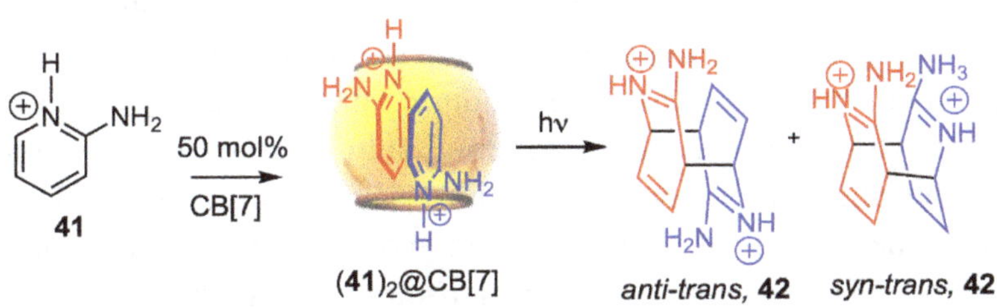

Scheme 14. CB[7] promoted [2+2]-photodimerization of aminopyridines in solution. Reproduced from Ref. [11] with permission from Royal Society of Chemistry.

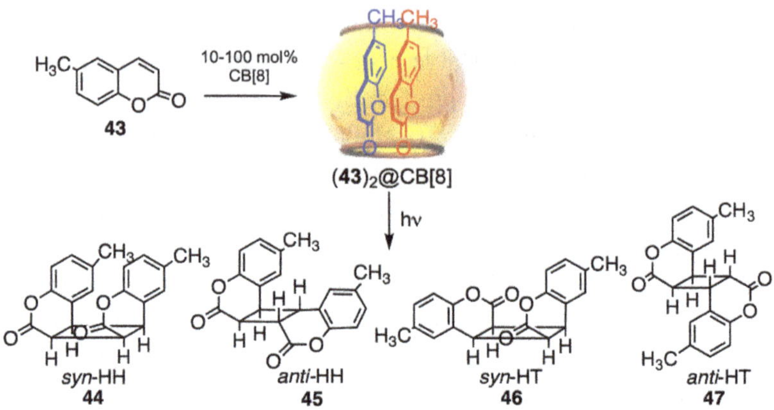

Scheme 15. Photodimerization of 6-methylcoumarin catalyzed by CB[8]. Reproduced from Ref. [11] with permission from Royal Society of Chemistry.

crystallized, giving valuable insights into the role of confinement in influencing the excited state chemical reactivity of coumarin derivatives. Analysis of the single-crystal XRD revealed that the 1:2 CB[8]-**43** host–guest complex had a head-to-tail (HT) orientation (Figure 5a). If this orientation was preferred during photoreaction, one will expect the formation of HT photodimers (*syn*-HT **46** and *anti*-HT **47**; Scheme 15) as the major product. But the formation of the *syn*-photoproducts (i.e., *syn*-HH *and syn*-HT) can be explained with a combination of effects, viz. the difference in orientation of the guest molecules within CB[8] in solution as compared to CB[8] in crystalline environment and/or the higher reactivity of the HH orientation of the host–guest complex in solution in the presence of CB[8] than the HT orientation, leading to *syn*-HH **44/** *syn*-HT **46** ratio of 70:30. The reaction was also carried out efficiently in water under sunlight, leading to comparable selectivity using Hg lamp irradiations [93, 95]. Exclusive formation of *syn*-dimers was observed during the photodimerization of **43** in CB[8], while *anti*-dimers are observed only in water and in isotropic media. This difference in the photoproduct stereochemistry allowed to narrow down the location of photodimerization of **43** when sub-stochiometric amount of CB[8] was employed. The *anti*-dimers were not observed upon employing even at 10 mol% (or higher) of CB[8] indicating that the photodimerization likely occurred within the CB[8] cavity.

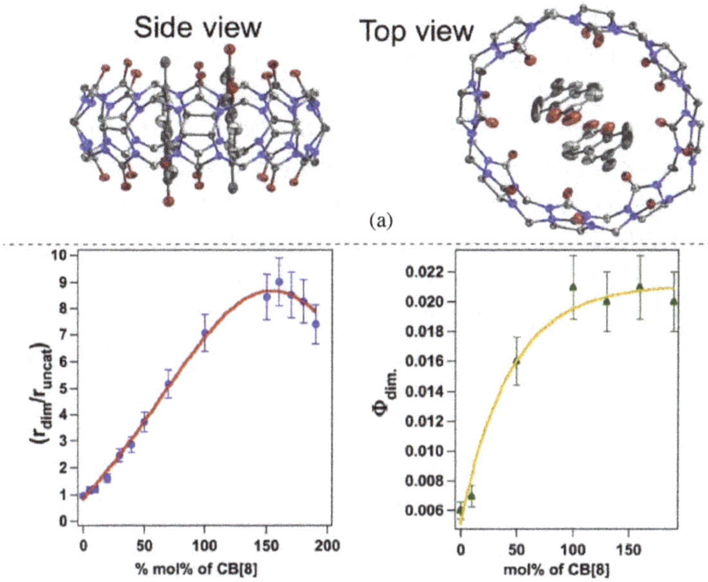

Figure 5. (a) Single-crystal XRD of $(\mathbf{43})_2$@CB[8] complex. (b) Photodimerization rate increase upon varying the mol% of CB[8]. r_{uncat} is the rate of photodimerization in the absence of CB[8] in water and r_{dim} is the rate of photodimerization in the presence of CB[8]. (c) Photodimerization quantum yield, Φ_{dim} with varying the mol% of CB[8]. The single-crystal XRD structure rendered from CCDC 814052 cif file and adapted from Ref. [95] with permission from Royal Society of Chemistry. Figures b and c were reproduced from Ref. [94] with permission from Canadian Journal of Chemistry.

Photochemical and photophysical parameters were evaluated to quantify the supramolecular photocatalytic efficiency (Figures 5 and 6). The kinetics of photodimerization was followed at various mol% CB[8] and the reaction rate (r_{dim}) was found to increase upon increasing the mol% of CB[8]. Compared to the uncatalyzed reaction, the maximum rate of photodimerization in the presence of CB[8] was ~9.0 times (Figure 5b) higher than uncatayzed reaction, i.e., $r_{dim}/r_{uncat} \approx 9.0$. The efficiency of dimerization was lowered upon further increase in CB[8] concentration, which was attributed to the formation of aggregates of CB[8] at high concentration as well as the formation of 1:1 host–guest complex that decreased the amount of 1:2 host–guest complex in solution that is responsible for the enhanced rate of photodimerization. There was an increase in quantum yield of photodimerization (Φ_{dim}) when there was an increase in mol% of CB[8].On inspecting Figure 5c, it is clear that the Φ_{dim} levels off beyond 70 mol% of CB[8], and there was no noticeable change in Φ_{dim} upon increasing the mole equivalents of CB[8] beyond 100 mol%. Hence, beyond 70 mol% of CB[8] the photoreaction occurred at the same efficiency, i.e., the *syn*-photodimer formed per unit of time to the number of photons absorbed was similar beyond 70 mol% of CB[8]. The rate of photodimerization (r_{dim}) depended on the relative population of 1:2 H–G complex, 1:1 H–G complex, and uncomplexed **43**, which in turn were dependent on the relative amounts of the host and guest. The fluorescence studies on **43** in the absence of CB[8] exhibited a structureless emission centered around 411 nm. The supramolecular complex of **43** within CB[8] exhibited an increase in the emission intensity with a prominent red shift with the emission maximum centered around 443 nm (Figure 6a). The fluorescence lifetime of **43** was found to increase in the presence of CB[8]. The uncomplexed **43** had a fluorescence lifetime (τ_f) of <0.1 ns (Figure 6c), in the presence of CB[8], a 1:1 CB[8]-**43** host–guest complex had a lifetime of $\tau_f = 3.7$ ns and 1:2 CB[8]-**43** host–guest complex had a $\tau_f = 0.7$ ns (Figure 6c).

Phosphorescence at 77K (Figure 6d) in water was observed in the presence CB[8] which exhibited the involvement of a triplet excited state of **43**. Hence in addition to a dynamic host–guest system, where uncomplexed **43**, 1:1, and 1:2 CB[8]-**43** host–guest complexes co-existed in water, photoexcitation led to the corresponding singlet and triplet excited states presenting a dynamic mixture (uncomplexed and complexed guests) in solution (Scheme 16). The enhanced fluorescence emission intensity and longer lifetimes was attributed to the confinement of **43** within the hydrophobic CB[8] cavity. Similarly, the same confinement of **43** within CB[8] was reasoned to be responsible for the observed phosphorescence at 77 K. In solution phase the excited **43** formed a mixture of photoproducts **44–47** as there was no control over the excited state reactivity/dynamics. But the confined environment of CBs not only altered the excited state lifetimes, but also siphoned the excited-state energy toward photochemical reactivity leading to photoproduct selectivity.

There was also no appreciable difference in the reaction conversion of photodimerization of **43** in the presence of CB[8] under nitrogen, air, or

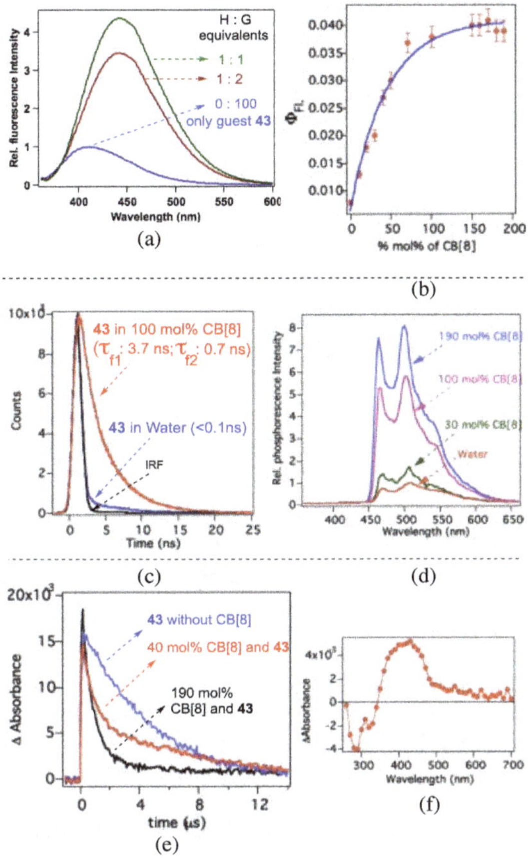

Figure 6. (a) Steady-state emission of **43** at 298 K (λex = 320 nm) with various equivalents of CB[8]. (b) Change in fluorescence quantum yield, Φ_{FL} with varying the mol% of CB[8]. (c) Fluorescence decay of **43** at 298 K (λex = 340 nm; λem = 411 nm for decay in water and λem = 443 nm for 100 mol% of CB[8]); IRF = instrument response function. (d) Phosphorescence of **43** at 77 K in water recorded from a 30–60 ms time window with various mol% of CB[8] (relative to the concentration of **43**); λex = 345 nm. (e) Transient absorption decay traces recorded at 420 nm of ^{3}(**43**)* in the absence and presence of different amounts of CB[8]. (f) Transient absorption spectrum of CB[8]- **43** host–guest complex (25 mM CB[8] and 50 mM **43**). Figures b, c, and d were reproduced from Ref. [94] with permission from Canadian Journal of Chemistry. Figures a, e, and f were reproduced from Ref. [95] with permission from Royal Society of Chemistry.

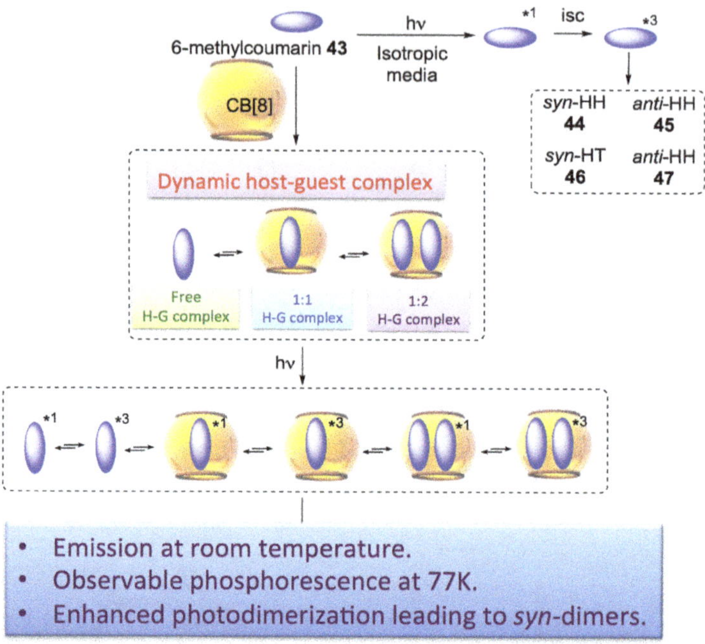

Scheme 16. Interplay of supramolecular complexation dynamics and its effects on excited-state photophysics and photochemistry. Reproduced from Ref. [94] with permission from Canadian Journal of Chemistry.

oxygen-saturated conditions. This implied two distinct possibilities on how the excited states of **43** were influenced by complexation within CB[8]. In the first possibility, a singlet excited state of **43** was conjectured to be involved in the photodimerization pathway mediated by CB[8]. In the second possibility, a triplet excited state of **43** mediated the photodimerization upon encapsulation within CB[8] and this triplet state of **43** encapsulated by CB[8] was protected from external quencher(s) namely, ground-state molecular oxygen for a longer time compared to uncomplexed **43** outside the supramolecular cavity, leading to enhanced photoreactivity. Laser flash photolysis studies were performed under photoreaction conditions (Table 3) to understand the triplet state dynamics involved during supramolecular photocatalysis.

A transient absorption spectrum (Figure 6f) of deoxygenated aqueous solution of **43** was recorded using 308 nm laser excitation (Table 3) and the

Table 3. Triplet lifetimes (τ) and pre-exponential factors (A) of transient absorption decays of $^3(\mathbf{43})^*$ at 420 nm with various mol% of CB[8].[a]

Entry	mol% CB	τ_1 (μs)	A_1	τ_2 (μs)	A_2
1	0	4.6	25	—	—
2	10	5.4	14	0.75	2.6
3	70	12	2.5	0.75	20
4	100	—	—	0.74	36
5	190	—	—	0.75	37

The decays were fitted to: ΔAbsorbance (t) = $A_1\exp(-t/\tau_1) + A_2\exp(-t/\tau_2)$. Data reproduced from Ref. [95] with permission from Royal Society of Chemistry.

[a]Laser pulse (308 nm, pulse width 15 ns).

triplet absorption decay was recorded at different concentrations of CB[8] (Figure 6e; Table 3). The triplet excited state of **43** was found to decay mono-exponentially with a lifetime (τ_1) of 4.6 μs (Table 3, entry 1) in the absence of CB[8]. The triplet excited state of **43** encapsulated within CB[8] exhibited a mono-exponential decay with a lifetime (τ_2) of ~0.75 μs in the presence of 100 mol% of CB[8] or higher. In between 10 and 100 mol% of CB[8] a bi-exponential triplet kinetic decay was observed with a short component (τ_2 ~ 0.75 μs) and long component with lifetime (τ_1) varying between 5.4 and 12 μs (Table 3, entries 2 and 3). Increasing concentration of CB[8] resulted in an increase in contribution of the short component (τ_2) that was also exposed by the increase of the pre-exponential factor (Table 3). The changing lifetime (τ_1) was justified based on efficient self-quenching of triplet excited **43** with a rate constant of 4.1×10^9 M^{-1} s^{-1}. The fraction of uncomplexed **43** decreased in aqueous solution with increasing CB[8] concentrations that was reflected in the decrease in A_1 values (Table 3); this can be attributed to the supramolecular encapsulation of **43** within CB[8]. This reduction in the effective concentration of uncomplexed **43** upon addition of CB[8] resulted in a decreased self-quenching rate of triplet excited state of **43**. An increase in transient lifetime as displayed by τ_1 values was observed due to the decrease in the rate of self-quenching on increasing the concentration of CB[8]. The 1:2 CB[8]-**43** H–G complex on photoirradiation was short-lived and hence was not detected during transient decay kinetic study which can attributed to the enhanced deactivation of the excited state by proximity effect. The triplet excited state of **43** that was uncomplexed in aqueous solution was quenched effectively with a high rate constant of $2 \times 10^9\ M^{-1}s^{-1}$, but the triplet excited state of **43** encapsulated within CB[8] was not effectively quenched by molecular oxygen. The encapsulation of the triplet excited state of **43** within CB[8] of the protected it from quenching by molecular oxygen. This was rationalized to be responsible for the similar conversions observed under oxygen- and nitrogen-saturated atmospheres. Saturation kinetics gave a turnover frequency of 3.4 min^{-1}, with sigmoidal dependence on coumarin concentration (with constant CB[8] concentration) that implied that the overall catalytic process was cooperative in nature (Figure 7a), hence indicating the enzyme-mimetic nature of supramolecular photocatalysis. The sigmoidal dependence was further investigated by a Hill plot that gave a Hill constant of 1.8 implying a positive allosteric effect in the catalytic process. A representative catalytic cycle involving CB[8] mediating the photodimerization of **43** was proposed (Figure 7b) based on the above photochemical, photophysical, and kinetic studies, in which a sequential formation of a 1:1 CB[8]-**43** host–guest complex was followed by the formation of 1:2 CB[8]-**43** host–guest complex. To analyze the catalytic process involved in the photodimerization involving CB[8], thermodynamic binding constant measurements were done using fluorescence titrations and stopped-flow spectrometry was used to understand the kinetic aspects of the host–guest complexation. The thermodynamic binding analysis for the formation of 1:1 and 1:2 CB[8]-**43** host–guest complex was found to be K_{a1}: $1.3 \times 10^4\ M^{-1}$ and K_{a2}: $2 \times 10^6\ M^{-1}$ respectively [94].

The control studies showed that there was no product formation inhibition during the photocatalysis mediated by CB[8] [96]. Depending on the difference in the fluorescence intensity observed for **43** with various mol% of CB[8], stopped-flow kinetic measurements were performed to determine the host–guest complexation rate constants. Based on the stopped-flow data, the formation of the 1:1 CB[8]-**43** host–guest complex was kinetically fast and formation of the 1:2 CB[8]-**43** host–guest complex was slow and was likely to serve as the rate limiting step during the CB[8] supramolecular photocatalysis. The reason rationalized for the slow second step, i.e., the formation of the 1:2 H–G complex, in spite of it being thermodynamically favorable, was due to the longer time required

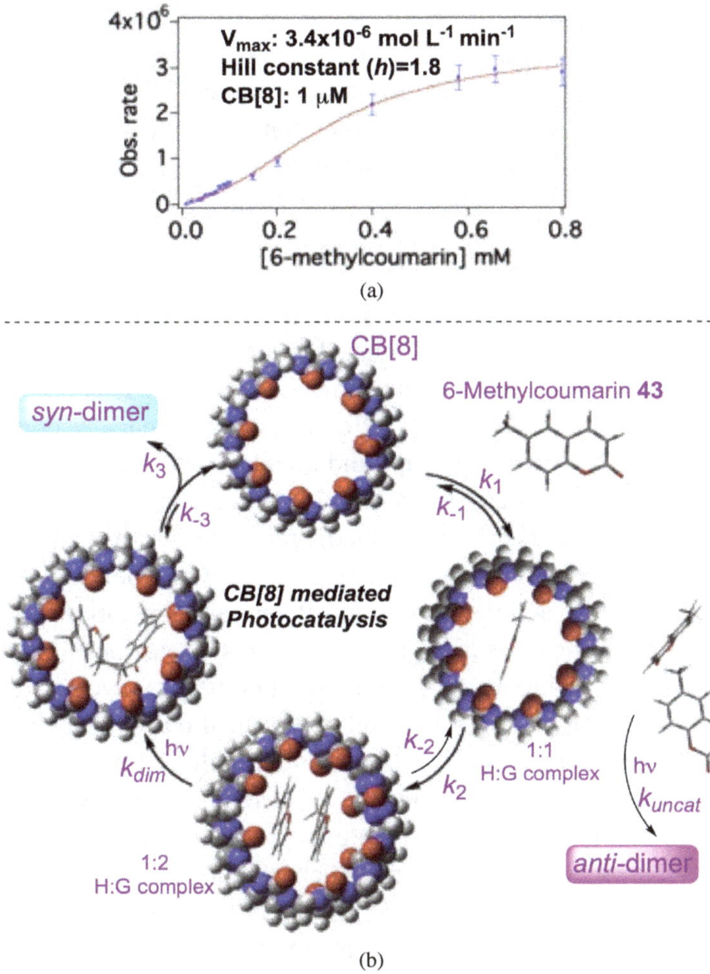

Figure 7. (a) Saturation kinetics of CB[8]-mediated photocatalysis of **43**. (b) Supramolecular photocatalysis of 6-methylcoumarin **43** mediated by CB[8]. Figures a and b reproduced from Ref. [93] with permission from Royal Society of Chemistry.

for the first coumarin guest molecule to alter the shape of the CB[8] cavity to accommodate the second coumarin guest molecule that was reflected in the allosteric response observed in the saturation kinetic measurements. This induced shape change in the presence of the guest which mediated the supramolecular photocatalysis by CB[8] resemble catalysis promoted by enzymes. The catalytic cycle with CB[8] was generalized in Figure 4c with the sequential formation of 1:2 host–guest complex that reacts efficiently on photoexcitation. While the basic mechanism of action of these systems will not be much different one can envision a catalytic cycle involving other supramolecular assemblies (*vide infra*).

1.3.3.2 *Cyclodextrins*

Cyclodextrins (CDs) are cyclic oligosaccharides that are made of six to eight united of d-glucopyranose units, that are linked by 1,4-glycosidic bonds (Figure 3) [64]. Cyclodextrins structurally resemble a cone or truncated funnel. CDs consist of two different portals, a narrow portal that consists of primary hydroxyl groups and a wider portal that contains secondary hydroxyl groups. The hydrophobic character of the central cavity in CDs is due to the presence of D-glucopyranoside units made of carbon atoms with an ether linkage. This cavity gives the CDs their unique complexation ability in aqueous solution. Potential applications of CDs involve their critical role as catalysts in biomimetic reactions, drug delivery systems, and organic reactions [64]. With numerous investigations by distinguished groups on controlling photochemical reactivity by employing CDs as a reaction flask, in this section we will highlight only a few selected examples of cyclodextrin and its derivatives involved in light-induced transformations. Readers are encouraged to look at other dedicated reviews in this area [16]. Paternó–Büchi reaction of adamantan-2-ones **48** with fumaronitrile **49** (Scheme 17) in the presence of β-cyclodextrins (β-CDs) was reported by Turro and coworkers [97]. The irradiation of 5-substituted adamantan-2-ones **48** with fumaronitrile in aqueous media with β-CDs resulted in the formation of oxetanes favoring the *syn* isomer **50**, in contrast to the preferential formation of *anti* isomer **50** in isotropic media. There were no significant changes observed in the presence of α- and γ-CD. The interaction of the carbonyl group with the β-CD host (represented by the host–guest complexes A-C in Scheme 17) helped in rationalizing the facial selectivity within β-CDs [97].

[4+4]-Photocyclodimerization of 2-anthracenecarboxylate **51** by using γ-cyclodextrin (γ-CD) as a chiral template (Scheme 18) was reported by Inoue and coworkers [98]. The large cavity volume of γ-CD

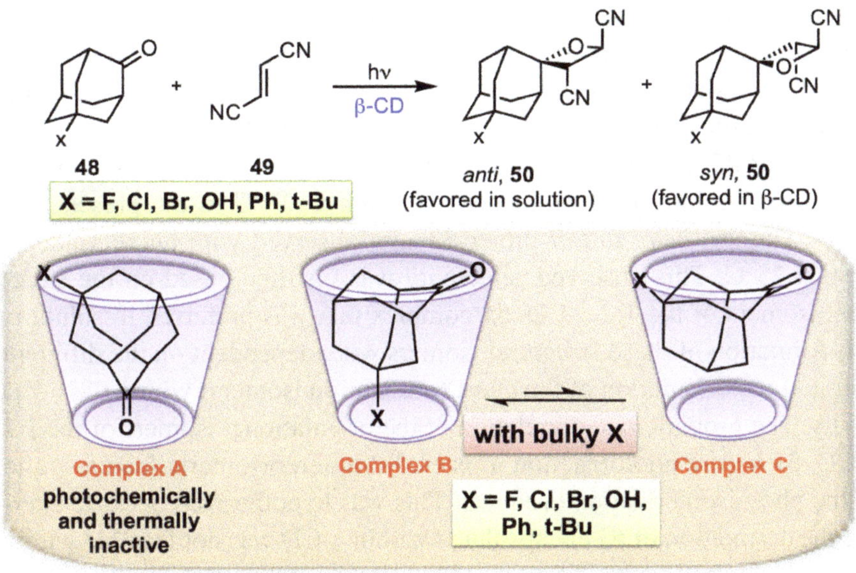

Scheme 17. Paternó–Büchi reaction of adamantan-2-ones **48** with fumaronitrile **49** mediated of β-cyclodextrin. Reproduced from Ref. [11] with permission from Royal Society of Chemistry.

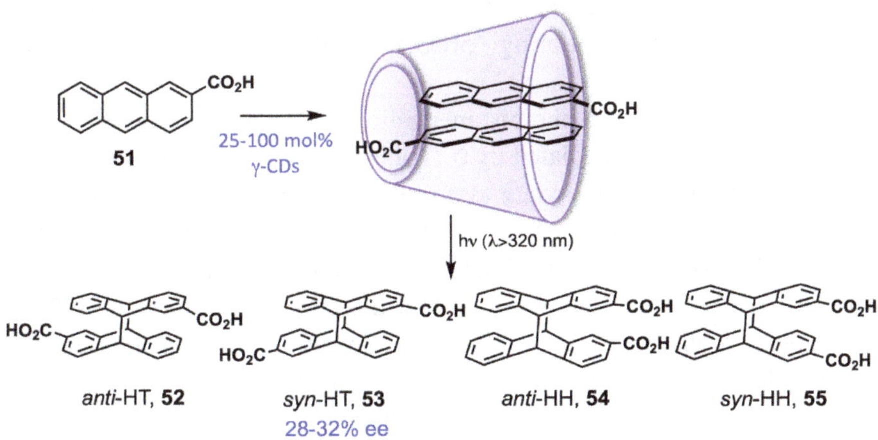

Scheme 18. [4+4]-Photocyclodimerization of 2-anthracenecarboxylate **51** catalyzed by γ-cyclodextrin (γ-CD). Reproduced from Ref. [11] with permission from Royal Society of Chemistry.

allowed encapsulation of two aromatic molecules within its cavity making them desirable as templates for bimolecular photochemical transformation. A very stable 1:2 inclusion complex with γ-CDs was detected with thermodynamic binding constants with $K_{a1} = 161\ M^{-1}$ and $K_{a2} = 3.85 \times 10^4\ M^{-1}$ at 25°C. Four photodimers were observed in the photoinduced cyclodimerization of **51** in aqueous solution at 25°C in the presence of γ-CD. The *syn-head-to-tail* dimer **53** was observed with an ee value of 32% at 25°C. This observed selectivity was justified based on the structural isomers of the 1:2 γ-CD–**51** complex that was preferred in solution. The formation of these structural isomers were dependent on the different longitudinal orientation of the guest (orientation isomers) within the γ-CD cavity. The ground-state population of the orientational isomers of the 1:2 γ-CD–**51** before photoreaction impacted the stereoisomeric features/ratio of the photoisomerization products. This was hypothesized because altering the orientation of **51** encapsulated within γ-CD was not feasible within its short-excited state lifetime. The enantioselectivity in the chiral cyclodimer(s) arose from the difference in stability/activation energy difference of the diastereomeric pair of orientational isomers of the 1:2 γ-CD–**51** complex. The ee at 25°C in the chiral cyclodimer **53** was 28%–32% and was dependent on the loading level of the cyclodextrin (25–100 mol%). On reducing the temperature to 0°C, the ee value increased to 41% in the cyclodimer **53** that highlighted the role of the temperature during the enantiodifferentiating process within the supramolecular container.

Enantiodifferentiating photoisomerization of (*Z*)-cyclooctene **56** to chiral (*E*)-cyclooctene **57** mediated by 6-*O*-substituted cyclodextrins functionalized with aromatic carboxylate as a sensitizer (Scheme 19) was reported by Inoue and coworkers [99]. In order to fine-tune the space/bulkiness of the sensitizer (benzoates, isomeric phthalates, and tethered benzamides) a series of α-, β-, and γ-CDs. They observed that the photostationary state (**57/56** ratio) obtained on sensitization of **56** with conventional alkyl/alkoxy benzoates (that lacked the CD cavity) in hydrocarbon solvent was ~0.25 and that observed on sensitization with 6-*O*-substituted cyclodextrins in 1:1 water–methanol was between 0.4 and 0.8. For the *E*-isomer **57** low ee values were observed with α-CDs (3% ee) and γ-CDs (5% ee); this was rationalized due to the mismatch in size of the CD cavity. Contrary to this, the β-CD

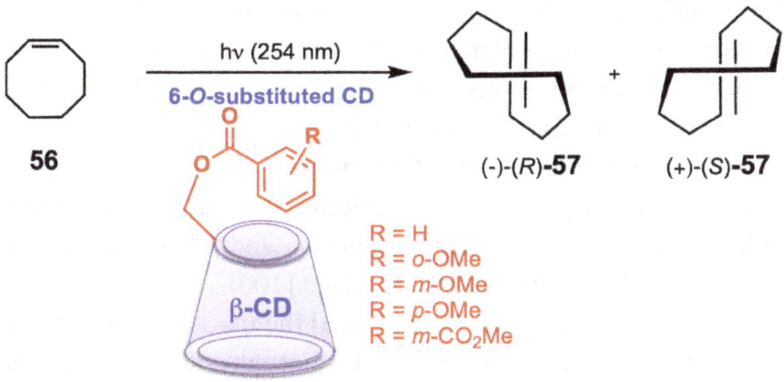

Scheme 19: Enantiodifferentiating photoisomerization of (*Z*)-cyclooctene to chiral (*E*)-cyclooctene induced and sensitized by 6-*O*-substituted-*β*-cyclodextrins. Reproduced from Ref. [11] with permission from Royal Society of Chemistry.

derivatives gave high ee values (ee values up to 24%) in **57** due to cavity size matching with the reactive substrate.

1.3.3.3 *Cavitands*

Cavitands are supramolecular hosts based on the different aromatic functionalities that have been synthesized in order to modulate their supramolecular characteristics [65, 82, 83]. In this section we will limit ourselves to a deep cavity cavitand known and octa acid (OA) that was synthesized by Gibb and coworkers [65]. OAs have been successfully used to control various photochemical transformations and photophysical events [82, 83, 100]. Even though the shape of OA resembles cyclodextrins (Figure 3), OA is a deep cavity cavitand due to two stacked aromatic units with a total of eight carboxylic acid groups, hence making them water soluble with a well-defined hydrophobic pocket. OA feature a shorter portal of ~5.4 Å and a larger portal of ~11 Å. The cavity has a depth of ~7.5 Å (Figure 3). The guest molecules which are organic enter the deep cavity through the larger OA portal. Based on the type of the guest(s), 1:1 and/or 2:2 host–guest complexes were observed with OAs. These were also probed for controlling the light-initiated processes. In this section we will discuss a couple of reactions which use OA to control the photochemical reactivity.

In principle OA could be employed for supramolecular photocatalysis. Ramamurthy and coworkers have established that the OA capsules are no passive spectators during photochemical transformations, but are active in enabling the electron and energy transfer reactions with appropriately designed guest molecules [100]. This opened up new opportunities to design and develop supramolecular systems that can communicate with one another, promoting molecular recognition and photocatalysis making them fertile grounds for future developments [100].

Ramamurthy and coworkers established the use of OA for controlling the excited state photochemistry and photophysics of encapsulated organic molecules [82]. In this section we will discuss a couple of photochemical processes (Schemes 20 and 21) that were controlled within the confinement of OA. Acenaphthylene **58** and OA formed a 2:2 host–guest complex ($\mathbf{58}_2$@OA_2) in water (pH ~ 9, borate buffer) [82]. Due to the preferential pre-orientation of the aromatic guests in the OA, direct irradiation of the 2:2 complex resulted in the exclusive formation of *syn*-**59** (Scheme 20). Eosin was used to perform the triplet sensitized irradiation

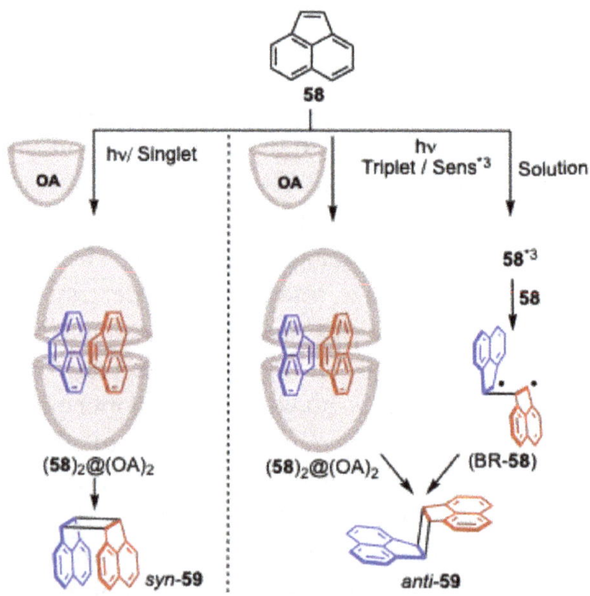

Scheme 20. Photodimerization of acenaphthylene **58** in presence of octa acid (OA). Reproduced from Ref. [11] with permission from Royal Society of Chemistry.

which resulted in *syn*-**59** and *anti*-**59** (*syn*-**59**: *anti*-**59** = 3:2). The dynamics of host–guest complexation and excited state lifetimes helped in rationalizing the contrasting selectivities observed from singlet and triplet reactivity of **58** in the presence of OA. The short singlet lifetime of **58** (0.35 ns) did not allow the pre-oriented configuration to change its orientation/structural geometric features that led to the *syn*-**59** photoproduct. For the triplet sensitized dimerization, the long triplet lifetime of **58** (6 ms) allowed the substrate excimers to escape from the OA capsule; this led to the reorientation of their geometry resulting in the formation of the *anti*-**59** photoproduct (Scheme 20). This study highlighted the pre-orientation of substrates using the confinement of cavitands in a reaction conducive configuration is responsible for the observed selectivity as well as restricting molecular movements of substrates within the cavity of the OA that was dependent on the excited state lifetimes of the reactive substrates. It also highlighted how the dynamic aspects of the host–guest complexation are related to the excited state lifetimes of the reactants and the way their kinetics influenced the photoproduct selectivity.

Norrish Type I and Type II reactions of optically pure α-alkyl deoxybenzoins **60** within the OA nanocapsule (Scheme 21) was reported by

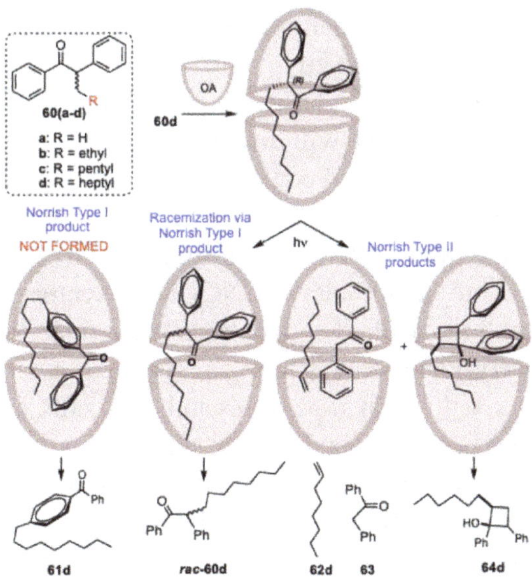

Scheme 21. Photolysis of α-alkyl deoxybenzoins (α-ADBs) within a water-soluble OA. Reproduced from Ref. [11] with permission from Royal Society of Chemistry.

Ramamurthy and coworkers [83]. The role of alkyl substituents was elucidated by investigating the different alkyl substituted deoxybenzoins **60a-d**. Photoirradiation of a 2:1 host–guest complex with OA (**60**@OA_2) in borate buffer (Scheme 21) was compared to the photoreactivity in isotropic media (benzene). The photoreaction of **60b-d** in isotropic media (benzene) resulted in the formation of both Norrish Type I and Type II products, but **60a** gave only Type I photoproduct. In the presence of OA in borate buffer solution Type I rearrangement product was observed with **60a-b**, Type II products were observed with **60c-d**. On employing optically pure isomers, the optical activity after photoirradiation was dependent on the length of alkyl chain in **60**. OA capsules were able to promote the conformation control that influenced the product distribution, and the length of the alkyl chain determined the free space available within the confined environment. The shorter alkyl substituents in **60**, the racemization product was attributed to the free movement of α-alkyl benzyl radical or sliding of benzoyl radical within the OA cavity. The long-chain alkyl substituents in **60**, the reactants adopted a confirmation within the OA that was suitable for Type II reaction.

1.3.3.4 *CNCs*

The interaction of supramolecular chemistry and co-ordination chemistry was manipulated by various groups to design water-soluble CNCs (Figure 3) that served as molecular reaction flasks for chemical transformation [66]. In this section we will discuss pioneering work done on one such reaction flask by Fujita and coworkers [66–68], they employed capping ligands such as ethylenediamine derivatives to enforce the 90° *cis*-geometry around square-planar-co-ordinated Pd^{II} and Pt^{II} ions. Self-assembled co-ordination cages (CNC) were synthesized by mixing end-capped metal ions with the tridentate, triangular ligands such as 1,3,5-tris(4-pyridyl)triazine and utilized as molecular reaction flasks. The structure of CNCs built with triazine ligands as walls of co-ordination cages that were secured by metal ions at the corners of the octahedron (Figure 3). This created an exclusive structural feature with a cavity that could be used for confinement and catalysis.

Highly stereoselective [2+2]-photodimerization of acenaphthylenes **58** in the presence of self-assembled Pd-nanocage, Pd-CNC1 (Figure 3;

Scheme 22) was reported by Fujita and coworkers [67]. Photoirradiation of acenaphthylenes (3 h) **58a-b** in the presence of Pd-CNC1 resulted in the formation of *syn*-HT dimers as the exclusively photoproduct (>98%). There was no observable decomposition of CNC during the course of the photoreaction, and it was found to be very effective. There was no reaction observed for similar irradiation times in the absence of the cage where a mixture of *syn* (19%) and *anti* (17%) dimers was observed at high concentrations of acenaphthylene **58**. Even though they are not catalytic, CNCs were demonstrated as viable molecular flasks that sequestered and pre-oriented two substrates in close proximity, thereby promoting dimerization of olefins in a selective fashion.

Photo-induced hydration of internal arylalkynes with catalytic amounts of CNCs was reported by Fujita and coworkers [68] (Scheme 23).

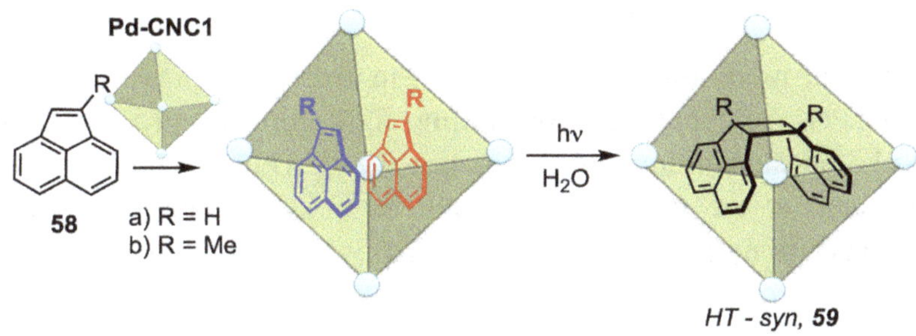

Scheme 22. Stereoselective intermolecular photodimerization of acenaphthylenes in presence of co-ordination nanocage (CNC). Reproduced from Ref. [11] with permission from Royal Society of Chemistry.

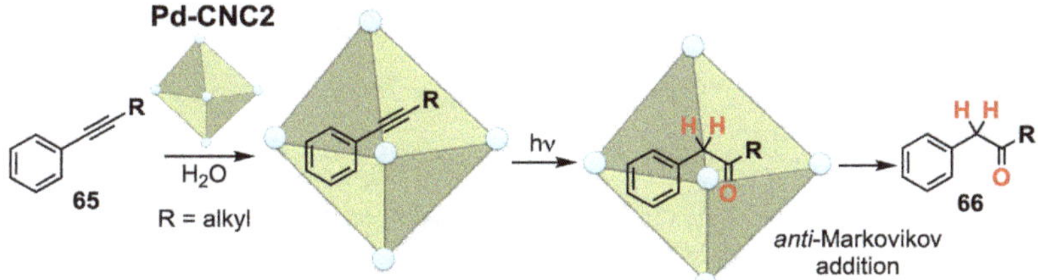

Scheme 23. Photohydration reaction of arylalkyne in presence of catalytic amount co-ordination nanocage (CNC) leading to anti-Markovnikov photoproduct. Reproduced from Ref. [11] with permission from Royal Society of Chemistry.

1-Phenyl-1-hexyne (**65**, 20 equiv.) on photoirradiation in the presence of Pd-nanocage, Pd-CNC2 (Figure 3; Scheme 23) resulted in the formation of the *anti*-Markovnikov hydration product **66** (benzyl ketone) specifically through an electron transfer mechanism. A highly active phenyl alkyne radical cation intermediate was generated by the electron transfer from the alkyne to the electron-deficient CNC. A stabilized benzylic radical was generated when the phenyl alkyne radical cation intermediate reacted with a water molecule (from the solvent) regioselectivity resulting in the *anti*-Markovnikov hydration product. The nanocage furnished as a UV-filter for the photoproduct, hence preventing the benzyl ketone photoproduct from further photochemical transformation (α-cleavage/hydrogen abstraction reactions that are expected from the photoproduct).

1.3.3.5 *Combining supramolecular assemblies to enhance reactivity and selectivity*

Using two or more assemblies in unison in order to improve the role of confining environments in controlling light-initiated processes. The [4+4]-photocycloaddition α-CD functionalized with an anthracene carboxylic acid at the 6-position through an ester linkage viz., 6-*O*-(2-anthracenecarbonyl)-α-CD **67**, (Scheme 24) was reported by Kim, Inoue,

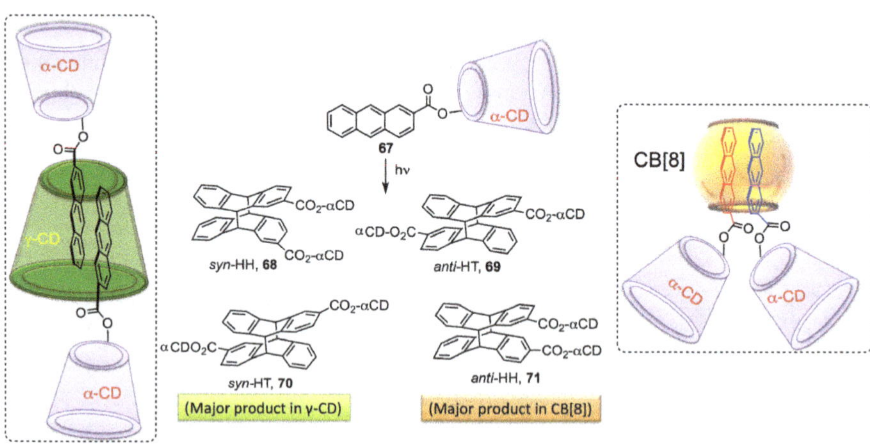

Scheme 24: Combining supramolecular assemblies to influence excited state chemistry of cyclodextrin-appended anthracene carboxylic acid derivative **67**. Reproduced from Ref. [11] with permission from Royal Society of Chemistry.

and coworkers [101]. The photodimerization efficiency and selectivity of **67** functionalized α-CD templated within either γ-CD or CB[8] was investigated. The mode of binding of anthracene motif inside the supramolecular cavity of both CB[8] and γ-CD was dictated by the α-CD functionality in **67** which served as a bulky hydrophilic substituent; this was reflected in the dramatic inversion of the HH:HT ratio at –20°C at 210 MPa. The major photoproduct observed in the case of γ-CD was the *syn*-HT **70** with a HH:HT product ratio of 2:98 (Scheme 24). With CB[8] *anti*-HH **71** was observed as the major product with a HH:HT ration of 99:1. On removing the α-CD chiral auxiliary by base hydrolysis, optically enriched parent dimers were obtained. The photoreactivity with γ-CD resulted in 91% ee value in the hydrolyzed *syn*-HT dimer and an ee value of 2% in the hydrolyzed *anti*-HH dimer at –20°C at 210 MPa. Contrarily in the case of CB[8] an ee value of 18% and 8% were observed for the hydrolyzed *syn*-HT and *anti*-HH dimers at –20°C at 210 MPa, respectively.

1.3.3.6 *Photocatalysis mediated by metallosupramolecular assemblies*

The use of supramolecular host–guest chemistry and metal-mediated photochemistry to control both the enantiodifferentiation of substrates within a chiral host and accelerate the photoreaction without using a sensitizer was cleverly exploited by Inoue and coworkers [77]. The γ-CD was functionalized at the 6-position with diamino substituents (DA-γ-CD), and their role was evaluated in promoting the photodimerization of anthracene carboxylic acid **51** in the presence of Cu(II) ions (Scheme 25). The DA-γ-CD oriented the two anthracene carboxylic acid molecules in a HH fashion in the presence of Cu(II) ions, which led to the formation of the optically active *anti*-HH dimer **54**. An ee value of 70% in the *anti*-HH dimer **54** was observed on photoirradiation of **51** in the presence of 0.1 equivalents of DA-γ-CD and 0.5 equivalents of Cu(II) ions. On decreasing the loading level of DA-γ-CD an ee value of 43% was observed in the *anti*-HH dimer **54**. Hence, the metallosupramolecular photocatalysis was established as an effective substitute to the hosts/templates appended with sensitizers (electron/energy transfer sensitizers).

Channeling photochemical reactivity through metal ion complexation has been cleverly manipulated by sequestering them using molecular

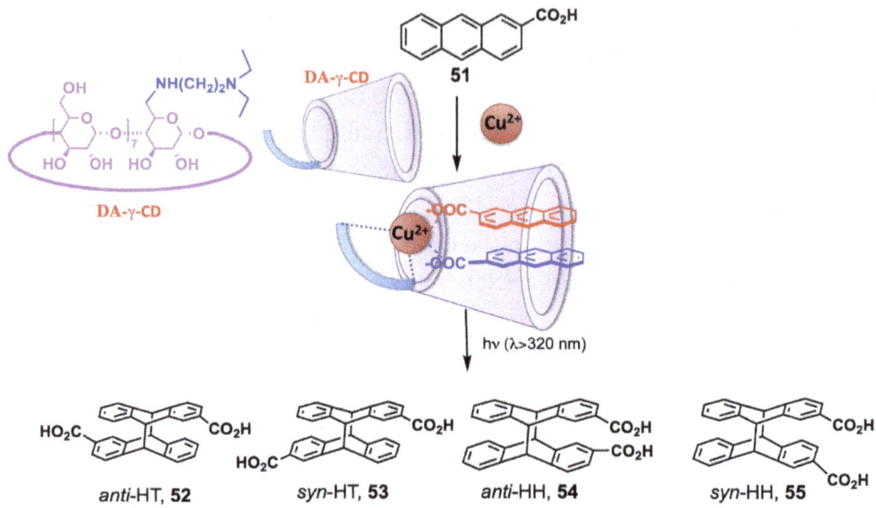

Scheme 25. Metallosupramolecular photodimerization of **51** mediated by diamino-functionalized γ-CD derivatives (DA-γ-CD) in the presence of $Cu(ClO_4)_2$ in aqueous methanol. Reproduced from Ref. [11] with permission from Royal Society of Chemistry.

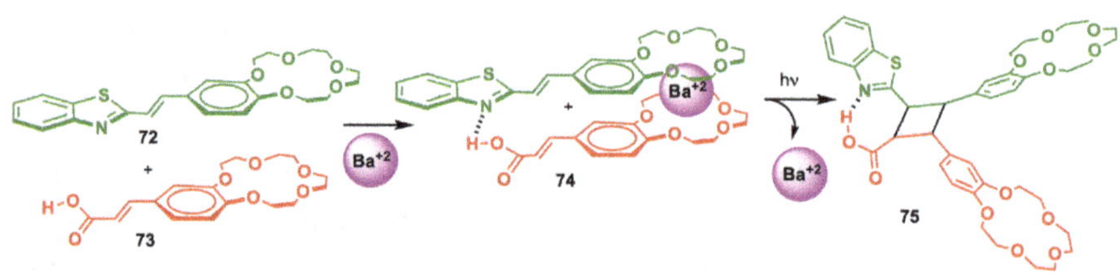

Scheme 26. Metallosupramolecular photochemistry: Intermolecular [2+2]-photocycloaddition of supramolecular assembly **74** anchored by Ba^{2+} leading to formation of unsymmetrical cycloadduct **75**. Reproduced from Ref. [11] with permission from Royal Society of Chemistry.

recognition motifs (e.g., crown ethers). The intermolecular [2+2]-photocycloaddition of supramolecular assembly of **74** formed from **72** and **73** (Scheme 26) was reported by Saltiel and coworkers [79]. Stabilization of the self-assembled **74** was reported to be via hydrogen bonding, π–π interactions, and sandwich Ba^{+2} complex formation between crown ether units in **72** and **73**. The stabilization due to these non-bonding interactions oriented the double bonds at an optimal distance for photochemical dimerization that occurred efficiently in the presence of Ba^{+2} ions leading to the formation of unsymmetrical cyclobutane adduct **75**. The importance of pre-organization of substrates in the presence of metal cation(s) that controlled photocycloaddition in solution leading to the exclusive formation of cyclobutane **75** in high yields was highlighted in this report.

1.3.3.7 *Supramolecular photocatalysis assisted by mechanical grinding*

Tailoring the thermodynamic and kinetic aspects to achieve an efficient dynamic exchange of reactants and product is critical for supramolecular photocatalysis in solution. Manipulating the supramolecular confinement in the solid state offers avenues to overcome many of the issues faced in solution-phase reactivity. Regulating reaction rates in the solid state via crystal engineering is very alluring as the slow diffusion limits the number of unwanted side reactions that are often encountered during photo-induced processes. This also opens up an opportunity to use mechanical force to alter the supramolecular arrangement in addition to altering/promoting the orientation of the guest molecule(s). The mechanical force can also be utilized as a mechanism for exchanging the reactants and the photoproducts with the supramolecular host. The enhancement of reactivity by this strategy involves mechanical grinding of the photoactive reactants with templates/additives. After photoirradiation, the molecular templates are re-grinded, hence recycling the solid-state mixture allowing the supramolecular template/additives bind to other reactants (exchanging the reactant and the product with the supramolecular host by mechanical force) hence allowing the photochemical transformation in a sustainable manner. This process is desirable as the phototransformation is done under solvent-free conditions.

The mechanical grinding of cucurbit[8]uril (CB[8]; 50 mol%) or other additives (e.g., thiourea or adamantine) with 6-methylcoumarin **43** resulted in enhancement in the photodimerization efficiency with exclusive formation of head-to-head photodimer **44** (Scheme 27) [102]. Upon

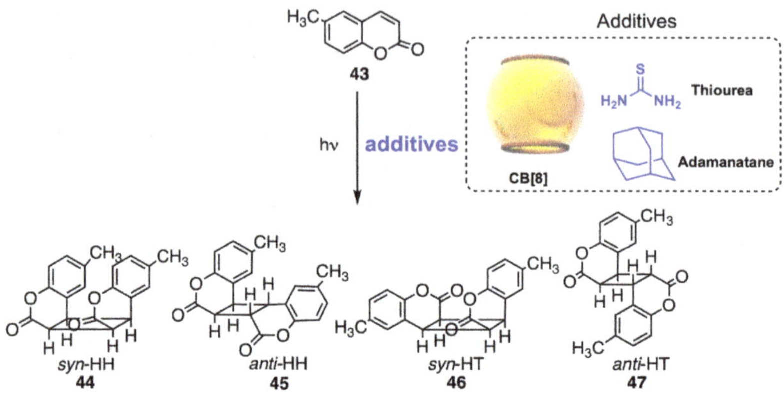

Scheme 27. Photodimerization of 6-methylcoumarin catalyzed by CB[8]. Reproduced from Ref. [11] with permission from Royal Society of Chemistry.

mechanical grinding with CB[8], a red shift in the fluorescence of **43** was observed in the solid state. The red-shifted fluorescence emission in solid state is similar to the one observed for CB[8]-**43** host–guest complex in solution at 298 K (Figure 3) [94, 102]. The solid-state phosphorescence emission for mechanically ground **43** with CB[8] at 77K was very similar to the phosphorescence spectrum observed for **43**@CB[8] host–guest complex at 77 K [94, 102]. A head-to-head orientation of coumarins with the distance between the double bonds being 4.695 Å was revealed from the single-crystal X-ray structure of **43**, and no observable photodimerization even after 200 h of irradiation [103]. A control reaction in the presence of H-bonding thiourea as additive gave selectively *syn*-HH **44**. Control reaction by mechanical grinding of **43** alone also resulted in similar conversions; this was in contrast to the reported behavior [103] for photodimerization in the solid state. This enhancement in the photoreactivity of **43** upon mechanical grinding was rationalized due to a combination of factors, viz. (a) the crystalline imperfections which are likely to be formed during the mechanical grinding and/or (b) the formation of a completely new amorphous phase during the mechanical grinding was not ruled out by powder XRD analysis.

1.4 Organocatalytic Photochirogenesis

1.4.1 Sensitizing H-bonding Templates for Supramolecular Photocatalysis

Enhanced photochemical reactivity and selectivity of caged substrates within molecular flasks/nano-containers were discussed in the previous section. This strategy of using molecular templates that anchor substrates through non-covalent interactions (e.g., H-bonding interactions) can also be extended to enhance photochemical reactivity and selectivity. Also, stereodifferentiation can be attained in the photoproduct due to the chiral

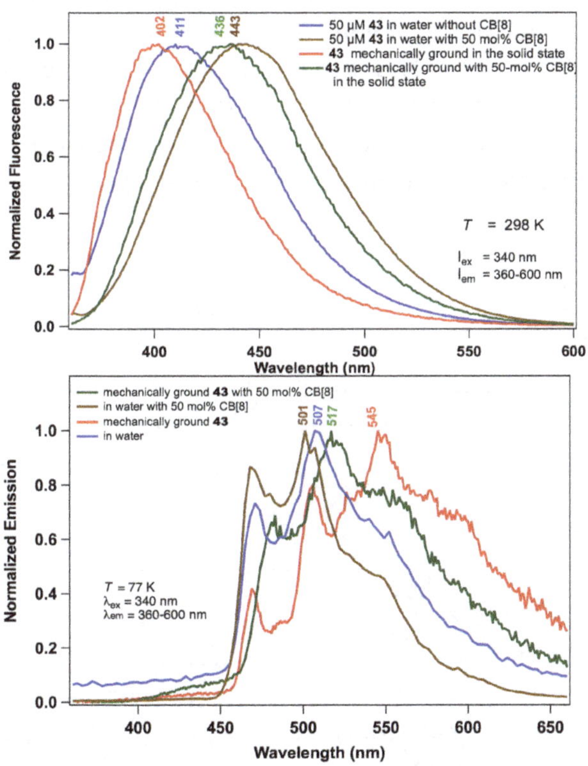

Figure 8. (Top) Normalized emission of **43** in the presence and absence of CB[8] in water and in the solid state at 298 K. (Bottom) Normalized emission spectra of **43** in the presence and absence of CB[8] in water and in the solid state at 77 K. Adapted from Ref. [102] and reproduced with permission of Elsevier.

environment presented by these templates with help of an asymmetric unit that can be incorporated into the template. This characteristic feature to achieve stereoselectivity in small chiral molecules during photoreactions is quite challenging. To enhance chemical reactivity/selectivity in light-initiated reactions, one of the methodologies employed by chemist is where a hydrogen-bonding template (Figure 3; Schemes 28–31) is appended with a sensitizer that permits energy transfer or electron transfer to a reactive substrate. Hydrogen-bonding interactions play the key role in bringing the reactive substrate closer to the sensitizer in order to enable the energy/electron transfer, while the architecture of the template allows facial/stereodifferentiation in the reaction of interest [58].

Bach and coworkers established the [2+2]-photocycloaddition reactions of 4-substituted-2-quinolone **76** by linking Kemp triacid lactam template (+)-KTL1 or (–)-KTL2 to the substrate (Scheme 28) [69]. Low selectivity observed in the photoproduct **77a** (37% ee, 89% yield, –15°C) with (–)-KTL2 host due to the weak association between **76a** and the H-bonding template with menthol ring that showed the deficiency in

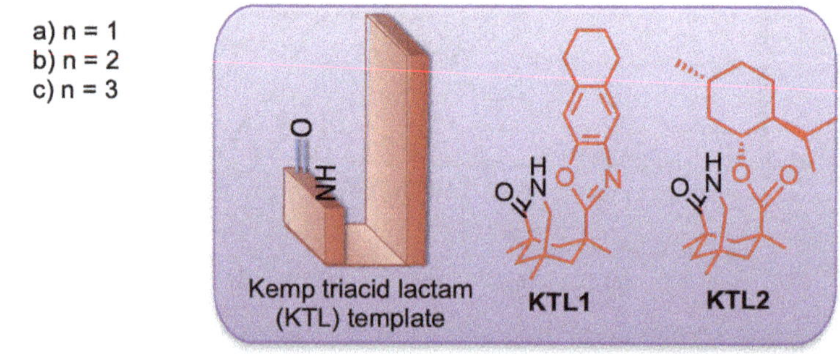

Scheme 28. Enantioselective intramolecular [2+2]-photocycloaddition reaction of substituted 2-quinolone **76**. Reproduced from Ref. [11] with permission from Royal Society of Chemistry.

stereo-discrimination. Nevertheless, with 2.6 equivalents of (+)-KTL1 enhanced enantioselectivity (93% ee, 77% yield at –60°C) was observed in **77a**. The photoreaction was also accomplished with the longer alkyl-chain quinolone **76c** in 1.2 equivalents of chiral host (+)-KTL1 in toluene at –15°C. The photoproduct **77c** showed high enantioselectivity (88% ee, 88% yield). Although not catalytic, this report showed that enantiotopic faces of the photoreactive substrates can be differentiated with the help of chiral lactam host consisting of Kemp triacid through hydrogen bonding.

To broaden the mode for catalytic turnover, with triplet sensitizers were affixed with the chiral Kemp lactam templates for facilitating energy transfer (Scheme 29) [70, 71]. Intramolecular [2+2]-photocycloaddition of quinolone **76b** was used to assess the Kemp lactam host KTL3 with xanthone sensitizer and KTL4 with benzophenone sensitizer (Scheme 29). Low enantioselectivity (39% ee) was observed in photoproduct **78b** at –25°C in trifluorotoluene when catalytic loading of KTL4 (10 mol%) was

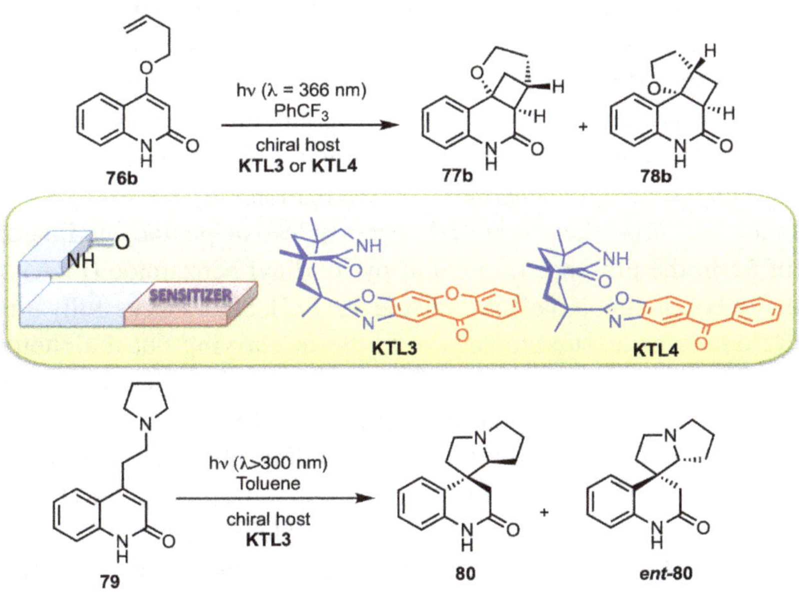

Scheme 29. Enantioselective intramolecular [2+2]-photocycloaddition sensitized by chiral template. Top: Energy transfer sensitization of quinolones **76b**. Bottom: Electron transfer sensitization of quinolone **79**. Reproduced from Ref. [11] with permission from Royal Society of Chemistry.

low. However, when the benzophenone moiety in KTL4 was replaced with xanthone as the triplet sensitizer in KTL3 (Scheme 29) high enantioselectivity (90% ee with 5 mol% catalytic loading of KTL3) in the photoproduct **78b** was observed. Increasing the catalyst loading from 5 to 20 mol% enhanced the enantioselectivity with 94% ee value (78% yield). The regiomeric ratio was 79:21 for the photoproducts **77b**:**78b**. It was rationalized that the absorptivity (optical density at the irradiating wavelengths) and low triplet energy of benzophenone when compared to xanthone accounted difference in stereoselectivity between the two templates.

The chiral sensitizers appended to Kemp lactam host can also be used for electron-transfer-mediated phototransformations. A photoinduced electron transfer (PET) reaction between quinolone **79** (Scheme 29, bottom) and benzophenone-appended Kemp lactam host KTL4 was investigated by Bach and coworkers [72]. By using benzophenone-appended Kemp lactam host KTL4 (0.3 eq.) high enantioselectivity (70% ee, 64% yield, 1 h irradiation) was attained in the photoproduct **80** in toluene at −60°C. The high selectivity was attained due to the two-point hydrogen bonding to substrate and KTL4 along with the ability of benzophenone moiety to act as a sensitizer for the PET reaction. Further, these benzophenone-appended templates provided shielding for enantiofacial differentiation throughout the photochemical transformation [72].

Inoue and coworkers reported enantioselective photocyclodimerization of **51** in the presence of a chiral pyrrolidinyl benzamide (PBA) that acted as a H-bonding template (Scheme 30) [73], after successfully utilizing cyclodextrins as supramolecular hosts for carrying out the photocyclodimerization of **51** (Schemes 18, 24, and 25) [98]. They employed amide and primary alcohol functionalities in PBA templates PBA1 and PBA2 that provided two-point hydrogen bonding sites for stereodifferentiation. Preferential formation of *anti*-HT **52** and *anti*-HH **54** at −50°C was observed among photoproducts **52–55** while irradiating **51** in the presence of PBA template in dichloromethane at two different temperatures (25°C and −50°C). In the presence of PBA1 and PBA2 significant difference in the ee values was observed in the case of chiral photoproducts *syn*-HT **53** and *anti*-HH **54**. In the presence of PBA2 template (*trans* geometry between amide and alcohol functionality) low ee (<3% ee) of

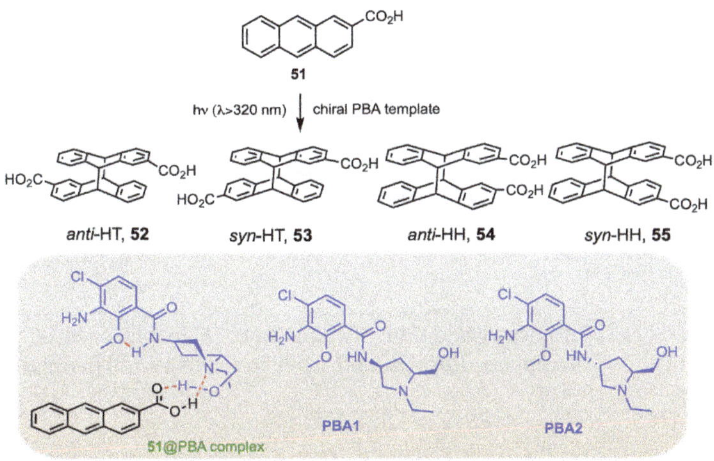

Scheme 30. [4+4]-Photocyclodimerization of 2-anthracenecarboxylate **51** in the presence of pyrrolidinyl benzamide (PBA) template. Reproduced from Ref. [11] with permission from Royal Society of Chemistry.

photoproducts was observed. However, in the presence of PBA1 template (*cis* geometry between amide and alcohol functionality), photoproduct *syn*-HT **53** was observed with an ee value of 25%–43% and *anti*-HH **55** was observed with an ee value of 10%–43% depending on the temperature. This enhancement in the selectivity of photoproducts was justified due to the steric shielding/stereofacial differentiation showed by two templates (PBA1 and PBA2) while interacting with **51**.

Krische and coworkers [74] reported the enantioselective [2+2]-photocycloaddition reaction of quinolones **76b** (Scheme 31) with triplet sensitizer consisting of chiral pyridine-isophthalamide (PIT) template that is appended with benzophenone. An increased efficiency of the photoreaction of quinolone **76b** was observed with the use of 0.25–2 equivalents of the PIT template in $CDCl_3$ at −70°C. Nevertheless, the ee values (20% ee) of the photoproduct **78b** were unaffected with the sub-stoichiometric or stoichiometric loading levels of the PIT template. This was explained based on the underlying fact that the host–guest complex is formed between the PIT template and the substrate **76b**.

Bassani and coworkers [75, 76] showed the [2+2]-photodimerization of cinnamate ester **81** functionalized with diaminotriazine moeity

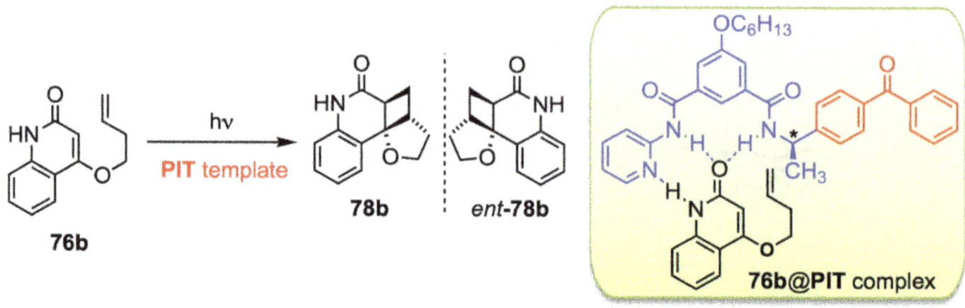

Scheme 31. [2+2]-Photocycloaddition of quinolone **76** in presence of pyridine-isophthalamide (PIT) template. Reproduced from Ref. [11] with permission from Royal Society of Chemistry.

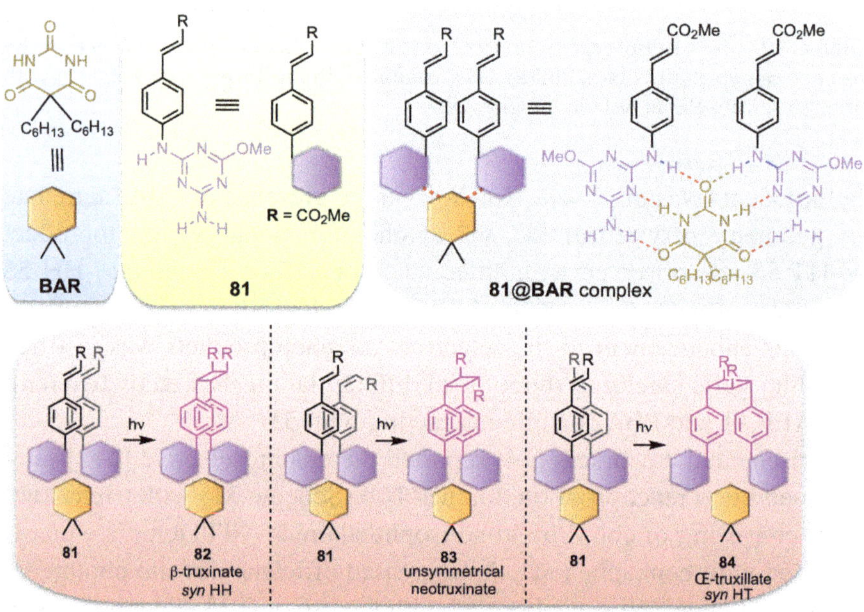

Scheme 32. [2+2]-Photodimerization of cinnamates in presence of a supramolecular hydrogen-bonded assembly. Reproduced from Ref. [11] with permission from Royal Society of Chemistry.

(Scheme 32). Barbiturate (BAR) served as a scaffold perceiving the diaminotriazine functionality through pre-orienting the cinnamate ester, thus increasing the reactivity and selectivity in the photoproduct. NMR binding experiment revealed a strong hydrogen bonding between BAR template and **81** with binding constants $K_1 = 620\ M^{-1}$ and $K_2 = 410\ M^{-1}$ (1:2 host–guest binding isotherm curve). In the absence of the BAR template, substrate **81** underwent double-bond photoisomerization (λ~350 nm) in dichloromethane with an *Z*:*E* ratio of 2:1 along with the formation of cyclobutane dimers. Irradiation of **81** in the presence of 0.5 equivalents of BAR template led to the formation of photoproducts **82–84**, while cyclobutane dimers were observed only on prolonged irradiations. The regiochemistry in the photoproducts **82–84** was rationalized based on the orientation of the cinnamate chromophore that reacted in the presence of the BAR template. The face-to-face orientation of the two *E*-**81** molecules led to the formation of the ternary complex with BAR template yielding the β-truxinate isomer **82**. On the other hand, the face-to-face orientation of one *E*-**81** and one Z-**81** yielded unsymmetrical neotruxinate isomer **83**. The crossed double bonds in the two *E*-**81** molecules due to supramolecular complexation with the BAR template yielded the -truxillate dimer **84**.

1.4.2 Organo-photocatalysis Mediated by Hydrogen Bonding Templates

Although the chiral templates controlling photochemical reactions have been successfully established, the stereoelectronic effects to govern the photocatalytic activity in asymmetric photoreactions was still underdeveloped compared to other supramolecular analogs. This is in part as a result of the change in the excited state behavior when stereoelectronic effects are introduced to chemical systems. This made the purpose of using catalyst developed based on stereoelectronic effects for photoreactions very challenging. Sivaguru and coworkers [104, 105] addressed this issue of asymmetric photoreactions by employing atropisomeric thiourea (TUA) to fine-tune enantioselectivity in photochemical reactions using stereoelectronic effects. A binaphthyl-based **TUA** template was employed to evaluate the efficiency of light absorption and photochemical transformations that has good enantiocontrol (Scheme 33). The organocatalysts/**TUA** template was fine-tuned sterically and electronically for influencing the [2+2]-photocycloaddition of 4-alkenyl substituted coumarins **85**, leading to enantioenriched photoproduct **86**. In case of coumarin derivative **85a**, high enantioselectivity (77%–96% ee) was observed in the photoproduct

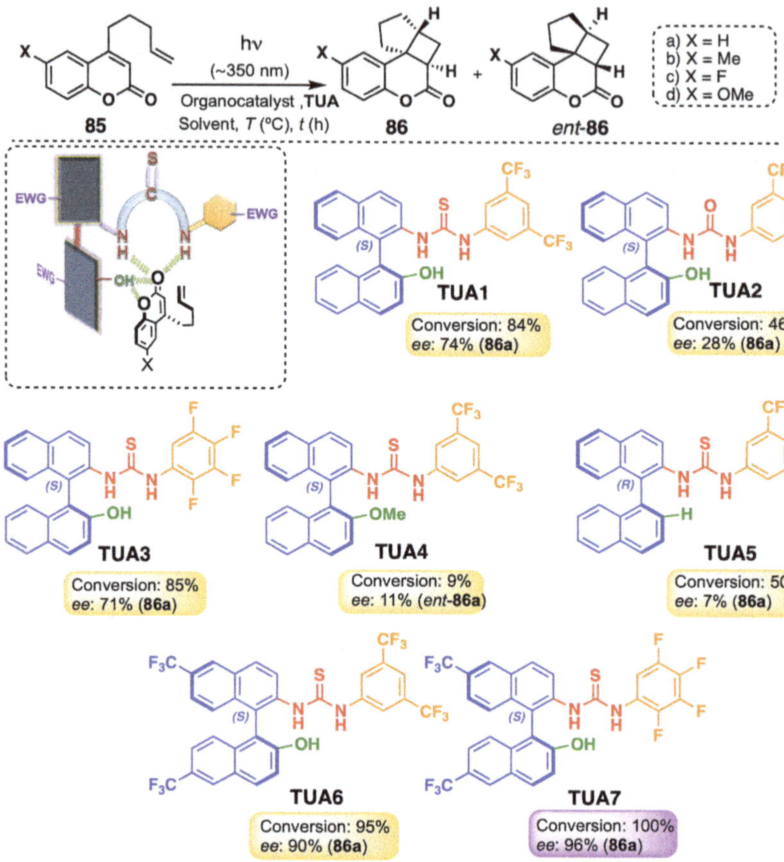

Scheme 33. Enantioselective intramolecular [2+2]-photocycloaddition of coumarin **85** catalyzed by thiourea templates **TUA1–TUA7**. Reprinted with permission from Ref. [104]. Copyright 2014. John Wiley and Sons.

86a (Scheme 33) at low loading (1–10 mol%; Table 4) of catalyst **TUA7**. The adaptability of the catalyst was further proved by performing the reaction in large scale that gave 94% ee with 77% isolated yield in the photoproduct (Table 4, entry 4).

To understand both the ground state and excited state interactions with the substrate the H-bonding ability of the catalysts was systematically changed in **TUA1–TUA7**. There was a significant impact on both the photochemical reactivity and enantioselectivity in product when the

Table 4. Enantioselective intramolecular [2+2]-photocycloaddition of **85a** with various mol% of catalyst **TUA7** in 1:1 toluene/m-xylene at −78 (±2)°C.[a–d]

TUA7 (mol%)	Time	1:2[b]	%ee **86a**[b]
—	4 h	88:12	Racemic
100	30 min[c]	0:100	96
30	30 min	0:100	94
10	30 min	0:100 (77)[d]	94 [92][d]
5	30 min	41:59	89
5	1 h	07:93	88
1	30 min	64:36	77
1	1 h	52:48	75
1	2 h	30:70	74
1	4 h	17:83	69

[a]Samples degassed by N_2 bubbling for 8–10 min before irradiations.
[b]By GC analysis.
[c]Similar ee and conversions at 2 and 3 h irradiations (error ±3%).
[d]Large-scale reaction (2 h), isolated yield = 77%, quantitative conversion.

stereoelectronic features of **TUA** (Scheme 33) were varied. To study about the effect of stereoelectronic effect, **TUA1** was kept as a reference for the photoreactions of unsubstituted coumarin **85a**. High conversions and enantioselectivity was shown by **TUA1** in product **86a** (84% conversion and 74% ee), while **TUA2** gave lower conversions (46%) and enantioselectivity (28%). This abrupt drop in the conversion and ee values while changing the thiourea **TUA1** to urea **TUA2** clearly showcased the influence of thiourea functionality in the photocycloaddition process. Upon changing the EWGs on the aryl ring from 3,5-trifluoromethyl substituents in **TUA1** to tetrafluoro-substituted phenyl ring in **TUA3** there was a minimal change in the ee value and conversions. It was also revealed that the H-bonding on the bi-naphthyl substituent also played a critical part in the ee and conversions. By replacing hydroxyl group in **TUA1** with methoxy substituent in **TUA4** and hydrogen in **TUA5** yielded very low ee with both **TUA4** and **TUA5**, indicating the importance of the OH group in influencing the stereoselection and reaction efficiency. The hydrogen-bonding interactions from the naphthyl substituent was further enhanced by evaluating **TUA6** and **TUA7** containing electron withdrawing CF_3 groups at the 6- and 6′-positions of the binaphthyl backbone. High enantioselectivity of the photoproduct was observed with both **TUA6** and **TUA7** (90% ee for **TUA6** and 96% ee for **TUA7**; Scheme 33).

Comprehensive photophysical studies were carried out to have a better understanding of the importance of **TUA** catalyst in promoting the photoreaction. In spite of the catalyst's excited-state energy being lower than that of the reactants the photoreaction advanced with **TUA1** and **85a**. The hydrogen-bonding interactions between **TUA1** and coumarin substrate playing a crucial role in promoting the reaction gave the idea of a dual catalytic cycle (Figure 9) based on the loading level of the catalyst. The light absorbing species can be manipulated with the loading level of the catalyst **TUA1** (at high catalyst loading) or the **TUA1–85a** complex (at low catalyst loading). With high concentrations of the substrate (i.e., at low catalyst loading), the **TUA1–85a** complex (that displayed a bathochromic shift in its absorption spectrum) was excited and reacted to form the corresponding photoproduct (Figure 9). While at low concentrations of the substrate (i.e., at high catalyst loading), **TUA1** absorbs light (as a result of higher molar absorptivity of **TUA1** compared to the **TUA1–85a** complex and/or **85a**) and reacts. The quenching of excited **TUA1** by substrate **85a** was observed and it was explained with the formation of an exciplex where the excited-state energy of **TUA1** was shared with **85a** (Figure 9). This exciplex reacted efficiently to form the photoproduct. Thus, this strategy illustrated the model to avoid reliance on photocatalysts or sensitizers possessing higher excited-state energies to promote photochemical reaction through formation of static and dynamic complexes. Further the strategy provided a platform for catalyst development via stereoelectronic effects that had a direct impact on both ground state complexation and excited state reactivity.

Sivaguru and coworkers also used thiourea catalysts for the intermolecular [2+2]-photochemical reactions [106]. Irradiation of coumarin **87a** with the thiourea catalysts **TUT9-12** changed the reaction yield due to the hydrogen bonding (Scheme 34). This allowed **87a** to react efficiently with tetramethylethylene **88** yielding the [2+2]-photoproduct **89**. For example,

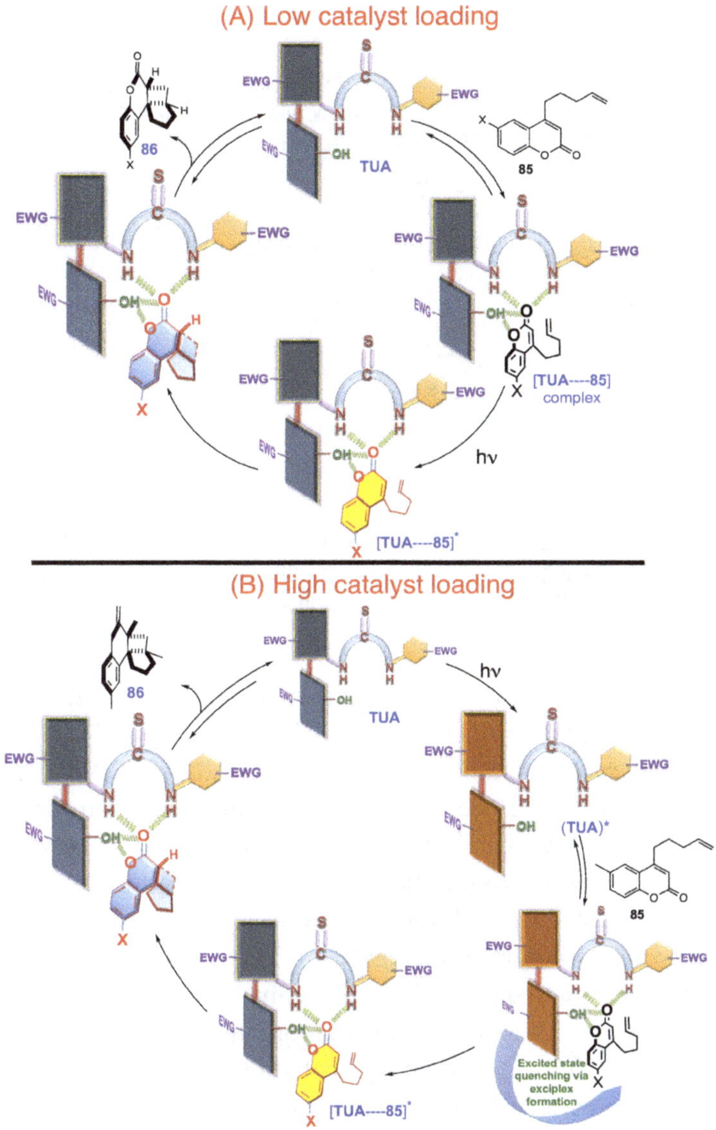

Figure 9. Enantioselective organo-photocatalysis with atropisomeric thiourea catalysts. Catalytic cycles at low (a) and high (b) catalytic loading levels. Reprinted with permission from Ref. [104]. Copyright 2014. John Wiley and Sons.

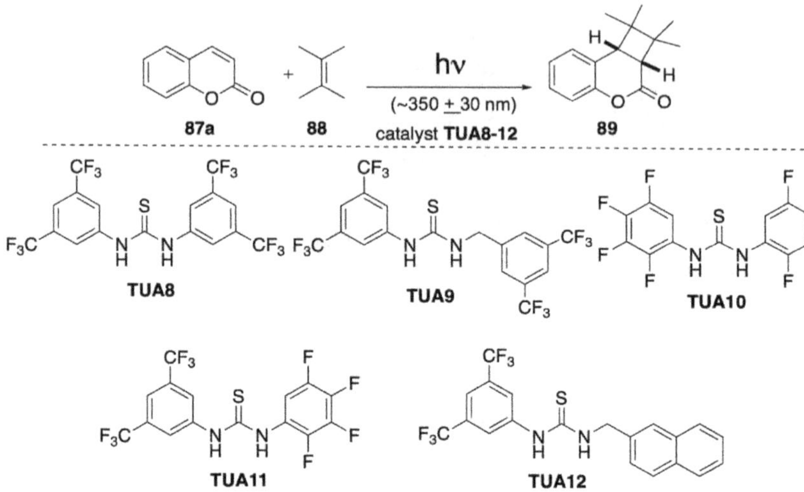

Scheme 34. Thiourea-catalyst-mediated intermolecular [2+2]-photochemical reaction of coumarin **87a** with tetramethyl ethylene **88**. Reprinted with permission from Ref. [106]. Copyright 2014. John Wiley and Sons.

the thiourea catalyst presence of the thiourea catalyst **TUT12** affects the conversion of the photoproduct giving a 75% isolated yield while providing <4% conversion in the absence of the catalyst **TUT12**. Although intramolecular photocycloaddition of coumarins was controlled by the amount of the thiourea catalyst (via an exciplex or a ground-state complex) [104], leading to two mechanistic pathways (Figure 9), the intermolecular [2+2]-photocycloaddition of coumarins mediated by thiourea catalyst provides minimized aggregation (Figure 10), enhanced intersystem crossing, and altered excited-state lifetime(s). The bathochromic shift of absorptivity in the thiourea–coumarin complex(es) allowed the excited selectivity [104, 106]. The photophysical studies explained the different excited state characteristics of the complex.

MacGillivray and coworkers [107] reported the solid-state photoreactivity through mechanical grinding in the presence of catalytic amounts of additives/supramolecular receptor(s) (Scheme 35) [107]. The [2+2]-photodimerization of trans-1,2-bis(4-pyridyl)ethylene (DPE) **91** in solid state was examined by utilizing a ditopic supramolecular catalyst DCR **90** through mechanical grinding [107]. The [2+2]-photodimerization of **91** in the presence of **90** (20 mol%) yielded quantitative amount of

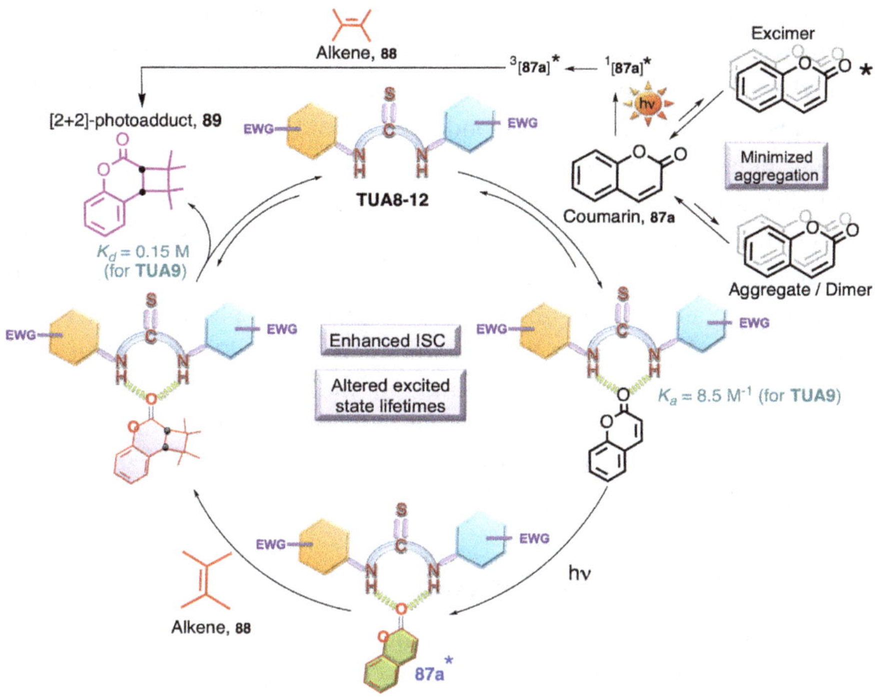

Figure 10. Catalytic cycle for thiourea-catalyst-mediated intermolecular [2+2]-photochemical reaction of coumarin **87a** with tetramethylethylene **88**. Reprinted with permission from Ref. [106]. Copyright 2014. John Wiley and Sons.

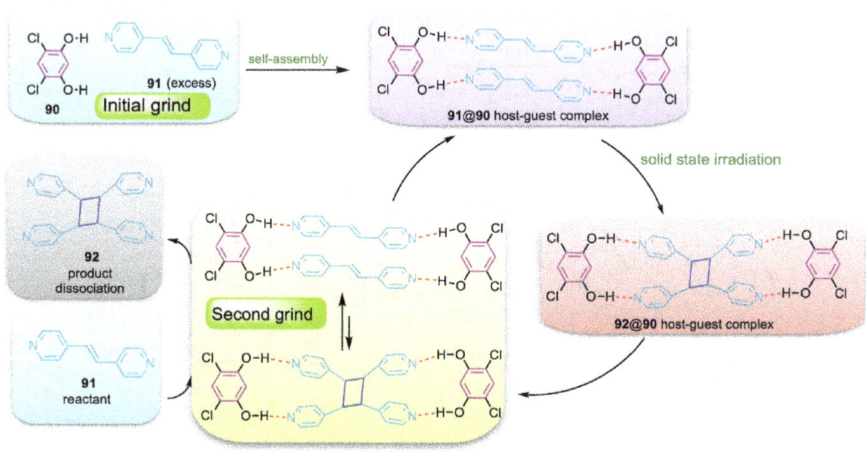

Scheme 35. Solid-state [2+2]-photodimerization of BPE in presence of a ditopic H-bonding host 4,6-dichloro-resorcinol (DCR) **90** through mechanical grinding. Reprinted.

tetrakis(4-pyridyl) cyclobutane (4,4′-tpcb) **92** photoproduct. The reaction was performed through dry mechanical grinding so that turnover was attained even in the absence of solvent. The formation of reactive **91@90** host–guest complex (hydrogen-bonded complex) as a result of the initial grinding helped in the formation of photoproduct **92** due to the shorter double-bond distance (Scheme 35). The photoproduct **92** formed was linked to the catalyst **91** and a second grinding was required to dissociate the receptor (catalyst **90**) from the photoproduct **92**. The free catalyst **90** then bound to unreacted substrates to enhance the photoreactivity and turnover. The photoproduct **92** was obtained in good yields when sub-stoichiometric amounts (10, 20, and 50 mol%) of catalyst **90** were employed. This report established a classic way of controlling and promoting photochemical transformations by combining supramolecular photocatalysis and mechanical stimuli.

1.5 Lewis-acid-mediated Photocatalysis

One of the earlier reports on photodimerization of *trans,trans*-dibenzylideneacetone **93** (DBA) in solution/solid state using uranyl salts was reported by Praetorius and Korn [108]. Later in 1975, Kemp and coworkers [109] reported that the orientation of truxillic-type dimer **94** could be controlled by uranyl salts as well as other salts such as tin chloride

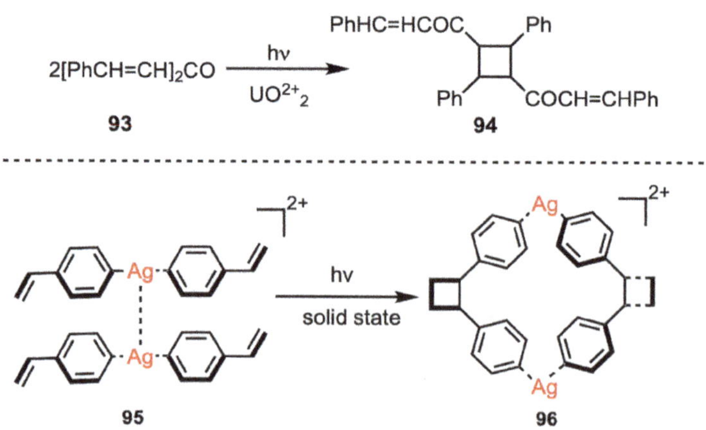

Scheme 36. Metal-mediated-solid state photodimerization (Top); templating olefin dimerization by silver(I) salts (Bottom).

(Scheme 36, top) for photodimerization. Other than uranyl or tin salts, silver salts ($Ag^+ X^-$ and $X^- = BF_4$, CF_3COO, CF_3SO_3, NO_3, ClO_4, ClO_3) have also been reported as a template for pyridyl containing olefins for controlling their photodimerization. The first report on using Ag^+ salts as template was reported by Schroder and coworkers[110] (Scheme 36, bottom) by employing bispyridyl ethylene **95** as a substrate for photodimerization. The investigation also pointed to the interactions between the lone pair of the nitrogen substituent and Ag^+, and π–π interactions between the pyridyl rings that led to pre-organization of C=C bonds by bringing them at a proximity distance for reactivity.

For many decades Lewis acid has been used as a catalyst for several thermal reactions (i.e., Diels–Alder reaction). However, Lewis-acid-promoted photochemical reaction has not been well investigated as their thermal counterparts. One of the earliest works on Lewis-acid-mediated photocatalysis was performed by Fredrick Lewis and coworkers [111] (Scheme 37) in the early 1980s where they utilized the photocycloaddition and photodimerization of coumarin derivatives (Scheme 37, top) as a

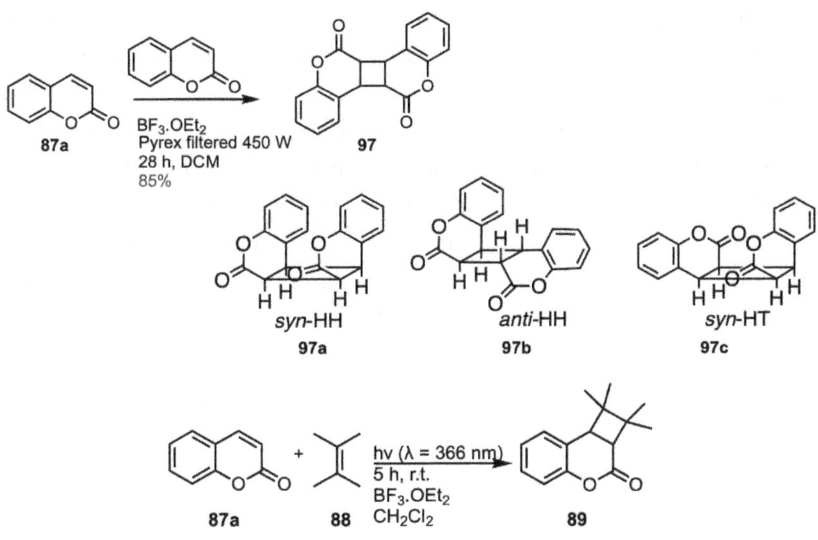

Scheme 37. Lewis-acid-mediated solution-phase photodimerization of coumarin (Top); cross [2+2]-photocycloaddition of 2,3-dimethyl-2-butene to coumarin (Bottom).

model system. The authors proposed that the complexation of reactive substrates with Lewis acid led to an increased ISC rate along with a bathochromic shift of the absorption profiles indicating the lowering of the excited-state energies.

Lewis and coworkers also performed a [2+2]-photocycloaddition (Scheme 37, bottom) of coumarin **87a** with tetramethylethylene **88** (10 eq.) in the presence of Lewis acid ($BF_3{\cdot}OEt_2$) [112]. In the presence of Lewis acid, the cycloadduct was observed in high yield. Their study also showed that Lewis acid influenced the photophysics of the substrates leading to faster deactivation of the excited state(s). In a similar fashion, they also reported a [2+2]-photodimerization of coumarins in the presence of Lewis acid (Scheme 37, top). The mechanism for the coumarin dimerization involved z complexation between coumarin and Lewis acid (BF_3) that was excited under specific wavelength and subsequently adding to the ground-state coumarin leading to the formation of the corresponding product (Scheme 38).

During the [2+2] cycloaddition (photocycloaddition or dimerization) the possibility of formation of up to four stereogenic centers were in a single step makes this photochemical transformation an efficient tool for organic synthesis [47]. An efficient yet underreported way to manipulate the stereochemical outcome is by employing Lewis acid to control photochemical transformations which broadly classified as Lewis-acid-mediated photocatalysis. Recently, Lewis-acid-mediated photocatalysis has been widely explored by the groups of Thorsten Bach (Scheme 39) and Tehshik Yoon (Scheme 41) to promote photochemical transformations with high stereoselectivity. Bach and coworkers [12, 113] employed a versatile chiral aluminum-based Lewis acid and performed enantioselective photoreaction on coumarin systems and found out that $AlBr_3$-activated

$$\text{Coumarin} + BF_3 \rightleftharpoons [\text{Coumarin}.\ BF_3]$$

$$[\text{Coumarin}.\ BF_3] \xrightarrow{h\nu} [\text{Coumarin}.\ BF_3]^*$$

$$[\text{Coumarin}.\ BF_3]^* + \text{Coumarin} \longrightarrow \text{Product} + BF_3$$

Scheme 38. Mechanism for Lewis-acid-mediated coumarin photodimerization.

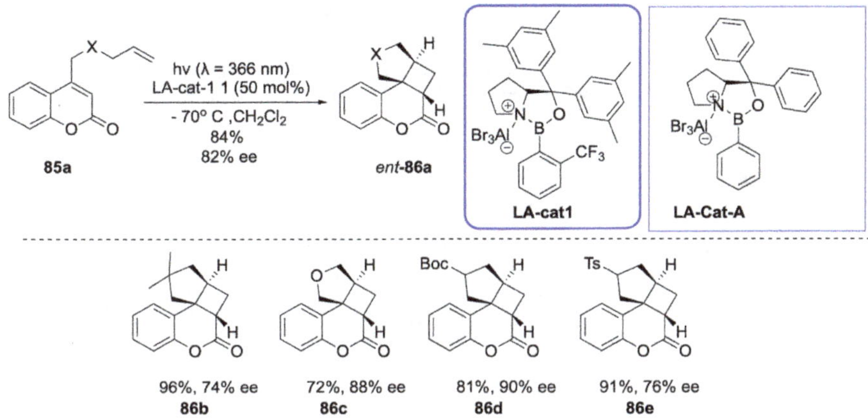

Scheme 39. Intramolecular enantioselective [2+2] cycloaddition of coumarin **85a** and the structures of the related photoproducts. Boc = *tert*-butyoxy carbonyl, Ts = *para*-toluene sulfonyl.

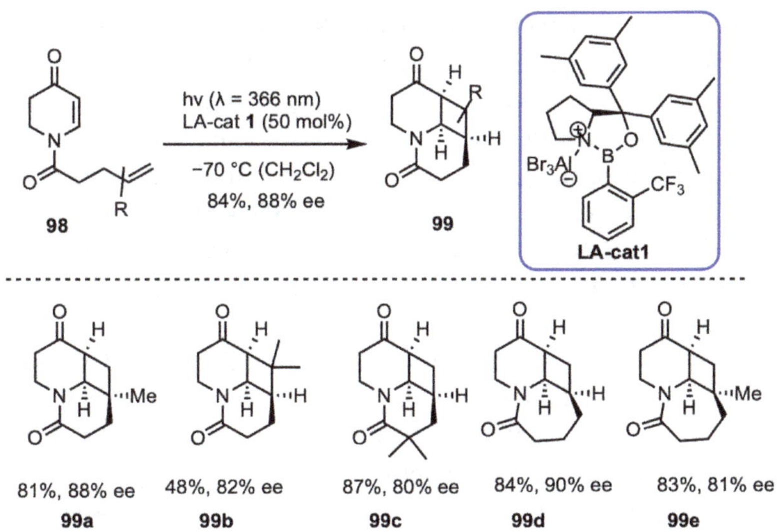

Scheme 40. Lewis-acid-catalyzed Intramolecular enantioselective [2+2] cycloaddition of enone **98** to product **99** and the structures of the related products.

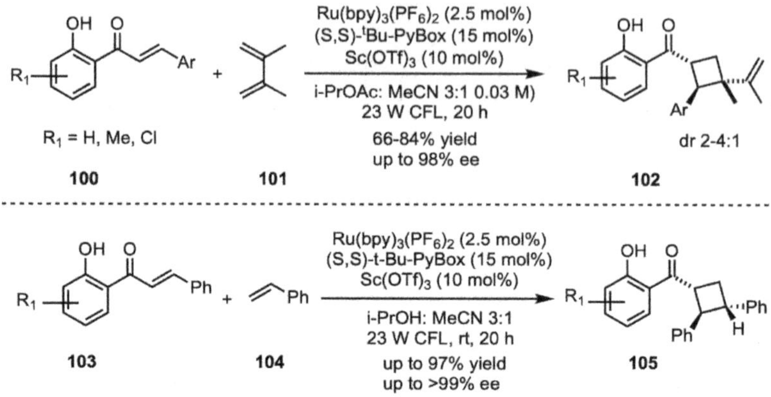

Scheme 41. Lewis-acid-mediated enantioselective [2+2] cycloaddition of hydroxy-chalcones (Top); Lewis-acid-mediated cross enantioselective [2+2] cycloaddition of hydroxychalcones (Middle); Lewis-acid-mediated enantioselective [2+2] cycloaddition of cinnamate esters (Bottom).

oxazaborolidines are a suitable catalyst for studying Lewis-acid-mediated enantioselective photoreactions (Scheme 39). The parent triphenyl-substituted oxaborolidine catalyst (LAcatA; Scheme 39) gave an ee (~32% ee) and proved that high enantioselectivity and chirality turnover can be achieved by optimizing the condition and the amount of catalyst loading. They also performed the photochemical reaction in the absence of Lewis acid under standard conditions and reported 28% yield (racemic product) after 5 h of irradiation. Due to the difficulty in controlling the unselective background reaction, it is noteworthy to find a chiral catalyst for photochemical cycloadditions.

Bach and coworkers [113] reported Lewis-acid-mediated intramolecular [2+2]-photocycloaddition of 4-(pent-4-enyl) coumarin **85a**, where they manipulated the catalytic activity by employing catalyst LACatA (Scheme 39) and obtained the photocycloadduct with high yield (up to 97%). They employed an intramolecular substrate 4-(Pent-4-enyl) coumarin **85a** considering the fact that enantioselectivity determining bond formation step would be much faster in such system than intermolecular addition. By employing oxaborolidine catalyst **LA-cat-1**, 4-(pent-4-enyl) coumarin **85a** underwent enantioselective [2+2] cycloaddition and yielded *ent*-**86a** with 82% ee (84% yield) and the reaction was found to proceed via the formation of triplet intermediate(s).

In 2013, Bach and coworkers [114] reported the enantioselective [2+2]-photocycloaddition (Scheme 40) of dihydropyridone **98** (enone derivative) in which enone substrate underwent photocycloaddition at 366 nm and yielded **99** with 88% ee (84% yield). The mechanism involved the co-ordination between Lewis acid and enone that led to a bathochromic shift of the π-π transition. The LA–substrate complex was thus selectively excited to yield photoproduct **99** (Scheme 40).

Yoon and coworkers [115] reported the photoexcitation energy transfer mediated enantioselective [2+2]-photocycloaddition involving 2′-hydroxychalcone **100** and diene **101** in the presence of Lewis acid (Scheme 41, top). They also observed that enones formed a complex with Lewis acid that resulted in a bathochromic shift of the absorbance. This enabled them to carry out the [2+2]-photocycloaddition under visible light. Yoon and coworkers decided to expand this strategy to enantioselective photocatalytic cycloaddition reactions involving hydroxy chalcone **103** and styrene **104** (Scheme 41, bottom) where they reported that the complex alters the triplet energy of the hydroxy chalcones (for hydroxy chalcone, ET = 54 kcal mol^{-1} to Lewis-acid-bound complex, ET= 32 kcal mol^{-1}) that allowed triplet energy transfer from the photocatalyst to the substrate, leading to the formation of the photoproduct with very high enantioselectivity (~99% ee).

In order to generalize the Lewis-acid-mediated enantioselective reactions, Yoon and coworkers [116] have also explored the

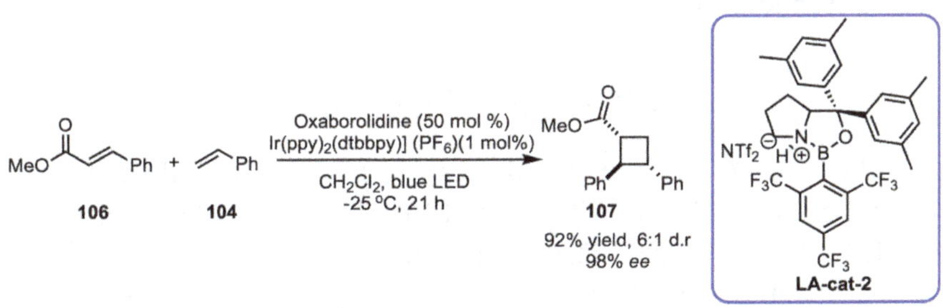

Scheme 42. Lewis-acid-mediated enantioselective [2+2] cycloaddition of cinnamate esters.

Lewis-acid-mediated [2+2]-photocycloaddition (Scheme 42) of simple cinnamate ester **106** in the presence of **LA-cat-2** (50 mol%). Irradiation under blue LED of the LA–substrate complex with styrene **104** using Ir-photocatalyst led to photoproduct **107** with 92% yield with 98% ee and dr ratio of 6:1. The role of Lewis acid was explored by performing computational experiments where they reported that the Lewis acids can stabilize the triplet state of the substrate that increased the thermodynamic driving force for energy transfer in addition to modulating the absolute position of the substrate frontier molecular orbitals for efficient reactivity.

1.6 Conclusions

Chiral enrichment of molecules with light offers unparalleled access to building complex structural skeletons. Developing novel strategies leading to photochirogenesis provides avenues to build sustainable, safe, and ecologically friendly approaches that will be of interest to the scientific community and the society. However, realizing the immense potential of excited state photoreactions leading to photochirogenesis requires an intricate understanding of both photophysical processes and photochemical events. This is because manipulating the stereochemical outcome of a photochemical reaction depends on the in-depth understanding of nature of the excited states involved, their finite excited state lifetimes as well as associated processes (radiation and radiationless processes) some of which might be nearly barrierless. These pose tremendous challenges and present a wide array of opportunities to develop new strategies for stereocontrol in photochemical reactions in solution, which has remained largely unaddressed.

1.7 Acknowledgment

The authors thank the NSF for the generous support of their research (CHE-1955524, CHE-1811795, CHE-1465075, CHE-1213880 and NSF CAREER CHE-0748525). The authors thank his group coworkers from prior years Dr. Barry Pemberton, Dr. Anoklase Jean-Luc Ayitou, Dr. Elango Kumarasamy, Dr. Josepha Jesuraj, Dr. Nandini Vallavoju, Dr. Ramya Raghunathan, Dr. Akila Iyer, Dr. Anthony Clay, and Dr. Sunil Kumar

Kandapa for their intellectual and experimental contributions that led to the development of asymmetric photoreaction methodologies in Prof. Sivaguru's group.

References

1. Ramamurthy, V., and Inoue, Y. (2004) *Molecular and Supramolecular Photochemistry: Chiral Photochemistry*. Marcel Dekker, New York.
2. Wang, W., Clay, A., Krishnan, R., Lajkiewicz Neil, J., Brown Lauren, E., Sivaguru, J., and Porco John, A. (2017) Total syntheses of the isomeric again natural products foveoglin A and perviridisin B: selective excited-state intramolecular proton-transfer photocycloaddition. *Angew. Chem. Int. Ed., 56*, 14479–14482.
3. Wen, S., Boyce, J.H., Kandappa, S.K., Sivaguru, J., and Porco, J.A. (2019) Regiodivergent photocyclization of dearomatized acylphloroglucinols: asymmetric syntheses of (−)-nemorosone and (−)-6-epi-garcimultiflorone A. *J. Am. Chem. Soc.*, **141**, 11315–11321.
4. Inoue, Y. (1992) Asymmetric photochemical reactions in solution. *Chem. Rev.*, **92**, 741–770.
5. Gamlin, J.N., Jones, R., Leibovitch, M., Patrick, B., Scheffer, J.R., and Trotter, J. (1996) The ionic auxiliary concept in solid state organic photochemistry. *Acc. Chem. Res.*, **29**, 203–209.
6. Garcia-Garibay, M.A. (2003) Engineering carbene rearrangements in crystals: from molecular information to solid-state reactivity. *Acc. Chem. Res.*, **36**, 491–498.
7. Pemberton, B.C., Raghunathan, R., Volla, S., and Sivaguru, J. (2012) From containers to catalysts: supramolecular catalysis within cucurbiturils. *Chem. Eur. J.*,**18**, 12178–12190.
8. Ramamurthy, V. (2015) Photochemistry within a water-soluble organic capsule. *Acc. Chem. Res.*, **48**, 2904–2917.
9. Sivaguru, J., Natarajan, A., Kaanumalle, L. S., Shailaja, J., Uppili, S., Joy, A., and Ramamurthy, V. (2003) Asymmetric photoreactions within zeolites: role of confinement and alkali metal ions. *Acc. Chem. Res.*, **36**, 509–521.
10. Sivaguru, J., Solomon, M.R., Poon, T., Jockusch, S., Bosio, S.G., Adam, W., and Turro, N.J. (2008) The reaction of singlet oxygen with enecarbamates: a mechanistic playground for investigating chemoselectivity, stereoselectivity, and vibratioselectivity of photooxidations. *Acc. Chem. Res.*, **41**, 387–400.
11. Vallavoju, N., and Sivaguru, J. (2014) Supramolecular photocatalysis: combining confinement and non-covalent interactions to control light initiated reactions. *Chem. Soc. Rev.*, **43**, 4084–4101.
12. Brimioulle, R., Lenhart, D., Maturi, M.M., and Bach, T. (2015) Enantioselective catalysis of photochemical reactions. *Angew. Chem. Int. Ed.*, **54**, 3872–3890.

13. Sakamoto, M. (1997) Absolute asymmetric synthesis from achiral molecules in the chiral crystalline environment. *Chem. Eur. J.*, **3**, 684–689.
14. Sakamoto, M., Iwamoto, T., Nono, N., Ando, M., Arai, W., Mino, T., and Fujita, T. (2003) Memory of chirality generated by spontaneous crystallization and asymmetric synthesis using the frozen chirality. *J. Org. Chem.*, **68**, 942–946.
15. Sakamoto, M. (2006) Spontaneous chiral crystallization of achiral materials and absolute asymmetric photochemical transformation using the chiral crystalline environment. *J. Photochem. Photobiol. C: Photochem. Rev.*, **7**, 183–196.
16. Ramamurthy, V., and Sivaguru, J. (2016) Supramolecular photochemistry as a potential synthetic tool: photocycloaddition. *Chem. Rev.*, **116**, 9914–9993.
17. Kumarasamy, E., Ayitou, A.J.-L., Vallavoju, N., Raghunathan, R., Iyer, A., Clay, A., Kandappa, S.K., and Sivaguru, J. (2016) Tale of twisted molecules. atropselective photoreactions: taming light induced asymmetric transformations through non-biaryl atropisomers. *Acc. Chem. Res.*, **49**, 2713–2724.
18. Kumarasamy, E., Raghunathan, R., Sibi, M.P., and Sivaguru, J. (2015) Nonbiaryl and heterobiaryl atropisomers: molecular templates with promise for atropselective chemical transformations. *Chem. Rev.*, **115**, 11239–11300.
19. Ayitou, A.J.-L., and Sivaguru, J. (2009) Light-induced transfer of molecular chirality in solution: enantiospecific photocyclization of molecularly chiral acrylanilides. *J. Am. Chem. Soc.*, **131**, 5036–5037.
20. Ayitou, A.J.-L., Ugrinov, A., and Sivaguru, J. (2009) 6π-photocyclization of *O*-tert-butylacrylanilides. N-substitution dictates the regiochemistry of cyclization. *Photochem. Photobiol. Sci.*, **8**, 751–754.
21. Jesuraj, J.L., and Sivaguru, J. (2010) Photochemical type II reaction of atropchiral benzoylformamides to point chiral oxazolidin-4-ones. Axial chiral memory leading to enantiomeric resolution of photoproducts. *Chem. Commun.*, **46**, 4791–4793.
22. Ayitou, A., Pemberton, B.C., Kumarasamy, E., Vallavoju, N., and Sivaguru, J. (2011) Fun with photons: selective light induced reactions in solution and in water soluble nano-containers. *CHIMA*, **65**, 202–209.
23. Ayitou, A.J.-L., and Sivaguru, J. (2011) Reactive spin state dependent enantiospecific photocyclization of axially chiral α-substituted acrylanilides. *Chem. Commun.*, **47**, 2568–2570.
24. Ayitou, A.J.-L., Vallavoju, N., Ugrinov, A., and Sivaguru, J. (2011) Enantiospecific 6π-photocyclization of atropisomeric α-substituted acrylanilides in the solid-state: role of crystalline confinement on enantiospecificity. *Photochem. Photobiol. Sci.*, **10**, 1380–1383.
25. Kumarasamy, E., Jesuraj, J.L., Omlid, J.N., Ugrinov, A., and Sivaguru, J. (2011) Light-induced enantiospecific 4π-ring closure of axially chiral 2-pyridones: enthalpic and entropic effects promoted by H-bonding. *J. Am. Chem. Soc.*, **133**, 17106–17109.

26. Ayitou, A.J.L., Fukuhara, G., Kumarasamy, E., Inoue, Y., and Sivaguru, J. (**2013**) Enantiospecific photochemical transformations under elevated pressure. *Chem. Eur. J.*, **19**, 4327–4334.
27. Kumarasamy, E., and Sivaguru, J. (2013) Light-induced stereospecific intramolecular [2+2]-cycloaddition of atropisomeric 3,4-dihydro-2-pyridones. *Chem. Commun.*, **49**, 4346–4348.
28. Raghunathan, R., Kumarasamy, E., Iyer, A., Ugrinov, A., and Sivaguru, J. (2013) Intramolecular Paternò-Büchi reaction of atropisomeric α-oxoamides in solution and in the solid-state. *Chem. Commun.*, **49**, 8713–8715.
29. Ayitou, A.J.-L., Clay, A., Kumarasamy, E., Jockusch, S., and Sivaguru, J. (2014) Enantiospecific photochemical 6π-ring closure of α-substituted atropisomeric acrylanilides — role of alkali metal ions. *Photochem. Photobiol. Sci.*, **13**, 141–144.
30. Clay, A., Kumarasamy, E., Ayitou, A.J.L., and Sivaguru, J. (2014) Photochemistry of atropisomers: non-biaryl atropisomers for stereospecific phototransformations. *Chem. Lett.*, **43**, 1816–1825.
31. Iyer, A., Jockusch, S., and Sivaguru, J. (2014) Dictating photoreactivity through restricted bond rotations: cross-photoaddition of atropisomeric acrylimide derivatives under UV/visible-light irradiation. *J. Phys. Chem. A.* 2014, **118**, 10596–10602.
32. Kumarasamy, E., Raghunathan, R., Jockusch, S., Ugrinov, A., and Sivaguru, J. (2014) Tailoring atropisomeric maleimides for stereospecific [2 + 2] photocycloaddition — photochemical and photophysical investigations leading to visible-light photocatalysis. *J. Am. Chem. Soc.*, **136**, 8729–8737.
33. Raghunathan, R., Kumarasamy, E., Jockusch, S., Ugrinov, A., and Sivaguru, J. (2016) Engaging electronic effects for atropselective [5+2]-photocycloaddition of maleimides. *Chem. Commun.*, **52**, 8305–8308.
34. Vallavoju, N., Sreenitya, A., Ayitou, A., Jockusch, S., Sunoj, R.B., and Sivaguru, J. (2016) Photoreactions with a twist: atropisomerism driven divergent reactivity of enones with UV and visible light. *Chem. Eur. J.*, **22**, 11339–11348.
35. Jacobs, H.J.C., Gielen, J.W.J., and Havinga, E. (1981) Effects of wavelength and conformation on the photochemistry of vitamin D and related conjugated trienes. *Tetrahedron Lett.*, **22**, 4013–4016.
36. Whitesell, J.K., Minton, M.A., and Tran, V.D. (1989) The non-equilibration of excited rotamers (NEER) principle. Ground-state conformational bias in triene photocyclizations. *J. Am. Chem. Soc.*, **111**, 1473–1476.
37. Clayden, J. (1997) Non-biaryl atropisomers: new classes of chiral reagents, auxiliaries, and ligands? *Angew. Chem. Int. Ed.*, **36**, 949–951.
38. Curran, D.P., Qi, H., Geib, S.J., and DeMello, N.C. (1994) Atroposelective thermal reactions of axially twisted amides and imides. *J. Am. Chem. Soc.*, **116**, 3131–3132.
39. Curran, D.P., Hale, G.R., Geib, S.J., Balog, A., Cass, Q.B., Degani, A.L.G., Hernandes, M.Z., and Freitas, L.C.G. (1997) Rotational features of carbon-nitrogen bonds in axially chiral O-tert-butyl anilides and related molecules. potential sub-

strates for the 'prochiral auxiliary' approach to asymmetric synthesis. *Tetrahedron: Asymmetry*, **8**, 3955–3975.

40. Tanaka, K., and Toda, F. (2000) Solvent-free organic synthesis. *Chem. Rev.*, **100**, 1025–1074.
41. Ford, W.T. (1971) Cycloaddition of benzyne to substituted cyclopentadienes and cyclopentadienyl Grignard reagents. *J. Org. Chem.*, **36**, 3979–3987.
42. Naito, T., Tada, Y., and Ninomiya, I. (1984) Asymmetric photocyclization of N-α,β-unsaturated acylanilides. *Heterocycles*, **22**, 237–240.
43. Ayitou, A.J.-L., Jesuraj, J.L., Barooah, N., Ugrinov, A., and Sivaguru, J. (2009) Enantiospecific photochemical Norrish/Yang type II reaction of nonbiaryl atropchiral α-oxoamides in solution: axial to point chirality transfer. *J. Am. Chem. Soc.*, **131**, 11314–11315.
44. Natarajan, A., Mague, J.T., and Ramamurthy, V. (2005) Asymmetric induction during yang cyclization of α-oxoamides: the power of a covalently linked chiral auxiliary is enhanced in the crystalline state. *J. Am. Chem. Soc.*, **127**, 3568–3576.
45. Wang, R., Chen, C., Duesler, E., Mariano, P.S., and Yoon, U.C. (2004) β-lactam-forming photochemical reactions of N-trimethylsilylmethyl- and N-tributylstannylmethyl-substituted α-ketoamides. *J. Org. Chem.*, **69**, 1215–1220.
46. Chesta, C.A., and Whitten, D.G. (1992) Photocyclization of α-keto amides in homogeneous solution and aqueous cyclodextrin media. The role of zwitterions and diradicals in photoinduced electron transfer reactions. *J. Am. Chem. Soc.*, **114**, 2188–2197.
47. Poplata, S., Tröster, A., Zou, Y.-Q., and Bach, T. (2016) Recent advances in the synthesis of cyclobutanes by olefin [2 + 2] photocycloaddition reactions. *Chem. Rev.*, **116**, 9748–9815.
48. Chang, X.-P., Zheng, Y., Cui, G., Fang, W.-H., and Thiel, W. (2016) Photocycloaddition reaction of atropisomeric maleimides: mechanism and selectivity. *Phys. Chem. Chem. Phys.*, **18**, 24713–24721.
49. Alder, A., Bühler, N., and Bellus, D. (1982) A note on intramolecular photochemical cycloaddition of N-substituted dimethacrylimides. *Helv. Chim. Acta*, **65**, 2405–2412.
50. Maruyama, K., and Ishitoku, T. (1980) Photochemical cyclization of acrylimide derivatives. *Chem. Lett.*, **9**, 359–360.
51. Sylvie Le, B., Jean-Pierre, P., and Olivier, P. (1993) Intramolecular [2+2] photocycloaddition of oxoesters and oxoamides. *Tetrahedron Lett.*, **34**, 635–638.
52. Clay, A., Vallavoju, N., Ugrinov, A., Krishnan, R., and Sivaguru, J. (2016) Metal free visible light mediated photocatalysis: controlling intramolecular [2+2] photocycloaddition of enones through axial chirality. *J. Org. Chem.*, **81**, 7191–7200.
53. Yang, N. C., Nussim, M., Jorgenson, M. J., and Murov, S. (1964) Photochemical reactions of carbonyl compounds in solution The Paterno-Büchi reaction. *Tetrahedron Lett.*, **5**, 3657–3664.
54. Vogt, F., Jödicke, K., Schröder, J., and Bach, T. (2009) Paternò-Büchi reactions of silyl enol ethers and enamides. *Synthesis*, **2009**, 4268–4273.

55. Davies, D.M.E., Murray, C., Berry, M., Orr-Ewing, A.J., and Booker-Milburn, K.I. (2007) Reaction optimization and mechanism in maleimide [5 + 2] photocycloaddition: a dual approach using tunable UV lasers and time-dependent DFT. *J. Org. Chem.*, **72**, 1449–1457.
56. Breslow, R. (1995) Biomimetic chemistry and artificial enzymes: catalysis by design. *Acc. Chem. Res.*, **28**, 146–153.
57. Vriezema, D.M., Comellas, A.M., Elemans, J.A.A.W., Cornelissen, J.J.L.M., Rowan, A.E., and Nolte, R.J.M. (2005) Self-assembled nanoreactors. *Chem. Rev.*, **105**, 1445–1490.
58. Svoboda, J., and König, B. (2006) Templated photochemistry: toward catalysts enhancing the efficiency and selectivity of photoreactions in homogeneous solutions. *Chem. Rev.*, **106**, 5413–5430.
59. Pemberton, B.C., Raghunathan, R., Volla, S., and Sivaguru, J. (2012) From containers to catalysts: supramolecular catalysis within cucurbiturils. *Chem. Eur. J.*, **18**, 12178–12190.
60. Sivaguru, J., Natarajan, A., Kaanumalle, L.S., Shailaja, J., Uppili, S., Joy, A., and Ramamurthy, V. (2003) Asymmetric photoreactions within zeolites: role of confinement and alkali metal ions. *Acc. Chem. Res.*, **36**, 509–521.
61. Behrend, R., Meyer, E., and Rusche, F. (1905) Condensation products from glycouril and formaldehyde. *Liebigs Ann. Chem.*, **339**, 1–37.
62. Lee, J.W., Samal, S., Selvapalam, N., Kim, H.-J., Kim, K. (2003) Cucurbituril homologues and derivatives: new opportunities in supramolecular chemistry. *Acc. Chem. Res.*, **36**, 621–630.
63. Masson, E., Ling, X., Joseph, R., Kyeremeh-Mensah, L., and Lu, X. (2012) Cucurbituril chemistry: a tale of supramolecular success. *RSC Adv.*, **2**, 1213–1247.
64. Szejtli, J. (1998) Introduction and general overview of cyclodextrin chemistry. *Chem. Rev.*, **98**, 1743–1754.
65. Gibb, C.L.D., and Gibb, B.C. (2004) Well-defined, organic nanoenvironments in water: the hydrophobic effect drives a capsular assembly. *J. Am. Chem. Soc.*, **126**, 11408–11409.
66. Yoshizawa, M., Klosterman, J.K., and Fujita, M. (2009) Functional molecular flasks: new properties and reactions within discrete, self-assembled hosts. *Angew. Chem. Int. Ed.*, **48**, 3418–3438.
67. Yoshizawa, M., Takeyama, Y., Kusukawa, T., and Fujita, M. (2002) Cavity-directed, highly stereoselective [2+2] photodimerization of olefins within self-assembled coordination cages. *Angew. Chem. Int. Ed.*, **41**, 1347–1349.
68. Murase, T., Takezawa, H., and Fujita, M. (2011) Photo-driven anti-Markovnikov alkyne hydration in self-assembled hollow complexes. *Chem. Commun.*, **47**, 10960–10962.
69. Bach, T., Bergmann, H., and Harms, K. (2000) Enantioselective intramolecular [2+2]-photocycloaddition reactions in solution. *Angew. Chem. Int. Ed.*, **39**, 2302–2304.

70. Müller, C., Bauer, A., and Bach, T. (2009) Light-driven enantioselective organocatalysis. *Angew. Chem. Int. Ed.*, **48**, 6640–6642.
71. Müller, C., Bauer, A., Maturi, M.M., Cuquerella, M.C., Miranda, M.A., and Bach, T. (2011) Enantioselective intramolecular [2 + 2]-photocycloaddition reactions of 4-substituted quinolones catalyzed by a chiral sensitizer with a hydrogen-bonding motif. *J. Am. Chem. Soc.*, **133**, 16689–16697.
72. Bauer, A., Westkamper, F., Grimme, S., and Bach, T. (2005) Catalytic enantioselective reactions driven by photoinduced electron transfer. *Nature*, **436**, 1139–1140.
73. Mizoguchi, J.-i., Kawanami, Y., Wada, T., Kodama, K., Anzai, K., Yanagi, T., and Inoue, Y. (2006) Enantiodifferentiating photocyclodimerization of 2-anthracenecarboxylic acid using a chiral N-(2-hydroxymethyl-4- pyrrolidinyl)benzamide template. *Org. Lett.*, **8**, 6051–6054.
74. Cauble, D.F., Lynch, V., and Krische, M.J. (2003) Studies on the enantioselective catalysis of photochemically promoted transformations: "sensitizing receptors" as chiral catalysts. *J. Org. Chem.*, **68**, 15–21.
75. Bassani, D.M., Darcos, V., Mahony, S., and Desvergne, J.-P. (2000) Supramolecular catalysis of olefin [2 + 2] photodimerization. *J. Am. Chem. Soc.*, **122**, 8795–8796.
76. Darcos, V., Griffith, K., Sallenave, X., Desvergne, J.-P., Guyard-Duhayon, C., Hasenknopf, B., and Bassani, D.M. (2003) Supramolecular control of [2 + 2] photodimerization via hydrogen bonding. *Photochem. Photobiol. Sci.*, **2**, 1152–1161.
77. Ke, C., Yang, C., Mori, T., Wada, T., Liu, Y., and Inoue, Y. (2009) Catalytic enantiodifferentiating photocyclodimerization of 2-anthracenecarboxylic acid mediated by a non-sensitizing chiral metallosupramolecular host. *Angew. Chem. Int. Ed.*, **48**, 6675–6677.
78. Cibulka, R., Vasold, R., and König, B. (2004) Catalytic photooxidation of 4-methoxybenzyl alcohol with a flavin–zinc(II)-cyclen complex. *Chem. Eur. J.*, **10**, 6223–6231.
79. Fedorova, O., Fedorov, Y.V., Gulakova, E., Schepel, N., Alfimov, M., Goli, U., and Saltiel, J. (2007) Supramolecular photochemical synthesis of an unsymmetrical cyclobutane. *Photochem. Photobiol. Sci.*, **6**, 1097–1105.
80. Yang, C., and Inoue, Y. (2014) Supramolecular photochirogenesis. *Chem. Soc. Rev.*, *ASAP*, 10.1039/C1033CS60339C.
81. Bibal, B., Mongin, C., and Bassani, D.M. (2014) Template effects and supramolecular control of photoreactions in solution. *Chem. Soc. Rev.*, 10.1039/C3CS60366K.
82. Kaanumalle, L.S., and Ramamurthy, V. (2007) Photodimerization of acenaphthylene within a nanocapsule: excited state lifetime dependent dimer selectivity. *Chem. Commun.*, 1062–1064.
83. Kulasekharan, R., Maddipatla, M.V.S.N., Parthasarathy, A., and Ramamurthy, V. (2013) Role of free space and conformational control on photoproduct selectivity of optically pure α-alkyldeoxybenzoins within a water-soluble organic capsule. *J. Org. Chem.*, **78**, 942–949.
84. Connors, K.A. (1987) *Binding Constants*. John Wiley, New York.

85. Freeman, W.A., Mock, W.L., and Shih, N.-Y. (1981) Cucurbituril. *J. Am. Chem. Soc.*, **103**, 7367–7368.
86. Kim, J., Jung, I.-S., Kim, S.-Y., Lee, E., Kang, J.-K., Sakamoto, S., Yamaguchi, K., and Kim, K. (2000) New cucurbituril homologues: syntheses, isolation, characterization, and X-ray crystal structures of cucurbit[n]uril (n) 5, 7, and 8). *J. Am. Chem. Soc.*, **122**, 540–541.
87. Day, A., Arnold, A. P., Blanch, R.J., and Snushall, B. (2001) Controlling factors in the synthesis of cucurbituril and its homologues. *J. Org. Chem.*, **66**, 8094–8100.
88. *Special Issue on Cucurbiturils* (2011). Israel J. Chem. WILEY-VCH Verlag, Vol. 51, pp. 485–678.
89. Jon, S.Y., Ko, Y.H., Park, S.H., Kim, H.-J., and Kim, K. (2001) A facile, stereoselective [2 + 2] photoreaction mediated by cucurbit[8]uril. *Chem. Commun.*, 1938–1939.
90. Pattabiraman, M., Natarajan, A., Kaliappan, R., Mague, J.T., and Ramamurthy, V. (2005) Template directed photodimerization of trans-1,2-bis(n-pyridyl)ethylenes and stilbazoles in water. *Chem. Commun.*, 4542–4544.
91. Maddipatla, M.V.S.N., Kaanumalle, L.S., Natarajan, A., Pattabiraman, M., and Ramamurthy, V. (2007) Preorientation of olefins toward a single photodimer: cucurbituril-mediated photodimerization of protonated azastilbenes in water. *Langmuir*, **23**, 7545–7554.
92. Wang, R., Yuan, L., and Macartney, D.H. (2006) Cucurbit[7]uril mediates the stereoselective [4+4] photodimerization of 2-aminopyridine hydrochloride in aqueous solution. *J. Org. Chem.*, **71**, 1237–1239.
93. Pemberton, B.C., Barooah, N., Srivatsava, D.K., and Sivaguru, J. (2010) Supramolecular photocatalysis by confinement — photodimerization of coumarins within cucurbit[8]urils. *Chem. Commun.*, **46**, 225–227.
94. Pemberton, B.C., Kumarasamy, E., Jockusch, S., Srivastava, D.K., and Sivaguru, J. (2011) Photophysical aspects of 6-methylcoumarin-cucurbit[8]uril host-guest complexes. *Can. J. Chem.*, **89**, 310–316.
95. Pemberton, B.C., Singh, R.K., Johnson, A.C., Jockusch, S., Da, S.J.P., Ugrinov, A., Turro, N.J., Srivastava, D.K., and Sivaguru, J. (2011) Supramolecular photocatalysis: insights into cucurbit[8]uril catalyzed photodimerization of 6-methylcoumarin. *Chem. Commun.*, **47**, 6323–6325.
96. Pemberton, B.C. (2012) From nanocontainer to nanocatalyst: Mechanistic studies of [2+2] photodimerization of coumarin derivatives within cucurbit[8]uril. PhD Thesis, North Dakota State University.
97. Chung, W.S., Turro, N.J., Silver, J., and Le Noble, W.J. (1990) Modification of face selectivity by inclusion in cyclodextrins. *J. Am. Chem. Soc.*, **112**, 1202–1205.
98. Nakamura, A., and Inoue, Y. (2003) Supramolecular catalysis of the enantiodifferentiating [4 + 4] photocyclodimerization of 2-anthracenecarboxylate by γ-cyclodextrin. *J. Am. Chem. Soc.*, **125**, 966–972.

99. Inoue, Y., Wada, T., Sugahara, N., Yamamoto, K., Kimura, K., Tong, L.-H., Gao, X.-M., Hou, Z.-J., and Liu, Y. (2000) Supramolecular photochirogenesis. 2. enantiodifferentiating photoisomerization of cyclooctene included and sensitized by 6-O-modified cyclodextrins. *J. Org. Chem.*, **65**, 8041–8050.
100. Jagadesan, P., Mondal, B., Parthasarathy, A., Rao, V.J., and Ramamurthy, V. (2013) Photochemical reaction containers as energy and electron-transfer agents. *Org. Lett.*, **15**, 1326–1329.
101. Yang, C., Mori, T., Origane, Y., Ko, Y.H., Selvapalam, N., Kim, K., and Inoue, Y. (2008) Highly stereoselective photocyclodimerization of α-cyclodextrin-appended anthracene mediated by γ-cyclodextrin and cucurbit[8]uril: a dramatic steric effect operating outside the binding site. *J. Am. Chem. Soc.*, **130**, 8574–8575.
102. Pemberton, B.C., Ugrinov, A., and Sivaguru, J. (2013) Evaluating photodimerization of 6-methylcoumarin mediated by cucurbit[8]uril through mechanical grinding — supramolecular effects of additives. *J. Photochem. Photobiol. A. Chem.*, **255**, 10–15.
103. Gnanaguru, K., Ramasubbu, N., Venkatesan, K., and Ramamurthy, V. (1985) A Study on the photochemical dimerization of coumarins in the solid state. *J. Org. Chem.*, **50**, 2337–2346.
104. Vallavoju, N., Selvakumar, S., Jockusch, S., Sibi, M.P., and Sivaguru, J. (2014) Enantioselective organo-photocatalysis mediated by atropisomeric thiourea derivatives. *Angew. Chem. Int. Ed.*, **53**, 5604–5608.
105. Vallavoju, N., Selvakumar, S., Jockusch, S., Prabhakaran, M.T., Sibi, M.P., and Sivaguru, J. (2014) Evaluating thiourea architecture for intramolecular [2+2] photocycloaddition of 4-alkenylcoumarins. *Adv. Synt. Catal.*, **356**, 2763–2768.
106. Vallavoju, N., Selvakumar, S., Pemberton, B.C., Jockusch, S., Sibi, M.P., and Sivaguru, J. (2016) Organophotocatalysis: insights into the mechanistic aspects of thiourea-mediated intermolecular [2+2] photocycloadditions. *Angew. Chem. Int. Ed.*, **55**, 5446–5451.
107. Sokolov, A.N., Bučar, D.-K., Baltrusaitis, J., Gu, S.X., MacGillivray, L.R. (2010) Supramolecular catalysis in the organic solid state through dry grinding. *Angew. Chem. Int. Ed.*, **49**, 4273–4277.
108. Praetorius, P., and Korn, F. (1910) Belichtung ungesättigter Ketone bei Gegenwart von Uranylsalzen. *Ber. Dtsch. Chem. Ges.*, **43**, 2744–2746.
109. Alcock, N.W., Herron, N., Kemp, T.J., and Shoppee, C.W. (1975) Orientation of photo-dimerization by metal ions: the crystal structure of bis(dibenzylideneacetone) uranyl dichloride. *J. Chem. Soc., Chem. Commun.*, 785–786.
110 Blake, A.J., Champness, N.R., Chung, S.S.M., Li, W.-S., and Schröder, M. (1997) In situ ligand synthesis and construction of an unprecedented three-dimensional array with silver(I): a new approach to inorganic crystal engineering. *Chem. Commun.*, 1675–1676.

111. Lewis, F.D., and Barancyk, S.V. (1989) Lewis acid catalysis of photochemical reactions. 8. Photodimerization and cross-cycloaddition of coumarin. *J. Am. Chem. Soc.*, **111**, 8653–8661.
112. Lewis, F.D., Howard, D.K., and Oxman, J.D. (1983) Lewis acid catalysis of coumarin photodimerization. *J. Am. Chem. Soc.*, **105**, 3344–3345.
113. Guo, H., Herdtweck, E., and Bach, T. (2010) Enantioselective Lewis acid catalysis in intramolecular [2+2] photocycloaddition reactions of coumarins. *Angew. Chem. Int. Ed.*, **49**, 7782–7785.
114. Brimioulle, R., and Bach, T. (2013) Enantioselective Lewis acid catalysis of intramolecular enone [2+2] photocycloaddition reactions. *Science*, **342**, 840.
115. Blum, T.R., Miller, Z.D., Bates, D.M., Guzei, I.A., and Yoon, T.P. (2016) Enantioselective photochemistry through Lewis acid–catalyzed triplet energy transfer. *Science*, **354**, 1391.
116. Daub, M.E., Jung, H., Lee, B.J., Won, J., Baik, M.-H., and Yoon, T.P. (2019) Enantioselective [2+2] cycloadditions of cinnamate esters: generalizing Lewis acid catalysis of triplet energy transfer. *J. Am. Chem. Soc.*, **141**, 9543–9547.

https://doi.org/10.1142/9789811259227_0008

8 Modeling of Chirogenesis: Best Practices and Applications

Peeter Burk* **and Jaanus Burk**

Institute of Chemistry, University of Tartu, Tartu, Estonia
*peeter.burk@ut.ee

State of the art for density functional theory (DFT) modeling of chiral systems, including chirality transfer, is reviewed. Reasonable modeling protocol is provided, and the possible caveats highlighted. Examples of DFT modeling of chirality transfer in last 10 years are revised.

1.1 Introduction

The term chirality is used to refer to the handedness of an object, i.e., when the object is not superimposable with its mirror image. The word chiral was first used by Lord Kelvin in 1894 and is derived from the Greek word χείρ (*cheir*), meaning "hand", which is the canonical example of an object with this property [1]. Chirality is widespread in nature at all length scales, ranging from the astronomical where circularly polarized light (CPL) emanates from large regions of space [2] to macroscopic objects like crystals, living bacterial colonies [3] and shells of some gastropod mollusks [4] to molecules and down to the interactions between fundamental particles [5].

In the context of a geometrically rigid body such as crystal, a structure is considered chiral, if it lacks mirror plane symmetry and is therefore not superimposable onto its mirror image. This definition implies that for chiral molecules there exist both left- and right-handed forms (the molecule and its nonsuperimposable mirror image), which are called enantiomers, if there are no combinations of mirror, rotation and inversion symmetry operations that transform the molecular structure into itself [6].

Molecular chirality was first observed in the early 1800s, by detecting the optical activity of enantiomers, i.e., their ability to rotate linearly polarized light by equal magnitudes but in opposite directions. The phenomenon of optical activity of the chemical compounds was demonstrated in the early 19th century by scientists such as Mitscherlich, Biot, and others [7]. The origin of optical activity in molecular chirality was discovered by Louis Pasteur in 1848 [8, 9], who found that crystals of sodium ammonium tartrate tetrahydrate had chiral shapes and the dissolution of those enantiomorphous crystals yielded solutions that rotated polarized light by equal angles but in opposite directions. Solutions obtained from equimolar amounts of the enantiomorphous crystals did not rotate light. Pasteur's experiment was the first to establish a link between enantiomorphism at the macroscopic level (crystal shape) and at the molecular level and to establish the link to optical activity [10]. This laid the foundation of the field of stereochemistry.

The most important manifestation of molecular chirality is found in the biomolecules essential for life on Earth. Most such molecules (amino acids, sugars, DNA, etc.) have chiral structures and, therefore, can have two enantiomeric forms. However, in life on Earth these molecules are homochiral, i.e., chiral biomolecules exist only in one enantiomeric form in nature [11, 12]. So, only the L-enantiomers of amino acids are present in naturally occurring proteins and the molecular backbones of DNA and RNA contain only D-sugars. Replacing some of the amino acids in proteins or replacing the sugar subunits in DNA with their mirror images would drastically change the in vivo properties of these biomolecules, rendering them incompatible with life [13, 14]. The origin of homochirality in life on Earth has been the subject of intense research over the past hundred years as it may provide clues to the processes that led to the origin of life itself [12, 15–17]. The reasons for this biomolecular homochirality on Earth are not known, but several of contributing factors have been proposed. These include, but are not limited to, tiny energy differences between enantiomers induced by parity violation in the weak force [18], stochastic fluctuations in enantiomeric excess (ee) [19], and chiral amplification of enantiomeric excess on inorganic surfaces [20]. The most important consequence of the biomolecular homochirality of life for human society is that chirality influences the physiological response of living organisms to ingested chiral compounds such as pharmaceuticals [21]. The tragic birth defects resulting from ingestion of a racemic (enantiomerically equimolar) mixture of D- and L-thalidomide by pregnant women in the late 1950s led to our understanding of the need for stereochemical control of chiral drugs [22, 23]. The understanding of the differences in physiological impact of enantiomers and established regulations for chiral drug development have led to a steady growth in the production of single enantiomer drugs [24–27]. This has led to a market with 4200 billion USD annual sales of enantiopure drugs in 2005 and its growth at a rate of 15% since 2010 [28, 29]. It is estimated that ~95% of all drugs marketed by 2020 will be chiral [30, 31].

Chirality research was focused on asymmetric synthesis of single enantiomer compounds in the 20th century, and decades of significant work has been recognized by the 2001 Nobel Prize in chemistry [32–34]. In the 21st century, however, the research on chirality has expanded and now includes fields such as chiral nanostructures [35, 36], chiral surfaces [37–39], molecular motors [40–42], and spintronics [43] One of the most active fields has been supramolecular chirality [44, 45], i.e., the supramolecular chirogenesis, formation of complex chiral structures by self-assembly.

1.2 Chirogenesis

Supramolecular chirogenesis, a blend of supramolecular chemistry and molecular chirality, plays a prominent role in the functioning of a large variety of natural (such as heme proteins, DNA double helix, and secondary α-helix structure of proteins) and artificial systems, endowing it with importance in divergent fields of applied sciences and modern technologies and also for the applicability in a range of areas such as asymmetric catalysis, assigning absolute configuration, memory devices, chiral recognition, and developing biomimetic processes [45–70]. Generally, in supramolecular chirogenesis, noncovalent interactions (e.g., hydrogen bonding, electrostatic, van der Waals, steric, and π–π interactions) are found to be the key modes of chiral communication. However, these interactions are also affected by several internal (steric and electronic effects, bonding strength, etc.) and external (solvent polarity, phase transition, temperature, etc.) factors and consequently exert influence on the overall chirality induction process [45–70]. Therefore, an in-depth understanding of supramolecular chirogenesis is essential to have a clear insight into its fundamental mechanisms and the various controlling factors, which would eventually lead to a variety of practical applications of chiral supramolecular science. This subject has intrigued the modern scientific community and it may become an essential tool for academic and industrial research in the near future.

Over the past few decades, researchers have been trying to comprehend various supramolecular principles to correlate with a diverse collection of complex chiral assemblies and modern artificial systems [45–70]. Among a great variety of chromophoric compounds, metalloporphyrin tweezers have been widely used for chirogenic processes due to their interesting physicochemical and tunable spectral properties, remarkable stabilities, easily accessible synthetic protocols, and abilities to construct molecular assemblies [71–81]. The tweezer methodology has been exploited to avail the porphyrins as molecular probes. In the host–guest supramolecular assembly consisting of an achiral metalloporphyrin tweezer and a chiral bidentate substrate, a stereospecific twist between the two metalloporphyrin chromophores is observed owing to chirality transfer from the chiral substrate to the achiral tweezer [50–56]. As a consequence, a bisignate electronic circular dichroism (ECD) response (exciton couplet) is generated in the porphyrin's Soret band region at about 400 nm, while the spatial orientation between the two porphyrins dictates the sign of the ECD response [50–56, 82–88]. Since the absolute configuration of the substrate (alongside with various other factors) determines the inter-porphyrin twisting, the tweezer methodology is an impressive and effective chiroptical tool for probing the molecular chirality of a range of chiral substrates [50–56, 89–127]. Such a metalloporphyrin tweezer receptor usually contains two substrate binding sites (generally metal ions) parted by a conformationally flexible or rigid spacer. The identity of metal ions inside the porphyrin core has a large impact on the thermodynamics and kinetics of the host–substrate bindings. Such features of the metalloporphyrin tweezer make it a perfect probe for comprehensively understanding supramolecular chirogenesis [50–56, 71–81]. In the last few decades, various metalloporphyrin tweezers have been developed and exploited by several research groups for assigning the absolute configuration to a wide range of chiral diamines, conjugates of primary and secondary amines, amino alcohols, secondary alcohols, α-chiral carboxylic acids, etc. [50–56, 89, 90, 125–127].

1.3 Exciton Coupling Theory

Exciton coupling theory is one of the most convincing and frequently used approach for rationalizing the spectral properties (absorption and CD spectra) of multi-chromophoric systems, and it has been widely employed to interpret the conformational and configurational properties of various organic molecules and natural products [50–63, 71–88]. Exciton coupling takes place when two (or more) independent chromophores interact through space. Such interaction generates a difference in the absorption spectrum, depending on the inter-chromophoric orientation and their spatial arrangement. So, in a co-facially oriented multi-chromophoric aggregate (often referred to as H-aggregation), the spectra shift toward higher energy (blue shift), while for chromophores arranged in a nearly parallel head-to-tail manner, the spectral bands shift toward lower energy compared to its monomeric counterpart (red shift). Another important type of exciton coupling involves obliquely arranged chromophores in space, which shows band splitting in the absorption spectra depending upon the distance and spatial arrangement of the participating chromophores. Because of such splitting (also known as Davydov splitting), the corresponding CD displays a bisignate signal. However, the sign of the CD signal depends on the torsional angle, which is defined as the twist angle between two effective transition moments of the chromophores [82–88]. Therefore, at least two strongly absorbing chromophores, having C_2 or lower symmetry, are required to be arranged spatially to promote the chirogenic processes [82–85, 88]. In this regard, metalloporphyrin tweezers are the important structural motifs for our quest toward chiroptical sensitivity, versatility, and applicability. In most cases, binding of a chiral substrate with an achiral metalloporphyrin tweezer (host) forms a conformationally rigid helical system, which generates exciton coupling between two porphyrin chromophores [50–56, 83–85]. As a consequence, the chirality of the substrate is transferred to the metalloporphyrin tweezer (host) by introducing a preferential chiral twist in the porphyrinic arrangement with the torsional angle between the dipole moments of each porphyrin, leading to a bisignate CD spectrum, which is denoted either as positive (clockwise arrangement of the interacting dipoles) or negative (anticlockwise orientation of the interacting dipoles) exciton coupling circular dichroism (ECCD) [50–56, 83–85]. The helicity of the inter-porphyrin chromophores is directly related to the sign of the spectrum. This nonempirical approach, ECCD, developed by Harada and Nakanishi [83] has been employed for the absolute stereochemical assignments of a wide range of substrates (e.g., carboxylic acids, epoxides, amines, alcohols, and diols) [89–127].

1.4 Detection of Chirality

As the modeling of chirality transfer heavily involves prediction of the different spectra, e.g., for absolute configuration assignment a short overview of such experimental methods is warranted here. Chiroptical spectroscopies, which emerge from the difference in absorption and speed of right CPL and left CPL upon interaction with a chiral molecule result in phenomena known as birefringence and circular dichroism (CD), respectively. The outcome of birefringence is the rotation of the plane-polarized wave when traveling through a chiral system, commonly known as optical rotation. Originally these measurements were performed only at 589 nm, the D-line of sodium, resulting in the αD values characteristic of any chiral molecule. Later, the measurement of the optical rotation at different wavelengths provided optical rotatory dispersion (ORD), a more reliable methodology for structural elucidation [51, 128]. CD can be measured in the infrared region, as is the case for vibrational CD (VCD), or in the ultraviolet–visible (UV–Vis) region of the spectrum, with spectroscopies such as ECD, circularly polarized luminescence, or Raman optical activity

[129]. In general, the chiroptical responses of enantiomers are of equal magnitude, but of the opposite sign to each other [130].

The generation of chiroptical responses is very general, ranging from monodisperse to polydisperse systems, individual or collective chirality, of organic, organometallic, or inorganic nature, or even a combination of them. The absolute configuration determination has been the most frequent application of chiroptical responses after their discovery. The general applicability of such methods is, however, wider, and now also includes the conformational assignment of the responding systems, detection of chiral or achiral molecules, as well as the characterization of self-assemblies [131]. The use of more than one chiroptical response is often required to confidentially determine the structure of a particular system [132].

The advance of chiroptical methods, which are the key spectroscopic means of probing and reporting the presence of chiral motive(s) within the molecular framework, came because of simultaneous advances in instrumentation as well as computer technology and algorithms that allow implementation of *ab initio* theoretical simulations. In a timely manner, scientists have made remarkable advances in the intelligent design of chiral systems capable of manifesting chirality in different ways: by selecting a chromophore providing chiral signals in a specific spectral region or by fine-tuning the chiroptical response via methodical selection of the preferred symmetry and geometry of the system in a given media.

Historically, the synthesis and isolation of chiral molecules have routinely been followed by measurements of OR at the single sodium-D-line wavelength (589 nm) and ECD spectral signature. To achieve a reliable stereochemical structural elucidation, nowadays, a simultaneous use of more than one chiroptical method is strongly encouraged: ORD, ECD, VCD, and ROA [133, 134].

The above-mentioned chiroptical methods examine chiral systems under variable conditions and with different forms of energy, hence making available unique sensitivities to different structural features. So, UV–Vis linearly polarized light is used in the case of ORD, while CPL is used in the case of ECD. VCD extends the functionality of ECD to the mid-infrared regime where normal mode vibrational transitions are observed. Like VCD, ROA is a form of vibrational optical activity that is sensitive to chirality associated with all fundamental vibrational normal modes. ROA measures a small difference in the intensity of vibrational Raman scattering from chiral molecules exposed to right- or left-polarized circular incident light. It is considered as important probe of the structure and behavior of biomolecules in aqueous solution [135, 136]. As such, different methods require different solvation environments, different concentrations, different consideration of solute–solute versus solute–solvent interactions, as well as intra versus intermolecular interactions.

Usually, merits and limitations of a given chiroptical method can be identified when different methods are applied simultaneously to cross-check a given stereochemical structural objective. As mentioned above, while OR at 589 nm is routinely reported, for structural elucidation it is advisable to use the overall ORD pattern that extends over multiple wavelengths. ECD requires a UV–Vis chromophore in order to produce a chiroptical response, while in case of VCD each of 3N-6 normal mode vibrations has potential to serve as a chirality probe of N-atomic molecule. VCD typically accesses multiple well-defined bands, but it is less sensitive than ECD as reflected in the need for higher concentrations in orders of magnitude due to the lower response intensity [137]. Chiroptical responses depend on the spatial orientation of chromophoric groups, moieties near the stereogenic centers, and therefore on the overall molecular flexibility. As VCD is highly conformation sensitive, it can be occasionally hindered by conformational averaging that leads to the attenuation of the bands in the case of flexible molecules [136].

1.5 Theoretical Prediction of the Chiroptical Responses

The theoretical prediction of the chiroptical responses is not trivial, and therefore different methodologies have been developed for their prediction. *Ab initio* calculations have been extensively used for small- and medium-sized molecular systems, not far above 1000 heavy (non-hydrogen) atoms. The specific level of theory must be chosen considering the chiroptical response under evaluation and the size and nature of the chiroptical system.

The exponential growth of computational resources has been the key driving force that has forever changed the way chemists approach the stereochemical structural elucidation, without the need for any additional chiral reagents, chemical derivatization, or reference system [138]. So, simulations related to identifying stable geometries and predicting corresponding chiroptical responses from first principles (*ab initio*) have become an essential tool for investigating chiral molecular systems. Many research studies have demonstrated that molecular chiroptical properties can be computed reliably, starting from first principles quantum theory, particularly density functional theory (DFT).

The progress made in the computational chiroptics has been documented by several recent review articles [135, 139–144]. Modeling is used to predict and visualize stable conformations, to assign absolute configurations, to simulate and analyze chiroptical data, as well as to provide a basis for understanding their true origin. Additionally, simulations can provide supporting evidence and insight into the origin of chiral induction phenomena. However, one should keep in mind also other side of the coin, as one needs to be careful not to treat molecular modeling as a black-box tool or take the quality of simulated data for granted [135, 137, 143]. *Ab initio*-based chiroptical predictions (see Figures 1 and 2) should begin

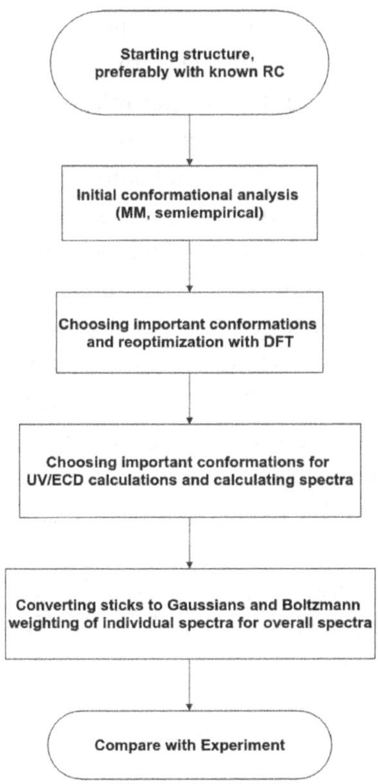

Figure 1. Typical steps in electronic circular dichroism (ECD) calculations.

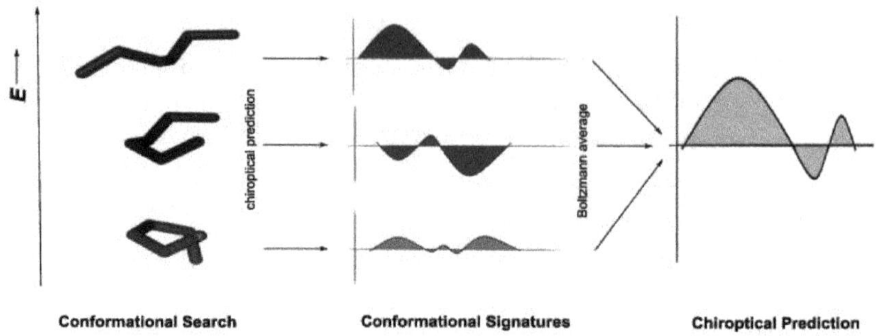

Figure 2. Outcomes for different steps of the chiroptical prediction. Adapted with permission from Ref. [144].

with building virtual models of systems under study (the connectivity of atoms and handedness of stereogenic centers) and for each model conformational survey aiming to identify important molecular geometries must be conducted. Empirical data from experiment (e.g., NMR spectroscopy or X-ray crystallography) could yield the relative configuration (RC) of selected stereogenic centers which can considerably reduce the number of necessary virtual models in this initial step. An illustrative example of valuable a priori RC determination is the study of marine natural product (+)-bistramide C, which has 10 stereogenic centers [145]. Specifically, a study based on NMR has reduced the candidate pool of 1024 possible stereoisomers to only 16 diastereomers needed for ECD theoretical consideration. In the absence of RC data, for a system with n elements of chirality, one must consider all $2n-1$ diastereomers, as the enantomeric forms provide mirror image chiroptical responses.

The most common approach to identify the local minima is to start with the survey of potential energy surface (PES) by using molecular mechanics (MM)-based molecular dynamics (MD) or Monte Carlo algorithms with a careful selection of parameterized force field based on atomtypes and moieties (e.g., MMFFs, OPLS3, AMBER, etc.). It is of critical importance that conformational search is conducted in a thorough manner in order to avoid any carryover errors from this initial step that might have negative consequences on the final chiroptical results and are challenging to track down. Even if the conformer is not highly populated, it could intrinsically provide a strong chiroptical response and therefore have influence on the overall predicted spectral signature. So, the use of multiple algorithms and possibly applicable force fields are sometimes suggested [135, 142, 143] to explore variation in all rotatable bonds and overcome energy barriers associated with ring-puckering modes. However, it should be kept in mind that such redundant application of force fields and algorithms gives no warranty for avoiding the errors.

The stable MM-identified geometries within ~10–20 kcal mol^{-1} energy window [146], depending on the intrinsic degrees of freedom, are subsequently fully optimized under appropriate quantum mechanical levels of theory. For larger molecular systems, it is not uncommon to first apply semi-empirical AM1 or PM6 methods, followed by full geometry optimization at a higher *ab initio* level of theory. Again, such repeated optimization can be beneficial for selecting smaller pool of configurations for more time-consuming calculations but can also introduce significant errors by excluding some important structures. DFT, which includes a large variety of different levels of approximation and theory, has been benchmarked as a standard computational method, yielding the good average performance for energy minimization and chiroptical predictions. Contrary to common statement "one size fits all", in choosing the appropriate DFT-based level of theory, there is no simple classification of molecular system by size or atom type that allows for single universal reliable approach. Every simulations method is prone to errors [147]. Therefore, it is often advised to resort to exploring different combinations of DFT functionals and basis sets in the search of consistencies in Boltzmann populations of stable conformations as well as, at a later stage, corresponding chiroptical responses. Simulation results that are more consistent among each other and, most importantly, that more closely correlate with the experimental data are considered as results delivering a higher confidence level. Identification of true PES minima resulting from the geometry optimization is verified by the absence of negative vibrational frequencies at the same level of theory. Unfortunately, there is no test which would assure that the found minimum is also a global one for a system under study.

Two suggested families of functionals that can be explored are hybrid and range-separated ones. Popular hybrid functional with increasing amounts of "exact" HF exchange integral are B3LYP (20% HF), PBE0 (25%), M06 (27%), BH&HLYP (50%), and M06-2X (54%). Range-separated functionals such as Coulomb-attenuated CAM-B3LYP and ωB97X functionals often outperform the hybrid ones in ECD predictions and are therefore considered standard. One instructive illustrative example can be found in Refs. [143, 148]. It must be emphasized that relative Gibbs free energies, dipole, and rotational strength can be very sensitive to the choice of functional. The literature provides several cautionary tales of studies which demonstrate that popularly used B3LYP is not necessarily the most accurate functional for predicting structures and energies of standard organic molecules, especially when noncovalent interactions come into play [135, 143, 148–151]. Sometimes, for more precise calculations of conformational preferences even more sophisticated methods such as CCSD(T) might be needed [152].

While 6-31G* can be used as the most fundamental basis set to get initial simulation-based insight, more advanced double-zeta- or triple- as well as augmented basis sets (aug-TZVP, aug-cc-pVDZ) or those with more polarization functions (6-311++G(2d,2p)) are recommended for producing better quality chiroptical outcomes. The diffuse functions must be included within the basis set selection for anionic species. Los Alamos National Laboratory 2 Double-Zeta (LANL2DZ) basis set [153] has been suggested as extensively tested effective core potential to model metal atoms [154]. Benchmark investigation demonstrated that CAM-B3LYP with the aug-cc-pVDZ basis set can be a useful method considering both accuracy and reasonable computational time for prediction of the Boltzmann average optical rotation that matches the experimental, and thereby the absolute configuration of chiral molecules [155]. An illustrative case for which both geometry optimization and time-dependent DFT (TDDFT) calculation of chiroptical property can be highly sensitive to the used basis set is the ORD-based investigation of isocytoxazone [156].

Moreover, when noncovalent interactions might be important, a reliable prediction of conformational energies requires the inclusion of dispersion, either as a correction of normal hybrid or double-hybrid functionals [157–159]. For example, 2-(3-fluoro-(p-tolyl)propyl)naphthalene may exist in unfolded or folded conformations, depending on the presence of an intramolecular π-stacking. Folded conformations, however, are totally overlooked by B3LYP calculations, while they turn out to be the most stable ones when B2GPPLYP-D3 is used for geometry optimizations in vacuo. As the two conformational families (unfolded and folded) have very different calculated ECD spectra, their relative population has a large impact on the average calculated ECD spectrum [135].

Aside from the influence of the functional and basis set the solvent model should be accounted to match the media used for experimental chiroptical measurements. Solvent may affect the relative stability of conformers, especially with propensities for nonbonding interactions. In most cases, continuum solvent models such as polarizable continuum model (PCM) and conductor-like screening model (COSMO) are satisfactory. According to few case studies, the reliability of PCM predictions of solvent effects on optical rotations is dependent on the chosen solvent [160]. If spectral predictions are not optimal and changing the level of theory does not improve the level of correlation, one may choose to add a few explicit solvent molecules to account for specific solute–solvent and solute–solute (dimer, trimer) interactions. For examples of such a procedure, we refer the interested readers to the literature [161–165].

While the choice of functional, basis set, and solvent model is important, vibrational effects can also play a notable role in the simulated chiroptical response, especially in the case of ORD and ECD. Predicted electronic chiroptical data in which a molecule's environment and inclusion of vibrational effects have been considered have generally better agreement with experimental data [166]. An illustrative example is the ORD-based structural elucidation of (R)-methyloxirane by means of a novel computational protocol, involving MD trajectory. Specifically, this method takes into account vibrational averaging and solvent effects, leading for the first time to a quantitative agreement (both sign and absolute value) between computed and experimental OR values at several frequencies [167]. This research area will remain highly active as new computational methods are being developed.

Chiroptical properties are subsequently computed from first principles and Boltzmann averaged based on predicted Gibbs free energies if more than one stable structures ($\Delta E < 1$ kcal mol^{-1}) are identified. The variability of *ab initio* predicted chiroptical signals, depending on conformer, underscores the importance of through conformational search and geometry optimization as a fundamental and one of the most important steps in carrying out absolute configuration assignment [156]. In general, the weaker the experimental chiroptical response and/or the more conformationally flexible the molecular system under study, the more likely will be its dependence on the calculation method, including functional, basis set, and solvation model. As mentioned above, TDDFT has been established to be main state-of-the-art technique to obtain accurate numerical results for chiroptical predictions. In addition, methods that have been used to provide a trustworthy chiroptical spectroscopic response are time-dependent HF [168] and more computationally demanding coupled cluster method. For larger systems with up to 1000 atoms, the simplified TDDFT (sTDDFT) method has been recently introduced as good combination between computational cost and satisfactory accuracy for ECD prediction [169].

Typically, when ECD and VCD spectra are simulated, the computed "stick" spectra corresponding to vertical electronic and vibrational transitions are broadened with Gaussian and Lorentzian functions, respectively. ECD spectra often require a global shift of the excitations, because of the tendency of TDDFT to underestimate excitation energies [135, 170]. Therefore, the excitation energies are typically shifted by +0.2–+0.45 eV. On the other hand, due to the several approximations made for calculating VCD signals, vibrational energies are typically overestimated and need to be downshifted by factors recommended depending on the given basis set and functional.

It might appear that *ab initio* predictions of multiple chiroptical spectroscopic methods are likely to give redundant structural information. However, at the very least, simulations of multiple chiroptical responses serve as independent verifications of molecular structures. In general, the weaker the computed chiroptical spectrum, the stronger the dependence on the level of theory (functional, basis set, and solvation model selection). When a given chiroptical spectroscopic method gives ambiguous results, the use of more than one chiroptical spectroscopic method may provide missing information and increase the overall confidence level of the stereochemical assignment.

It is worth mentioning that in the past decade, it has been brought to light that traditional solely qualitative correlation between experimental and *ab initio* predicted spectral profiles can provide limiting and even misleading analyses as such approach overlooks the valuable stereochemical information embedded within the corresponding electronic absorption (EA) as well as electronic dissymmetry factor (EDF) spectra. Therefore, one of the recent trends pioneered by Polavarapu *et al.* is to correlate theoretical versus experimental ECD, EA, and EDF spectra to enhance applicability as well as increase confidence level of stereochemical structural elucidations via electronically based chiroptical spectroscopy. As such, quantitative correlations between experimental and theoretical electronic signatures represent the suggested overall *ab initio* protocol. Algorithms have been implemented in general-purpose programs such as CDSpecTech which helps generate, cross-correlate, and quantitatively score the degree of overall between theoretical and experimental EA, ECD, and EDF spectra. We refer readers to further literature for recent case studies which demonstrate the scope of applicability and benefits of applying the novel quantitative ECD approach toward reliable stereochemical elucidations [171–173].

Considering now the options for modeling chirality transfer, several additional caveats emerge [55]. The systems under study are usually bigger than average chiroptical molecules considered so far. Besides that obvious size problem and the possibility of different attachment sites and combinations of 1:1, 1:2, etc. complexes, there is also the question related to interactions involved in chirality transfer systems — van der Waals contacts, electrostatic and steric interactions, π–π interactions, and hydrogen bonding. These are an order of magnitude weaker that covalent bonding are therefore also more demanding toward the level of theory to get the results right "for the right reason," not just due to fortuitous error-cancellation. This all contributes to relatively limited use of *ab initio* calculations for studying chirality transfer. Still, there is a growing pool of studies overcoming these problems.

1.6 Application of Theoretical Modeling to Chirogenesis

CD spectra for C21-alkylated nickel(II) complexes of N-confused porphyrin have been calculated by Chmielewski *et al.* using TDDFT (B3LYP with 6-31G** and LANLDZ basis to confirm the absolute configuration assignment and hypothesis that the system remains chiral after removal of the metal ion from the macrocyclic crevice by comparing experimental and theoretical spectra [174]. The adduct formation of protonated bis(N-confused porphyrin) (BNCP, 3,3′-bis(meso-tetratolyl-2-aza-21-carbaporphyrin) with chiral anions, carboxylic acids, and alcohols was studied in solution by means of ^{1}H NMR and CD spectroscopic analysis and DFT methods by the same group [175]. A comparison of the experimental and calculated CD spectra was employed for the assignment of the absolute configuration of atropo-enantiomers.

The host–guest binding in bis[Zn^{2+}(salphen)] complex (salphen=N,N′-phenylenebis-(salicylimine)) with carboxylic acids was studied by Wezenberg *et al.* [176]. DFT optimization using the B3P86/6-31G*/LanLDZ level of theory was first used to identify the most favorable location of the acidic proton, whose position could not be resolved by X-ray diffraction, and rationalize the stability of formed complex. Single-point DFT calculations (single-point B3P86) were then carried out to determine the energy of the structures that were found in the solid state and the preferential diastereoisomer formation was confirmed.

New type of bis-Zn(salphen) hosts (salphen=N,N′-1,2-phenylenebis-(salicylimine)) in complexes with a number of potentially ditopic and monotopic chiral guests (O and N donors) was studied by Escarcega-Bobadilla *et al.* [177] to examine the chiral induction effects. DFT analysis at CAM-B3LYP/6-31G** (LanLDZ for Zn and Cl) level of the 1:1 host–guest complex between bis-salphen-Zn_3 and cinchonidine was performed to establish the preferred conformation of the host upon binding of (−)-cinchonidine. The computational ECD spectrum for the (M)-(−) diastereoisomer was in good agreement with the experimental data, showing a first (larger) negative Cotton effect followed by a smaller second positive Cotton effect. The electronic transitions responsible for these effects have host-based π–π* character, as supported by frontier orbital analysis.

Pintre *et al.* [101] studied a bisporphyrin tweezer receptor and its metalation with Zn(II) and Rh(III) cations. The supramolecular chirality induction process that takes place when the metalated bisporphyrin receptors coordinate to enantiopure 1,2-diaminocyclohexane in two different solvents, toluene, and dichloromethane, was explored. A simplified analog of studied bisporphyrin-Zn_2, in which three meso-phenyl substituents in each porphyrin unit and the benzyl ether in the linker are removed, its 1:1 and 2:1 complexes with 1,2-diaminocyclohexane, was used to compute (BP86/TZP) the gas-phase energy of their conformers and find out the preferential conformation. Also, TDDFT was used to calculate EA and CD spectra, which were in a nice agreement with experimental ones at both the BP86/TZP and the SAOP/TZP levels.

In situ-generated multicomponent assembly that creates diastereomeric tris(pyridine) metal complexes incorporating chiral secondary alcohols for determining alcohol identity and *ee* was studied by You *et al.* [178]. DFT calculations (B3LYP functional and the 6-31G(d)) were employed to predict diastereomeric ratios and the preferential twist based on the configuration of the newly created stereocenter. CD spectra were calculated in vacuo at the CAM-B3LYP/SVP level; other functionals (B3LYP) and larger basis sets (TZVP), as well as the inclusion of a solvent (PCM model for acetonitrile), did not substantially modify the results. Calculated positions and signs of CD signals matched nicely the experimental ones. Orbital and population analyses were used to rationalize the obtained experimental spectra.

Monoamide-linked monozinc bisporphyrinate was designed and synthesized by Zhang *et al.* [179] and used as a host to complex with chiral amino acid esters. DFT and TDDFT (B3LYP/6-31G*) calculations were performed to elucidate the chiral induction process. According to DFT and TDDFT calculations, the positive band experimentally found for the host–guest complex mainly originates from π–π^* transition due to orbital characters of the corresponding starting and arriving states while the negative band is contributed by both π–π^* transitions and metal-to-ligand charge-transfer. It was shown that there is no direct contribution from the chiral guest.

Novel oxalic amide-linked bisporphyrinate, which shows chiral recognition ability for amino acid ethyl esters, has been designed and synthesized by Jiang *et al.* [180]. The mechanism of chiral recognition has also been investigated by X-ray crystallography, UV–Vis spectra, ^{1}H NMR, and DFT/TDDFT calculations at the B3LYP/6-31G* level. The calculated CD spectrum agreed with the experimental result, especially the sign and shape. Comparing with the observed CD bands, the calculated CD bands have slight blue-shift.

Fang *et al.* [105] studied chirality transfer from chiral monoamines to *m*-phthalic diamide-linked zinc bisporphyrinates. DFT calculations (B3LYP/6-31G*) were used to investigate the energies of different conformers and their geometries. Calculations were also used to rationalize, how the studied system can function not only as chirality probes for chiral monoamines but also distinguish between aryl and alkyl substituents of monoamines.

A calix[4]arene host equipped with two bis-[Zn(salphen)] complexes, which self-assemble into a capsular complex in the presence of a chiral diamine guest with 2:1 ratio between the host and the guest, was studied by Martinez-Rodriguez *et al.* [181]. Effective chirality transfer from the diamine to the calix–salen hybrid host is observed by CD spectroscopy. Density functional theory calculations were used to investigate the origin of the stability of the host–guest system and the experimental CD spectrum compared with those calculated for both possible diastereoisomers showing that the M,M isomer is the one that is preferentially formed. Two-layer ONIOM method (B3LYP/6-31G(d,p):PM6) was used to cope with the size of the system for optimization and TDDFT was used to calculate the ECD spectra.

A series of new photo-responsive amino acid-derived azobenzenedicarboxylic acid derivatives were synthesized by Sogawa *et al.* [182]. DFT calculations at the B3LYP/6-31G(d,p) level were used to clarify the structures of dimers corresponding to the experimental spectra recorded in DMF.

An apparently rigid dibenzothiophene-bridged zinc(II)/magnesium(II) bisporphyrin host has been explored to determine the absolute configuration of a series of amino alcohols, while the effect of the metal ions has been thoroughly scrutinized by Dhamija *et al.* [119]. DFT calculations using B97D functional and 6-31+G(d,p) basis set for C, H, N, and O and LANL2DZ for Zn and S were used to uncover the reasons behind the stabilization of various conformations of 1:2 host–guest complexes, namely the influences of the H bonding and steric crowding toward the relative stabilities of endo–endo and exo–endo conformers, which both can be important depending on the size of substituent. Theoretical spectra for selected complexes were also calculated using TDDFT method (ωB97X-D functional, the basis set was 6-31G*, LANL2DZ for Zn, Mg, and S; dichloromethane was used for solvent correction) and were in a good correlation with the experimental spectra.

DFT calculations at B3LYP/6-31G* level (starting from experimentally determined crystal structures) were used by Hu *et al.* to gain more insight into the chirality transfer mechanism in *m*-phthalic diamide-linked bisporphyrin complex with a series of chiral monoamines and diamines [183]. The calculations suggested that, in the host–guest complexes, the nature of the bulkiest substituent group in monoamine makes a difference. When it is an alkyl group, the steric repulsion causes one conformer to be energetically more favorable, which leads to a clockwise twist, while an aryl group can form π–π interactions with the phenyl ring present in the linker, and causes another conformer to be energetically favored, which leads to an anticlockwise twist. When the linker was modified with a 2-naphthalenecarboxamide [184] substituent similar conclusions were reached. Similar results were obtained by the same authors for *m*-phthalic diamide-linked bisporphyrin complexes with chiral monoalcohols [185]. The study suggests that in this bisporphyrin system the interactions between the linker and guests are crucial.

Lu *et al.* [186] have studied intra- and inter-molecular chirogenesis events on a family of chiral bisporphyrin hosts constructed by connecting two zinc(II) porphyrinates to a homochiral (R)-(+)- or (S)-(–)-1,10-bi-2-naphthol with ester or ether linkages and its consecutive intermolecular binding with 4,4′-bipyridyl, as a typical achiral ditopic guest. It was noted that the intensity of CD signals decreases significantly for all of the supramolecular complexes in comparison with the chiral bisporphyrin hosts alone, though the CD sign does not change. The observation is rationalized by DFT molecular modeling (at B97D/LANL2DZ level), which suggests that upon the formation of 1:1 cyclic supramolecular complexes the chiral twist direction of the two porphyrin moieties does not change, while the Zn–Zn distance increases remarkably.

The combination of spectroscopic methods, DFT/TDDFT calculations, X-ray crystal structure, and microscopy techniques were used by Wan *et al.* [187] to study the chirality transfer process during the self-assembly for pincer-type Pt^{II} complexes. The study demonstrated that an achiral Pt^{II} complex can be induced to undergo symmetry breaking self-assembly by a chiral doping approach and to ultimately aggregate into chiral supramolecular polymers. Only dimeric and tetrameric structures were optimized at PBE0-D3/6-31G*/LANLDZ level using PCM solvation model, and the TDDFT (SAOP/TZP/COSMO) calculations were performed on the dimeric structure to simulate its CD spectrum due to computation constraints. The calculations provided an explanation for the solvent-dependent chirality change. Two dimeric conformers have been found and optimized, with quite different calculated dipole moments. As the relative stabilities of the conformers should be affected by the solvent via electrostatic interactions, it was proposed that the conformer with a larger dipole moment would be more stable in polar solvent, whereas the one with smaller dipole moment would be more stable in less-polar environment. The calculated CD spectra were similar to the experimentally observed CD spectra.

Bendzińska-Berus *et al.* [188] studied various aspects of incorporation of a sterically demanding trityl (CPh_3) group into a series of N-acylated amino acids, N-acylated Gly amides, and chiral Gly–Gly dipeptide derivatives. Structural analysis carried out with the use of complementary experimental methods and supported by DFT calculations (at TD-CAM-B3LYP/6-311++G(d,p)//B3LYP/6-311++G(d,p) level) to determine the mechanism of chirality transmission from the stereogenic center to the trityl reporter group. Calculations were mainly used to study different conformations of studied molecules and identify intramolecular interactions responsible for conformational preferences. The sensitivity in chirogenesis was found to be a function of steric requirements of the amino acid backbone, the inductor–reporter distance, and the environment polarity.

DFT calculations were used by Savchenko *et al.* [189] to characterize the self-assembly scenario of photosensitive star-shaped azobenzenes with benzene-1,3,5-tricarboxamide core and azobenzene arms using the M06-L/aug-cc-pVTZ functional and basis set for dimers and density-functional-based tight-binding with parameter set mio-1-1 [190] for larger stacks of azobenzenes. The results suggest that during the formation of ordered phases, these molecules prefer to build one-dimensional columns with the equilibrium distance between the star cores, in which stars are held together via weak interactions. These noncovalent interactions include among others the hydrogen bonding, whose length shortens with increasing the column length. The formed aggregates are also characterized by the helical twist and chirality.

The complexation and chiroptical behavior of the host–guest complexes using a new chiral Zn(II) bisporphyrin tweezer host and a series of achiral aliphatic diamine guests varying the chain length were studied by Saha *et al.* [123]. (1R,2R)-cyclohexanediamine linked two Zn(II) porphyrin moieties covalently, thereby producing a strong chiral field around the bisporphyrin tweezer framework. The chiral tweezer exhibited upon complexation with achiral guest large changes in the UV–Vis spectra and CD exciton couplets due to a sudden change in the porphyrin disposition, which is controlled by the host–guest stoichiometry as well as the chain lengths of the diamine guest. To obtain deeper insights into the sudden changes of interporphyrin disposition just by changing the length of the achiral diamines, authors have performed a series of computational studies by using MD simulations followed by full *ab initio* geometry optimization [ωB97XD/6-31+G* for C,H,N,O and LANL2DZ for Zn, implicit solvent (iefpcm = CHCl3)] and TDDFT CD prediction. The theoretical conformational and ECD analysis provided a rational for the inversion of experimental CD of the complexes with shorter diamines — the changes in porphyrin–porphyrin spatial disposition upon complexation lead to inversion of sign of the CD signal.

The chirality transfer from 1-phenylethylamine to a 5-amino-1,3-phthalic acid diamide-linked zinc bisporphyrinate was investigated by Zhang *et al.* [191] through crystallographic and DFT studies at the level of ωB97XD/6-31G*. Calculations were used to elucidate the energies of conformers of 1:1 and 1:2 complexes.

The intermolecular interaction active between polyaniline and 10-camphorsulfonic acid was studied by Mishra *et al.* [192]. Theoretical DFT UCAM-B3LYP results allowed them to prove that the 800-nm CD signal is due to the polyaniline alone without any 10-camphorsulfonic acid involvement, also allowing to ascertain the tight connection between the latter CD signal and the polyaniline oligomer helical-like structure. It was demonstrated that both the process of supramolecular-chiral-induction, and possibly in a wider mode the supramolecular induction of asymmetry (chiral handedness), which are due to the presence of a suitable "chiral-catalyst," are electronic (quantum mechanical) in nature, and are not governed and controlled by any "shape/geometric" factor (key-and-lock model).

The induced CD spectra of six zinc porphyrin complexes with chiral guests were simulated by Osadchuk *et al.* [193] using different DFT methods for benchmarking purposes. The best agreement between theoretical and experimental results for the Soret (B) band absorption region was achieved with the ωB97X-D, CAM-B3LYP, and M06-2X functionals with implicit inclusion of solvent effects (SMD model). Also, a good correlation between the simulated and experimental spectra was obtained with the DZVP basis sets; however, a more accurate simulation of the length and velocity rotational strengths needed larger TZVP basis sets. Additionally, it was shown that the conformation of the chiral guest influences the chirogenic mechanism.

Computational and experimental study of the linear optical properties and supramolecular structure of a rhodamine heterodimer assembled on a DNA scaffold was reported by Rukin *et al.* [194]. Computationally, starting from low-cost DFT/TDDFT simulations (at CAM-B3LYP-D/6-311G(d,p) level, the solvation was taken into account at PCM level) of the bare dimer the measured ± CD sign sequence of the S_1/S_2 bands to a specific chiral conformation of the heterodimer was assigned. In the monomers, as typical for rhodamine dyes, the electric transition dipole of the lowest π–π^* transition is parallel to the long axis of the xanthene planes. Authors show that in the heterodimer the sign sequence of the two CD bands is related to the orientation of these long axes. To account explicitly for environment effects, MD simulations were used for characterizing the supramolecular structure of the two optical isomers tethered on DNA. Average absorption and CD-profiles were modeled using *ab initio* TDDFT calculations at the geometries sampled along a few nanosecond MD run. The absorption profiles computed for both optical isomers were in good agreement with the experimental absorption spectrum and did not allow one to discriminate between them. The computed averaged CD profiles provided the orientation of monomers in the enantiomer that is dominant under the experimental conditions.

The exciton chirality (EC) rule was reexamined by Swathi *et al.* [195], and it was shown that this rule is not universally applicable by calculating CD spectra for rigid linear phenyleneethynylene dye chiral assemblies with the CAM-B3LYP functional and 6-31G(d) as basis set in the gas phase. The results clearly indicated that the EC rule holds for aggregates with repulsive interactions (so-called simple-helix) but fails for aggregates with attractive interactions (so-called creeper helix). From these studies, it appears that in supramolecular chiral aggregates, the sign of the CD couplet depends not only on the handedness of the aggregate, but also on the sign of the interaction energy between neighboring dipoles. The EC rule does not account for the sign of interactions and is valid only for aggregates in which adjacent dipoles have repulsive interaction energy.

Ustrnul *et al.* [196] studied the supramolecular chirogenesis in zinc porphyrins by complexation with enantiopure hemicucurbit[n]urils ($n = 6$, 8). It was demonstrated that zinc porphyrins can be effectively used to sense hemicucurbituril chirality. The used macrocycles have large association constants with porphyrins thanks to their suitable preorganization. DFT calculations (PBE/def2-SVP for optimization and CAM-B3LYP/def2-SVP for spectra) were used for rationalization of the mechanism of chirality transfer and the mode of interaction of cycHCs with porphyrins.

A chirality transfer in a series of [2]pseudorotaxanes through the intermolecular noncovalent interaction of a pair of chiral binaphthalene crown ethers with achiral secondary ammonium salts with different chromophores was studied by Wang *et al.* [197]. It was found that the binaphthalene groups can induce new CD signals only in the [2]pseudorotaxane structures between the host crown ethers and the guest molecule with the anthryl group. DFT (at B3LYP-D3/6-31G(d,p)//SMD(DCM) level) and TDDFT (at CAM-B3LYP-D3/6-31G(d,p)//SMD(DCM) level) calculations (the SMD solvation model with the dichloromethane solvent was used in all calculations) indicated that the generation of the new CD signal is mainly due to the intermolecular chiral induction between the anthryl group in the guest and the naphthalene groups in the host.

1.7 Conclusions

As can be seen from the presented articles, theoretical modeling can be used to gain versatile information that helps to get deeper understanding of chirality transfer mechanisms. Over the years the application of the modeling has steadily grown and has been applied to wider aspects of chirality: first to predict the energies of most important conformation, then also their CD spectra, and finally to the interactions present in the host–guest complex and also the origin of lines in ECD spectrum. Undoubtedly, we will see growth of the application of modeling in the future as the computational resources are steadily becoming more available and the methodologies used continue to develop into more robust tools that can increasingly be used also by nonexperts.

References

1. Kelvin, L. (1894) The molecular tactics of a crystal. *Oxford Univ. J. Sci. Club.*, **18**, 3–57.
2. Kwon, J., Tamura, M., Lucas, P.W., Hashimoto, J., Kusakabe, N., Kandori, R., Nakajima, Y., Nagayama, T., Nagata, T., and Hough, J.H. (2013) Near-infrared circular polarization images of NGC 6334-V. *Astrophys. J. Lett.*, **765** (1).
3. Ben-Jacob, E., and Levine, H. (2001) The artistry of nature. *Nature*, **409** (6823), 985–986.
4. Grande, C., and Patel, N.H. (2009) Nodal signalling is involved in left-right asymmetry in snails. *Nature*, **457** (7232), 1007–1011.
5. Bonner, W.A. (2000) Parity violation and the evolution of biomolecular homochirality. *Chirality,* **12** (3), 114–126.
6. Hermann, K. (2016) Crystallography and surface structure: an introduction for surface scientists and nanoscientists, in *Crystallography and Surface Structure: An Introduction for Surface Scientists and Nanoscientists*, Wiley-VCH Verlag Gmbh & Co., Weinheim, Germany.
7. Mauskopf, S. (2006) A history of chirality, in *Chiral Analysis* (Busch K.W., Busch M.A., eds.), Elsevier, Amsterdam, pp. 3–24.
8. Gal, J. (2013) Molecular chirality in chemistry and biology: historical milestones. *Helv. Chim. Acta*, **96** (9), 1617–1657.
9. Pasteur, L. (1848) Mémoire sur la relation qui peut exister entre la forme cristalline et la composition chimique, et sur la cause de la polarisation rotatoire. *Comp. Rend. Hebd Séanc. Acad. Sci.*, **26**, 535–538.
10. Kahr, B., and Shtukenberg, A.G. (2017) Seeing molecular configuration in twisted crystal form. *Isr. J. Chem.*, **57** (1), 31–38.
11. Bada, J.L. (1995) Origins of homochirality. *Nature*, **374** (6523), 594–595.
12. Bonner, W.A. (1991) The origin and amplification of biomolecular chirality. *Orig. Life Evol. Biosph.*, **21** (2), 59–111.
13. Keszthelyi, L. (1987) Asymmetries of nature and the origin of biomolecular handedness. *BioSystems*, **20** (1), 15–19.
14. Avetisov, V.A., Goldanskii, V.I., and Kuz'min, V.V. (1991) Handedness, origin of life and evolution. *Phys. Today*, **44** (7), 33–41.

15. Blackmond, D.G. (2010) The origin of biological homochirality. *Cold Spring Harb. Perspect. Biol.*, **2** (5).
16. Guijarro, A., and Yus, M. (2008) *The Origin of Chirality in the Molecules of Life*. [Internet], The Royal Society of Chemistry.
17. Ávalos, M., Babiano, R., Cintas, P., Jiménez, J.L., and Palacios, J.C. (2010) Homochirality and chemical evolution: new vistas and reflections on recent models. *Tetrahedron Asymmetry*, **21** (9–10), 1030–1040.
18. Quack, M. (2002) How important is parity violation for molecular and biomolecular chirality? *Angew Chem. Int. Ed.*, **41** (24), 4618–4630.
19. Plasson, R., Kondepudi, D.K., Bersini, H., Commeyras, A., and Asakura, K. (2007) Emergence of homochirality in far-from-equilibrium systems: mechanisms and role in prebiotic chemistry. *Chirality*, **19** (8), 589–600.
20. Fraser, D.G., Fitz, D., Jakschitz, T., and Rode, B.M. (2011) Selective adsorption and chiral amplification of amino acids in vermiculite clay-implications for the origin of biochirality. *Phys. Chem. Chem. Phys.*, **13** (3), 831–838.
21. Mane, S. (2016) Racemic drug resolution: a comprehensive guide. *Anal. Methods*, **8** (42), 7567–7586.
22. Kelsey, F.O. (1988) Thalidomide update: regulatory aspects. *Teratology*, **38** (3), 221–226.
23. Tomaszewski, J., and Rumore, M.M. (1994) Stereoisomeric drugs: FDA'S policy statement and the impact on drug development. *Drug Dev. Ind. Pharm.*, **20** (2), 119–139.
24. Sahajwalla, C. (2004) Chirality in drug design and development, in *Chirality in Drug Design and Development* (Reddy I., Mehvar R., eds.). CRC Press, FL, USA.
25. Agranat, I., Wainschtein, S.R., and Zusman, E.Z. (2012) The predicated demise of racemic new molecular entities is an exaggeration. *Nat. Rev. Drug Discov.*, **11** (12), 972–973.
26. Caner, H., Groner, E., Levy, L., and Agranat, I. (2004) Trends in the development of chiral drugs. *Drug Discov. Today*, **9** (3), 105–110.
27. Murakami, H. (2007) From racemates to single enantiomers — chiral synthetic drugs over the last 20 years, in *Topics in Current Chemistry* (Sakai, K., Hirayama, N., and Tamura, R., eds.), vol. 269 Springer-Verlag Berlin, Berlin, Germany pp. 273–299.
28. Erb, S. (2006) Single-enantiomer drugs poised for further market growth. *Pharm. Technol.*, **30**, 14–18.
29. Technologies, R.I. (2017) *Chiral Chemicals Market Research Report*.
30. Challener, C.A. (2016) Going green with biocatalysis. *Pharm. Technol.*, **40** (8), 24–25.
31. Global Industry Analysts, I. (2016) *Chiral Technology — A Global Strategic Business Report*.

32. Noyori, R. (2002) Asymmetric catalysis: science and opportunities (nobel lecture). *Angew. Chem. Int. Ed.*, **41** (12), 2008–2022.
33. Sharpless, K.B. (2002) Searching for new reactivity (nobel lecture). *Angew. Chem. Int. Ed.*, **41** (12), 2024–2032.
34. Knowles, W.S. (2002) Asymmetric hydrogenations (Nobel lecture 2001). *Angew. Chem. Int. Ed.,* **41** (12), 1998–2007.
35. Wang, Y., Xu, J., Wang, Y., and Chen, H. (2013) Emerging chirality in nanoscience. *Chem. Soc. Rev.*, **42** (7), 2930–2962.
36. Kumar, J., Thomas, K.G., and Liz-Marzán, L.M. (2016) Nanoscale chirality in metal and semiconductor nanoparticles. *Chem. Commun.*, **52** (85), 12555–12569.
37. Zaera, F. (2008) Chiral modification of solid surfaces: a molecular view. *J. Phys. Chem. C*, **112** (42), 16196–16203.
38. Gellman, A.J., Horvath, J.D., and Buelow, M.T. (2001) Chiral single crystal surface chemistry. *J. Mol. Catal. A Chem.*, **167** (1–2), 3–11.
39. Ernst, K.H. (2012) Molecular chirality at surfaces. *Phys. Status Solidi Basic Res.*, **249** (11), 2057–2088.
40. Wang, J., and Feringa, B.L. (2011) Dynamic control of chiral space in a catalytic asymmetric reaction using a molecular motor. *Science (80-)*, **331** (6023), 1429–1432.
41. Feringa, B.L. (2001) In control of motion: from molecular switches to molecular motors. *Acc. Chem. Res.*, **34** (6), 504–513.
42. Schalley, C.A., Beizai, K., and Vögtle, F. (2001) On the way to rotaxane-based molecular motors: studies in molecular mobility and topological chirality. *Acc. Chem. Res.*, **34** (6), 465–476.
43. Naaman, R., and Waldeck, D.H. (2015) Spintronics and chirality: spin selectivity in electron transport through chiral molecules, in *Annual Review of Physical Chemistry* (Johnson, M.A. and Martinez, T.J., eds.) vol. 66, Annual Reviews, Palo Alto, CA, pp. 263–281.
44. Elemans, J.A.A.W., De Cat, I., Xu, H., and De Feyter, S. (2009) Two-dimensional chirality at liquid–solid interfaces. *Chem. Soc. Rev.*, **38** (3), 722–736.
45. Liu, M., Zhang, L., and Wang, T. (2015) Supramolecular chirality in self-assembled systems. *Chem. Rev.*, **115** (15), 7304–7397.
46. Rest, C., Kandanelli, R., and Fernández, G. (2015) Strategies to create hierarchical self-assembled structures via cooperative non-covalent interactions. *Chem. Soc. Rev.*, **44** (8), 2543–2572.
47. Chen, L.J., Yang, H.B., and Shionoya, M. (2017) Chiral metallosupramolecular architectures. *Chem. Soc. Rev.*, **46** (9), 2555–2576.
48. Chen, Z., Wang, Q., Wu, X., Li, Z., and Jiang, Y.B. (2015) Optical chirality sensing using macrocycles, synthetic and supramolecular oligomers/polymers, and nanoparticle based sensors. *Chem. Soc. Rev.*, **44** (13), 4249–4263.
49. Miyake, H., and Tsukube, H. (2012) Coordination chemistry strategies for dynamic helicates: time-programmable chirality switching with labile and inert metal helicates. *Chem. Soc. Rev.*, **41** (21), 6977–6991.

50. Lu, H., and Kobayashi, N. (2016) Optically active porphyrin and phthalocyanine systems. *Chem. Rev.*, **116** (10), 6184–6261.
51. Pescitelli, G., Di Bari, L., and Berova, N. (2014) Application of electronic circular dichroism in the study of supramolecular systems. *Chem. Soc. Rev.*, **43** (15), 5211–5233.
52. Valderrey, V., Aragay, G., and Ballester, P. (2014) Porphyrin tweezer receptors: binding studies, conformational properties and applications. *Coord. Chem. Rev.*, **258–259** (1), 137–156.
53. Wolf, C., and Bentley, K.W. (2013) Chirality sensing using stereodynamic probes with distinct electronic circular dichroism output. *Chem. Soc. Rev.*, **42** (12), 5408–5424.
54. Berova, N., Pescitelli, G., Petrovic, A.G., and Proni, G. (2009) Probing molecular chirality by CD-sensitive dimeric metalloporphyrin hosts. *Chem. Commun.*, (40), 5958–5980.
55. Hembury, G.A., Borovkov, V.V., and Inoue, Y. (2008) Chirality-sensing supramolecular systems. *Chem. Rev.*, **108** (1), 1–73.
56. Borovkov, V.V., Hembury, G.A., and Inoue, Y. (2004) Origin, control, and application of supramolecular chirogenesis in bisporphyrin-based systems. *Acc. Chem. Res.*, **37** (7), 449–459.
57. Schwartz, E.S., Le Gac, J.J., Cornelissen, L.M.R., Nolte, J.M., and Rowan, A.E. (2010) Macromolecular multi-chromophoric scaffolding. *Chem. Soc. Rev.*, **39** (5), 1576–1599.
58. Shen, J., and Okamoto, Y. (2016) Efficient separation of enantiomers using stereoregular chiral polymers. *Chem. Rev.*, **116** (3), 1094–1138.
59. Jo, H.H., Lin, C.Y., and Anslyn, E.V. (2014) Rapid optical methods for enantiomeric excess analysis: from enantioselective indicator displacement assays to exciton-coupled circular dichroism. *Acc. Chem. Res.*, **47** (7), 2212–2221.
60. Escárcega-Bobadilla, M.V., and Kleij, A.W. (2012) Artificial chirogenesis: a gateway to new opportunities in material science and catalysis. *Chem. Sci.*, **3** (8), 2421–2428.
61. Canary, J.W., Mortezaei, S., and Liang, J. (2010) Transition metal-based chiroptical switches for nanoscale electronics and sensors. *Coord. Chem. Rev.*, **254** (19–20), 2249–2266.
62. Kane-Maguire, L.A.P., and Wallace, G G. (2010) Chiral conducting polymers. *Chem. Soc. Rev.*, **39** (7), 2545–2576.
63. Canary, J.W. (2009) Redox-triggered chiroptical molecular switches. *Chem. Soc. Rev.*, **38** (3), 747–756.
64. Lee, C.C., Grenier, C., Meijer, E.W., and Schenning, A.P.H.J. (2009) Preparation and characterization of helical self-assembled nanofibers. *Chem. Soc. Rev.*, **38** (3), 671–683.
65. Miyata, M., Tohnai, N., and Hisaki, I. (2007) Crystalline host-guest assemblies of steroidal and related molecules: diversity, hierarchy, and supramolecular chirality. *Acc. Chem. Res.*, **40** (8), 694–702.

66. Palmans, A.R.A., and Meijer, E.W. (2007) Amplification of chirality in dynamic supramolecular aggregates. *Angew. Chem. Int. Ed.*, **46** (47), 8948–8968.
67. Yashima, E., Ousaka, N., Taura, D., Shimomura, K., Ikai, T., and Maeda, K. (2016) Supramolecular helical systems: helical assemblies of small molecules, foldamers, and polymers with chiral amplification and their functions. *Chem. Rev.*, **116** (22), 13752–13990.
68. Rosaria, L., D'Urso, A., Mammana, A., and Purrello, R. (2008) Chiral memory: induction, amplification, and switching in porphyrin assemblies. *Chirality*, **20** (3–4), 411–419.
69. Ogoshi, H., and Mizutani, T. (1998) Multifunctional and chiral porphyrins: Model receptors for chiral recognition. *Acc. Chem. Res.*, **31** (2), 81–89.
70. Xing, P., and Zhao, Y. (2018) Controlling supramolecular chirality in multicomponent self-assembled systems. *Acc. Chem. Res.*, **51** (9), 2324–2334.
71. Chan, A.K.W., and Yam, V.W.W. (2018) Precise modulation of molecular building blocks from tweezers to rectangles for recognition and stimuli-responsive processes. *Acc. Chem. Res.*, **51** (12), 3041–3051.
72. Klärner, F.G., and Schrader, T. (2013) Aromatic interactions by molecular tweezers and clips in chemical and biological systems. *Acc. Chem. Res.*, **46** (4), 967–978.
73. Khan, F.S.T., Guchhait, T., Sasmal, S., and Rath, S.P. (2017) Hydroxo-bridged diiron(III) and dimanganese(III) bisporphyrins: modulation of metal spins by counter anions. *Dalt. Trans.*, **46** (4), 1012–1037.
74. Martín, N., and Pérez, E.M. (2010) Molecular tweezers for fullerenes. *Pure Appl. Chem.*, **82** (3), 523–533.
75. Dolenský, B., Havlík, M., and Král, V. (2012) Oligo tröger's bases–new molecular scaffolds. *Chem. Soc. Rev.*, **41** (10), 3839–3858.
76. Hardouin-Lerouge, M., Hudhomme, P., and Sallé, M. (2011) Molecular clips and tweezers hosting neutral guests. *Chem. Soc. Rev.*, **40** (1), 30–43.
77. Mondal, P., and Rath, S.P. (2020) Cyclic metalloporphyrin dimers: conformational flexibility, applications and future prospects. *Coord. Chem. Rev.*, **405**.
78. Lai, D.F., Khan, S.T., and Rath, S.P. (2018) Multiheme proteins: effect of heme-heme interactions. *Dalt. Trans.*, **47** (41), 14388–14401.
79. Guchhait, T., Sasmal, S.F., Khan, S.T., and Rath, S.P. (2017) Oxo- and hydroxo-bridged diiron(III) porphyrin dimers: inorganic and bio-inorganic perspectives and effects of intermacrocyclic interactions. *Coord. Chem. Rev.*, **337**, 112–144.
80. Sil, D., and Rath, S.P. (2015) An ethane-bridged porphyrin dimer as a model of di-heme proteins: inorganic and bioinorganic perspectives and consequences of heme-heme interactions. *Dalt. Trans.*, **44** (37), 16195–16211.
81. Khan, F.S.T., Pandey, A.K., and Rath, S.P. (2018) Modulation of iron spin in ethane-bridged diiron(III) porphyrin dimer: anion dependent spin state switching. *J. Chem. Sci.*, **130** (7).
82. Berova, N., Nakanishi, K., and Woody, R.W. (2000) *Circular Dichroism: Principles and Applications, 2nd ed.* (Berova, N., Nakanishi, K., Woody, R.W., eds.) Wiley-VCH, New York.

83. Harada, N., and Nakanishi, K. (1972) The exciton chirality method and its application to configurational and conformational studies of natural products. *Acc. Chem. Res.*, **5** (8), 257–263.
84. Telfer, S.G., McLean, T.M., and Waterland, M.R. (2011) Exciton coupling in coordination compounds. *Dalt. Trans.*, **40** (13), 3097–3108.
85. Pescitelli, G., Di Bari, L., and Berova, N. (2011) Conformational aspects in the studies of organic compounds by electronic circular dichroism. *Chem. Soc. Rev.*, **40** (9), 4603–4625.
86. Beychok, S. (1966) Circular dichroism of biological macromolecules. *Science (80-)* **154** (3754), 1288–1299.
87. Kasha, M., Rawls, H.R., and El-Bayoumi, M.A. (1965) The exciton model in molecular spectroscopy. *Pure Appl. Chem.*, **11** (3–4), 371–392.
88. Harada, N., and Nakanishi, K. (1983) *Circular Dichroic Spectroscopy: Exciton Coupling in Organic Stereochemistry.* [Internet] Oxford University Press, Oxford.
89. Borovkov, V.V., Lintuluoto, J.M., Sugiura, M., Inoue, Y., and Kuroda, R. (2002) Remarkable stability and enhanced optical activity of a chiral supramolecular bis-porphyrin tweezer in both solution and solid state. *J. Am. Chem. Soc.*, **124** (38), 11282–11283.
90. Borovkov, V.V., Lintuluoto, J.M., and Inoue, Y. (2001) Supramolecular chirogenesis in zinc porphyrins: mechanism, role of guest structure, and application for the absolute configuration determination. *J. Am. Chem. Soc.*, **123** (13), 2979–2989.
91. Gholami, H., Anyika, M., Zhang, J., Vasileiou, C., and Borhan, B. (2016) Host–guest assembly of a molecular reporter with chiral cyanohydrins for assignment of absolute stereochemistry. *Chem. A Eur. J.*, **22** (27), 9235–9239.
92. Tanasova, M., Anyika, M., and Borhan, B. (2015) Sensing remote chirality: Stereochemical determination of β-, γ-, and δ-chiral carboxylic acids. *Angew. Chem. Int. Ed.*, **54** (14), 4274–4278.
93. Li, X., Burrell, C.E., Staples, R.J., and Borhan, B. (2012) Absolute configuration for 1, n-glycols: a nonempirical approach to long-range stereochemical determination. *J. Am. Chem. Soc.*, **134** (22), 9026–9029.
94. Li, X., and Borhan, B. (2008) Prompt determination of absolute configuration for epoxy alcohols via exciton chirality protocol. *J. Am. Chem. Soc.*, **130** (48), 16126–16127.
95. Li, X., Tanasova, M., Vasileiou, C., and Borhan, B. (2008) Fluorinated porphyrin tweezer: a powerful reporter of absolute configuration for erythro and threo diols, amino alcohols, and diamines. *J. Am. Chem. Soc.*, **130** (6), 1885–1893.
96. Proni, G., Pescitelli, G., Huang, X., Nakanishi, K., and Berova, N. (2003) Magnesium tetraarylporphyrin tweezer: a CD-sensitive host for absolute configurational assignments of α-chiral carboxylic acids. *J. Am. Chem. Soc.*, **125** (42), 12914–12927.
97. Huang, X., Fujioka, N., Pescitelli, G., Koehn, F.E., Williamson, R.T., Nakanishi, K., and Berova, N. (2002) Absolute configurational assignments of secondary amines by CD-sensitive dimeric zinc porphyrin host. *J. Am. Chem. Soc.*, **124** (35), 10320–10335.

98. Proni, G., Pescitelli, G., Huang, X., Quraishi, N.Q., Nakanishi, K., and Berova, N. (2002) Configurational assignment of α-chiral carboxylic acids by complexation to dimeric Zn-porphyrin: host-guest structure, chiral recognition and circular dichroism. *Chem. Commun.*, **2** (15), 1590–1591.
99. Kurtán, T., Nesnas, N., Li, Y.Q., Huang, X., Nakanishi, K., and Berova, N. (2001) Chiral recognition by CD-sensitive dimeric zinc porphyrin host. 1. Chiroptical protocol for absolute configurational assignments of monoalcohols and primary monoamines. *J. Am. Chem. Soc.*, **123** (25), 5962–5973.
100. Kurtán, T., Nesnas, N., Koehn, F.E., Li, Y.Q., Nakanishi, K., and Berova, N. (2001) Chiral recognition by CD-sensitive dimeric zinc porphyrin host. 2. Structural studies of host-guest complexes with chiral alcohol and monoamine conjugates. *J. Am. Chem. Soc.*, **123** (25), 5974–5982.
101. Pintre, I.C., Pierrefixe, S., Hamilton, A., Valderrey, V., Bo, C., and Ballester, P. (2012) Influence of the solvent and metal center on supramolecular chirality induction with bisporphyrin tweezer receptors. Strong metal modulation of effective molarity values. *Inorg. Chem.*, **51** (8), 4620–4635.
102. Etxebarria, J., Vidal-Ferran, A., and Ballester, P. (2008) The effect of complex stoichiometry in supramolecular chirality transfer to zinc bisporphyrin systems. *Chem. Commun.*, **(45)**, 5939–5941.
103. Takeda, S., Hayashi, S., Noji, M., and Takanami, T. (2019) Chiroptical protocol for the absolute configurational assignment of alkyl-substituted epoxides using bis (zinc porphyrin) as a CD-sensitive bidentate host. *J. Org. Chem.*, **84** (2), 645–652.
104. Hayashi, S., Yotsukura, M., Noji, M., and Takanami, T. (2015) Bis(zinc porphyrin) as a CD-sensitive bidentate host molecule: Direct determination of absolute configuration of mono-alcohols. *Chem. Commun.*, **51** (55), 11068–11071.
105. Fang, X., Han, Z., Xu, C., Li, X., Wang, Y., and Hu, C. (2015) Discrimination between alkyl and aryl substituents of chiral monoamines by m-phthalic diamide-linked zinc bisporphyrinates. *Dalt. Trans.*, **44** (28), 12511–12515.
106. Jiang, J., Fang, X., Liu, B., and Hu, C. (2014) M-phthalic diamide-linked zinc bisporphyrinate: spontaneous resolution of its crystals and its application in chiral recognition of amino acid esters. *Inorg. Chem.*, **53** (7), 3298–3306.
107. Yoon, H., Lee, C.H., and Jang, W.D. (2012) Absolute stereochemical determination of chiral carboxylates using an achiral molecular tweezer. *Chem. A Eur., J.* **18** (39), 12479–12486.
108. Hou, J.L., Yi, H.P., Bin Shao, X., Li, C., Wu, Z.Q., Jiang, X.K., Wu, L.-Z., Tung, C.-H., and Li, Z.-T. (2006) Helicity induction in hydrogen-bonding-driven zinc porphyrin foldamers by chiral C60-incorporating histidines. *Angew. Chem. Int. Ed.*, **45** (5), 796–800.
109. Muraoka, T., Kinbara, K., and Aida, T. (2006) Mechanical twisting of a guest by a photoresponsive host. *Nature*, **440** (7083), 512–515.
110. Ouedraogo, Y.P., Huang, L., Torrente, M.P., Proni, G., Chadwick, E., Wehmschulte, R.J., and Nesnas, N. (2013) A direct stereoselective preparation of a fish pheromone

and application of the zinc porphyrin tweezer chiroptical protocol in its stereochemical assignment. *Chirality*, **25** (9), 575–581.
111. Ishii, Y., Soeda, Y., and Kubo, Y. (2007) Chirality induction in a cation-driven assembly using a crowned metalloporphyrin. *Chem. Commun.*, (28), 2953–2955.
112. Brahma, S., Ikbal, S.A., Dey, S., and Rath, S.P. (2012) Induction of supramolecular chirality in di-zinc(ii) bisporphyrin via tweezer formation: synthesis, structure and rationalization of chirality. *Chem. Commun.*, **48** (34), 4070–4072.
113. Brahma, S., Ikbal, S.A., and Rath, S.P. (2014) Synthesis, structure, and properties of a series of chiral tweezer-diamine complexes consisting of an achiral zinc(II) bisporphyrin host and chiral diamine guest: induction and rationalization of supramolecular chirality. *Inorg. Chem.*, **53** (1), 49–62.
114. Dhamija, A., Ikbal, S.A., and Rath, S.P. (2016) Induction and rationalization of supramolecular chirality in the tweezer-diamine complexes: insights from experimental and DFT studies. *Inorg. Chem.*, **55** (24), 13014–13026.
115. Brahma, S., Ikbal, S.A., Dhamija, A., and Rath, S.P. (2014) Highly enhanced bisignate circular dichroism of ferrocene-bridged Zn(II) bisporphyrin tweezer with extended chiral substrates due to well-matched host-guest system. *Inorg. Chem.*, **53** (5), 2381–2395.
116. Ikbal, S.A., Brahma, S., and Rath, S.P. (2015) Step-wise induction, amplification and inversion of molecular chirality through the coordination of chiral diamines with Zn(ii) bisporphyrin. *Chem. Commun.*, **51** (5), 895–898.
117. Ikbal, S.A., Brahma, S., and Rath, S.P. (2014) Transfer and control of molecular chirality in the 1:2 host-guest supramolecular complex consisting of Mg(II)bisporphyrin and chiral diols: the effect of H-bonding on the rationalization of chirality. *Chem. Commun.*, **50** (90), 14037–14040.
118. Ikbal, S.A., Dhamija, A., Brahma, S., and Rath, S.P. (2016) A nonempirical approach for direct determination of the absolute configuration of 1,2-diols and amino alcohols using Mg(II)bisporphyrin. *J. Org. Chem.*, **81** (13), 5440–5449.
119. Dhamija, A., Saha, B., and Rath, S.P. (2017) Metal-center-driven supramolecular chirogenesis in tweezer amino alcohol complexes: structural, spectroscopic, and theoretical investigations. *Inorg. Chem.*, **56** (24), 15203–15215.
120. Dhamija, A., Saha, B., Chandel, D., Malik, H., and Rath, S.P. (2020) Molecule to supramolecule: chirality induction, inversion, and amplification in a Mg(II)porphyrin dimer templated by chiral diols. *Inorg. Chem.*, **59** (1), 801–809.
121. Ikbal, S.A., Saha, B., and Rath, S.P. (2015) Stoichioinetry-controlled supramolecular chirality induction in magnesium(II) porphyrin dimer by amino alcohols: mechanistic insights and effect of ligand bulkiness. *J. Ind. Chem. Soc.*, **92** (12), 2001–2014.
122. Saha, B., Ikbal, S.A., and Rath, S.P. (2020) Complexation of chiral Zinc(II) Porphyrin tweezer with chiral guests: control, discrimination and rationalization of supramolecular chirality. *Inorg. Chem.*, **59** (11), 7795–7809.
123. Saha, B., Petrovic, A.G., Dhamija, A., Berova, N., and Rath, S.P. (2019) Complexation of chiral Zinc(II) porphyrin tweezer with achiral aliphatic diamines

revisited: molecular dynamics, electronic CD, and 1H NMR analysis. *Inorg. Chem.*, **58** (17), 11420–11438.

124. Saha, B., Ikbal, S.A., Petrovic, A.G., Berova, N., and Rath, S.P. (2017) Complexation of chiral Zinc-porphyrin tweezer with achiral diamines: induction and two-step inversion of interporphyrin helicity monitored by ECD. *Inorg. Chem.*, **56** (7), 3849–3860.
125. Borovkov, V.V., Fujii, I., Muranaka, A., Hembury, G.A., Tanaka, T., Ceulemans, A., Kobayashi, N., and Inoue, Y. (2004) Rationalization of supramolecular chirality in a bisporphyrin system. *Angew. Chem. Int. Ed.*, **43** (41), 5481–5485.
126. Borovkov, V.V., Lintuluoto, J.M., Hembury, G.A., Sugiura, M., Arakawa, R., and Inoue, Y. (2003) Supramolecular chirogenesis in zinc porphyrins: interaction with bidentate ligands, formation of tweezer structures, and the origin of enhanced optical activity. *J. Org. Chem.*, **68** (19), 7176–7192.
127. Lintuluoto, J.M., Borovkov, V.V., and Inoue, Y. (2002) Direct determination of absolute configuration of monoalcohols by bis(magnesium porphyrin). *J. Am. Chem. Soc.*, **124** (46), 13676–13677.
128. Petrovic, A., Berova, N., and Alonso-Gómez, J. (2015) *Absolute Configuration and Conformational Analysis of Chiral Compounds via Experimental and Theoretical Prediction of Chiroptical Properties: ORD, ECD, and VCD.*, Wiley, Weinheim.
129. Nakanishi, K., Berova, N., Polavarapu, P.L., and Woody, R.W. (2013) *Comprehensive Chiroptical Spectroscopy: Instrumentation, Methodologies, and Theoretical Simulations.* [Internet] Comprehensive Chiroptical Spectroscopy, Wiley-VCH Verlag, Hoboken, NJ.
130. Barron, L.D. (2004) *Molecular Light Scattering and Optical Activity. 2nd ed.* Cambridge University Press, New York, NY.
131. Bentley, K.W., and Wolf, C. (2014) Comprehensive chirality sensing: development of stereodynamic probes with a Dual (chir)optical response. *J. Org. Chem.*, **79** (14), 6517–6531.
132. Rivera-Fuentes, P., Alonso-Gómez, J.L., Petrovic, A.G., Santoro, F., Harada, N., Berova, N., and Diederich, F. (2010) Amplification of chirality in monodisperse, enantiopure alleno-acetylenic oligomers. *Angew. Chem. Int. Ed.*, **49** (12), 2247–2250.
133. Polavarapu, P.L. (2008) Why is it important to simultaneously use more than one chiroptical spectroscopic method for determining the structures of chiral molecules? *Chirality*, **20** (5), 664–672.
134. Polavarapu, P.L. (2012) Molecular structure determination using chiroptical spectroscopy: where we may go wrong? *Chirality*, **24** (11), 909–920.
135. Berova, N., Polavarapu, P.L., Nakanishi, K., and Woody, R.W. (2012) *Comprehensive Chiroptical Spectroscopy: Instrumentation, Methodologies, and Theoretical Simulations.* [Internet] Comprehensive Chiroptical Spectroscopy, Wiley, Hoboken, NJ.

136. Lassen, P.R., Skytte, D.M., Hemmingsen, L., Nielsen, S.F., Freedman, T.B., Nafie, L.A., and Christensen, S.B. (2005) Structure and absolute configuration of nyasol and hinokiresinol via synthesis and vibrational circular dichroism spectroscopy. *J. Nat. Prod.*, **68** (11), 1603–1609.
137. Petrovic, A.G., Navarro-Vazquez, A., and Lorenzo Alonso-Gomez, J. (2010) From relative to absolute configuration of complex natural products: interplay between NMR, ECD, VCD, and ORD assisted by *ab initio* calculations. *Curr. Org. Chem.*, **14** (15), 1612–1628.
138. Noumeur, S.R., Helaly, S.E., Jansen, R., Gereke, M.T., Stradal, E.B., Harzallah, D., and Stadler, M. (2017) Preussilides A-F, bicyclic polyketides from the endophytic fungus *Preussia similis* with antiproliferative activity. *J. Nat. Prod.*, **80** (5), 1531–1540.
139. Pecul, M., and Ruud, K. (2005) The ab initio calculation of optical rotation and electronic circular dichroism, in *Advances in Quantum Chemistry* (Sabin, J.R., and Brandas, E., eds.) Advances In Quantum Chemistry, vol. 50, Elsevier Academic Press Inc., San Diego, CA pp. 185–212.
140. Crawford, T.D. (2006) Ab initio calculation of molecular chiroptical properties. *Theor. Chem. Acc.*, **115** (4), 227–245.
141. Polavarapu, P.L. (2007) Renaissance in chiroptical spectroscopic methods for molecular structure determination. *Chem. Rec.*, **7** (2), 125–136.
142. Autschbach, J. (2009) Computing chiroptical properties with first-principles theoretical methods: background and illustrative examples. *Chirality*, **21** (1E), E116–E152.
143. Pescitelli, G., and Bruhn, T. (2016) Good computational practice in the assignment of absolute configurations by TDDFT calculations of ECD spectra. *Chirality*, **28** (6), 466–474.
144. Ozcelik, A., Pereira-cameselle, R., Ulrih, N.P., Petrovic, A.G., and Lorenzo Alonso-Gómez, J. (2020) Chiroptical sensing: a conceptual introduction. *Sensors (Switzerland)*, **20** (4), 974.
145. Zuber, G., Goldsmith, M.R., Hopkins, T.D., Beratan, D.N., and Wipf, P. (2005) Systematic assignment of the configuration of flexible natural products by spectroscopic and computational methods: the bistramide C analysis. *Org. Lett.*, **7** (23), 5269–5272.
146. Johnson, J.L., Raghavan, V., Cimmino, A., Moeini, A., Petrovic, A.G., Santoro, E., Superchi, S., Berova, N., Evidente, A., and Polavarapu, P.L. (2018) Absolute configurations of chiral molecules with multiple stereogenic centers without prior knowledge of the relative configurations: a case study of inuloxin C. *Chirality*, **30** (11), 1206–1214.
147. Brémond, E., Savarese, M., Adamo, C., and Jacquemin, D. (2018) Accuracy of TD-DFT geometries: a fresh look. *J. Chem. Theory Comput.*, **14** (7), 3715–3727.
148. Berardozzi, R., Guido, C., Capozzi, A.M., Cardellicchio, C., Di Bari, L., and Pescitelli, G. (2015) Circular dichroism and TDDFT investigation of chiral fluorinated aryl benzyl sulfoxides. *Eur. J. Org. Chem.*, **2015** (25), 5554–5562.

149. Bruhn, T., Witterauf, F., Götz, D.C.G., Grimmer, C.T., Würtemberger, M., Radius, U., and Bringmann, G. (2014) C,C- and N,C-coupled dimers of 2-aminotetraphenylporphyrins: regiocontrolled synthesis, spectroscopic properties, and quantum-chemical calculations. *Chem. A Eur. J.*, **20** (14), 3998–4006.
150. Le Guennic, B., Hieringer, W., Görling, A., and Autschbach, J. (2005) Density functional calculation of the electronic circular dichroism spectra of the transition metal complexes [M(phen) 3] 2+ (M = Fe, Ru, Os). *J. Phys. Chem. A*, **109** (21), 4836–4846.
151. Brémond, É., Savarese, M., Su, N.Q., Pérez-Jiménez, Á.J., Xu, X., Sancho-García, J.C., and Adamo, C. (2016) Benchmarking density functionals on structural parameters of small-/medium-sized organic molecules. *J. Chem. Theory Comput.*, **12** (2), 459–465.
152. Gerbig, D., Desch, S., and Schreiner, P.R. (2018) Making glycine methyl ester chiral. *Chem. A Eur. J.*, **24** (46), 11904–11907.
153. Li, C.-Z., Jiang, X.-K., Zhanting, L., Gao, X., and Wang, Q.-R. (2007) Foldamer-based molecular recognition. *Chin. J. Org. Chem.*, **27**, 188–196.
154. Ferrand, Y., Kendhale, A.M., Kauffmann, B., Grélard, A., Marie, C., Blot, V., Pipelier, M., Dubreuil, D., and Huc, I. (2010) Diastereoselective encapsulation of tartaric acid by a helical aromatic oligoamide. *J. Am. Chem. Soc.*, **132** (23), 7858–7859.
155. Howard, J.C., Sowndarya, S.V., Ansari, I.M., Mach, T.J., Baranowska-ŁAczkowska, A., and Crawford, T.D. (2018) Performance of property-optimized basis sets for optical rotation with coupled cluster theory. *J. Phys. Chem. A*, **122** (28), 5962–5969.
156. Mazzeo, G., Giorgio, E., Zanasi, R., Berova, N., and Rosini, C. (2010) Absolute configuration through the DFT simulation of the optical rotation. Importance of the correct selection of the input geometry: a caveat. *J. Org. Chem.*, **75** (13), 4600–4603.
157. Grimme, S., and Steinmetz, M. (2013) Effects of London dispersion correction in density functional theory on the structures of organic molecules in the gas phase. *Phys. Chem. Chem. Phys.*, **15** (38), 16031–16042.
158. Kruse, H., Goerigk, L., and Grimme, S. (2012) Why the standard B3LYP/6-31G* model chemistry should not be used in DFT calculations of molecular thermochemistry: understanding and correcting the problem. *J. Org. Chem.*, **77** (23), 10824–10834.
159. Goerigk, L., and Grimme, S. (2014) Double-hybrid density functionals. *Wiley Interdiscip. Rev. Comput. Mol. Sci.*, **4** (6), 576–600.
160. Mennucci, B., Cappelli, C., Cammi, R., and Tomasi, J. (2011) Modeling solvent effects on chiroptical properties. *Chirality*, **23** (9), 717–729.
161. Petrovic, A.G., Vantomme, G., Negrõn-Abril, Y.L., Lubian, E., Saielli, G., Menegazzo, I., Cordero, R., Proni, G., Nakanishi, K., Carofiglio, T., and Berova, N.

(2011) Bulky melamine-based Zn-porphyrin tweezer as a CD probe of molecular chirality. *Chirality*, **23** (9), 808–819.
162. Troche-Pesqueira, E., Pérez-Juste, I., Navarro-Vázquez, A., and Cid, M.M. (2013) A β-cyclodextrin-resveratrol inclusion complex and the role of geometrical and electronic effects on its electronic induced circular dichroism. *RSC Adv.*, **3** (26), 10242–10250.
163. Giner-Casares, J.J., and Liz-Marzán, L.M. (2014) Plasmonic nanoparticles in 2D for biological applications: toward active multipurpose platforms. *Nano Today*, **9** (3), 365–377.
164. Tang, L., Li, S., Xu, L., Ma, W., Kuang, H., Wang, L., and Xu, C. (2015) Chirality-based Au@Ag nanorod dimers sensor for ultrasensitive PSA detection. *ACS Appl. Mater. Interfaces*, **7** (23), 12708–12712.
165. Perera, A.S., Thomas, J., Poopari, M.R., and Xu, Y. (2016) The clusters-in-a-liquid approach for solvation: new insights from the conformer specific gas phase spectroscopy and vibrational optical activity spectroscopy. *Front. Chem.*, **4**, 9.
166. Orlova, A.V., Andrade, R.R., Da Silva, C.O., Zinin, A.I., and Kononov, L.O. (2014) Polarimetry as a tool for the study of solutions of chiral solutes. *ChemPhysChem.*, **15** (1), 195–207.
167. Lipparini, F., Egidi, F., Cappelli, C., and Barone, V. (2013) The optical rotation of methyloxirane in aqueous solution: a never ending story? *J. Chem. Theory Comput.*, **9** (4), 1880–1884.
168. Míguez-Lago, S., Llamas-Saiz, A.L., Magdalena Cid, M., and Alonso-Gõmez, J.L. (2015) A covalent organic helical cage with remarkable chiroptical amplification, *Chem. A Eur. J.* **21** (50), 18085–18088.
169. Kumar, J., Eraña, H., López-Martínez, E., Claes, N., Martín, V.F., Solís, D.M., Bals, S., Cortajarena, A.L., Castilla, J., and Liz-Marzán, L.M. (2018) Detection of amyloid fibrils in Parkinson's disease using plasmonic chirality. *Proc. Natl. Acad. Sci. U S A.* **115** (13), 3225–3230.
170. Guerrero-Martínez, A., Auguié, B., Alonso-Gómez, J.L., Džolič, Z., Gómez-Graña, S., Žinić, M., Magdalena, M., and Liz-Marzán, L.M. (2011) Intense optical activity from three-dimensional chiral ordering of plasmonic nanoantennas. *Angew. Chem. Int. Ed.*, **50** (24), 5499–5503.
171. Covington, C.L., and Polavarapu, P.L. (2013) Similarity in dissymmetry factor spectra: a quantitative measure of comparison between experimental and predicted vibrational circular dichroism. *J. Phys. Chem. A,* **117** (16), 3377–3386.
172. Junior, F.M.S., Covington, C.L., De Amorim, M.B.L., Velozo, S.M., Kaplan, M.A.C., and Polavarapu, P.L. (2014) Absolute configuration of a rare sesquiterpene: (+)-3-ishwarone. *J. Nat. Prod.*, **77** (8), 1881–1886.
173. Johnson, J.L., Nair, D.S., Pillai, S.M., Johnson, D., Kallingathodi, Z., Ibnusaud, I., and Polavarapu, P.L. (2019) Dissymmetry factor spectral analysis can provide useful

diastereomer discrimination: chiral molecular structure of an analogue of (–)-crispine A. *ACS Omega*, **4** (4), 6154–6164.

174. Chmielewski, P.J., Szterenberg, L., and Siczek, M. (2011) Enantiomer resolution of intrinsically chiral C21-alkylated N-confused porphyrin complexes. *Chem. A Eur. J.*, **17** (3), 1009–1020.
175. Chmielewski, P.J., Siczek, M., and Stepień, M. (2015) Bis(N-confused porphyrin) as a semirigid receptor with a chirality memory: a two-way host enantiomerization through point-to-axial chirality transfer. *Chem. A Eur. J.*, **21** (6), 2547–2559.
176. Wezenberg, S.J., Salassa, G., Escudero-Adán, E.C., Benet-Buchholz, J., and Kleij, A.W. (2011) Effective chirogenesis in a bis(metallosalphen) complex through host-guest binding with carboxylic acids. *Angew. Chem. Int. Ed.*, **50** (3), 713–716.
177. Escárcega-Bobadilla, M.V., Salassa, G., Martínez Belmonte, M., Escudero-Adán, E.C., and Kleij, A.W. (2012) Versatile switching in substrate topicity: supramolecular chirality induction in di- and trinuclear host complexes. *Chem. A Eur. J.*, **18** (22), 6805–6810.
178. You, L., Pescitelli, G., Anslyn, E.V., and Di Bari, L. (2012) An exciton-coupled circular dichroism protocol for the determination of identity, chirality, and enantiomeric excess of chiral secondary alcohols. *J. Am. Chem. Soc.*, **134** (16), 7117–7125.
179. Zhang, Y., Ma, B., Wang, Y., and Hu, C. (2014) The role of amino acid esters and the amide group in the chiral induction process by a novel monoamide-linked monozinc bisporphyrinate. *Zeitsch. Anorg. Allg. Chem.*, **640** (10), 2078–2085.
180. Jiang, J., Feng, Z., Liu, B., Hu, C., and Wang, Y. (2013) Chiral recognition of amino acid esters by a novel oxalic amide-linked bisporphyrin. *Dalt. Trans.*, **42** (21), 7651–7659.
181. Martínez-Rodríguez, L., Bandeira, N.A.G., Bo, C., and Kleij, A.W. (2015) Highly efficient chirality transfer from diamines encapsulated within a self-assembled calixarene-salen host. *Chem. A Eur. J.*, **21** (19), 7144–7150.
182. Sogawa, H., Terada, K., Miyagi, Y., Shiotsuki, M., Inai, Y., Masuda, T., and Sanda, F. (2015) Photoinduced formation of an azobenzene-based CD-active supramolecular cyclic dimer. *Chem. A Eur. J.*, **21** (18), 6747–6755.
183. Hu, T., Hu, C., Wang, Y., Young, D.J., and Lang, J.P. (2018) Stoichiometrically controlled chirality inversion in zinc bisporphyrinate-monoamine complexes. *Dalt. Trans.*, **47** (15), 5503–5512.
184. Hu, T., Liu, T., Zhang, Z., Wang, Y., Yang, Y., Young, D.J., Hu, C., and Langa, J.-P. (2019) Precise control of chirality transfer by adjusting the alkyl substituents of guests. *Dye Pigment.*, **160**, 692–699.
185. Hu, Q., Zhuo, C., Wang, Y., Hu, C., and Lang, J. (2017) Chirality transfer from chiral monoamines to an m-phthalic diamide-linked zinc bisporphyrinate with a benzylamide substituent. *Inorg. Chem.*, **56** (17), 10204–10214.
186. Lu, W., Li, L., Yang, H., Zhao, L., Qi, D., Bian, Y., and Jiang, J. (2017) Intramolecular chirality induction and intermolecular chirality modulation in BINOL bridged bisporphyrin hosts. *Dye Pigment.*, **137**, 608–614.

187. Wan, Q., Xiao, X.S., To, W.P., Lu, W., Chen, Y., Low, K.H., and Che, C.-M. (2018) Counteranion- and solvent-mediated chirality transfer in the supramolecular polymerization of luminescent platinum(II) complexes. *Angew. Chem. Int. Ed.*, **57** (52), 17189–17193.
188. Bendzińska-Berus, W., Jelecki, M., Kwit, M., and Rychlewska, U. (2019) Transfer of chirality in: N-triphenylacetylamino acids and chiral derivatives of N-triphenylacetyl Gly-Gly dipeptide and control of their assembly with steric constraints. *CrystEngComm.*, **21** (22), 3420–3430.
189. Savchenko, V., Koch, M., Pavlov, A.S., Saphiannikova, M., and Guskova, O. (2019) Stacks of azobenzene stars: self-assembly scenario and stabilising forces quantified in computer modelling. *Molecules*, **24** (23), 1–18.
190. Elstner, M., Porezag, D., Jungnickel, G., Elsner, J., Haugk, M., Frauenheim, T., Suhai, S., and Seifert, G. (1998) Self-consistent-charge density-functional tight-binding method for simulations of complex materials properties. *Phys. Rev. B.*, **58** (11), 7260–7268.
191. Zhang, Z., Hu, C., Wang, Y., and Lang, J.P. (2019) Crystallographic and DFT studies on host-guest complexes consisting of zinc bisporphyrinates and 1-phenylethylamine. *J. Coord. Chem.*, **72** (5–7), 1156–1170.
192. Mishra, S., Kumar, A., Venkatesan, M., Pigani, L., Pasquali, L., and Fontanesi, C. (2020) Exchange interactions drive supramolecular chiral induction in polyaniline. *Small Methods*, **4** (10).
193. Osadchuk, I., Borovkov, V., Aav, R., and Clot, E. (2020) Benchmarking computational methods and influence of guest conformation on chirogenesis in zinc porphyrin complexes. *Phys. Chem. Chem. Phys.*, **22** (19), 11025–11037.
194. Rukin, P.S., Komarova, K.G., Fresch, B., Collini, E., and Remacle, F. (2020) Chirality of a rhodamine heterodimer linked to a DNA scaffold: an experimental and computational study. *Phys. Chem. Chem. Phys.*, **22** (14), 7516–7523.
195. Swathi, K., Sissa, C., Painelli, A., and Thomas, K.G. (2020) Supramolecular chirality: a caveat in assigning the handedness of chiral aggregates. *Chem. Commun.*, **56** (59), 8281–8284.
196. Ustrnul, L., Kaabel, S., Burankova, T., Martõnova, J., Adamson, J., Konrad, N., Burk, P., Borovkov, V., and Aav, R. (2019) Supramolecular chirogenesis in zinc porphyrins by enantiopure hemicucurbit[n] urils (n = 6, 8). *Chem. Commun.*, **55** (96), 14434–14437.
197. Wang, H.J., Zhang, H.Y., Zhang, H.Y., Liu, G., Dai, X., Wu, H., and Liu, Y. (2020) Guest-induced supramolecular chirality transfer in [2]pseudorotaxanes: experimental and computational study. *Org. Biomol. Chem.*, **18** (38), 7649–7655.

https://doi.org/10.1142/9789811259227_0009

9 Chirogenesis in Materials Science and Other Applications

Fan Yi,[*,¶] **Qiang He,**[*,¶] **Wajahat Ali**[†] **and Yue Sun**[*,‡,§]

*Hubei Key Laboratory of Catalysis and Materials Science, College of Chemistry and Material Sciences, South-Central Minzu University, Wuhan, China
†State Key Laboratory of Separation Membrane and Membrane Process, School of Chemistry, Tiangong University, Tianjin, China
‡Department of Chemistry, University of Baltistan, Skardu, Pakistan

Chirogenesis has aroused extensive attention as it is directly correlated with numerous biological systems and materials science. This chapter collected the chirogenesis in materials science ranging from chirogenesis in 0D and 1D materials to 3D materials, and other applications. Hopefully, this chapter will guide the design and functionality of chiral structures. The field of chirogenesis in sciences and relevant sciences are significantly promising in understanding the origins of life in nature, as well as in developing smart materials.

§Corresponding author: sunyue@tiangong.edu.cn
¶The authors equally contributed to this work

1.1 Introduction

Chirogenesis is an emerging branch of materials science that shows great promise for developing new material systems with unique properties [1, 2]. Chirogenesis, and related effects that occur on surfaces, also have great relevance for applications in many fields [3–5]. In this chapter, we summarize the most recent developments in the area of chirogenesis in materials science and other applications. We consider chirogenesis in 0D (zero dimensional), 1D, 2D, and 3D material systems [6–16]. Additionally, fundamental advances based on the use of chirogenesis in self-assembled soft materials and moiré stacking will be reviewed (Scheme 1).

1.2 Chirogenesis in 0D and 1D Materials

Plasmonic nanoparticles assemblies have highly tunable optical properties that play an important role in many fields [17–19]. Gold nanoparticles (AuNPs) are one of the most remarkable 0D noble metals, that have been extensively studied in plasmonic applications [20–23].

Recently, Xu and coworkers reported heterodimers based on different-sized AuNPs with sodium chloride (NaCl) and DNA [24]. Specifically, a circular dichroism (CD) band with a magnitude of (+)10 or (−)20 mdeg appeared near the surface plasmonic frequency of AuNPs upon the formation of NaCl or DNA triggered heterodimers (Figure 1). Notably, the chiral response was tunable and reversible, based on changing the size of the precursors and the temperature. They further studied the factors

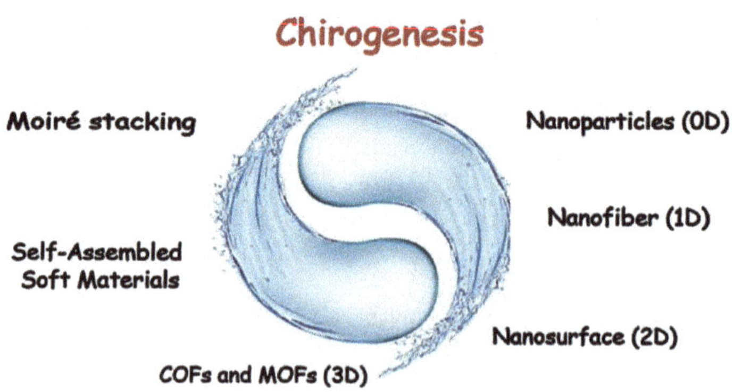

Scheme 1. Chirogenesis in materials science and other applications.

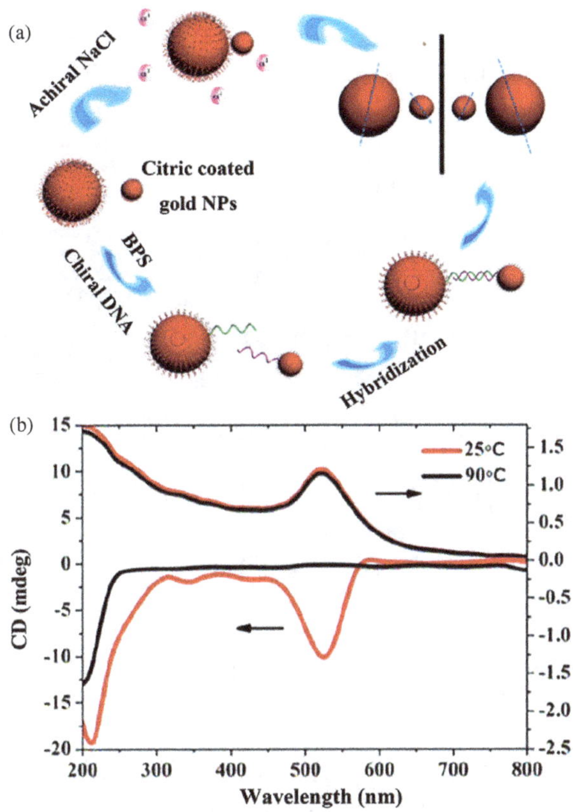

Figure 1. (a) Schematic strategy for chiral heterodimers; (b) reversible ultraviolet (UV)/vis and CD spectra of heterodimers formed by DNA at different temperatures. Reprinted (adapted) with permission from Ref. [24]. Copyright (2013) American Chemical Society.

influencing the formation of NPs dimers and mechanisms for these processes have since been proposed. On the basis of space simulations, the chiral signals of heterodimers have been attributed to the combined effects of the dihedral angle between plasmonically coupled NPs and two adjacent NPs, leading to breaking of the symmetry of the uniform spheres. These studies have provided a method for better understanding the origin of chirality in NPs-based assemblies and also offer opportunities for creating ways of engineering nanostructures in the field of chirogenesis.

Chirogenesis in 1D materials is relevant to applications in the fields of nanocomposites, nanoelectronics, and energy harvesting. The

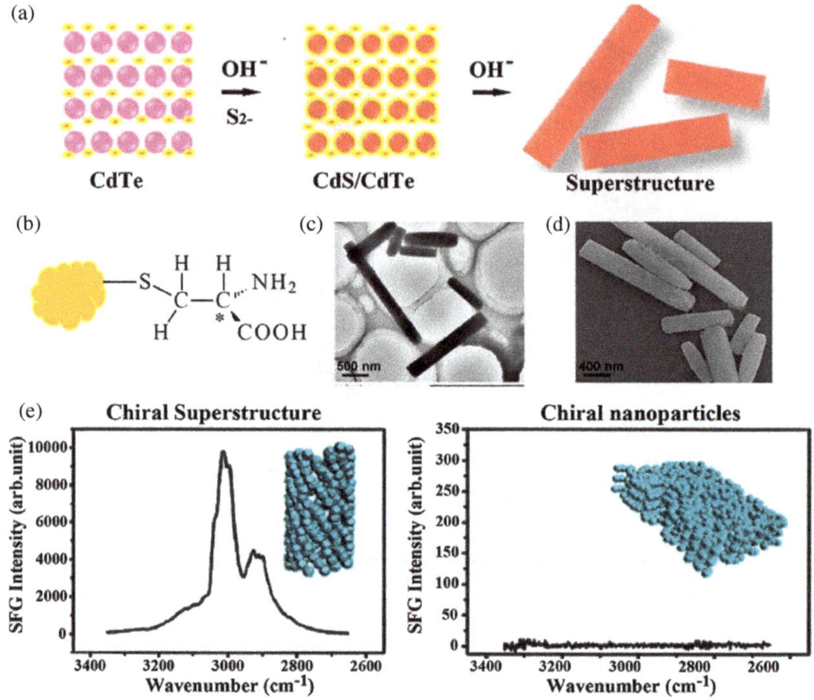

Figure 2. (a) Synthetic CdS/CdTe nanoparticles and chiral superstructure formed by self-assembly; (b) CdS/CdTe NPs surface modified with Cys; (c) transmission electron microscope images of superstructures; (d) scanning electron microscope images of superstructures; (e) sum frequency generation spectra of chiral functional CdS/CdTe NPs and monodisperse CdS/CdTe nanoparticles. Reprinted (adapted) with permission from Ref. [25]. Copyright (2019) American Chemical Society.

incorporation of chirogenesis into 1D materials can improve the mechanical and electrical properties of composites and enhance modulation efficiency in optical modulators and photoelectric devices. Chiral NPs can spontaneously self-assemble into stationary structures, which are of interest in terms of applied and fundamental investigations because of their unique properties and ordered collective behaviors. Wang and coworkers used L-cysteine (Cys) as a chiral molecule to generate functional surfactants to cap CdS/CdTe NPs. They further used a functional surfactant as a linkage for constructing ordered assemblies of NPs (Figure 2) [25]. In terms of light absorption and CD effects, Cys ligands attached to the nanoparticles exhibit related

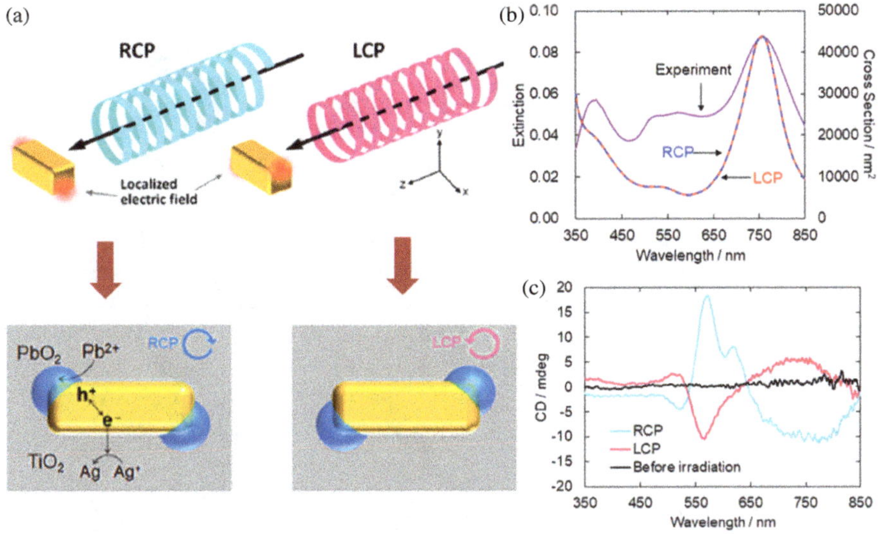

Figure 3. (a) Chiral nanostructure fabricated by PICS with circularly polarized light as the chiral source; (b) experimental and simulated extinction spectra of the TiO_2 substrate with gold nano-cuboids irradiated from the back side; (c) CD spectra of the TiO_2 with gold nano-cuboids before and after PbO_2 deposition. Reprinted (adapted) with permission from Ref. [26]. Copyright (2018) American Chemical Society.

changes, in molecular solution used to control the procedures at different pH. Notably, during the spontaneous self-assembly process, synthesized chiral NPs organized into ordered rod-shaped NPs based on monitoring of the CD response of SPs in different cultivating times. Experimental studies and theoretical calculations have further confirmed that ordered stacking of the chiral NPS and stabilizer, lead to collective optical activity.

Tatsuma and Saito used circularly polarized light (CPL) as a chiral trigger to introduce chirality into plasmonic nanostructures (Figure 3) [26]. Gold nano-cuboids as precursors were irradiated with CPL to localize the electric fields at specific sites of the cuboids depending on the light handedness, resulting in the deposition of dielectric moieties [27, 28]. Plasmonic nanostructures with high chirality have been developed by this approach. Macroscopic CD signals appeared within the wavelength range of the RCP-light-irradiated substrate (530–800 nm), whereas those of the LCP-light-irradiated substrate were opposite over the same range. This

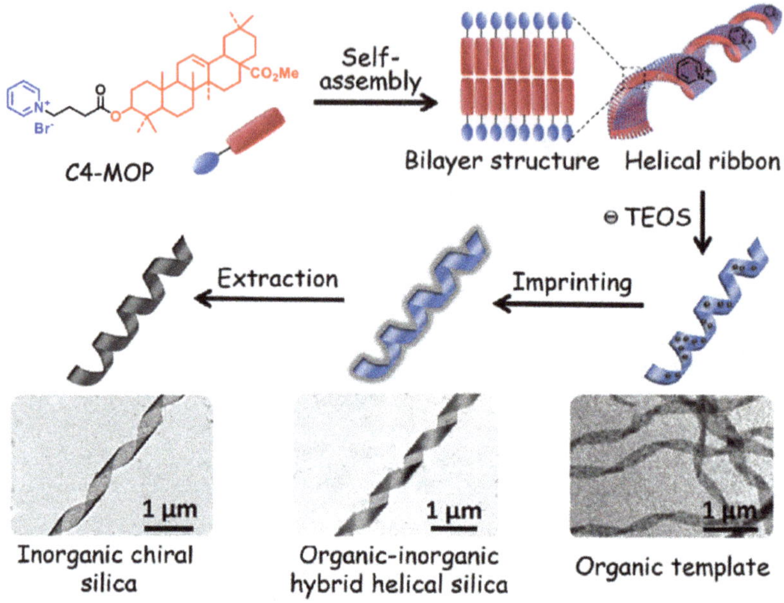

Figure 4. Chirality imprinting of inorganic silica with C4-MOP. Reproduced from Ref. [29] with permission from RSC 2018.

study suggested that chiral structures are spontaneously formed by the irradiation of LCP and RCP in solution.

Ju and coworkers imprinted triterpenoid-regulated helical ribbons to produce a chiral silica material [29]. Oleanolic acid (OA) with fused structures consisting of six-membered rings bearing carboxylic and hydroxyl groups as the core building block for constructing chirogenesis. The hydroxyl group was decorated with a cationic pyridinium ion allowing for interactions with the silica precursor (TEOS) bearing anionic charges [30, 31]. Additionally, the carboxylic group was modified with a methyl ester group, to increase the hydrophobicity of the OA units. These design features produced an ideal amphiphilic molecule C4-MOP with a flexible hydrophilic headgroup and rigid hydrophobic skeleton. C4-MOP could be transformed into uniform helical ribbons by adjusting the solvent composition and concentration. Chiral silica materials were successfully generated by silica mineralization of a sol-gel, driven by electrostatic interactions between the negatively charged tetraethyl orthosilicate, and the positively charged pyridinium cation [29]. Notably, Iwasa *et al.* provided unambiguous evidence of superconductivity in a chiral structure [32]. Such superconductivity can be realized by ionic gating in individual chiral nanotubes. In the superconducting state, the nonreciprocal signal was greatly enhanced.

1.3 Chirogenesis in 2D Materials

1.3.1 Chiral Perovskites

In chiral perovskite materials, the combination of chirogenesis and perovskite scaffolds offers a new approach to developing smart optoelectronic devices [33–36]. This section summarizes recent advances in chiral perovskite materials and their related optoelectronic performance [37, 38].

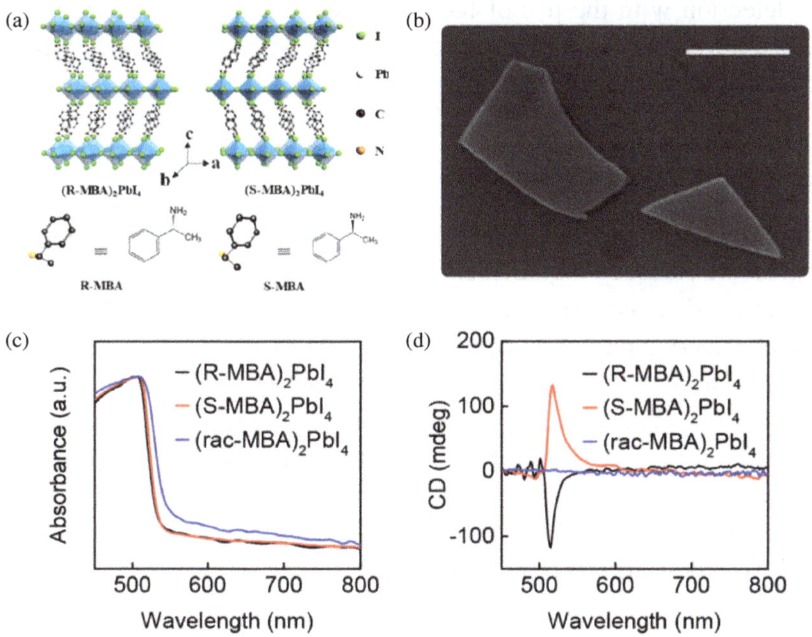

Figure 5. (a) Structure of chiral perovskites (R- and S-MBA)$_2$PbI$_4$; (b) scanning electron microscope (SEM) image of (R-MBA)$_2$PbI$_4$; (c) ultraviolet (UV)-vis spectra of 2D perovskites; (d) CD spectra of 2D perovskites. Reprinted (adapted) with permission from Ref. [39]. Copyright (2019) American Chemical Society.

Ma and coworkers used R-MBA and S-MBA as chiral ligands to synthesize chiral 2D perovskite (R- and S-MBA)$_2$PbI$_4$ (Figure 5) [39]. The surface modified with (R-MBA)$_2$PbI$_4$ was smooth and flat because of the layered nature of the 2D perovskites. The perovskite crystals were dissolved in DMF solution, and the CD spectra were measured by spin coating the solution onto quartz substrates to confirm the chirality of the perovskites. Strong and mirrored CD signals were observed from the films of (R-MBA)$_2$PbI$_4$ and (S-MBA)$_2$PbI$_4$. These data confirmed that the incorporation of chiral molecules can induce chirality in 2D perovskites. The as-synthesized crystals had good phase purity and crystalline quality, as confirmed by low-temperature photoluminescence and X-ray diffraction studies. As the temperature was increased, the degree of circularly polarized photoluminescence also decreased. These results suggest that lattice distortion formed by the chiral molecules contributes to the observed chirality in the samples. Moreover, the researchers achieved CPL detection with the use of these pure chiral 2D perovskites. Further studies showed that desired functionalities can be integrated into chiral 2D perovskites with layered materials.

1.3.2 Chiral Liquid Crystals

Chiral liquid crystals (CLCs) have drawn considerable interest in the field of optoelectronics owing to their potential for applications, such as in information displays, and unique properties including optical rotation, CD, and selective reflection of polarized light [40, 41].

Graphene-based CLCs have been the focus of a number of investigations [42, 43]. Chirality is a fundamental requirement for constructing a spiral morphology. Akagi *et al.* obtained a series of graphite films from helical polyacetylene [44, 45] by heating the helical polyacetylene film at a high temperature to induce carbonization and graphitization [46]. In this way, they produced stable graphitic films under atmospheric conditions.

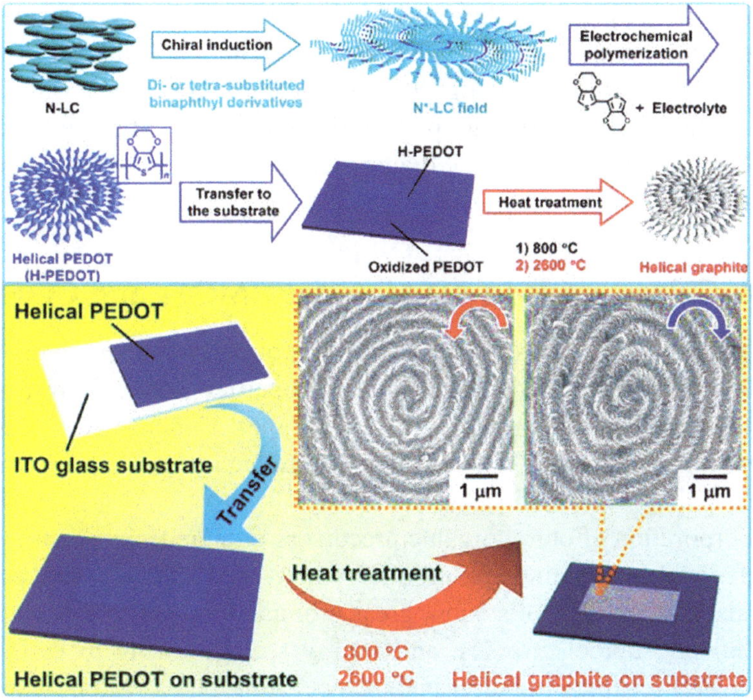

Figure 6. Synthetic process for obtaining helical graphite using CLCs as building blocks. Reprinted (adapted) with permission from Ref. [47]. Copyright (2016) American Chemical Society.

Notably, precise regulation of the twist degree, horizontal and vertical alignment, and screw direction of the helical graphite was achieved by altering the helical pitch and orientation state (Figure 6) [47]. Recently, the same researchers used polymers to generate graphite films with various morphologies. Graphite and carbon whiskers were simply aligned with the use of self-organized helical poly(*m*-phenylene) derivatives. A spiral graphite film was readily prepared, which contained left- and right-handed helical morphologies composed of fibril bundles [48].

Carbonization of precursor-helical PEDOT films was achieved by electrochemical polymerization in a chiral reaction solution. After the heating process, the helical morphology of the H-PEDOT precursor was maintained in the resultant spiral graphite and carbon films. This well-defined helical graphite film had unique diamagnetic, electrical, and electrochemical, properties.

1.4 Chirogenesis in 3D Materials (MOFs and COFs)

There is a need for rational design principles for microstructured materials with advanced physical and mechanical properties [49]. Chemists have made use of node rotation to generate two types of innovative 3D chiral covalent organic frameworks (CCOFs) [50, 51] and metal-organic frameworks (MOFs) [49, 52, 53].

Incorporation of chiral organic precursors into predesigned structures through covalent bonding can produce a new class of crystalline porous materials [54, 55]. CCOFs show great promise for applications such as drug delivery, optoelectronics, and energy storage owing to their high porosity and tunable structural parameters. 3D CCOFs warrant exploration owing to their unique confinement effects and large surface areas. For example, Cui and coworkers synthesized the first 3D CCOFs by imine condensation. After postsynthetic oxidation of the imine linkages, the framework was transformed into amide-linked CCOFs, which retained their porosity and crystallinity. Chiral high-performance liquid chromatography (HPLC) with the use of natural polymers as chiral stationary phases has become an important system for chiral separation. Thus, two CCOFs have been used to pack HPLC columns.

To examine the separation performance toward chiral molecules, 1-phenyl-2-propanol was injected into a CCOF-based column. A small decrease in the retention time for the eluted enantiomer was found as the mass increased (Figure 7) [56]. Notably, the two CCOFs as chiral stationary phases could be used to separate racemic alcohols. Importantly, the oxidized CCOFs had superior separation performance compared with that of the pristine framework. Hence, CCOFs are a powerful platform for

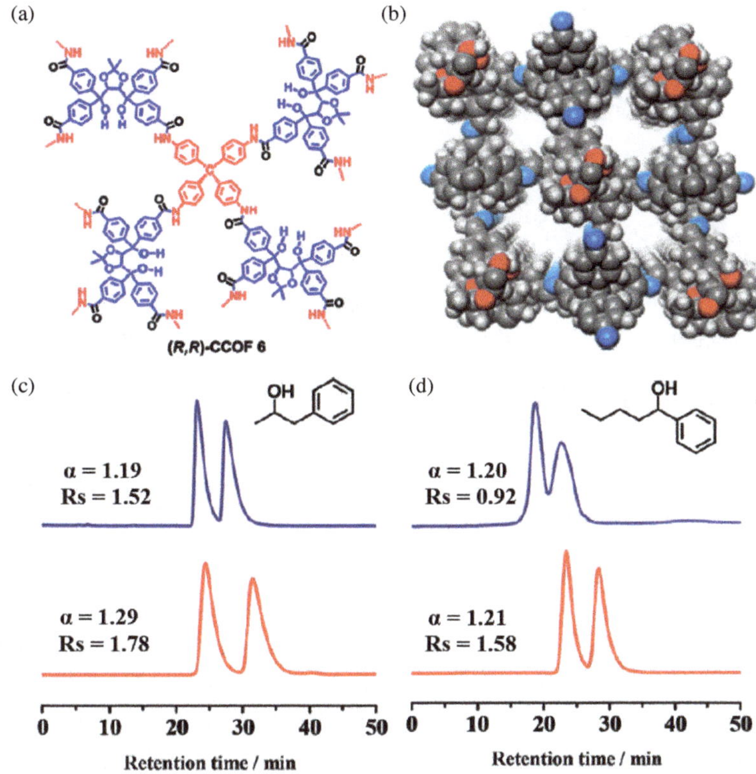

Figure 7. (a) Structure of COFs; (b) structural representations of 1D channels; (c) separation of racemic 1-phenyl-2-propanol and 1-phenyl-1-pentanol by functional HPLC. Reprinted (adapted) with permission from Ref. [56]. Copyright (2018) American Chemical Society.

chiral resolution that will open new avenues of research in materials engineering and design.

MOFs, predetermined by coordination driven self-assembly, can be systematically decorated by tuning the organic bridging linkers. Precise control of pore surfaces and sizes is important for selective recognition in MOF-based porous materials [57–59]. The pore surfaces of MOFs can be designed by immobilization of different recognition groups, such as chiral

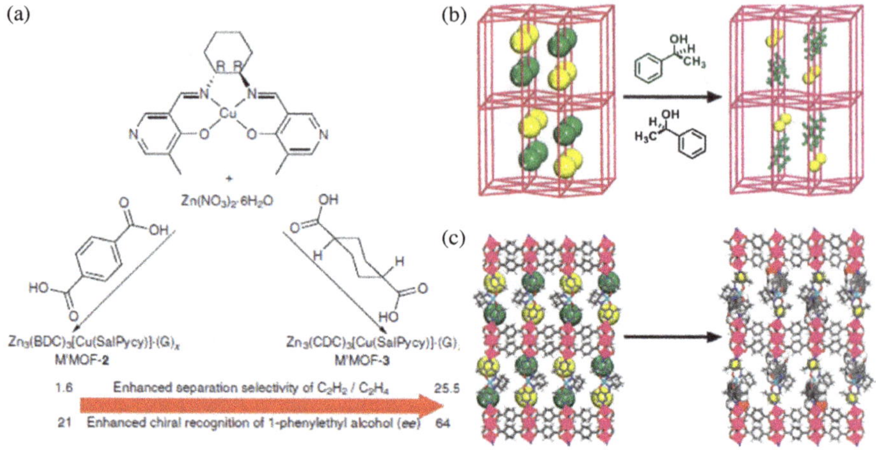

Figure 8. (a) Synthesis of M′MOF; (b) hexagonal primitive network topology; (c) M′MOF-3 framework with chiral pore cavities. Reproduced from Ref. [65] with permission from Springer Nature 2011.

pockets and open metal sites, to recognize small molecules [60–64]. Separation of ethylene and acetylene is an important industrial process because the compounds are essential chemical reagents. Current separation approaches, such as extraction of cracked olefins and partial hydrogenation of acetylene are energy consuming and costly. Chen and coworkers reported tunable microporous enantiopure mixed metal-organic framework (M′MOF) materials for the separation of C_2H_2 and C_2H_4 (Figure 8). High selectivities were achieved, suggesting potential for applications of M′MOFs to practical separation of C_2H_2/C_2H_4 [65].

1.5 Chirogenesis in Other Applications

1.5.1 Chirogenesis in Self-assembled Soft Materials

Chirogenesis has been used to form scaffold materials for cell cultures and in three-dimensional tissue engineering based on self-assembled hydrogels. Chiral gels with supramolecular nanostructures can operate as templates to form chiral inorganic nanomaterials [66–68]. Chirogenesis in self-assembled soft materials can enhance site and spatial selection.

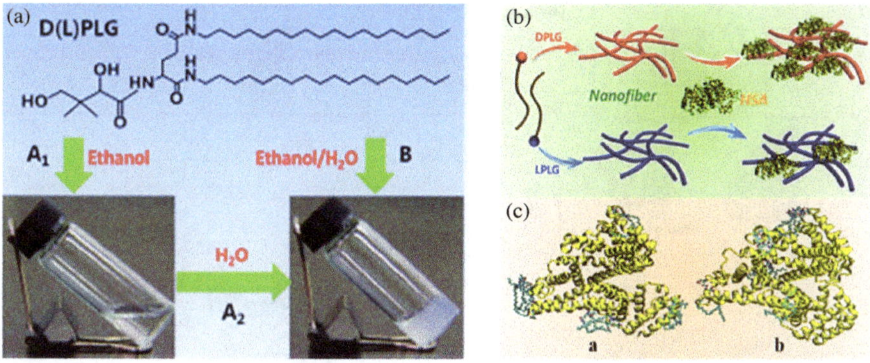

Figure 9. (a) Molecular structure and two ways to form gel; (b) different adhesion abilities of nanomaterials; (c) simulation conformations of the gelators and HSA. Reprinted (adapted) with permission from Ref. [69]. Copyright (2014) American Chemical Society.

Liu and coworkers designed DPLG and LPLG (L-glutamide and D- or L-pantolactone) as two gelators to study interactions with proteins [69]. Notably, the two gelators formed gels in a mixed solvent. The corresponding gels were characterized with CD, ultraviolet (UV)–vis spectroscopy, X–ray diffraction, and atomic force microscopy. The supramolecular chirality resulted in nanofiber structures with many similar properties but opposite signals. Additionally, the chirality of the nanofibrous structure markedly affected protein adhesion. The researchers used a quartz crystal microbalance technique to study the adsorption of human serum albumin (HSA) onto the nanofibrous structures. These results indicated that the supramolecular nanostructure of DPLG has a stronger adhesive ability than that of LPLG. Thus, chiral nanostructures can amplify minor differences in the interactions between D-/L- substances and biological molecules. Additionally, molecular dynamic simulations were used to propose a model to describe the differences and confirm the relationship between the gelators and proteins.

1.5.2 Chiral Metamaterials via Moiré Stacking

Chirality can be transferred from the macro- to nanoscale, without the requirement of chiral building blocks, based on methods related to

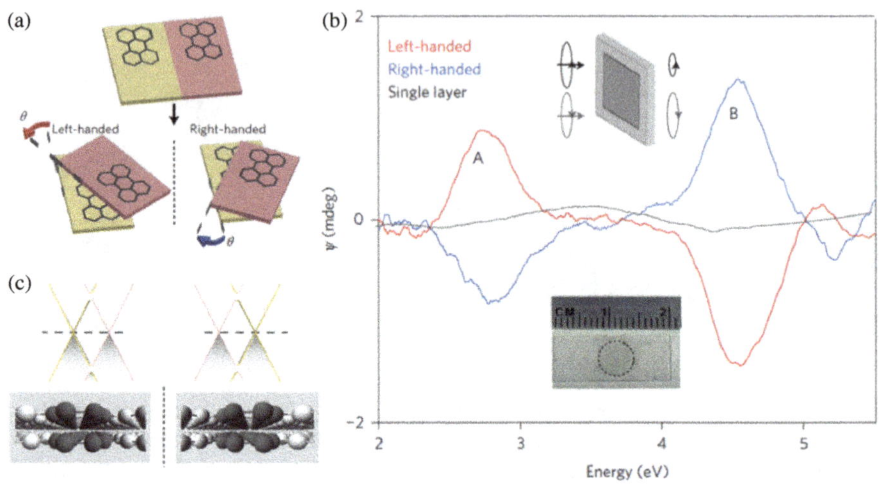

Figure 10. (a) Chiral stacking process for generating left- and right-handed twisted bilayer graphene films; (b) electronic band structures and cross-sectional schematics with spiral atomic arrangements of left- and right-handed twisted bilayer graphene films; (c) ellipticity spectra, or CD spectra. Reproduced from Ref. [75] with permission from Springer Nature 2010.

supramolecular chirality [70, 71]. For example, as shown in Figure 10, macroscale rotation generates structural chirality at the nanoscale [72]. One approach to defining nanoscale chiral features by manipulation on the microscale is through moiré stacking. Moiré stacking relates to the superimposition of two similar or identical periodic lattices with rotational displacement, leading to remarkable chiral structures [73, 74].

The realization of spatially uniform chiral films may provide a powerful way of developing nanodevices with unique chiral properties [75–77]. Park and coworkers reported on positioning of layer-by-layer assemblies with precise control of the interlayer rotation (θ) and polarity by a chiral stacking approach, which resulted in tunable chiral properties [78]. The authors generated left- and right-handed bilayer graphene with two-atom-thick chiral films. The film had one of the highest intrinsic ellipticity values, and a remarkably strong CD, where the sign was tuned by θ and the solution polarity. These chiral properties originate from the large in-plane magnetic moment associated with the interlayer optical transition. Furthermore, by producing three-layer graphene films with controlled CD signal, the authors programmed the chiral properties of atomically thin films in a layer-by-layer fashion.

1.6 Conclusions and Outlook

Chirogenesis, generally associated with self-assembly processes, can be induced in systems containing achiral or chiral molecules. Recently, chirogenesis has drawn great interest and growth in applications of chirogenesis clearly demonstrating great potential for novel opportunities and changing the way we conceive materials science. We show that chirogenesis is a useful tool for both understanding materials science and tailoring materials for specific applications. Therefore, we foresee further advances in this field focused on the use of chirogenesis beyond the molecular level, translating chirogenesis to the macroscopic world, such as in chiral self-assembled materials with new types of multifunctional components. Undoubtedly, this will have an impact on the development of a next generation smart responsive devices and molecular machines on surfaces. Therefore, further efforts in chirogenesis research should integrate novel concepts from biology, supramolecular chemistry, and materials science.

1.7 Acknowledgments

This work was supported from the Open Project of Hubei Key Laboratory of Wudang Local Chinese Medicine Research, Hubei University of Medicine (Grant No. WDCM2019003) and the Start-up Funding from South-Central University for Nationalities (No. YZZ19005).

References

1. Escárcega-Bobadilla, M.V., and Kleij, A.W. (2012) Artificial Chirogenesis: A Gateway to New Opportunities in Material Science and Catalysis. *Chem. Sci.*, **3**, 2421–2428.
2. Hembury, G.A., Borovkov, V.V., and Inoue, Y. (2008) Chirality-Sensing Supramolecular Systems. *Chem. Rev.*, **108**, 1–73.
3. Borovkov, V.V., Hembury, G.A., and Inoue, Y. (2004) Origin, Control, and Application of Supramolecular Chirogenesis in Bisporphyrin-Based Systems. *Acc. Chem. Res.*, **37**, 449–459.

4. Dydio, P., and Reek, J.N.H. (2014) Supramolecular Control of Selectivity in Transition-Metal Catalysis Through Substrate Preorganization. *Chem. Sci.*, **5**, 2135–2145.
5. Manera, M.G., Ferreiro-Vila, E., Cebollada, A., García-Martín, J.M., García-Martín, A., Giancane, G., Valli, L., and Rella, R. (2012) Ethane-Bridged Zn Porphyrins Dimers in Langmuir–Schäfer Thin Films: Spectroscopic, Morphologic, and Magneto-Optical Surface Plasmon Resonance Characterization. *J. Phys. Chem. C.*, **116**, 10734–10742.
6. Ishikawa, M., Maruyama, Y., Ye, J.Y., and Futamata, M. (2002) Single-Molecule Imaging and Spectroscopy Using Fluorescence and Surface-Enhanced Raman Scattering. *J. Biomed. Phys.*, **28**, 573–585.
7. Xiao, Z., Levy-Nissenbaum, E., Alexis, F., Lupták, A., Teply, B.A., Chan, J.M., Shi, J., Digga, E., Cheng, J., Langer, R., and Farokhzad, O.C. (2012) Engineering of Targeted Nanoparticles for Cancer Therapy Using Internalizing Aptamers Isolated by Cell-Uptake Selection. *ACS Nano.*, **6**, 696–704.
8. Slocik, J.M., Govorov, A.O., and Naik, R.R. (2011) Plasmonic Circular Dichroism of Peptide-Functionalized Gold Nanoparticles. *Nano Lett.*, **11**, 701–705.
9. MacQuarrie, S., Thompson, M.P., Blanc, A., Mosey, N.J., Lemieux, R.P., and Crudden, C.M. (2008) Chiral Periodic Mesoporous Organosilicates Based on Axially Chiral Monomers: Transmission of Chirality in the Solid State. *J. Am. Chem. Soc.*, **130**, 14099–14101.
10. Stoumpos, C.C., Cao, D.H., Clark, D.J., Young, J., Rondinelli, J.M., Jang, J.I., Hupp, J.T., and Kanatzidis, M.G. (2016) Ruddlesden–Popper Hybrid Lead Iodide Perovskite 2D Homologous Semiconductors. *Chem. Mater.*, **28**, 2852–2867.
11. Filip, M.R., Eperon, G.E., Snaith, H.J., and Giustino, F. (2014) Steric Engineering of Metal-Halide Perovskites with Tunable Optical Band Gaps. *Nat. Commun.*, **5**, 5757.
12. Qi, X., Zhang, Y., Ou, Q., Ha, S.T., Qiu, C.-W., Zhang, H., Cheng, Y.-B., Xiong, Q., and Bao, Q. (2018) Photonics and Optoelectronics of 2D Metal-Halide Perovskites. *Small*, **14**, 1800682.
13. Byun, J., Cho, H., Wolf, C., Jang, M., Sadhanala, A., Friend, R.H., Yang, H., and Lee, T.-W. (2016) Efficient Visible Quasi-2D Perovskite Light-Emitting Diodes. *Adv. Mater.*, **28**, 7515–7520.
14. Meza, L.R., Zelhofer, A.J., Clarke, N., Mateos, A.J., Kochmann, D.M., and Greer, J.R. (2015) Resilient 3D Hierarchical Architected Metamaterials. *Proc. Nat. Acad. Sci.*, **112**, 11502.
15. Vignolini, S., Yufa, N.A., Cunha, P.S., Guldin, S., Rushkin, I., Stefik, M., Hur, K., Wiesner, U., Baumberg, J.J., and Steiner, U. (2012) A 3D Optical Metamaterial Made by Self-Assembly. *Adv. Mater.*, **24**, OP23–OP27.
16. Bisht, K.K., Parmar, B., Rachuri, Y., Kathalikattil, A.C., and Suresh, E. (2015) Progress in the Synthetic and Functional Aspects of Chiral Metal–Organic Frameworks. *CrystEngComm*, **17**, 5341–5356.
17. Oh, S.S., Demetriadou, A., Wuestner, S., and Hess, O. (2013) On the Origin of Chirality in Nanoplasmonic Gyroid Metamaterials. *Adv. Mater.*, **25**, 612–617.

18. Wang, R.-Y., Wang, H., Wu, X., Ji, Y., Wang, P., Qu, Y., and Chung, T.-S. (2011) Chiral Assembly of Gold Nanorods with Collective Plasmonic Circular Dichroism Response. *Soft Matter*, **7**, 8370–8375.
19. Noguez, C., and Garzón, I.L. (2009) Optically Active Metal Nanoparticles. *Chem. Soc. Rev.*, **38**, 757–771.
20. Fan, Z., and Govorov, A.O. (2010) Plasmonic Circular Dichroism of Chiral Metal Nanoparticle Assemblies. *Nano Lett.*, **10**, 2580–2587.
21. Li, Z., Zhu, Z., Liu, W., Zhou, Y., Han, B., Gao, Y., and Tang, Z. (2012) Reversible Plasmonic Circular Dichroism of Au Nanorod and DNA Assemblies. *J. Am. Chem. Soc.*, **134**, 3322–3325.
22. Oh, H.S., Liu, S., Jee, H., Baev, A., Swihart, M.T., and Prasad, P.N. (2010) Chiral Poly(fluorene-alt-benzothiadiazole) (PFBT) and Nanocomposites with Gold Nanoparticles: Plasmonically and Structurally Enhanced Chirality. *J. Am. Chem. Soc.*, **132**, 17346–17348.
23. Ma, W., Kuang, H., Wang, L., Xu, L., Chang, W.-S., Zhang, H., Sun, M., Zhu, Y., Zhao, Y., Liu, L., Xu, C., Link, S., and Kotov, N.A. (2013) Chiral Plasmonics of Self-Assembled Nanorod Dimers. *Sci. Rep.*, **3**, 1934.
24. Yan, W., Ma, W., Kuang, H., Liu, L., Wang, L., Xu, L., and Xu, C. (2013) Plasmonic Chirogenesis from Gold Nanoparticles Superstructures. *J. Phys. Chem. C.*, **117**, 17757–17765.
25. Mao, X., Wang, Z., Zeng, D., Cao, H., Zhan, Y., Wang, Y., Li, Q., Shen, Y., and Wang, J. (2019) Self-Assembled Chiral Nanoparticle Superstructures and Identification of Their Collective Optical Activity from Ligand Asymmetry. *ACS Nano*, **13**, 2879–2887.
26. Saito, K., and Tatsuma, T. (2018) Chiral Plasmonic Nanostructures Fabricated by Circularly Polarized Light. *Nano Lett.*, **18**, 3209–3212.
27. Tatsuma, T., Nishi, H., and Ishida, T. (2017) Plasmon-Induced Charge Separation: Chemistry and Wide Applications. *Chem. Sci.*, **8**, 3325–3337.
28. Tian, Y., and Tatsuma, T. (2005) Mechanisms and Applications of Plasmon-Induced Charge Separation at TiO_2 Films Loaded with Gold Nanoparticles. *J. Am. Chem. Soc.*, **127**, 7632–7637.
29. Gao, Y., Hao, J., Liu, J., Liang, Y., Du, F., Hu, J., and Ju, Y. (2019) Imprinting Supramolecular Chirality on Silica from Natural Triterpenoid-Regulated Helical Ribbons. *Mater. Chem. Front.*, **3**, 308–313.
30. Liu, C., Jin, Q., Lv, K., Zhang, L., and Liu, M. (2014) Water Tuned the Helical Nanostructures and Supramolecular Chirality in Organogels. *Chem. Commun.*, **50**, 3702–3705.
31. Wu, X., Ji, S., Li, Y., Li, B., Zhu, X., Hanabusa, K., and Yang, Y. (2009) Helical Transfer through Nonlocal Interactions. *J. Am. Chem. Soc.*, **131**, 5986–5993.
32. Qin, F., Shi, W., Ideue, T., Yoshida, M., Zak, A., Tenne, R., Kikitsu, T., Inoue, D., Hashizume, D., and Iwasa, Y. (2017) Superconductivity in a Chiral Nanotube. *Nat. Commun.*, **8**, 14465.
33. Shi, Y., Duan, P., Huo, S., Li, Y., and Liu, M. (2018) Endowing Perovskite Nanocrystals with Circularly Polarized Luminescence. *Adv. Mater.*, **30**, 1705011.

34. Yuan, C., Li, X., Semin, S., Feng, Y., Rasing, T., and Xu, J. (2018) Chiral Lead Halide Perovskite Nanowires for Second-Order Nonlinear Optics. *Nano Lett.*, **18**, 5411–5417.
35. Long, G., Jiang, C., Sabatini, R., Yang, Z., Wei, M., Quan, L.N., Liang, Q., Rasmita, A., Askerka, M., Walters, G., Gong, X., Xing, J., Wen, X., Quintero-Bermudez, R., Yuan, H., Xing, G., Wang, X.R., Song, D., Voznyy, O., Zhang, M., Hoogland, S., Gao, W., Xiong, Q., and Sargent, E.H. (2018) Spin Control in Reduced-Dimensional Chiral Perovskites. *Nat. Photonics.*, **12**, 528–533.
36. Yang, Y., da Costa, R.C., Fuchter, M.J., and Campbell, A.J. (2013) Circularly Polarized Light Detection by a Chiral Organic Semiconductor Transistor. *Nat. Photonics.*, **7**, 634–638.
37. Cao, D.H., Stoumpos, C.C., Farha, O.K., Hupp, J.T., and Kanatzidis, M.G. (2015) 2D Homologous Perovskites as Light-Absorbing Materials for Solar Cell Applications. *J. Am. Chem. Soc.*, **137**, 7843–7850.
38. Liu, Y., Xiao, H., and Goddard, W.A. (2016) Two-Dimensional Halide Perovskites: Tuning Electronic Activities of Defects. *Nano Lett.*, **16**, 3335–3340.
39. Ma, J., Fang, C., Chen, C., Jin, L., Wang, J., Wang, S., Tang, J., and Li, D. (2019) Chiral 2D Perovskites with a High Degree of Circularly Polarized Photoluminescence. *ACS Nano*, **13**, 3659–3665.
40. Uchimura, M., Watanabe, Y., Araoka, F., Watanabe, J., Takezoe, H., and Konishi, G.-i. (2010) Development of Laser Dyes to Realize Low Threshold in Dye-Doped Cholesteric Liquid Crystal Lasers. *Adv. Mater.*, **22**, 4473–4478.
41. Zhai, F., Feng, Y., Zhou, K., Wang, L., Zheng, Z., and Feng, W. (2019) Graphene-Based Chiral Liquid Crystal Materials for Optical Applications. *J. Mater. Chem. C.*, **7**, 2146–2171.
42. Narayan, R., Kim, J.E., Kim, J.Y., Lee, K.E., and Kim, S.O. (2016) Liquid Crystals: Graphene Oxide Liquid Crystals: Discovery, Evolution and Applications. *Adv. Mater.*, **28**, 3044–3044.
43. Nych, A., Fukuda, J.-i., Ognysta, U., Žumer, S., and Muševič, I. (2017) Spontaneous Formation and Dynamics of Half-Skyrmions in a Chiral Liquid-Crystal Film. *Nat. Phys.*, **13**, 1215–1220.
44. Akagi, K., Piao, G., Kaneko, S., Sakamaki, K., Shirakawa, H., and Kyotani, M. (1999) ChemInform Abstract: Helical Polyacetylene Synthesized with a Chiral Nematic Reaction Field. *ChemInform*, **30**, 1683–1686
45. Goh, M., Matsushita, S., and Akagi, K. (2010) From Helical Polyacetylene to Helical Graphite: Synthesis in the Chiral Nematic Liquid Crystal Field and Morphology-Retaining Carbonisation. *Chem. Soc. Rev.*, **39**, 2466–2476.
46. Kyotani, M., Matsushita, S., Nagai, T., Matsui, Y., Shimomura, M., Kaito, A., and Akagi, K. (2008) Helical Carbon and Graphitic Films Prepared from Iodine-Doped Helical Polyacetylene Film Using Morphology-Retaining Carbonization. *J. Am. Chem. Soc.*, **130**, 10880–10881.
47. Subacius, D., Shiyanovskii, S.V., Bos, P., and Lavrentovich, O.D. (1997) Cholesteric Gratings with Field-Controlled Period. *Appl. Phys. Lett.*, **71**, 3323–3325.

48. Yan, B., Matsushita, S., and Akagi, K. (2016) An Advanced Method for Preparation of Helical Carbon and Graphitic Films Using a Carbonization Substrate. *Chem. Mater.*, **28**, 8781–8791.
49. Wu, W., Qi, D., Liao, H., Qian, G., Geng, L., Niu, Y., and Liang, J. (2018) Deformation Mechanism of Innovative 3D Chiral Metamaterials. *Sci. Rep.*, **8**, 12575.
50. Côté, A.P., Benin, A.I., Ockwig, N.W., Keeffe, M., Matzger, A.J., and Yaghi, O.M. (2005) Porous, Crystalline, Covalent Organic Frameworks. *Science*, **310**, 1166.
51. El-Kaderi, H.M., Hunt, J.R., Mendoza-Cortés, J.L., Côté, A.P., Taylor, R.E., Keeffe, M., and Yaghi, O.M. (2007) Designed Synthesis of 3D Covalent Organic Frameworks. *Science*, **316**, 268.
52. Kaskel, S. (2010) Metal–Organic Frameworks. Design and Application. Edited by Leonard R. MacGillivray. *Angew. Chem. Int. Ed.*, **49**, 9565.
53. Krawiec, P., Kramer, M., Sabo, M., Kunschke, R., Fröde, H., and Kaskel, S. (2006) Improved Hydrogen Storage in the Metal-Organic Framework Cu3(BTC)2. *Adv. Eng. Mater.*, **8**, 293–296.
54. Zhou, L., Zhou, Y., Li, M., Chen, P., and Wang, Y. (2000) Experimental and Modeling Study of the Adsorption of Supercritical Methane on a High Surface Activated Carbon. *Langmuir*, **16**, 5955–5959.
55. Sumida, K., Rogow, D.L., Mason, J.A., McDonald, T.M., Bloch, E.D., Herm, Z.R., Bae, T.-H., and Long, J.R. (2012) Carbon Dioxide Capture in Metal–Organic Frameworks. *Chem. Rev.*, **112**, 724–781.
56. Han, X., Huang, J., Yuan, Chen., Liu, Y., and Cui, Y. Chiral 3D Covalent Organic Frameworks for High Performance. *J. Am. Chem. Soc.*, **140**, 892–895.
57. Liquid Chromatographic Enantioseparation.
58. Kuznicki, S.M., Bell, V.A., Nair, S., Hillhouse, H.W., Jacubinas, R.M., Braunbarth, C.M., Toby, B.H., and Tsapatsis, M. (2001) A Titanosilicate Molecular Sieve with Adjustable Pores for Size-Selective Adsorption of Molecules. *Nature*, **412**, 720–724.
59. Chen, B., Xiang, S., and Qian, G. (2010) Metal–Organic Frameworks with Functional Pores for Recognition of Small Molecules. *Acc. Chem. Res.*, **43**, 1115–1124.
60. Ma, L., Falkowski, J.M., Abney, C., and Lin, W. (2010) A Series of Isoreticular Chiral Metal–Organic Frameworks as a Tunable Platform for Asymmetric Catalysis. *Nat. Chem.*, **2**, 838–846.
61. Britt, D., Furukawa, H., Wang, B., Glover, T.G., and Yaghi, O.M. (2009) Highly Efficient Separation of Carbon Dioxide by a Metal-Organic Framework Replete with Open Metal Sites. *Proc. Nat. Acad. Sci.*, **106**, 20637.
62. Shimomura, S., Higuchi, M., Matsuda, R., Yoneda, K., Hijikata, Y., Kubota, Y., Mita, Y., Kim, J., Takata, M., and Kitagawa, S. (2010) Selective Sorption of Oxygen and Nitric Oxide by an Electron-Donating Flexible Porous Coordination Polymer. *Nat. Chem.*, **2**, 633–637.
63. Devic, T., Horcajada, P., Serre, C., Salles, F., Maurin, G., Moulin, B., Heurtaux, D., Clet, G., Vimont, A., Grenèche, J.-M., Ouay, B.L., Moreau, F., Magnier, E., Filinchuk, Y., Marrot, J., Lavalley, J.-C., Daturi, M., and Férey, G. (2010) Functionalization in

Flexible Porous Solids: Effects on the Pore Opening and the Host–Guest Interactions. *J. Am. Chem. Soc.*, **132**, 1127–1136.
64. Seo, J.S., Whang, D., Lee, H., Jun, S.I., Oh, J., Jeon, Y.J., and Kim, K. (2000) A Homochiral Metal–Organic Porous Material for Enantioselective Separation and Catalysis. *Nature*, **404**, 982–986.
65. Morris, R.E., and Bu, X. (2010) Induction of Chiral Porous Solids Containing Only Achiral Building Blocks. *Nat. Chem.*, **2**, 353–361.
66. Xiang, S.-C., Zhang, Z., Zhao, C.-G., Hong, K., Zhao, X., Ding, D.-R., Xie, M.-H., Wu, C.-D., Das, M.C., Gill, R., Thomas, K.M., and Chen, B. (2011) Rationally Tuned Micropores Within Enantiopure Metal-Organic Frameworks for Highly Selective Separation of Acetylene and Ethylene. *Nat. Commun.*, **2**, 204.
67. Zhang, M., Qing, G., and Sun, T. (2012) Chiral Biointerface Materials. *Chem. Soc. Rev.*, **41**, 1972–1984.
68. Zhao, Y., Abdul Rahim, N.A., Xia, Y., Fujiki, M., Song, B., Zhang, Z., Zhang, W., and Zhu, X. (2016) Supramolecular Chirality in Achiral Polyfluorene: Chiral Gelation, Memory of Chirality, and Chiral Sensing Property. *Macromolecules*, **49**, 3214–3221.
69. Ben-Moshe, A., Wolf, S.G., Sadan, M.B., Houben, L., Fan, Z., Govorov, A.O., and Markovich, G. (2014) Enantioselective Control of Lattice and Shape Chirality in Inorganic Nanostructures Using Chiral Biomolecules. *Nat. Commun.*, **5**, 4302.
70. Qiu, Y., Chen, P., Guo, P., Li, Y., and Liu, M. (2008) Supramolecular Chiroptical Switches Based on Achiral Molecules. *Adv. Mater.*, **20**, 2908–2913.
71. Wu, Z., Liu, Y., Hill, E.H., and Zheng, Y. (2018) Chiral Metamaterials via Moiré Stacking. *Nanoscale*, **10**, 18096–18112.
72. Wu, Z., Chen, X., Wang, M., Dong, J., and Zheng, Y. (2018) High-Performance Ultrathin Active Chiral Metamaterials. *ACS Nano*, **12**, 5030–5041.
73. Kim, Y., Yeom, B., Arteaga, O., Jo Yoo, S., Lee, S.-G., Kim, J.-G., and Kotov, N.A. (2016) Reconfigurable Chiroptical Nanocomposites with Chirality Transfer from the Macro- to the Nanoscale. *Nat. Mater.*, **15**, 461–468.
74. Oster, G., Wasserman, M., and Zwerling, C. (1964) Theoretical Interpretation of Moiré Patterns. *J. Opt. Soc. Am.*, **54**, 169–175.
75. Li, N., Wu, W., and Chou, S.Y. (2006) Sub-20-nm Alignment in Nanoimprint Lithography Using Moiré Fringe. *Nano Lett.*, **6**, 2626–2629.
76. Shopsowitz, K.E., Qi, H., Hamad, W.Y., and MacLachlan, M.J. (2010) Free-Standing Mesoporous Silica Films with Tunable Chiral Nematic Structures. *Nature*, **468**, 422–425.
77. Edgar, C.D., and Gray, D.G. (2001) Induced Circular Dichroism of Chiral Nematic Cellulose Films, *Cellulose*, **8**, 5–12.
78. Arnaboldi, S., Cauteruccio, S., Grecchi, S., Benincori, T., Marcaccio, M., Biroli, A.O., Longhi, G., Licandro, E., and Mussini, P.R. (2019) Thiahelicene-Based Inherently Chiral Films for Enantioselective Electroanalysis. *Chem. Sci.*, **10**, 1539–1548.

https://doi.org/10.1142/9789811259227_bmatter

Index

Lightning Source UK Ltd.
Milton Keynes UK
UKHW020946170123
415487UK00002B/39

9 789811 259210